Operator Methods in Quantum Mechanics

Operator Methods in Quantum Mechanics

O.L. DE LANGE

and

R.E. RAAB

Department of Physics, University of Natal,
Pietermaritzburg, South Africa

CLARENDON PRESS · OXFORD

This book has been printed digitally and produced in a standard specification in order to ensure its continuing availability

OXFORD
UNIVERSITY PRESS

Great Clarendon Street, Oxford OX2 6DP

Oxford University Press is a department of the University of Oxford. It furthers the University's objective of excellence in research, scholarship, and education by publishing worldwide in

Oxford New York

Auckland Cape Town Dar es Salaam Hong Kong Karachi Kuala Lumpur Madrid Melbourne Mexico City Nairobi New Delhi Shanghai Taipei Toronto

With offices in

Argentina Austria Brazil Chile Czech Republic France Greece Guatemala Hungary Italy Japan South Korea Poland Portugal Singapore Switzerland Thailand Turkey Ukraine Vietnam

Published in the United States by Oxford University Press Inc., New York

ISBN 978-0-19-853961-2

PREFACE

The purpose of this volume is two-fold: to provide an introduction to the use of operator methods in quantum mechanics and to serve as a reference work on this topic. As such it should be suitable for use as a complement to senior and graduate courses in quantum mechanics. Although this is primarily a book on applications of operator methods, we have attempted to make the presentation self-contained by including an introductory chapter on the formalism of quantum mechanics and by giving some background material at various points in the text.

In the early days of quantum mechanics two apparently different methods were used to solve physical problems: the matrix mechanics of Heisenberg, Born, and Jordan, and the wave mechanics of Schrödinger. An abstract formulation of quantum mechanics, that includes the wave and matrix theories as particular cases, was developed by Dirac.

It soon emerged that, at least in those circumstances where it could be applied, wave mechanics was to be the preferred version of the theory. Thus in 1928 Darwin commented on Dirac's approach that 'the methods of non-commutative algebra are harder to follow, and certainly much more difficult to invent, than are operations of types long familiar to analysis'. He argued that 'it is surely better to present the theory in a mathematical form that dates from the time of Laplace and Legendre, if only because the details of the calculus have been so much more thoroughly explored'.* This familiarity with and preference for analysis manifested itself in an emphasis on the wave-mechanical approach in most books on quantum mechanics.

There have been counter-trends to this development. Modern texts on quantum mechanics adopt the abstract formulation and they stress the special nature of wave mechanics.† Furthermore, it has often been noted in the literature that the machinery of wave mechanics (such as choice of coordinate system, separation of variables, boundary conditions, single-valuedness) can obscure the underlying quantum-mechanical principles and complicate the analysis. Although there has been since 1928 much research on operator methods to solve quantum-mechanical problems, many of these results remain scattered in the literature and therefore somewhat inaccessible to the non-specialist. These circumstances have prompted us to write the present volume.

Much of this book is based on standard algebraic methods such as shift operators and the virial, hypervirial, and Hellmann–Feynman theorems. To

---

* Darwin, C.G. (1928). The wave equations of the electron. *Proceedings of the Royal Society*, A **118**, 654–80.

† See Refs 1–3 of Chapter 1.

construct shift operators we frequently use the factorization method that originated with Weyl, Dirac, and Schrödinger, and was extensively developed by Infeld and Hull, and Miller.*

For many problems the abstract shift operators yield first-order differential operators when expressed in a suitable representation (usually the coordinate representation, sometimes the momentum representation). Consequently one has a simple means of constructing normalized wavefunctions by what Schrödinger termed the 'manufacturing process'. As Infeld and Hull, and Miller have shown, this technique is an attractive method of studying the special functions of mathematical physics. We use this method extensively to determine normalized wavefunctions from first principles, and in this way the reader encounters Legendre, Hermite, Laguerre, and Gegenbauer polynomials and Bessel functions. In these calculations it is natural to obtain the results in terms of hypergeometric functions and we have consistently done so as we believe it has pedagogical value.

The reader who wishes to use this book as an introduction to operator methods can be selective, and in Section 1.12 we have made some suggestions of material for a first reading. It is with such a reader in mind, and conscious of J.E. Littlewood's remark that 'two trivialities omitted can add up to an impasse', that we have usually presented the analysis in considerable detail.

A word on subject matter that we have omitted in this book. First, there are topics such as the Wigner–Eckart theorem, perturbation theory, and scattering theory that we decided to omit because there are other books which contain comprehensive treatments of these subjects. Of these we recommend A. Böhm's *Quantum mechanics*† for the excellence of its presentation and its emphasis on general algebraic methods. Second, because this is a book on operator methods, physical applications are usually not discussed in any detail; for these the reader is frequently referred to the literature.

We hope that what we have presented does not obscure the elegance and simplicity of operator methods and that the reader may be encouraged to invent novel ways of applying such techniques.

The entire manuscript was typed by Sandy Diederichs, and we are much indebted to her for undertaking this arduous task. We also thank Paul Jackson for his encouragement during the writing of this book.

Pietermaritzburg, South Africa O.L. de L.
July 1991 R.E.R.

---

* See Refs 1 and 7 of Chapter 3.

† See Ref. 3 of Chapter 1.

CONTENTS

1 THE MATHEMATICAL FORMALISM OF QUANTUM MECHANICS

2 THE HARMONIC OSCILLATOR

3 OTHER ONE-DIMENSIONAL SYSTEMS

4 ANGULAR MOMENTUM

5 SPHERICALLY SYMMETRIC POTENTIALS

6 APPLICATIONS

7 FACTORIZATIONS WITH APPLICATION TO THE MOMENTUM REPRESENTATION

8 THE ISOTROPIC HARMONIC OSCILLATOR IN AN ANGULAR MOMENTUM BASIS

9 THE COULOMB PROBLEM IN AN ANGULAR MOMENTUM BASIS

10 THE COULOMB PROBLEM IN THE BASIS $|H\mathscr{A}_z L_z\rangle$

11 RELATIVISTIC QUANTUM MECHANICS

12 RELATIVISTIC SPHERICALLY SYMMETRIC PROBLEMS

1

THE MATHEMATICAL FORMALISM OF QUANTUM MECHANICS

Quantum mechanics was created mainly to account for microscopic phenomena that occur on a scale of order 10^{-10} m. It has turned out to be an astonishingly successful theory which applies on a scale down to at least 10^{-18} m. Quantum theory has become established as part of the essential core of physics.

The purpose of this introductory chapter is two-fold. Firstly, in Sections 1.1–1.7 we review the formalism of elementary non-relativistic quantum mechanics. In this review we attempt to give enough of the terminology and basics of the subject so that a reader who is not already familiar with quantum mechanics will be able to follow the applications considered later in the book. For the reader who finds our discussion too brief, the literature on quantum mechanics is vast and we recommend in particular Refs 1–5. The second purpose of this chapter is to present the basic tools of operator methods as they are used in this book: thus in Sections 1.8–1.11 we discuss commutation relations, shift operators, and the virial, hypervirial, and Hellmann–Feynman theorems.

1.1 Hilbert space

In the standard formulation of quantum theory the (pure [1]) states of a physical system correspond to vectors in a Hilbert space over the complex numbers [5]. The observables and dynamical variables of the system are represented by linear operators which transform the state vectors. Throughout this book we adopt Dirac's notation: thus a state vector is denoted by the *ket* $|\psi\rangle$. This ket provides a complete description of the physical state.

The space of kets (state space) has the properties:*

(i) The sum of any two kets is also a ket, and the commutative and associative laws of addition are valid.

(ii) There exists a unique ket $|\text{null}\rangle$ (the null vector) such that†

$$|\psi\rangle + |\text{null}\rangle = |\psi\rangle. \tag{1.1}$$

(iii) If c is a complex number, $c|\psi\rangle\ (\equiv |c\psi\rangle)$ is a ket.

---

* It may be helpful here to bear in mind the more familiar example of vector analysis in three-dimensional space.

† We avoid using the symbol $|0\rangle$ for the null vector since it is convenient to let $|0\rangle$ denote the ground state of a system (see Chapters 2 and 3). We often denote $|\text{null}\rangle$ simply by 0, as in (1.3).

(iv) The difference between any two kets is defined as

$$|\psi\rangle - |\phi\rangle = |\psi\rangle + (-1)|\phi\rangle.$$

(v) Multiplication by a number is distributive

$$c(|\psi\rangle + |\phi\rangle) = c|\psi\rangle + c|\phi\rangle. \tag{1.2}$$

(vi) A set of kets $|\psi_n\rangle$ ($n = 1, 2, \ldots, D$) is linearly independent if

$$\sum_{n=1}^{D} c_n|\psi_n\rangle = 0 \tag{1.3}$$

implies $c_n = 0$ for all n. The dimension D of the space is the maximum number of linearly independent kets. If no such maximum number exists, the space is infinite dimensional. For the present we restrict ourselves to spaces for which D is finite or denumerably infinite.

(vii) For every pair of kets $|\phi\rangle$ and $|\psi\rangle$ there exists a unique complex number denoted by the bracket

$$(|\phi\rangle, |\psi\rangle) \equiv \langle\phi|\psi\rangle. \tag{1.4}$$

This complex number is known as the *scalar product* or *inner product* of $|\phi\rangle$ and $|\psi\rangle$.

(viii) The scalar product has the properties

$$\begin{gathered}\langle\phi|\psi\rangle = \langle\psi|\phi\rangle^*, \qquad \langle\phi|c\psi\rangle = c\langle\phi|\psi\rangle,\\ \langle\phi|\psi + \chi\rangle = \langle\phi|\psi\rangle + \langle\phi|\chi\rangle,\end{gathered} \tag{1.5}$$

where the asterisk denotes complex conjugation, and

$$\langle\psi|\psi\rangle \geq 0. \tag{1.6}$$

The equality in (1.6) holds if and only if $|\psi\rangle$ is the null vector. We refer to the real number $\langle\psi|\psi\rangle$ as the *norm* of $|\psi\rangle$ and to the condition (1.6) as *non-negativity of the norm.* (This condition is essential for the probabilistic interpretation of quantum mechanics [1–3] and it is frequently used in this book.) If

$$\langle\psi|\psi\rangle = 1 \tag{1.7}$$

then $|\psi\rangle$ is a *normalized* ket. If

$$\langle\phi|\psi\rangle = 0 \tag{1.8}$$

and neither $|\phi\rangle$ nor $|\psi\rangle$ is the null vector then the kets are *orthogonal.*

(ix) There exist *complete orthonormal sets of bases.* Each such set consists of the kets $|n\rangle$ ($n = 1, 2, \ldots, D$) which are *orthonormal**

* The Kronecker δ function is defined by $\delta_{mn} = 0$ if $m \neq n$, $\delta_{mn} = 1$ if $m = n$.

$$\langle m|n\rangle = \delta_{mn}, \tag{1.9}$$

and which have the property that any ket of the space can be expanded as

$$|\psi\rangle = \sum_{n=1}^{D} |n\rangle\langle n|\psi\rangle. \tag{1.10}$$

It is clear, even from the material presented so far, that quantum mechanics is a highly 'counter-intuitive' theory. For example, regarding the assertion that the state of a physical system is completely described by a ket, it has been emphasized that 'This axiom goes strongly against intuition. However well known it may have become, it nevertheless remains utterly strange. What has reality to do with a linear space, generally infinite-dimensional, using complex numbers as coordinates? The axiom contains in an almost aggressive way the mystery of the intelligibility of reality in terms of mathematics, through deeply unintelligible ways.... The only justification for the rule lies in its incredible success, although some more intrinsic reasons have been sought for it. It originated from the accumulated wisdom of history as construed through a long comparison between imagination and experiment and never, up to now, has it been found either inadequate or faulty.' [6]

1.2 Operators

An operator A on the state space assigns to each* ket another ket of that space. This is indicated symbolically by

$$A|\psi\rangle = |\phi\rangle, \tag{1.11}$$

where $|\psi\rangle$ and $|\phi\rangle$ are kets of the same space. An operator is completely specified if its action on every ket of the space is known. Operators may be added together and multiplied by complex numbers and by other operators to yield new operators. Specifically, the sum and product of two operators are defined by

$$(A + B)|\psi\rangle = A|\psi\rangle + B|\psi\rangle, \tag{1.12}$$

$$(AB)|\psi\rangle = A(B|\psi\rangle). \tag{1.13}$$

Two operators are said to be equal if they have the same effect on all kets: that is, if

$$A|\psi\rangle = B|\psi\rangle \tag{1.14}$$

for all $|\psi\rangle$ then $A = B$. The unit operator 1 leaves all kets and operators unchanged, and the zero operator 0 maps all kets into the null vector

$$0|\psi\rangle = |\text{null}\rangle. \tag{1.15}$$

---

* We assume for the moment that the operators are bounded (see below).

An operator A is *linear* if

$$A(|\psi\rangle + |\phi\rangle) = A|\psi\rangle + A|\phi\rangle \tag{1.16}$$

and

$$A(c|\psi\rangle) = cA|\psi\rangle, \tag{1.17}$$

for all $|\psi\rangle$ and $|\phi\rangle$ and all complex numbers c. If A and B are linear it follows from (1.12) and (1.13) that $A + B$ and AB are linear. It also follows that a linear operator is completely specified if its action on each member of a complete orthonormal set of kets is known.

The scalar product

$$\langle\phi|A|\psi\rangle \tag{1.18}$$

is the *matrix element* of A between $|\psi\rangle$ and $|\phi\rangle$. If all the matrix elements of the linear operators A and B are equal then $A = B$.

From now on we consider only linear operators. These obey all the laws of algebra except the commutative law; that is, $AB \neq BA$ in general.

The *adjoint operator* $A^\dagger$ of A is defined by

$$\langle\phi|A^\dagger|\psi\rangle = \langle\psi|A|\phi\rangle^* \tag{1.19}$$

for all $|\psi\rangle$ and $|\phi\rangle$. It follows that

$$(A^\dagger)^\dagger = A, \tag{1.20}$$

$$(cA)^\dagger = c^* A^\dagger, \tag{1.21}$$

$$(A + B)^\dagger = A^\dagger + B^\dagger, \tag{1.22}$$

$$(AB)^\dagger = B^\dagger A^\dagger. \tag{1.23}$$

An operator A that acts on the entire state space is said to be *self-adjoint* or *Hermitian* if

$$\langle\phi|A|\psi\rangle = \langle\psi|A|\phi\rangle^* \tag{1.24}$$

for all $|\psi\rangle$ and $|\phi\rangle$. This is often written simply as

$$A^\dagger = A. \tag{1.25}$$

It should be noted that the Hermitian property is an attribute of the Hilbert space, as emphasized in (1.24), and that formal manipulations based on (1.25) can be misleading (see Appendix F and Sections 5.4 and 10.3).

An operator is said to be *unitary* if

$$A^\dagger A = AA^\dagger = 1. \tag{1.26}$$

From (1.4), (1.5), (1.19), and (1.26) we have

$$(A|\phi\rangle, A|\psi\rangle) = (|\phi\rangle, |\psi\rangle). \tag{1.27}$$

Thus scalar products are invariant under unitary transformations. Note that the product of two Hermitian operators is Hermitian if and only if the operators commute, whereas the product of two unitary operators is always unitary.

It is useful to introduce a space of *bras* that is dual to the space of kets [1, 2]: thus to each ket $|\psi\rangle$ there corresponds a bra $\langle\psi|$. (The terminology is associated with the notation for the bracket (1.4).) The operator $|\psi\rangle\langle\phi|$, known as the *outer product* of $|\psi\rangle$ and $\langle\phi|$, is defined by

$$(|\psi\rangle\langle\phi|)|\chi\rangle = |\psi\rangle(\langle\phi|\chi\rangle) \tag{1.28}$$

for all kets $|\chi\rangle$. An outer product is postulated to act to the left on a bra. Thus

$$\langle\chi|(|\psi\rangle\langle\phi|) = (\langle\chi|\psi\rangle)\langle\phi|. \tag{1.29}$$

Equations (1.28) and (1.29) can be regarded as the associative law for a triple product of vectors. It can be shown that the outer product is a linear operator and that

$$(|\psi\rangle\langle\phi|)^\dagger = |\phi\rangle\langle\psi|. \tag{1.30}$$

Thus, in particular, $|\psi\rangle\langle\psi|$ is a Hermitian operator.

From (1.10) is obtained the *closure relation* (or resolution of the identity)

$$\sum_{n=1}^{D} |n\rangle\langle n| = 1. \tag{1.31}$$

This relation makes it possible to express any operator as

$$A = \sum_{n'}\sum_{n} |n'\rangle\langle n'|A|n\rangle\langle n|. \tag{1.32}$$

Here the complex numbers $\langle n'|A|n\rangle$ are a *matrix representation* of A. Note that the matrix representation of $A^\dagger$ is obtained by transposing the matrix representation of A and taking the complex conjugate of each element.

If

$$A|\psi\rangle = a|\psi\rangle \tag{1.33}$$

where a is a complex number, then $|\psi\rangle$ is an *eigenket* of A with *eigenvalue a*. Equation (1.33) is an eigenvalue equation for A, and the set of eigenvalues is the *spectrum* of A. If there exist g linearly independent eigenkets of A with the same eigenvalue, this eigenvalue is said to be g-fold *degenerate*. For the present we suppose that the eigenvalues of A are non-degenerate ($g = 1$) and that the spectrum is discrete

$$A|\psi_n\rangle = a_n|\psi_n\rangle \qquad (n = 1, 2, \ldots). \tag{1.34}$$

If A is Hermitian it follows from (1.24) and (1.33) that the eigenvalues a_n are real and the eigenkets belonging to different eigenvalues are orthogonal. In the following we suppose that $|\psi_n\rangle$ are normalized: thus we have the orthonormality condition

$$\langle\psi_m|\psi_n\rangle = \delta_{mn}. \tag{1.35}$$

The above definition of an operator acting on the entire state space is too restrictive for many physical applications. It is necessary to extend this definition to operators which act only on a subspace of the state space; this subspace is known as the *domain* of the operator. It is also necessary to introduce the concept of a *bounded* operator; by definition, A is bounded if

$$|\langle\phi|A|\psi\rangle| \leq b < \infty \tag{1.36}$$

for all normalized kets $|\psi\rangle$ and $|\phi\rangle$. Boundedness of A ensures that the adjoint $A^\dagger$ is a bounded linear operator with the same domain as A [5]. Furthermore, it can be shown that any Hermitian operator defined on the entire space is bounded (Hellinger–Toeplitz theorem) [5].

Whereas in a finite-dimensional Hilbert space all operators are bounded, this is not so in an infinite-dimensional space. The action of an unbounded operator can be defined only on some subspace of the state space. Consequently the sum of two unbounded operators A and B is defined only in the common domain of A and B, and the product AB is defined only for kets $|\psi\rangle$ such that $B|\psi\rangle$ is in the domain of A. For an unbounded operator to be Hermitian, (1.24) must be valid for all $|\psi\rangle$ and $|\phi\rangle$ in the domain of A; if, in addition, the domains of A and $A^\dagger$ are identical, A is said to be self-adjoint.

Many operators which occur in quantum mechanics are unbounded* and 'this unfortunate fact of life is the source of much difficulty (and suffering) in quantum physics' [7]. An approach which has been proposed to avoid these difficulties is use of the rigged Hilbert space rather than the Hilbert space [3]. The rigorous treatment of these topics is beyond the scope of this book and we will treat unbounded operators heuristically. For the applications considered in this book such an approach will be sufficient: thus, for example, from unbounded operators we construct shift operators that generate all the excited states of a discrete spectrum.

1.3 Observables

A Hermitian operator A is an *observable* if its eigenkets $|\psi_n\rangle$ are a basis in state space [1]; that is, if an arbitrary ket can be written

$$|\psi\rangle = \sum_{n=1}^{D} |\psi_n\rangle\langle\psi_n|\psi\rangle. \tag{1.37}$$

(If D is finite, all Hermitian operators are observables; if D is infinite this is not necessarily so.) The corresponding expansion for a bra is

$$\langle\psi| = \sum_{n=1}^{D} \langle\psi|\psi_n\rangle\langle\psi_n|, \tag{1.38}$$

* For example, the eigenvalues of the position and momentum operators range from $-\infty$ to ∞ (Section 1.4) and hence these operators are unbounded.

and the closure relation is

$$\sum_{n=1}^{D} |\psi_n\rangle\langle\psi_n| = 1. \tag{1.39}$$

In quantum mechanics it is a postulate that every measurable physical quantity is described by an observable and that the only possible result of the measurement of a physical quantity is one of the eigenvalues of the corresponding observable [1]. Immediately after an observation of A which yields the eigenvalue a_n, the system is in the corresponding state $|\psi_n\rangle$. It is also a postulate that the probability of obtaining the result a_n when observing A on a system in the normalized state $|\psi\rangle$ is*

$$P(a_n) = |\langle\psi_n|\psi\rangle|^2. \tag{1.40}$$

The normalization of $|\psi\rangle$ and the closure relation ensure that

$$\sum_{n=1}^{D} P(a_n) = 1.$$

For an observable (1.32) reduces to

$$A = \sum_n a_n |\psi_n\rangle\langle\psi_n|, \tag{1.41}$$

which is the *spectral decomposition* of A. The *expectation value* of A with respect to $|\psi\rangle$ is

$$\langle A\rangle = \langle\psi|A|\psi\rangle. \tag{1.42}$$

From (1.40)–(1.42)

$$\langle A\rangle = \sum_n a_n P(a_n), \tag{1.43}$$

and thus (1.42) corresponds to an average measured value.

Let A and B be two observables and suppose that rapid successive measurements yield the results a_n and b_n respectively. If immediate repetition of the observations always yields the same results for all possible values of a_n and b_n then A and B are *compatible* (or non-interfering) observables. For any set of observables $A, B, \ldots$ it can be proved that any one of the conditions (i) $A, B, \ldots$ commute with each other, (ii) $A, B, \ldots$ are compatible, (iii) $A, B, \ldots$ possess a complete orthonormal set of simultaneous eigenkets,† implies the other two.

If A is an observable, so is the polynomial function

$$P_N(A) = \sum_{k=1}^{N} c_k A^k, \tag{1.44}$$

---

* This probability is determined empirically by observing A on each system of a large assemblage of N systems each in the state $|\psi\rangle$ [1–3].

† We are assuming non-degenerate eigenkets.

where c_k is real. More generally we define a *function of an observable* as follows. Let $f(x)$ be a real single-valued function* of the real variable x defined in the interval $\alpha \leq x \leq \beta$ which includes all the eigenvalues of A: $\alpha \leq a_n \leq \beta$ for all n. We define the linear operator $f(A)$ by the condition

$$f(A)|\psi_n\rangle = f(a_n)|\psi_n\rangle \tag{1.45}$$

for all $|\psi_n\rangle$. One can readily show that $f(A)$ is also an observable. If A has non-degenerate eigenvalues it can be proved that any observable B which commutes with A is a function of A.

If the eigenvalues of A are non-degenerate the normalized eigenkets $|\psi_n\rangle$ are unique to within a *phase factor*:† hence a well-defined initial state can be obtained by measuring A. If the eigenvalues of A are degenerate we can in principle identify additional observables $B, C, \ldots$ which commute with A and each other (but are not functions of A or each other), until we have a set of commuting observables for which there is no degeneracy. Then the simultaneous eigenkets $|a_n b_p c_q \ldots\rangle$ are unique to within a phase factor; they are a basis for which the orthonormality relations are

$$\langle a_{n'} b_{p'} \ldots | a_n b_p \ldots \rangle = \delta_{n'n} \delta_{p'p} \ldots . \tag{1.46}$$

The observables $A, B, C, \ldots$ constitute a *complete set of commuting observables* (c.s.c.o.). A well-defined initial state can be obtained by an observation of a c.s.c.o.

The Hilbert space and the c.s.c.o. one associates with a physical system are dependent on the model one adopts for that system. For example, the Hilbert space and c.s.c.o. for the non-relativistic Schrödinger model of the hydrogen atom are different from those in the Dirac theory of this atom (see Chapters 9 and 12). Also, for a given model there can be more than one c.s.c.o. (see Chapters 8–11).

Suppose A and B are two non-commuting observables. Write

$$[A, B] = \mathrm{i}C, \tag{1.47}$$

where

$$[A, B] \equiv AB - BA \tag{1.48}$$

is the *commutator* of A and B. From (1.21) and (1.23), C is Hermitian. It can be shown that [8]

$$\Delta A \Delta B \geq \tfrac{1}{2}|\langle\psi|C|\psi\rangle| \tag{1.49}$$

where

$$\Delta A = [\langle\psi|(A - \langle A\rangle)^2|\psi\rangle]^{1/2}, \tag{1.50}$$

* $f(x)$ need be defined only at $x = a_n$.

† The kets $|\psi\rangle$ and $e^{i\theta}|\psi\rangle$, where θ is any real number, yield the same physical results.

and similarly for ΔB. According to the *uncertainty principle* (1.49) it is not possible for two non-commuting observables to possess a complete set of simultaneous eigenkets; in particular, if C is a non-zero real number then A and B cannot possess any simultaneous eigenkets.

Suppose $|\psi_n\rangle$ and $|\phi_n\rangle$, respectively, are the eigenkets of the non- commuting observables A and B of a system. We can use either the set $\{|\psi_n\rangle\}$ or the set $\{|\phi_n\rangle\}$ as basis kets of the state space. The two bases are related by the transformation

$$|\phi_n\rangle = U|\psi_n\rangle \tag{1.51}$$

where

$$U = \sum_i |\phi_i\rangle\langle\psi_i|. \tag{1.52}$$

Orthonormality of the $|\phi_n\rangle$ and the closure relation for the $|\psi_n\rangle$ ensure that from (1.30) and (1.31) U is a unitary operator. Thus (1.34) can also be written

$$(UAU^\dagger)(U|\psi_n\rangle) = a_n(U|\psi_n\rangle), \tag{1.53}$$

and hence $U|\psi_n\rangle$ is an eigenket of

$$A' = UAU^\dagger \tag{1.54}$$

with eigenvalue a_n.

The Hilbert space discussed in Section 1.1 is adequate for quantum-mechanical problems involving observables with discrete spectra. For observables with continuous spectra it is necessary to use a space whose kets are not restricted to have finite norm. A rigorous discussion of this topic is difficult and we merely state the changes that must be made in some of the above results.

The eigenvalue equation (for a non-degenerate spectrum) is

$$A|\psi_a\rangle = a|\psi_a\rangle \tag{1.55}$$

where the eigenvalue a has a continuous range of values. The orthonormality condition (1.35) and the expansion (1.37) become

$$\langle\psi_{a'}|\psi_a\rangle = \delta(a' - a) \tag{1.56}$$

and

$$|\psi\rangle = \int \mathrm{d}a\, |\psi_a\rangle\langle\psi_a|\psi\rangle, \tag{1.57}$$

where the integration is over all values of a. Here the Dirac δ function satisfies [9]

$$\begin{gathered} \delta(x) = 0 \quad (x \neq 0) \\ \int \delta(x)\,\mathrm{d}x = 1 \end{gathered} \tag{1.58}$$

where the range of integration includes the origin. The closure relation is

$$\int da\,|\psi_a\rangle\langle\psi_a| = 1 \tag{1.59}$$

instead of (1.39). For a normalized ket $|\psi\rangle$ the probability that an observation of A will yield a result lying between a and $a + da$ is

$$P(a)\,da = |\langle\psi_a|\psi\rangle|^2\,da. \tag{1.60}$$

Equation (1.59) and the normalization (1.7) ensure that

$$\int P(a)\,da = 1. \tag{1.61}$$

If the spectrum has both discrete and continuous parts then the expansion of an arbitrary ket is

$$|\psi\rangle = \sum_n |\psi_n\rangle\langle\psi_n|\psi\rangle + \int da\,|\psi_a\rangle\langle\psi_a|\psi\rangle. \tag{1.62}$$

1.4 Position and momentum: operators and representations

We denote the observables associated with the position and momentum of a particle by $\mathbf{r}$ and $\mathbf{p}$ respectively. In Cartesian coordinates

$$\mathbf{r} = (r_1, r_2, r_3) = (x, y, z), \tag{1.63}$$

$$\mathbf{p} = (p_1, p_2, p_3) = (p_x, p_y, p_z). \tag{1.64}$$

It is a postulate of quantum mechanics that the Cartesian components of the operators $\mathbf{r}$ and $\mathbf{p}$ satisfy the *canonical commutation relations* (or *fundamental commutation relations*)

$$[r_i, r_j] = 0 \tag{1.65}$$

$$[p_i, p_j] = 0 \tag{1.66}$$

$$[r_i, p_j] = i\hbar\delta_{ij}, \tag{1.67}$$

for i and $j = 1, 2, 3$. Here $\hbar$ denotes Planck's constant divided by 2π. The p_i are momentum operators canonically conjugate to r_i.

Consider first the position operator. In non-relativistic quantum mechanics it is assumed that r_1, r_2, r_3 is a c.s.c.o. for a spinless* free particle; then there is only one linearly independent eigenket $|\mathbf{r}'\rangle$ of $\mathbf{r}$ belonging to the eigenvalue $\mathbf{r}'$. We write the eigenvalue equation as†

$$\mathbf{r}|\mathbf{r}'\rangle = \mathbf{r}'|\mathbf{r}'\rangle, \tag{1.68}$$

which is an abbreviation of the three equations

---

* Spin is discussed in Section 4.14.

† We simplify the notation of (1.55) by using $|\mathbf{r}'\rangle$ to denote $|\psi_{\mathbf{r}'}\rangle$.

$$r_i|\mathbf{r}'\rangle = r_i'|\mathbf{r}'\rangle, \tag{1.69}$$

where $|\mathbf{r}'\rangle = |r_1'r_2'r_3'\rangle$. The components r_i' vary continuously from $-\infty$ to ∞ [3]. These kets satisfy the orthonormality condition

$$\langle\mathbf{r}''|\mathbf{r}'\rangle = \delta^3(\mathbf{r}'' - \mathbf{r}') = \delta(r_1'' - r_1')\delta(r_2'' - r_2')\delta(r_3'' - r_3') \tag{1.70}$$

and the closure relation

$$\int \mathrm{d}^3r'|\mathbf{r}'\rangle\langle\mathbf{r}'| = 1, \tag{1.71}$$

where $\mathrm{d}^3r' = \mathrm{d}x'\,\mathrm{d}y'\,\mathrm{d}z'$ and the integrations are over all space.

The expansion of an arbitrary ket in terms of the basis $|\mathbf{r}'\rangle$ is

$$|\psi\rangle = \int \mathrm{d}^3r'|\mathbf{r}'\rangle\langle\mathbf{r}'|\psi\rangle. \tag{1.72}$$

A similar expansion applies to a bra $\langle\psi|$ (see (1.38)). The choice of the eigenkets $|\mathbf{r}'\rangle$ as a basis yields the *coordinate representation*. The expansion coefficient

$$\psi(\mathbf{r}') = \langle\mathbf{r}'|\psi\rangle \tag{1.73}$$

in (1.72) is the *coordinate-space wavefunction*: it is a complex function of the continuous real variable $\mathbf{r}'$. For a normalized ket $|\psi\rangle$ the probability that a measurement of the coordinates of a particle will yield results in the intervals r_i' to $r_i' + \mathrm{d}r_i'$ is

$$|\langle\mathbf{r}'|\psi\rangle|^2\,\mathrm{d}^3r' = |\psi(\mathbf{r}')|^2\,\mathrm{d}^3r'. \tag{1.74}$$

The normalization (1.7) and the closure relation (1.71) ensure that

$$\int |\psi(\mathbf{r}')|^2\,\mathrm{d}^3r' = 1. \tag{1.75}$$

According to (1.75), $|\psi(\mathbf{r}')|^2$ is zero at infinity and hence the probability of infinite values of $\mathbf{r}'$ is zero: the motion is finite and the particle is said to be in a *bound state*.

Multiplying (1.72) on the left by $\langle\phi|F(\mathbf{r})$ we have

$$\langle\phi|F(\mathbf{r})|\psi\rangle = \int \mathrm{d}^3r'\phi^*(\mathbf{r}')F(\mathbf{r}')\psi(\mathbf{r}'). \tag{1.76}$$

A similar expression can be deduced for the matrix elements of $\mathbf{p}$. It can be shown that [2]

$$\mathbf{p}|\psi\rangle = \int \mathrm{d}^3r'|\mathbf{r}'\rangle(-\mathrm{i}\hbar\nabla_{\mathbf{r}'})\langle\mathbf{r}'|\psi\rangle \tag{1.77}$$

where

$$\nabla_{\mathbf{r}'} \equiv \left(\frac{\partial}{\partial x'}, \frac{\partial}{\partial y'}, \frac{\partial}{\partial z'}\right). \tag{1.78}$$

From (1.77) we obtain

$$\langle\phi|\mathbf{p}|\psi\rangle = \int d^3r'\phi^*(\mathbf{r}')(-i\hbar\nabla_{\mathbf{r}'})\psi(\mathbf{r}'). \tag{1.79}$$

If $F(\mathbf{r}, \mathbf{p})$ denotes a power series in r_i, p_j and products of these operators, eqns (1.76) and (1.79) generalize to

$$\langle\phi|F(\mathbf{r}, \mathbf{p})|\psi\rangle = \int d^3r'\phi^*(\mathbf{r}')F(\mathbf{r}', -i\hbar\nabla_{\mathbf{r}'})\psi(\mathbf{r}'). \tag{1.80}$$

If $\langle\phi|$ is an eigenbra of $\mathbf{r}$ the integrations in (1.79) and (1.80) can be performed using (1.70) and (1.58), and we find

$$\langle\mathbf{r}'|\mathbf{p}|\psi\rangle = -i\hbar\nabla_{\mathbf{r}'}\langle\mathbf{r}'|\psi\rangle \tag{1.81}$$

and

$$\langle\mathbf{r}'|F(\mathbf{r}, \mathbf{p})|\psi\rangle = F(\mathbf{r}', -i\hbar\nabla_{\mathbf{r}'})\langle\mathbf{r}'|\psi\rangle. \tag{1.82}$$

Equations (1.80) and (1.82) are useful results. The right-hand side of (1.80) is the formula for matrix elements of F in the coordinate representation of wave mechanics; the relation (1.82) enables us to write down the differential equations satisfied by coordinate-space wavefunctions (see below).

For the momentum operator similar results apply: thus corresponding to (1.68)–(1.72) we have

$$\mathbf{p}|\mathbf{p}'\rangle = \mathbf{p}'|\mathbf{p}'\rangle, \tag{1.83}$$

$$\langle\mathbf{p}''|\mathbf{p}'\rangle = \delta^3(\mathbf{p}'' - \mathbf{p}'), \tag{1.84}$$

$$\int d^3p'|\mathbf{p}'\rangle\langle\mathbf{p}'| = 1, \tag{1.85}$$

$$|\psi\rangle = \int d^3p'|\mathbf{p}'\rangle\langle\mathbf{p}'|\psi\rangle, \tag{1.86}$$

where the integrations are over all momentum space. The choice of the eigenkets $|\mathbf{p}'\rangle$ as a basis yields the *momentum representation*, and the expansion coefficient

$$\bar{\psi}(\mathbf{p}') = \langle\mathbf{p}'|\psi\rangle \tag{1.87}$$

in (1.86) is the *momentum-space wavefunction*. The normalization condition for this wavefunction is

$$\int d^3p'|\bar{\psi}(\mathbf{p}')|^2 = 1. \tag{1.88}$$

The expansions (1.72) and (1.86) show that there are integral relations between the coordinate- and momentum-space wavefunctions,

$$\langle \mathbf{p}'|\psi\rangle = \int d^3r' \langle \mathbf{p}'|\mathbf{r}'\rangle\langle \mathbf{r}'|\psi\rangle \tag{1.89}$$

and

$$\langle \mathbf{r}'|\psi\rangle = \int d^3p' \langle \mathbf{r}'|\mathbf{p}'\rangle\langle \mathbf{p}'|\psi\rangle. \tag{1.90}$$

We can readily evaluate $\langle \mathbf{r}'|\mathbf{p}'\rangle$ as an explicit function of $\mathbf{r}'$ and $\mathbf{p}'$. Equations (1.81) and (1.83) show that this function satisfies the three first-order partial differential equations

$$-i\hbar\nabla_{\mathbf{r}'}\langle \mathbf{r}'|\mathbf{p}'\rangle = \mathbf{p}'\langle \mathbf{r}'|\mathbf{p}'\rangle. \tag{1.91}$$

It also satisfies the orthonormality condition

$$\int d^3p' \langle \mathbf{r}''|\mathbf{p}'\rangle\langle \mathbf{p}'|\mathbf{r}'\rangle = \langle \mathbf{r}''|\mathbf{r}'\rangle = \delta(\mathbf{r}'' - \mathbf{r}'). \tag{1.92}$$

The solution to (1.91) and (1.92) is

$$\langle \mathbf{r}'|\mathbf{p}'\rangle = (2\pi\hbar)^{-3/2} \exp(i\mathbf{p}'.\mathbf{r}'/\hbar), \tag{1.93}$$

where we have used the formula [10]

$$\int_{-\infty}^{\infty} e^{ikx}\, dx = 2\pi\delta(k), \tag{1.94}$$

and set a phase factor equal to unity. Equations (1.89), (1.90), (1.93), and (1.5) show that $\bar{\psi}(\mathbf{p}')$ and $\psi(\mathbf{r}')$ are Fourier transforms of each other:

$$\bar{\psi}(\mathbf{p}') = (2\pi\hbar)^{-3/2} \int d^3r' \exp(-i\mathbf{p}'.\mathbf{r}'/\hbar)\psi(\mathbf{r}') \tag{1.95}$$

and

$$\psi(\mathbf{r}') = (2\pi\hbar)^{-3/2} \int d^3p' \exp(i\mathbf{p}'.\mathbf{r}'/\hbar)\bar{\psi}(\mathbf{p}'). \tag{1.96}$$

Corresponding to (1.77) we have

$$\mathbf{r}|\psi\rangle = \int d^3p' |\mathbf{p}'\rangle(i\hbar\nabla_{\mathbf{p}'})\langle \mathbf{p}'|\psi\rangle, \tag{1.97}$$

where

$$\nabla_{\mathbf{p}'} \equiv \left(\frac{\partial}{\partial p_x'}, \frac{\partial}{\partial p_y'}, \frac{\partial}{\partial p_z'}\right). \tag{1.98}$$

Thus

$$\langle \mathbf{p}'|\mathbf{r}|\psi\rangle = i\hbar\nabla_{\mathbf{p}'}\langle \mathbf{p}'|\psi\rangle \tag{1.99}$$

and

$$\langle\phi|F(\mathbf{p},\mathbf{r})|\psi\rangle = \int \mathrm{d}^3p'\bar{\phi}^*(\mathbf{p}')F(\mathbf{p}', i\hbar\nabla_{\mathbf{p}'})\bar{\psi}(\mathbf{p}'). \tag{1.100}$$

The right-hand side of (1.100) is the formula for matrix elements of F in the momentum representation of wave mechanics.

If there exists more than one c.s.c.o. for a system, the wavefunctions will, of course, be different for each c.s.c.o. We will encounter several examples in later sections (see Section 2.9, 8.5, 9.4, 10.7, 11.3, 11.5).

1.5 Quantum dynamics

So far we have been considering a single instant of time. At time $t = 0$ the expectation value of an observable A is

$$\langle A(0)\rangle = \langle\psi(0)|A(0)|\psi(0)\rangle \tag{1.101}$$

and at time t it is

$$\langle A(t)\rangle = \langle\psi(t)|A(t)|\psi(t)\rangle, \tag{1.102}$$

where $|\psi(t)\rangle$ is a normalized ket describing the state of the system at time t and $A(t)$ is the operator representing the observable A at time t. (Here t is used as a parameter to label different kets in the same space which describe the system at various times, and to label operators which represent the same observable at different times.)

In this section we consider isolated (conservative) systems; that is, H in (1.106) is independent of t. The time dependence on the right-hand side of (1.102) can be distributed between $|\psi(t)\rangle$ and $A(t)$ in a convenient manner, provided only that this does not alter the experimentally significant quantity $\langle A(t)\rangle$. The simplest possibilities are [3]: (i) $A(t) = A(0)$, and $|\psi(t)\rangle$ is time dependent (the *Schrödinger picture*); (ii) $|\psi(t)\rangle = |\psi(0)\rangle$, and $A(t)$ is time dependent (the *Heisenberg picture*). Because the expectation value (1.102) must be independent of the picture used, we require

$$\langle\psi(t)|A(0)|\psi(t)\rangle = \langle\psi(0)|A(t)|\psi(0)\rangle. \tag{1.103}$$

We also require that the normalization of $|\psi(t)\rangle$ be independent of time. These two conditions are satisfied if

$$|\psi(t)\rangle = U(t)|\psi(0)\rangle \tag{1.104}$$

and

$$A(t) = U^\dagger(t)A(0)U(t), \tag{1.105}$$

where $U(t)$ is a unitary operator.

It is a postulate of quantum mechanics that the *time-evolution operator* $U(t)$ satisfies the differential equation

$$i\hbar \frac{dU(t)}{dt} = HU(t), \tag{1.106}$$

where H is a Hermitian operator termed the Hamiltonian (the observable associated with the total energy of the system) [2]. If H is independent of t the solution to (1.106) is

$$U(t) = \exp(-iHt/\hbar). \tag{1.107}$$

To determine the time dependence of kets and expectation values we require the effect of $U(t)$ on the basis kets. A simple case that we will frequently encounter occurs when H and $A(0)$, $B(0)$, ... are a c.s.c.o. Then the basis kets $|\psi_n\rangle$ are also eigenkets of H and we write*

$$H|\psi_n\rangle = E_n|\psi_n\rangle. \tag{1.108}$$

The time-independent kets $|\psi_n\rangle$ are known as *energy eigenkets* and E_n is the *energy eigenvalue*.

For a system that is initially in the state

$$|\psi(0)\rangle = \sum_n |\psi_n\rangle\langle\psi_n|\psi(0)\rangle, \tag{1.109}$$

eqns (1.107), (1.108), and (1.45) show that the state (1.104) at time t is

$$|\psi(t)\rangle = \sum_n |\psi_n\rangle\langle\psi_n|\psi(0)\rangle \exp(-iE_n t/\hbar). \tag{1.110}$$

In particular, if the initial state is one of the simultaneous eigenkets $|\psi_n\rangle$ of the c.s.c.o. then

$$|\psi(t)\rangle = |\psi_n\rangle \exp(-iE_n t/\hbar), \tag{1.111}$$

and the system remains in this simultaneous eigenstate. For the ket (1.111) the expectation value of any observable Ω is (see (1.103))

$$\langle\psi(t)|\Omega(0)|\psi(t)\rangle = \langle\psi_n|\Omega(0)|\psi_n\rangle. \tag{1.112}$$

Because (1.112) is independent of time the energy eigenkets are known as *stationary states*.

If we allow each side of (1.106) to operate on $|\psi(0)\rangle$ and use (1.104) we obtain the *Schrödinger equation* for the time evolution of $|\psi(t)\rangle$

$$i\hbar \frac{d}{dt}|\psi(t)\rangle = H|\psi(t)\rangle. \tag{1.113}$$

By taking the time derivative of both sides of (1.105) and using (1.106) we find

$$\frac{dA(t)}{dt} = \frac{i}{\hbar}[H, A(t)], \tag{1.114}$$

* We suppose the spectra are discrete and we use n to denote the set of indices that label the eigenvalues of the c.s.c.o.

which is the *Heisenberg equation* of motion for observables. An observable is a *constant of the motion* if it does not depend on t. Equation (1.114) shows that all constants of the motion commute with H and that every member of a c.s.c.o., $H, A, B, \ldots$, is a constant of the motion.

1.6 Combinations of physical systems

So far our discussion has dealt with a single physical system. In quantum mechanics the state space for a combination of uncorrelated [1] physical systems is a *direct-product space*.

Suppose $\mathscr{H}_1$ and $\mathscr{H}_2$ are two spaces of dimension D_1 and D_2 respectively. The space

$$\mathscr{H} = \mathscr{H}_1 \otimes \mathscr{H}_2 \tag{1.115}$$

is called the direct-product space if to each $|\psi\rangle$ in $\mathscr{H}_1$ and $|\chi\rangle$ in $\mathscr{H}_2$ there corresponds a ket*

$$|\psi\rangle \otimes |\chi\rangle \tag{1.116}$$

in $\mathscr{H}$, where the direct product (1.116) has the properties

(i)

$$c|\psi\rangle \otimes |\chi\rangle = |\psi\rangle \otimes c|\chi\rangle = c(|\psi\rangle \otimes |\chi\rangle), \tag{1.117}$$

(c is any complex number),

(ii)

$$\begin{aligned} |\psi\rangle \otimes (|\chi_1\rangle + |\chi_2\rangle) &= |\psi\rangle \otimes |\chi_1\rangle + |\psi\rangle \otimes |\chi_2\rangle, \\ (|\psi_1\rangle + |\psi_2\rangle) \otimes |\chi\rangle &= |\psi_1\rangle \otimes |\chi\rangle + |\psi_2\rangle \otimes |\chi\rangle, \end{aligned} \tag{1.118}$$

(iii) If $|\psi_n\rangle$ $(n = 1, 2, \ldots, D_1)$ and $|\chi_m\rangle$ $(m = 1, 2, \ldots, D_2)$ are bases in $\mathscr{H}_1$ and $\mathscr{H}_2$, then the set of $D_1 D_2$ kets $|\psi_n\rangle \otimes |\chi_m\rangle$ is a basis in $\mathscr{H}$.

The scalar product of $|\psi\rangle \otimes |\chi\rangle$ and $|\psi'\rangle \otimes |\chi'\rangle$ is defined as

$$(\langle\psi'|\psi\rangle)(\langle\chi'|\chi\rangle). \tag{1.119}$$

If A and B are linear operators in $\mathscr{H}_1$ and $\mathscr{H}_2$ respectively, then the linear operator

$$C = A \otimes B \tag{1.120}$$

in $\mathscr{H}$ is defined by

$$C(|\psi\rangle \otimes |\chi\rangle) = (A|\psi\rangle) \otimes (B|\chi\rangle). \tag{1.121}$$

* We will sometimes omit the symbol $\otimes$ in direct products such as (1.116).

It can be shown that if A, A', ... and B, B', ... are c.s.c.o.s in $\mathscr{H}_1$ and $\mathscr{H}_2$ respectively, then $A \otimes 1, A' \otimes 1, \ldots, 1 \otimes B, 1 \otimes B', \ldots$ is a c.s.c.o. in $\mathscr{H}$ [1].

Suppose $\mathscr{H}_1$ and $\mathscr{H}_2$ are the state spaces of two uncorrelated [1] physical systems. It is a postulate of quantum mechanics that the state space of the combination of these two systems is the direct-product space $\mathscr{H}$, and that the observables of the combined system are given in terms of the observables of each system by (1.120).

Examples of such combinations are the vibrating rotator,* the isotropic multi-dimensional oscillator (Section 2.10), a non-relativistic particle with spin (Section 4.14), and the two-body problem for non-identical particles (Section 5.1).

1.7 Wave mechanics

Suppose the Hamiltonian operator is a function of $\mathbf{r}$ and $\mathbf{p}$. By multiplying both sides of the Schrödinger equation (1.113) on the left by $\langle \mathbf{r}'|$ and using (1.82) and

$$\langle \mathbf{r}'|\frac{\mathrm{d}}{\mathrm{d}t}|\psi(t)\rangle = \frac{\partial}{\partial t}\langle \mathbf{r}'|\psi(t)\rangle = \frac{\partial}{\partial t}\Psi(\mathbf{r}', t),$$

we obtain Schrödinger's time-dependent wave equation

$$H(\mathbf{r}', -\mathrm{i}\hbar\nabla_{\mathbf{r}'})\Psi(\mathbf{r}', t) = \mathrm{i}\hbar\frac{\partial}{\partial t}\Psi(\mathbf{r}', t). \tag{1.122}$$

For a stationary state (see (1.111))

$$\Psi(\mathbf{r}', t) = \langle \mathbf{r}'|\psi_n\rangle \exp(-\mathrm{i}E_n t/\hbar), \tag{1.123}$$

and (1.122) reduces to Schrödinger's time-independent wave equation

$$H(\mathbf{r}', -\mathrm{i}\hbar\nabla_{\mathbf{r}'})\psi_n(\mathbf{r}') = E_n\psi_n(\mathbf{r}'), \tag{1.124}$$

where $\psi_n(\mathbf{r}') = \langle \mathbf{r}'|\psi_n\rangle$.

For a non-relativistic particle of mass M in a potential $V(\mathbf{r})$ the Hamiltonian operator is

$$H = \frac{1}{2M}\mathbf{p}^2 + V(\mathbf{r}), \tag{1.125}$$

and (1.124) is a second-order partial differential equation. Differential equations for other observables that depend on $\mathbf{r}$ and $\mathbf{p}$ can be written down in a similar manner. The solution of the simultaneous differential equations (for a c.s.c.o.) subject to certain boundary conditions [11] for ψ is the wave-mechanical method of solving quantum-mechanical problems. The wavefunc-

---

* For an excellent discussion of this topic see Ref. 3.

tions are calculated by applying techniques such as separation of variables (in a particular coordinate system) and series solutions to the differential equations [12]. Once the wavefunctions are known, the matrix elements (1.80) can be determined by integration.

An alternative and completely equivalent* formulation of wave mechanics can be given in terms of the momentum representation (Section 1.4). In this representation the time-dependent wave equation is

$$H(\mathrm{i}\hbar\nabla_{\mathbf{p}'}, \mathbf{p}')\bar{\Psi}(\mathbf{p}', t) = \mathrm{i}\hbar\frac{\partial}{\partial t}\bar{\Psi}(\mathbf{p}', t). \tag{1.126}$$

For most problems (1.126) is an integral equation [13]. The momentum representation is frequently used in scattering theory [3].

In the abstract formulation of quantum mechanics one is not concerned with any particular representation for operators or kets and the analysis of physical problems employs operator methods. The basic tools of these methods are commutation relations, special operators known as shift operators, and the virial, hypervirial, and Hellmann–Feynman theorems. In the remainder of this chapter we discuss these tools.

1.8 Commutation relations

We derive some general results which enable us to evaluate commutation relations of operators that are functions of **r** and **p**.

If†

$$f(z) = \sum_{n=-\infty}^{\infty} a_n(z - z_0)^n \tag{1.127}$$

then‡

$$[p_x, f(x)] = -\mathrm{i}\hbar\frac{\mathrm{d}f(x)}{\mathrm{d}x} \tag{1.128}$$

and

$$[x, f(p_x)] = \mathrm{i}\hbar\frac{\mathrm{d}f(p_x)}{\mathrm{d}p_x}. \tag{1.129}$$

Proof

To prove (1.128) we must show that for $n = 0, \pm 1, \pm 2, \ldots$

* The state spaces of the various representations are isomorphic to each other and to the abstract state space of quantum mechanics. As a result, scalar products are independent of which space one uses.

† Thus $f(z)$ is analytic except at an isolated point z_0: eqn (1.127) is the Laurent series for this function [14].

‡ The derivative in (1.128) means: replace the operator x with the number x' and differentiate $f(x')$ with respect to x'. Then replace x' with x in this derivative. Similarly for the derivatives in (1.129), (1.136), and (1.137).

$$[p_x, (x-c)^n] = -i\hbar n(x-c)^{n-1}, \tag{1.130}$$

where c is a constant. From (1.67)

$$[p_x, x-c] = -i\hbar \tag{1.131}$$

and hence (1.130) is true for $n = 1$. Suppose it is also true for a particular $n > 1$. Then (1.130) and (1.131) show that

$$[p_x, (x-c)^{n+1}] = -i\hbar(n+1)(x-c)^n. \tag{1.132}$$

Hence (1.130) is valid for $n = 1, 2, \ldots$.

Now

$$[p_x, (x-c)(x-c)^{-1}] = [p_x, 1] = 0. \tag{1.133}$$

From (1.133) and (1.131)

$$[p_x, (x-c)^{-1}] = i\hbar(x-c)^{-2}, \tag{1.134}$$

and hence (1.130) is true for $n = -1$. Suppose it is also valid for some $n < -1$. Then (1.130) and (1.134) show that

$$[p_x, (x-c)^{n-1}] = -i\hbar(n-1)(x-c)^{n-2}. \tag{1.135}$$

Hence (1.130) is true for $n = -1, -2, \ldots$. This proves (1.128): the proof of (1.129) is similar.

The commutation relations (1.128) and (1.129) are sufficient for all the one-dimensional problems discussed in this book (see Chapters 2 and 3). For three-dimensional problems we require the following generalizations: if $f(\mathbf{u})$ is an analytic function of u_1, u_2, u_3 (except perhaps at isolated points) then

$$[p_j, f(\mathbf{r})] = -i\hbar\frac{\partial f(\mathbf{r})}{\partial r_j} \tag{1.136}$$

and

$$[r_j, f(\mathbf{p})] = i\hbar\frac{\partial f(\mathbf{p})}{\partial p_j} \tag{1.137}$$

for $j = 1, 2, 3$. Thus, for example,*

$$p_j r_j = r_j p_j - i\hbar\frac{\partial r_j}{\partial r_j};$$

that is,

$$\mathbf{p}.\mathbf{r} = \mathbf{r}.\mathbf{p} - 3i\hbar. \tag{1.138}$$

Other commutators can be evaluated in a similar manner and the following

---

* We adopt the convention that a repeated index implies summation from 1 to 3.

list gives those that we shall use later:

$$[\mathbf{r}.\mathbf{p}, \mathbf{r}] = -i\hbar\mathbf{r} \tag{1.139}$$

$$[\mathbf{p}.\mathbf{r}, \mathbf{p}] = i\hbar\mathbf{p} \tag{1.140}$$

$$[\mathbf{r}.\mathbf{p}, r] = -i\hbar r \tag{1.141}$$

$$[\mathbf{p}.\mathbf{r}, p] = i\hbar p \tag{1.142}$$

$$[\mathbf{r}.\mathbf{p}, r^{-1}] = i\hbar r^{-1} \tag{1.143}$$

$$[\mathbf{p}.\mathbf{r}, p^{-1}] = -i\hbar p^{-1} \tag{1.144}$$

$$[\mathbf{r}.\mathbf{p}, \hat{\mathbf{r}}] = 0 \tag{1.145}$$

$$[\mathbf{p}.\mathbf{r}, \hat{\mathbf{p}}] = 0 \tag{1.146}$$

$$[\mathbf{r}, p^2] = 2i\hbar\mathbf{p} \tag{1.147}$$

$$[\mathbf{p}, r^2] = -2i\hbar\mathbf{r} \tag{1.148}$$

$$[\mathbf{p}, r] = -i\hbar\hat{\mathbf{r}} \tag{1.149}$$

$$[\mathbf{p}, r^{-1}] = i\hbar r^{-2}\hat{\mathbf{r}} \tag{1.150}$$

$$[p^2, r^{-1}] = 2i\hbar r^{-3}\mathbf{r}.\mathbf{p} = 2i\hbar\mathbf{p}.\mathbf{r}\, r^{-3} \tag{1.151}$$

$$[p^2, \hat{\mathbf{r}}] = 2i\hbar r^{-3}(\mathbf{r}.\mathbf{p}\mathbf{r} - r^2\mathbf{p}) \tag{1.152}$$

$$[p^2, r] = -2i\hbar r^{-1}(\mathbf{r}.\mathbf{p} - i\hbar), \tag{1.153}$$

where $r = (\mathbf{r}.\mathbf{r})^{1/2}$, $p = (\mathbf{p}.\mathbf{p})^{1/2}$, $\hat{\mathbf{r}} = r^{-1}\mathbf{r}$, and $\hat{\mathbf{p}} = p^{-1}\mathbf{p}$.

In calculations of commutators involving vector products we will also find the following formula useful:

$$(\mathbf{a} \times \mathbf{b})_i = \varepsilon_{ijk} a_j b_k \tag{1.154}$$

where ε_{ijk} is the totally antisymmetric tensor in three dimensions

$$\varepsilon_{ijk} = \begin{cases} 1 & ijk = \text{any even permutation of } 123, \\ -1 & ijk = \text{any odd permutation of } 123, \\ 0 & \text{if any two indices are equal.} \end{cases} \tag{1.155}$$

This tensor satisfies the identities

$$\varepsilon_{ijk}\varepsilon_{ijl} = 2\delta_{kl} \tag{1.156}$$

and

$$\varepsilon_{ijk}\varepsilon_{ilm} = \delta_{jl}\delta_{km} - \delta_{jm}\delta_{kl}. \tag{1.157}$$

1.9 Shift operators

An operator transforms a ket into another ket of the same space. A *shift operator* (or *ladder operator*) transforms between the eigenkets of observables.

Suppose that H (the Hamiltonian or some other observable) is a c.s.c.o. with eigenvalue equation

$$H|\psi\rangle = E|\psi\rangle. \tag{1.158}$$

The spectrum of H may be discrete or continuous. If there exist operators $\mathscr{L}^{\pm}$ such that

$$[H, \mathscr{L}^{\pm}] = \pm c\mathscr{L}^{\pm}, \tag{1.159}$$

where c is a positive constant, then

$$H\{\mathscr{L}^{\pm}|\psi\rangle\} = (E \pm c)\{\mathscr{L}^{\pm}|\psi\rangle\}. \tag{1.160}$$

Hence either $\mathscr{L}^{\pm}|\psi\rangle$ are eigenkets of H with eigenvalues $E \pm c$ or $\mathscr{L}^{\pm}|\psi\rangle = 0$. We denote $|\psi\rangle$ by $|E\rangle$, and write

$$\mathscr{L}^{\pm}|E\rangle = \alpha^{\pm}(E, c)|E \pm c\rangle, \tag{1.161}$$

where the numbers $\alpha^{\pm}$ are to be determined. Equations (1.161) are *shift operations*: $\mathscr{L}^{+}$ and $\mathscr{L}^{-}$, respectively, are *raising* and *lowering operators* for the eigenvalue E. The *coefficients* $\alpha^{\pm}$ play an important role in the applications of shift operators. Taking the norm of each side of (1.161) we have

$$|\alpha^{\pm}|^2\langle E \pm c|E \pm c\rangle = \langle E|(\mathscr{L}^{\pm})^{\dagger}\mathscr{L}^{\pm}|E\rangle, \tag{1.162}$$

and thus for normalized kets

$$|\alpha^{\pm}|^2 = \langle E|(\mathscr{L}^{\pm})^{\dagger}\mathscr{L}^{\pm}|E\rangle \tag{1.163}$$

$$\geq 0. \tag{1.164}$$

The condition (1.164) follows from the fact that (1.163) is a norm.

One way of deriving shift operators is to require that they *factorize* H; that is,

$$\mathscr{L}^{\mp}\mathscr{L}^{\pm} = H \pm \tfrac{1}{2}c. \tag{1.165}$$

Here the upper signs go together as do the lower signs, and (1.165) is a *two-way* factorization. By operating on the left on both sides of (1.165) with $\mathscr{L}^{\pm}$ we obtain (1.159). Shift operators are *mutually adjoint* if

$$(\mathscr{L}^{+})^{\dagger} = \mathscr{L}^{-}. \tag{1.166}$$

For mutually adjoint shift operators that factorize H, eqn (1.163) yields

$$|\alpha^{\pm}|^2 = E \pm \tfrac{1}{2}c. \tag{1.167}$$

Suppose there exists a normalizable ket $|E\rangle$ with $E > 0$, and consider the sequence of kets obtained by repeated application of $\mathscr{L}^{+}$ to $|E\rangle$

$$(\mathscr{L}^{+})^N|E\rangle \qquad (N = 1, 2, \ldots). \tag{1.168}$$

Let α_N^{+} denote the coefficient for the Nth such application. Equation (1.167) shows that

$$|\alpha_N^+|^2 = E + c(N-1) + \tfrac{1}{2}c \qquad (N = 1, 2, \ldots). \tag{1.169}$$

This satisfies (1.164) and hence the kets (1.168) are normalizable.

For the sequence

$$(\mathscr{L}^-)^N |E\rangle \qquad (N = 1, 2, \ldots) \tag{1.170}$$

eqn (1.167) gives

$$|\alpha_N^-|^2 = E - c(N-1) - \tfrac{1}{2}c \qquad (N = 1, 2, \ldots), \tag{1.171}$$

and there are two possibilities:

(i) If

$$E = c(n + \tfrac{1}{2}) \qquad (n = 0, 1, \ldots), \tag{1.172}$$

then (1.170) will terminate when $N = n + 1$. According to (1.164) and (1.171), each of the n kets in (1.170) will be normalizable.

(ii) If $E \neq c(n + \frac{1}{2})$ the sequence (1.170) will not terminate and when $N > 1 + (E - \frac{1}{2}c)/c$, (1.171) violates (1.164).

Hence the normalizable kets are $|\psi_n\rangle$ $(n = 0, 1, \ldots)$ and their eigenvalues are given by the discrete spectrum (1.172). The 'ground state' $|\psi_0\rangle$ satisfies the equation

$$\mathscr{L}^- |\psi_0\rangle = 0, \tag{1.173}$$

and all excited states are generated from $|\psi_0\rangle$ by repeated application of the raising operator:

$$|\psi_n\rangle = (\alpha_n^+ \ldots \alpha_2^+ \alpha_1^+)^{-1} (\mathscr{L}^+)^n |\psi_0\rangle. \tag{1.174}$$

For a continuous spectrum the shift operators in (1.161) will generate all states from the (non-denumerably infinite) set of states in E to $E + c$.

The preceding is the simplest description of shift operators and it applies, for example, to the energy eigenkets of the one-dimensional oscillator (see Chapter 2). There are several variations of the above:

(i) In (1.159), c need not be a constant. For example, one could have

$$[H, \mathscr{L}^\pm] = \mathscr{L}^\pm f^\pm(H), \tag{1.175}$$

where $f^\pm(H)$ are real functions of H. Then $\mathscr{L}^\pm$ transform $|E\rangle$ into $|E + f^\pm(E)\rangle$. Examples are the shift operators for the orbital angular momentum quantum number l (see (4.51) and (4.58)).

(ii) We will encounter cases where H, and consequently $\mathscr{L}^\pm$, are dependent on a parameter or quantum number k. Then $\mathscr{L}_k^\pm$ change the value of k in the eigenket $|Ek\rangle$ (see Sections 3.2 and 5.4). We will refer also to $\mathscr{L}_k^\pm$ as shift operators. In certain cases $\mathscr{L}_k^\pm$ transform between kets of different physical systems (see Sections 3.4–3.7).

(iii) Usually shift operators for energy are obtained by factorization of an operator other than the Hamiltonian (see Chapters 3, 5, 7, 10, 12).

(iv) Shift operators need not be mutually adjoint (see Chapters 5, 7, 10, 12).

(v) Instead of (1.165) one can consider the less restrictive condition

$$\mathscr{L}^+ \mathscr{L}^- = H - \tfrac{1}{2}c. \tag{1.176}$$

Such a *one-way* factorization is useful in obtaining generalized potentials and in supersymmetric quantum mechanics (see Chapters 3 and 5).

(vi) If the c.s.c.o. contains two or more observables it is sometimes possible to use dynamical variables instead of factorization to construct shift operators (see Chapters 8–10 and 12).

The factorization of H into shift operators for the discrete spectrum is a property of Hilbert space and is not restricted to any particular representation [15]. If H has a discrete spectrum it can be written (see (1.41))

$$H = \sum_n E_n |\psi_n\rangle\langle\psi_n|, \tag{1.177}$$

and thus the one-way factorization $\mathscr{L}^+ \mathscr{L}^- = H - E_0$ is provided by operators which have the spectral decompositions

$$\mathscr{L}^+ = \sum_n (E_{n+1} - E_0)^{1/2} |\psi_{n+1}\rangle\langle\psi_n|, \tag{1.178}$$

$$\mathscr{L}^- = \sum_n (E_{n+1} - E_0)^{1/2} |\psi_n\rangle\langle\psi_{n+1}|. \tag{1.179}$$

These mutually adjoint operators perform the shift operations

$$\mathscr{L}^+ |\psi_n\rangle = (E_{n+1} - E_0)^{1/2} |\psi_{n+1}\rangle, \tag{1.180}$$

$$\mathscr{L}^- |\psi_n\rangle = (E_n - E_0)^{1/2} |\psi_{n-1}\rangle. \tag{1.181}$$

The application of the shift operator method to discrete spectra rests on the assumption that there exists at least one normalizable ket. A simple way of checking this assumption is to construct a representative of $|\psi_n\rangle$. Suppose the discrete spectra are bounded below. Then $|\psi_0\rangle$ satisfies a set of operator equations*

$$\mathscr{L}_i^- |\psi_0\rangle = 0 \qquad (i = 1, 2, \ldots), \tag{1.182}$$

where $\mathscr{L}_i^-$ are lowering operators for the eigenvalues of the discrete spectra. For many tractable problems (1.182) expressed in a suitable representation (usually the coordinate representation) are a set of first-order partial differential equations which can be solved in a simple manner for $\psi_0(\mathbf{r}')$ (say). Because

$$\langle\psi_0|\psi_0\rangle = \int \mathrm{d}^3 r' |\psi_0(\mathbf{r}')|^2 \tag{1.183}$$

(see (1.72)), $|\psi_0\rangle$ is normalizable if $\psi_0(\mathbf{r}')$ is square-integrable.

* For example, for the Coulomb potential (1.182) consists of a set of three equations: see Sections 9.4 and 10.7.

It is, in any case, useful to calculate the normalized wavefunction $\psi_0(\mathbf{r}')$: repeated application of differential raising operators to this function is a simple method of determining all the normalized wavefunctions of the discrete spectrum. In this book coordinate- and momentum-space wavefunctions are obtained by this method.

An essential ingredient of the shift operator method is the calculation of the coefficients $\alpha^{\pm}$. If the shift operators are not mutually adjoint, then these coefficients cannot be determined using eigenvalue equations such as (1.158) alone; additional techniques are required, such as those described below.

1.10 The virial and hypervirial theorems

We use a simple scaling argument [16] to prove the virial theorem. Consider the Hamiltonian

$$\hat{H} = \frac{1}{2M}\mathbf{p}^2 + V(\xi\mathbf{r}), \tag{1.184}$$

where ξ is an arbitrary real parameter. Then

$$\left\langle \xi \frac{\partial \hat{H}}{\partial \xi}\bigg|_{\mathbf{r},\mathbf{p}} \right\rangle = \left\langle \xi \frac{\partial V(\xi\mathbf{r})}{\partial \xi} \right\rangle. \tag{1.185}$$

Here $\langle\ \rangle$ denotes an expectation value with respect to a normalizable* eigenket of $\hat{H}$. In terms of the conjugate operators $\mathbf{r}' = \xi\mathbf{r}$ and $\mathbf{p}' = \xi^{-1}\mathbf{p}$

$$\hat{H} = \xi^2 \frac{1}{2M}(\mathbf{p}')^2 + V(\mathbf{r}'). \tag{1.186}$$

Hence

$$\left\langle \xi \frac{\partial \hat{H}}{\partial \xi}\bigg|_{\mathbf{r}',\mathbf{p}'} \right\rangle = 2\left\langle \xi^2 \frac{1}{2M}(\mathbf{p}')^2 \right\rangle = 2\langle \hat{T} \rangle, \tag{1.187}$$

where

$$\hat{T} = \hat{H} - V(\mathbf{r}') \tag{1.188}$$

is the kinetic energy operator. The left-hand sides of (1.185) and (1.187) are equal by the Hellmann–Feynman theorem (1.203), and hence

$$2\langle \hat{T} \rangle = \sum_{i=1}^{3} \left\langle r_i \frac{\partial V(\xi\mathbf{r})}{\partial r_i} \right\rangle. \tag{1.189}$$

Equation (1.189) is the virial theorem for a non-relativistic particle. With

* The virial theorem can also be generalized to certain non-normalizable states [17].

$\xi = 1$ it has the convenient form

$$2\langle T \rangle = \sum_{i=1}^{3} \left\langle r_i \frac{\partial V(\mathbf{r})}{\partial r_i} \right\rangle. \tag{1.190}$$

In particular, if $V(\mathbf{r})$ is a homogeneous function of degree ν,

$$V(\xi\mathbf{r}) = \xi^{\nu} V(\mathbf{r}), \tag{1.191}$$

then

$$2\langle T \rangle = \nu\langle V \rangle. \tag{1.192}$$

The virial theorem can be generalized to relativistic equations (Section 12.5), to a charged particle in an electromagnetic field [16], and other problems [17].

The hypervirial theorem is an obvious result that was used in some of the earliest work on matrix mechanics [18]. Suppose A is Hermitian with respect to $|\psi\rangle$ and $W|\psi\rangle$, where $|\psi\rangle$ is an eigenket of A and W is some operator. Then

$$\langle\psi|AW|\psi\rangle = \langle\psi|W^{\dagger}A|\psi\rangle^* = a\langle\psi|W|\psi\rangle = \langle\psi|WA|\psi\rangle.$$

Hence we obtain the hypervirial theorem

$$\langle\psi|[A, W]|\psi\rangle = 0. \tag{1.193}$$

To apply this theorem one must, for a given A, make suitable choices for the operator W. For example, suppose A is the non-relativistic Hamiltonian (1.125). Choosing $W = \mathbf{r}.\mathbf{p}$ and using the commutation relations (1.136) and (1.147) we obtain (1.190). This proof of the virial theorem is not entirely general because H is not always Hermitian with respect to $\mathbf{r}.\mathbf{p}|\psi\rangle$ [19].

Equation (1.193) can be extended: if in the calculation leading to (1.193) we consider matrix elements instead of expectation values, we find

$$\langle\psi_j|[A, W]|\psi_k\rangle = (a_j - a_k)\langle\psi_j|W|\psi_k\rangle. \tag{1.194}$$

Thus, for example, with $W = p_i$ and A given by (1.125) this yields

$$\left\langle \psi_j \left| \frac{\partial V(\mathbf{r})}{\partial r_i} \right| \psi_k \right\rangle = -i\hbar^{-1}(E_j - E_k)\langle\psi_j|p_i|\psi_k\rangle. \tag{1.195}$$

1.11 The Hellmann–Feynman theorem and its extensions

Let $A(\xi)$ be an observable that depends on a continuous real parameter ξ. Suppose the spectrum of $A(\xi)$ is discrete and denote the eigenvalues and orthonormal eigenkets by $a_i(\xi)$ and $|\psi_i(\xi)\rangle$. Then*

$$\langle\psi_j|\psi_i\rangle = \delta_{ij} \tag{1.196}$$

* For convenience we suppress the dependence on ξ: thus A, a_i, and $|\psi_i\rangle$ denote $A(\xi)$, $a_i(\xi)$, and $|\psi_i(\xi)\rangle$.

and

$$A|\psi_i\rangle = a_i|\psi_i\rangle. \tag{1.197}$$

A small variation $\delta\xi$ changes these equations to

$$(\langle\psi_j| + \langle\delta\psi_j|)(|\psi_i\rangle + |\delta\psi_i\rangle) = \delta_{ij} \tag{1.198}$$

and

$$(A + \delta A)(|\psi_i\rangle + |\delta\psi_i\rangle) = (a_i + \delta a_i)(|\psi_i\rangle + |\delta\psi_i\rangle), \tag{1.199}$$

where

$$|\delta\psi_i\rangle \equiv \delta|\psi_i(\xi)\rangle = |\psi_i(\xi + \delta\xi)\rangle - |\psi_i(\xi)\rangle. \tag{1.200}$$

From (1.196) and (1.198) we have to first order

$$\langle\delta\psi_j|\psi_i\rangle + \langle\psi_j|\delta\psi_i\rangle = 0. \tag{1.201}$$

Multiplying (1.199) on the left by $\langle\psi_i| + \langle\delta\psi_i|$ and neglecting second-order terms we obtain

$$\langle\psi_i|\delta A|\psi_i\rangle + \langle\delta\psi_i|A|\psi_i\rangle + \langle\psi_i|A|\delta\psi_i\rangle = \delta a_i. \tag{1.202}$$

Because A is Hermitian, (1.197) and (1.201) show that the sum of the second and third terms in (1.202) is zero. Hence

$$\langle\psi_i(\xi)|\frac{\partial A(\xi)}{\partial\xi}|\psi_i(\xi)\rangle = \frac{\partial a_i(\xi)}{\partial\xi}. \tag{1.203}$$

Equation (1.203) is the Hellmann–Feynman theorem [20, 21]. This useful result provides a simple method of calculating certain expectation values and of determining the condition for a potential well to bind a particle (see Chapters 3, 6, 10, 12); it also has applications to solid state theory and quantum chemistry [22].

If we multiply (1.199) on the left by $\langle\psi_j| + \langle\delta\psi_j|$ and use (1.196)–(1.198), the Hermitian property of A, and (1.201), we obtain Epstein's extension of the Hellmann–Feynman theorem [21, 23]

$$\langle\psi_j(\xi)|\frac{\partial A(\xi)}{\partial\xi}|\psi_i(\xi)\rangle + \beta_{ji}[a_j(\xi) - a_i(\xi)] = \frac{\partial a_i(\xi)}{\partial\xi}\delta_{ij}, \tag{1.204}$$

where

$$\beta_{ji} = \left\langle\psi_j(\xi)\middle|\frac{\partial\psi_i(\xi)}{\partial\xi}\right\rangle \tag{1.205}$$

$$= -\beta_{ij}^* \tag{1.206}$$

and

$$\left|\frac{\partial\psi_i(\xi)}{\partial\xi}\right\rangle = \lim_{\delta\xi\to 0}\frac{\delta|\psi_i(\xi)\rangle}{\delta\xi}. \tag{1.207}$$

If $B(\xi)$ is a second operator that depends on ξ then

$$\delta\langle\psi_i|BA|\psi_i\rangle = \langle\delta\psi_i|BA|\psi_i\rangle + \langle\psi_i|BA|\delta\psi_i\rangle + \langle\psi_i|(\delta B)A + B(\delta A)|\psi_i\rangle. \tag{1.208}$$

But

$$\begin{aligned}\delta\langle\psi_i|BA|\psi_i\rangle &= \delta\langle\psi_i|B|\psi_i\rangle a_i\\ &= a_i(\langle\delta\psi_i|B|\psi_i\rangle + \langle\psi_i|B|\delta\psi_i\rangle + \langle\psi_i|\delta B|\psi_i\rangle)\\ &\quad + \langle\psi_i|B|\psi_i\rangle\delta a_i.\end{aligned} \tag{1.209}$$

Equating (1.208) and (1.209), and using the Hermitian property of A, we have

$$\langle\psi_i|B\delta A|\psi_i\rangle = \langle\psi_i|B|\psi_i\rangle\delta a_i + \langle\psi_i|[A, B]|\delta\psi_i\rangle.$$

Thus if B commutes with A we obtain another extension of the Hellmann–Feynman theorem, namely [24],

$$\langle\psi_i(\xi)|B(\xi)\frac{\partial A(\xi)}{\partial\xi}|\psi_i(\xi)\rangle = \langle\psi_i(\xi)|B(\xi)|\psi_i(\xi)\rangle\frac{\partial a_i(\xi)}{\partial\xi}. \tag{1.210}$$

From eqns (1.203) and (1.204) we can derive sum rules involving matrix elements of derivatives of $A(\xi)$ with respect to ξ. By considering a small variation $\delta\xi$ in (1.203) we find

$$\left\langle\frac{\partial\psi_i}{\partial\xi}\middle|\frac{\partial A}{\partial\xi}\middle|\psi_i\right\rangle + \left\langle\psi_i\middle|\frac{\partial A}{\partial\xi}\middle|\frac{\partial\psi_i}{\partial\xi}\right\rangle + \left\langle\psi_i\middle|\frac{\partial^2 A}{\partial\xi^2}\middle|\psi_i\right\rangle = \frac{\partial^2 a_i}{\partial\xi^2}. \tag{1.211}$$

Now

$$\left|\frac{\partial\psi_i}{\partial\xi}\right\rangle = \sum_j |\psi_j\rangle\left\langle\psi_j\middle|\frac{\partial\psi_i}{\partial\xi}\right\rangle = \sum_j \beta_{ji}|\psi_j\rangle. \tag{1.212}$$

(See (1.37) and (1.205).) Substituting (1.212) in (1.211) and using (1.204) and (1.206) to eliminate β_{ji} and β_{ji}^*, we obtain the sum rule

$$\frac{\partial^2 a_i}{\partial\xi^2} - \langle\psi_i|\frac{\partial^2 A}{\partial\xi^2}|\psi_i\rangle = 2\sum_j{}' \frac{\langle\psi_i|\frac{\partial A}{\partial\xi}|\psi_j\rangle\langle\psi_j|\frac{\partial A}{\partial\xi}|\psi_i\rangle}{a_i - a_j}. \tag{1.213}$$

Here the prime means that the term with $j = i$ is excluded from the sum. Proceeding in this way one can deduce a series of sum rules [25].

1.12 Discussion

This completes our introductory survey of the mathematical formalism of quantum mechanics. Before proceeding we summarize the results we will most frequently use.

(i) Non-negativity of the norm (see (1.6)).

(ii) The concept that an operator performs transformations in the state space (see (1.11)) and, in particular, of shift operations as transformations between the eigenkets of observables (see (1.161)).

(iii) Non-commutation of operators. The commutation relations (1.136) and (1.137), and others such as (4.1), are basic to all our considerations.

(iv) The Hermitian property (1.24) and results such as (1.20)–(1.23) for taking the adjoint of an operator.

(v) Orthonormality of the eigenkets of a Hermitian operator (in the sense of

TABLE 1.1(a) Coordinate-space wavefunctions

System	Section
1. One-dimensional harmonic oscilator	2.6
2. Coherent states of the oscillator	2.6
3. Charged oscillator in a uniform electric field	2.7
4. Charged particle in a uniform magnetic field	2.8
5. Isotropic multi-dimensional oscillator	2.10
6. Generalized one-dimensional harmonic potentials	3.1
7. One-dimensional Morse potential	3.4
8. One-dimensional Rosen–Morse potential	3.5
9. Pöschl–Teller potentials	3.6, 3.7
10. Spherical harmonics	4.10
11. Three-dimensional oscillator (angular momentum basis)	6.4, 8.5
12. Coulomb potential (angular momentum basis)	6.4, 9.4
13. Hulthén potential	6.4
14. Generalized three-dimensional oscillator potentials	5.8
15. Generalized Coulomb potentials	5.8
16. Sectionally constant potentials (angular momentum basis)	6.5–6.7
17. Three-dimensional Morse potential	5.10
18. Three-dimensional symmetric Rosen–Morse potential	5.10
19. Coulomb potential (parabolic basis)	10.7
20. Relativistic charged particle in a uniform magnetic field	11.5
21. Relativistic Coulomb problems	12.6
22. Dirac oscillator	12.10

TABLE 1.1(b) Momentum-space wavefunctions

System	Section
1. One-dimensional harmonic oscillator	2.6
2. Coherent states of the oscillator	2.6
3. Isotropic multi-dimensional oscillator	2.10
4. Three-dimensional oscillator (angular momentum basis)	7.6, 8.5
5. Coulomb potential (angular momentum basis)	7.6, 9.4
6. Dirac oscillator	12.10

(1.35) or (1.56)) and the property that the eigenvalues of such an operator are real.

(vi) The concept of the representative of a ket as a complex function of continuous real variables (see (1.73) and (1.87)).

The reader who is making a first encounter with operator methods may find it helpful to select material for a first reading of this book. A suitable starting point is the one-dimensional harmonic oscillator, for which the basic theory is given in Sections 2.1 and 2.2, and two simple examples are discussed in Sections 2.7 and 2.8. Other one-dimensional systems are considered in Section 3.2 where the factorization method is discussed in some detail. Four worked examples are presented in Sections 3.4–3.7, and it would be helpful if the reader studied at least one of these; all the calculations can be performed *ab initio*.

The other archetypal problem (after the oscillator) is the theory of angular momentum of a single particle (Sections 4.1–4.5). This can be followed by the important topic of spherically symmetric potentials (Sections 5.1–5.7, 5.9, and 5.10). Some applications to the evaluation of matrix elements are given in Sections 6.1–6.3.

For the reader who wishes to use operator methods to evaluate wavefunctions, we have attempted to make the calculations of these functions self-contained. Table 1.1 indicates where they are to be found.

References

1. Cohen-Tannoudji, C., Diu, B., and Laloë, F. (1977). *Quantum mechanics*, Vol. 1. Wiley, New York.
2. Sakurai, J.J. (1985). *Modern quantum mechanics*. Benjamin, Menlo Park, CA.
3. Böhm, A. (1979). *Quantum mechanics*. Springer, New York. See also Böhm, A. (1978). *The rigged Hilbert space and quantum mechanics*, Springer Lecture Notes in Physics, Vol. 78.
4. Jordan, T.F. (1986). *Quantum mechanics in simple matrix form*. Wiley, New York.
5. See, for example, Prugovečki, E. (1981). *Quantum mechanics in Hilbert space*, 2nd edn. Academic Press, New York. Jordan, T.F. (1969). *Linear operators for quantum mechanics*. Wiley, New York. Reed, M. and Simon, B. (1972). *Methods in modern mathematical physics, I. Functional analysis*. Academic Press, New York.
6. Omnès, R. (1990). From Hilbert space to common sense: A synthesis of recent progress in the interpretation of quantum mechanics. *Annals of Physics*, **201**, 354–447.
7. Biedenharn, L.C. and Louck, J.D. (1981). *The Racah–Wigner algebra in quantum theory*. Addison-Wesley, Reading, MA.
8. For the general proof of this result see Carruthers, P. and Nieto, M.M. (1968). Phase and angle variables in quantum mechanics. *Reviews of Modern Physics*, **40**, 411–40, and Ref. 7, pp. 307–53.
9. For a discussion of the δ function see Messiah, A. (1961). *Quantum mechanics*, Vol. 1. North-Holland, Amsterdam, and Gelfand, I.M. and Vilenkin, N. Ya (1968). *Generalized functions*, Vol. IV. Academic Press, New York.

10. See Schiff, L.I. (1955). *Quantum mechanics*, 2nd edn, Section 11. McGraw-Hill, New York.
11. For a discussion of the conditions on the wavefunction and the potential energy, see Prugovečki, Ref. 5.
12. See, for example, Ref. 10.
13. See Mott, N.F. and Sneddon, I.N. (1948). *Wave mechanics and its applications*. Oxford University Press, London. Lévy, M. (1950). Wave equations in momentum space. *Proceedings of the Royal Society*, **204 A**, 145–69.
14. Morse, P.M. and Feshbach, H. (1953). *Methods of theoretical physics*, Part 1, Ch. 4. McGraw-Hill, New York.
15. de la Peña, L. and Montemayor, R. (1980). Raising and lowering operators and spectral structure: A concise algebraic technique. *American Journal of Physics*, **48**, 855–60. Kauffmann, S.K. and Tupper, G.B. Ladder operators and factorization: A Hilbert space viewpoint. Unpublished manuscript.
16. Dongpei, Zhu (1986). Proof of the quantum virial theorem. *American Journal of Physics*, **54**, 267–70.
17. Papp, E. (1986). Quasiclassical approach to the virial theorem and to the evaluation of the ground-state energy. *Physics Reports*, **136**, 103–51.
18. Heisenberg, W. and Jordan, P. (1926). Anwendung der Quantenmechanik auf das Problem der anomalen Zeemaneffekte. *Zeitschrift für Physik*, **37**, 263–77. More extensive applications of the hypervirial theorem were made by Hirschfelder, J.O. (1960). Classical and quantum-mechanical hypervirial theorems. *Journal of Chemical Physics*, **33**, 1462–66, and Hirschfelder, J.O. and Coulson, C.A. (1962). Hypervirial theorems applied to molecular quantum dynamics. *Journal of Chemical Physics*, **36**, 941–6.
19. Thirring, W. (1979). *A course in mathematical physics*, Vol. 3. Springer, New York.
20. Hellmann, H. (1937). *Einführung in die Quantenchemie*. Deuticke, Vienna. Feynman, R.P. (1939). Forces in molecules. *Physical Review*, **56**, 340–3. Equation (1.203) is given in Pauli, W. (1933). *Handbuch der Physik*, Vol. 24, Part 1. Springer, Berlin, and it is sometimes referred to as the Pauli–Hellmann–Feynman theorem. See also Musher, J.I. (1966). Comment on some theorems of quantum chemistry. *American Journal of Physics*, **34**, 267–8.
21. A detailed review of this theorem has been given by Epstein, S.T. (1981). The Hellmann–Feynman theorem. In *The force concept in chemistry*, (ed. B.M. Deb), pp. 1–38. Van Nostrand Reinhold, New York.
22. See Ziesche, P., Kunze, K., and Milek, B. (1987). Generalisation of the Hellmann–Feynman theorem to Gamow states. *Journal of Physics A; Math. Gen.*, **20**, 2859–64 and references therein.
23. Epstein, S.T. (1954). Note on perturbation theory. *American Journal of Physics*, **22**, 613–14.
24. Balasubramanian, S. (1984). Note on Feynman's theorem. *American Journal of Physics*, **55**, 1143–4.
25. Singh, S.B. and Singh, C.A. (1989). Extensions of the Feynman–Hellmann theorem and applications. *American Journal of Physics*, **57**, 894–9.

2
THE HARMONIC OSCILLATOR

In both classical and quantum physics the harmonic oscillator is one of the most important physical systems [1–4]. Furthermore, because of the simplicity and symmetry of the Hamiltonian operator, the quantum-mechanical oscillator is particularly suited to analysis by abstract operator methods. For this reason the study of the one-dimensional oscillator is a useful introduction to the application of operator methods in quantum mechanics.

In Sections 2.1–2.4 we use these methods to determine energy eigenvalues and eigenkets, and matrix elements; the analysis is extended to isotropic multi-dimensional oscillators in Section 2.10. Quasi-classical states of the oscillator are discussed in Section 2.5. In Section 2.6 shift operators are used to calculate wavefunctions. In Sections 2.7–2.11 we discuss some applications of the theory of the oscillator.

2.1 Introduction

The simplest example of a classical one-dimensional oscillator consists of a particle of mass M acted on by a force $F_x = -kx_c$, where k is a positive constant and x_c is the x coordinate of the particle. The x component of the equation of motion is

$$\ddot{x}_c + \omega^2 x_c = 0, \tag{2.1}$$

and the Hamiltonian is

$$H = \frac{1}{2M} p_c^2 + \frac{1}{2} M\omega^2 x_c^2, \tag{2.2}$$

where

$$\omega = (k/M)^{1/2} \tag{2.3}$$

is the angular frequency of the oscillator, and

$$p_c = M\dot{x}_c \tag{2.4}$$

is the momentum.

From (2.1) and (2.4) we obtain the solutions

$$\begin{aligned} x_c(t) &= x_c(0) \cos \omega t + (M\omega)^{-1} p_c(0) \sin \omega t \\ &= A \cos(\omega t - \phi) \end{aligned} \tag{2.5}$$

and

$$\begin{aligned} p_c(t) &= -M\omega x_c(0) \sin \omega t + p_c(0) \cos \omega t \\ &= -M\omega A \sin(\omega t - \phi). \end{aligned} \tag{2.6}$$

Here A is the amplitude

$$A = [x_c^2(0) + (M\omega)^{-2} p_c^2(0)]^{1/2}, \tag{2.7}$$

and ϕ is the phase

$$\phi = \tan^{-1}\left(\frac{p_c(0)}{M\omega x_c(0)}\right). \tag{2.8}$$

The Hamiltonian of a one-dimensional quantum-mechanical oscillator is

$$H = \frac{1}{2M}p^2 + \frac{1}{2}M\omega^2 x^2, \tag{2.9}$$

where x and p are the position and momentum operators: they are linear Hermitian operators satisfying the canonical commutation relation

$$[x, p] = i\hbar. \tag{2.10}$$

In order to factorize H we consider the dimensionless operator*

$$a = \frac{1}{(2M\hbar\omega)^{1/2}}(ip + M\omega x). \tag{2.11}$$

Then

$$a^\dagger = \frac{1}{(2M\hbar\omega)^{1/2}}(-ip + M\omega x). \tag{2.12}$$

Equations (2.10)–(2.12) yield

$$[a, a^\dagger] = 1. \tag{2.13}$$

Because x and p commute with the unit operator 1, we also have

$$[a, 1] = [a^\dagger, 1] = 0. \tag{2.14}$$

The operators a, $a^\dagger$, and 1 are generators of a Lie algebra (the Heisenberg–Weyl algebra).

We also define the dimensionless Hermitian operator

$$N = a^\dagger a. \tag{2.15}$$

From (2.13) and (2.15)

$$[N, a] = -a \tag{2.16}$$

and

$$[N, a^\dagger] = a^\dagger. \tag{2.17}$$

* In Sections 2.1–2.3 the discussion is restricted to a given instant of time: the time dependence of operators and kets is considered in Section 2.4.

The operators a, $a^\dagger$, 1, and N are also generators of a Lie algebra (the oscillator algebra). The observables x, p, and H are given in terms of a, $a^\dagger$, and N by

$$x = \left(\frac{\hbar}{2M\omega}\right)^{1/2}(a^\dagger + a), \tag{2.18}$$

$$p = \mathrm{i}(\tfrac{1}{2}M\hbar\omega)^{1/2}(a^\dagger - a), \tag{2.19}$$

$$H = (aa^\dagger - \tfrac{1}{2})\hbar\omega = (a^\dagger a + \tfrac{1}{2})\hbar\omega = (N + \tfrac{1}{2})\hbar\omega. \tag{2.20}$$

Thus a and $a^\dagger$ provide a two-way factorization of H (see also Section 3.1).

2.2 Energy eigenvalues and eigenkets

Let $|n\rangle$ denote an eigenket* of N with eigenvalue n:

$$N|n\rangle = n|n\rangle. \tag{2.21}$$

Then

$$H|n\rangle = (n + \tfrac{1}{2})\hbar\omega|n\rangle, \tag{2.22}$$

and $|n\rangle$ is an energy eigenket with eigenvalue

$$E_n = (n + \tfrac{1}{2})\hbar\omega. \tag{2.23}$$

To determine n we consider the effect of a and $a^\dagger$ on $|n\rangle$. From (2.16), (2.17), and (2.21)

$$N\{a|n\rangle\} = (n - 1)\{a|n\rangle\}, \tag{2.24}$$

$$N\{a^\dagger|n\rangle\} = (n + 1)\{a^\dagger|n\rangle\}. \tag{2.25}$$

Thus a and $a^\dagger$ are lowering and raising operators for the quantum number n

$$a|n\rangle = \alpha^-|n - 1\rangle, \tag{2.26}$$

$$a^\dagger|n\rangle = \alpha^+|n + 1\rangle. \tag{2.27}$$

Taking the norm of both sides of (2.26) we have

$$|\alpha^-|^2\langle n - 1|n - 1\rangle = \langle n|a^\dagger a|n\rangle. \tag{2.28}$$

The right-hand side of (2.28) is equal to $n\langle n|n\rangle$. Thus for normalized kets $\alpha^- = n^{1/2}$ (a phase factor is set equal to unity) and hence

$$a|n\rangle = n^{1/2}|n - 1\rangle. \tag{2.29}$$

Similarly,

$$a^\dagger|n\rangle = (n + 1)^{1/2}|n + 1\rangle. \tag{2.30}$$

* We assume that such an eigenket exists.

The values of n are fixed by the requirement that the coefficients in the shift operations (2.26) and (2.27) must satisfy $|\alpha^\pm|^2 \geq 0$ (see Section 1.9). For positive integral values of n the lowering sequence $a|n\rangle$, $a^2|n\rangle, \ldots$ terminates at $|0\rangle$ because, from (2.29),

$$a|0\rangle = 0. \tag{2.31}$$

The raising sequence $a^\dagger|n\rangle, (a^\dagger)^2|n\rangle, \ldots$ does not terminate (see (2.30)). Hence the possible values of n are

$$n = 0, 1, 2, \ldots. \tag{2.32}$$

For any non-integral value of n the lowering sequence will produce kets which violate non-negativity of the norm in (2.28).

From (2.22) the energy of the ground state $|0\rangle$ is the zero-point energy [5]

$$E_0 = \tfrac{1}{2}\hbar\omega. \tag{2.33}$$

From the ground state eigenket we can obtain all other kets by repeated application of the raising operation (2.30)

$$|n\rangle = \frac{1}{(n!)^{1/2}}(a^\dagger)^n|0\rangle. \tag{2.34}$$

The kets $|n\rangle$ are orthonormal*

$$\langle n'|n\rangle = \delta_{n'n} \tag{2.35}$$

and complete [6]

$$\sum_{n=0}^{\infty} |n\rangle\langle n| = 1. \tag{2.36}$$

The set $\{|n\rangle\}$ is a basis for the Hilbert space of the one-dimensional harmonic oscillator: any state $|\psi\rangle$ can be expressed as

$$|\psi\rangle = \sum_{n=0}^{\infty} |n\rangle\langle n|\psi\rangle. \tag{2.37}$$

Similarly, any operator A acting on this space can be written

$$A = \sum_{n'=0}^{\infty}\sum_{n=0}^{\infty} |n'\rangle\langle n'|A|n\rangle\langle n|. \tag{2.38}$$

2.3 Matrix elements

The theory outlined above provides, in principle, a complete description of the quantum-mechanical properties of the one-dimensional harmonic oscilla-

* $|n\rangle$ and $|n'\rangle$ are orthogonal because they are non-degenerate eigenkets of a Hermitian operator with different eigenvalues (see Section 1.2).

tor. For example, matrix elements of certain functions of x and p can be evaluated from (2.18), (2.19), (2.29), (2.30), and (2.35). The simplest such matrix elements are

$$\langle n'|x|n\rangle = \left(\frac{\hbar}{2M\omega}\right)^{1/2}[n^{1/2}\delta_{n',n-1} + (n+1)^{1/2}\delta_{n',n+1}] \tag{2.39}$$

and

$$\langle n'|p|n\rangle = \mathrm{i}(\tfrac{1}{2}M\hbar\omega)^{1/2}[-n^{1/2}\delta_{n',n-1} + (n+1)^{1/2}\delta_{n',n+1}]. \tag{2.40}$$

Similarly,

$$\langle n'|x^2|n\rangle = \frac{\hbar}{2M\omega}\{[n(n-1)]^{1/2}\delta_{n',n-2} + (2n+1)\delta_{n'n} + [(n+1)(n+2)]^{1/2}\delta_{n',n+2}\} \tag{2.41}$$

and

$$\langle n'|p^2|n\rangle = \tfrac{1}{2}M\hbar\omega\{-[n(n-1)]^{1/2}\delta_{n',n-2} + (2n+1)\delta_{n'n} - [(n+1)(n+2)]^{1/2}\delta_{n',n+2}\}. \tag{2.42}$$

Equations (2.41) and (2.42) with $n = n'$ yield the expectation values*

$$\langle n|x^2|n\rangle = (n+\tfrac{1}{2})\frac{\hbar}{M\omega} \tag{2.43}$$

and

$$\langle n|p^2|n\rangle = (n+\tfrac{1}{2})M\hbar\omega. \tag{2.44}$$

For higher powers of x and p this procedure for calculating matrix elements becomes rather tedious. Closed formulae for certain matrix elements can be obtained by operator methods such as the hypervirial theorem [7]. A simple and elegant method, due to Morales and Flores-Riveros [8], is based on the generalized binomial theorem†

$$(A+B)^N = \sum_{N_1=0}^{N}\sum_{N_2=0}^{N-N_1}\left(-\frac{k}{2}\right)^{1/2(N-N_1-N_2)}\frac{N!}{N_1!N_2![\frac{1}{2}(N-N_1-N_2)]!}A^{N_1}B^{N_2} \tag{2.45}$$

$$= \sum_{N_1=0}^{N}\sum_{N_2=0}^{N-N_1}\left(\frac{k}{2}\right)^{1/2(N-N_1-N_2)}\frac{N!}{N_1!N_2![\frac{1}{2}(N-N_1-N_2)]!}B^{N_1}A^{N_2}. \tag{2.46}$$

---

* The simplest method of obtaining these expectation values is to apply the Hellmann–Feynman theorem (Section 1.11): thus (2.43) follows from $\langle \partial H/\partial\omega\rangle = \partial E/\partial\omega$.

† If $k = 0$, (2.45) and (2.46) reduce to the usual binomial theorem

$$(A+B)^N = \sum_{N_1=0}^{N}\frac{N!}{N_1!(N-N_1)!}A^{N_1}B^{N-N_1}.$$

Here A and B are operators such that

$$[A, B] = k, \tag{2.47}$$

where k is a constant, $N = 1, 2, \ldots$, and the summations are over values of N_1 and N_2 for which

$$N - N_1 - N_2 = 0, 2, 4, \ldots. \tag{2.48}$$

We first prove this theorem and then we discuss some of its applications.

We start with Cauchy's formula for the derivatives of an analytic function [9]

$$f^{(N)}(0) = \frac{N!}{2\pi i}\int_C \frac{dz}{z^{N+1}} f(z), \tag{2.49}$$

where C is a contour that encloses the origin and

$$f^{(N)}(z) = \frac{d^N f}{dz^N}. \tag{2.50}$$

Let

$$f(z) = \exp[z(A + B)], \tag{2.51}$$

where A and B are independent of z. Then

$$(A + B)^N = \frac{N!}{2\pi i}\int_C \frac{dz}{z^{N+1}} \exp[z(A + B)]. \tag{2.52}$$

The identity

$$\exp[z(A + B)] = \exp(-\tfrac{1}{2}z^2[A, B])\exp(zA)\exp(zB) \tag{2.53}$$

is valid for operators A and B that commute with $[A, B]$ (see Appendix A). From (2.52), (2.53), and (2.47)

$$(A + B)^N = \frac{N!}{2\pi i}\int_C \frac{dz}{z^{N+1}} \exp(zA)\exp(zB)\exp(-\tfrac{1}{2}kz^2). \tag{2.54}$$

By expanding $\exp(zA)$ and $\exp(zB)$, this becomes

$$(A + B)^N = \sum_{N_1=0}^{N}\sum_{N_2=0}^{N-N_1} \frac{N!}{N_1!N_2!} A^{N_1}B^{N_2} I, \tag{2.55}$$

where

$$I = \frac{1}{2\pi i}\int_C \frac{dz}{z^{N-N_1-N_2+1}} \exp(-\tfrac{1}{2}kz^2), \tag{2.56}$$

and we have used the fact that $I = 0$ if $N_1 + N_2 > N$. To evaluate I we use (2.49) and

$$\frac{\mathrm{d}^n}{\mathrm{d}u^n}\exp(-u^2)|_{u=0} = 0 \qquad (n \text{ odd})$$

$$= (-1)^{n/2}\frac{n!}{(\frac{1}{2}n)!} \qquad (n \text{ even}). \tag{2.57}$$

(See (2.158)–(2.160).) Hence

$$I = 0 \qquad (N - N_1 - N_2 \text{ odd}),$$

$$= \left(-\frac{k}{2}\right)^{1/2(N-N_1-N_2)}\left[\left(\frac{N-N_1-N_2}{2}\right)!\right]^{-1} \qquad (N - N_1 - N_2 \text{ even}). \tag{2.58}$$

Equations (2.55) and (2.58) yield (2.45); eqn. (2.46) follows by interchanging A and B, and replacing k with $-k$ in (2.45).

Let us use the above theorem to calculate the matrix elements* of x^N. (Knowledge of these matrix elements enables one to treat an anharmonic oscillator within perturbation theory to any order of anharmonicity.) We use (2.18) and (2.45) with $A = a^\dagger$ and $B = a$. Equation (2.47) is satisfied with $k = -1$. Multiply (2.45) on the left by $\langle n'|$ and on the right by $|n\rangle$, and note that

$$\langle n'|(a^\dagger)^{N_1}a^{N_2}|n\rangle = [n'(n'-1)\ldots(n'-N_1+1)]^{1/2}[n(n-1)\ldots \times (n-N_2+1)]^{1/2}\langle n'-N_1|n-N_2\rangle$$

$$= \left(\frac{n'!n!}{(n'-N_1)!(n-N_2)!}\right)^{1/2}\delta_{n'-N_1,n-N_2}. \tag{2.59}$$

(See (2.29) and (2.35).) Thus we obtain

$$\langle n'|x^N|n\rangle = \left(\frac{\hbar}{2M\omega}\right)^{N/2}(n'!n!)^{1/2}\sum_{N_1=0}^{N}\sum_{N_2=0}^{N-N_1} \times \frac{N!\delta_{n'-N_1,n-N_2}}{2^{1/2(N-N_1-N_2)}N_1!N_2![\frac{1}{2}(N-N_1-N_2)]![(n'-N_1)!(n-N_2)!]^{1/2}}. \tag{2.60}$$

From (2.59) we have the restrictions $N_1 \leq n'$ and $N_2 \leq n$. If† $n \geq n'$, the δ-function in (2.60) is non-zero for

$$N_2 = N_1 + n - n'. \tag{2.61}$$

Hence the sum over N_2 in (2.60) consists of a single term with N_2 given by (2.61). The results obtained from (2.60) depend on whether $n - n'$ and N are even or odd integers, and consequently there are four cases:

---

* Matrix elements of p^N and functions of p can be determined in a similar manner.

† Because $\langle n'|x^N|n\rangle = \langle n|x^N|n'\rangle^*$, this entails no loss of generality.

(i) If $n - n'$ is even, the matrix elements of even powers of x are

$$\langle n'|x^{2N}|n\rangle = \left(\frac{\hbar}{2M\omega}\right)^{N}(n'!n!)^{1/2}\sum_{N_1=0}^{K}$$
$$\times\frac{(2N)!}{2^{N-(1/2)(n-n')-N_1}N_1!(N_1+n-n')![N-\frac{1}{2}(n-n')-N_1]!(n'-N_1)!} \tag{2.62}$$

for $N \geq \frac{1}{2}(n-n')$, and

$$\langle n'|x^{2N}|n\rangle = 0 \tag{2.63}$$

for $N < \frac{1}{2}(n-n')$. Here K is the smaller of n' and $N - \frac{1}{2}(n-n')$. The value zero in (2.63) results from the restriction of $N - N_1 - N_2$ to positive values in (2.45).

(ii) If $n - n'$ is odd, the restriction to integer factorials in (2.60) means

$$\langle n'|x^{2N}|n\rangle = 0. \tag{2.64}$$

(iii) If $n - n'$ is odd, the matrix elements of odd powers of x are

$$\langle n'|x^{2N+1}|n\rangle = \left(\frac{\hbar}{2M\omega}\right)^{N+(1/2)}(n'!n!)^{1/2}\sum_{N_1=0}^{K'}$$
$$\times\frac{(2N+1)!}{2^{N-N_1-(1/2)(n-n'-1)}N_1!(N_1+n-n')![N-\frac{1}{2}(n-n'-1)-N_1]!(n'-N_1)!} \tag{2.65}$$

for $N \geq \frac{1}{2}(n-n'-1)$, and

$$\langle n'|x^{2N+1}|n\rangle = 0 \tag{2.66}$$

for $N < \frac{1}{2}(n-n'-1)$. Here K' is the smaller of n' and $N - \frac{1}{2}(n-n'-1)$.

(iv) If $n - n'$ is even,

$$\langle n'|x^{2N+1}|n\rangle = 0. \tag{2.67}$$

Consider next the matrix elements of a function $f(x)$ that can be expanded in the Taylor series

$$f(x) = \sum_{L=0}^{\infty}\left(\frac{1}{(2L)!}f^{(2L)}(0)x^{2L} + \frac{1}{(2L+1)!}f^{(2L+1)}(0)x^{2L+1}\right). \tag{2.68}$$

If $n - n'$ is even, eqns (2.62), (2.67), and (2.68) yield

$$\langle n'|f(x)|n\rangle = (n'!n!)^{1/2}\sum_{L=(1/2)(n-n')+N_1}^{\infty}\sum_{N_1=0}^{n'}\left(\frac{\hbar}{2M\omega}\right)^{L}$$
$$\times\frac{f^{(2L)}(0)}{2^{L-(1/2)(n-n')-N_1}N_1!(N_1+n-n')![L-\frac{1}{2}(n-n')-N_1]!(n'-N_1)!}. \tag{2.69}$$

With $N = L - \frac{1}{2}(n - n') - N_1$, eqn (2.69) becomes

$$\langle n'|f(x)|n\rangle = (n'!n!)^{1/2} \sum_{N=0}^{\infty} \sum_{N_1=0}^{n'} \left(\frac{\hbar}{2M\omega}\right)^{N+N_1+(1/2)(n-n')} \times \frac{f^{(2N+2N_1+n-n')}(0)}{2^N N_1!(N_1 + n - n')!N!(n' - N_1)!}. \tag{2.70}$$

In a similar manner, for $n - n'$ odd we also obtain (2.70).

Equation (2.70) is a useful formula for the evaluation of matrix elements [7, 8]. Consider, for example, the Gaussian-type operator

$$f(x) = \exp(-\alpha x^2), \tag{2.71}$$

where α is a constant. According to (2.57)

$$f^{(m)}(0) = 0 \qquad (m \text{ odd}) \tag{2.72}$$

$$= (-\alpha)^{m/2} \frac{m!}{(\frac{1}{2}m)!} \qquad (m \text{ even}). \tag{2.73}$$

From (2.70) and (2.72) with $m = 2N + 2N_1 + n - n'$, we see that

$$\langle n'|\exp(-\alpha x^2)|n\rangle = 0 \tag{2.74}$$

if $n - n'$ is odd. If $n - n'$ is even we substitute (2.73) in (2.70) and perform the sum over N in (2.70) using the expansion

$$\frac{1}{(1 + u)^{X+(1/2)}} = \frac{X!}{(2X)!} \sum_{N=0}^{\infty} \frac{(2N + 2X)!}{2^{2N} N!(N + X)!}(-u)^N \tag{2.75}$$

with

$$u = \hbar\alpha/M\omega \tag{2.76}$$

and $X = N_1 + \frac{1}{2}(n - n')$. Hence we obtain the matrix elements

$$\langle n'|\exp(-\alpha x^2)|n\rangle = \left(\frac{n'!n!}{1 + u}\right)^{1/2} \sum_{N=0}^{n'} \left(-\frac{u}{2(1 + u)}\right)^{N+(1/2)(n-n')} \times \frac{(2N + n - n')!}{N!(N + n - n')!(n' - N)![N + \frac{1}{2}(n - n')]!} \tag{2.77}$$

for $n - n'$ even.

The matrix elements calculated in this section are for a given instant of time, say $t = 0$. In terms of these the values at time t are given by (2.88).

2.4 Time dependence

The discussion in the previous sections is restricted to a single instant of time, and we now consider the time dependence of the quantum-mechanical oscillator. The time dependence of operators and kets depends on which picture we choose to adopt (see Section 1.5).

In the Schrödinger picture operators are time independent and hence the operators x, p, a, and $a^\dagger$ used above apply in this picture for all t. If $|n\rangle$ is an eigenket of the oscillator at time $t = 0$, then in the Schrödinger picture the eigenket at time t is (see Section 1.5)

$$|\psi_n(t)\rangle = \exp\left(-\frac{\mathrm{i}}{\hbar}E_n t\right)|n\rangle. \tag{2.78}$$

In the Heisenberg picture the eigenkets are time independent and hence $|n\rangle$, for example, is an eigenket for all t. Heisenberg operators satisfy (1.114). Thus for the oscillator,

$$M\dot{x}(t) = p(t) \tag{2.79}$$

and

$$\dot{p}(t) = -M\omega^2 x(t). \tag{2.80}$$

These equations can be decoupled to give

$$\ddot{x}(t) + \omega^2 x(t) = 0. \tag{2.81}$$

From (2.80) and (2.81) we obtain the solutions*

$$x(t) = x\cos\omega t + (M\omega)^{-1} p \sin\omega t \tag{2.82}$$

and

$$p(t) = -M\omega x \sin\omega t + p\cos\omega t. \tag{2.83}$$

The definition (2.11) is generalized to

$$a(t) = \frac{1}{(2M\hbar\omega)^{1/2}}[\mathrm{i}p(t) + M\omega x(t)]. \tag{2.84}$$

From (2.82)–(2.84),

$$a(t) = a\exp(-\mathrm{i}\omega t), \tag{2.85}$$

and hence

$$a^\dagger(t) = a^\dagger \exp(\mathrm{i}\omega t). \tag{2.86}$$

We can readily extend the results derived in Sections 2.1–2.3 to arbitrary instants of time. For example, from (2.85) and (2.86) we see that

$$N(t) = a^\dagger(t)a(t) = a^\dagger a \tag{2.87}$$

is a time-independent operator. For matrix elements use of the Schrödinger picture yields

* We use the notation $A(0) = A$: thus $x(0) = x$, etc.

$$\langle \psi_{n'}(t)|\, A\, |\psi_n(t)\rangle = \exp\left(-\frac{\mathrm{i}}{\hbar}(E_n - E_{n'})t\right)\langle n'|\, A\, |n\rangle, \tag{2.88}$$

where $\langle n'|\, A\, |n\rangle$ are the matrix elements at time $t = 0$. In the Heisenberg picture the left-hand side of (2.88) is

$$\langle n'|\, A(t)\, |n\rangle = \langle n'|\exp\left(\frac{\mathrm{i}}{\hbar}Ht\right)A \exp\left(-\frac{\mathrm{i}}{\hbar}Ht\right)|n\rangle,$$

which is equal to the right-hand side of (2.88).

It is instructive to compare properties of the classical and quantum-mechanical oscillators. There is a striking similarity between the operator solutions (2.82) and (2.83), and the classical solutions (2.5) and (2.6). However, this does not imply that the quantum-mechanical expectation values of x and p are necessarily equal to the classical solutions. For example, for energy eigenstates*

$$\langle n|\, x\, |n\rangle = 0 \tag{2.89}$$

and

$$\langle n|\, p\, |n\rangle = 0. \tag{2.90}$$

(See (2.39) and (2.40).) It is clear that energy eigenstates are not suitable for discussing the classical limit of the oscillator.

Other properties support this conclusion. Thus there is no limit in which $\langle n|\, H\, |n\rangle$ is equal to the classical energy†

$$E_{\mathrm{c}} = \tfrac{1}{2}M\omega^2 A^2. \tag{2.91}$$

Furthermore, one expects that for states which approximate a classical oscillator, the uncertainty product $\Delta x \Delta p$ should be close to the minimum value allowed by the uncertainty principle, namely (see (1.49))

$$\Delta x \Delta p = \tfrac{1}{2}\hbar. \tag{2.92}$$

For energy eigenstates (1.50), (2.39), and (2.43) yield

$$\Delta x = [(n + \tfrac{1}{2})\hbar/M\omega]^{1/2}. \tag{2.93}$$

Similarly,

$$\Delta p = [(n + \tfrac{1}{2})\hbar M\omega]^{1/2} \tag{2.94}$$

and therefore

$$\Delta x \Delta p = (n + \tfrac{1}{2})\hbar. \tag{2.95}$$

Except for the ground state the uncertainty product (2.95) exceeds the minimum value (2.92).

---

* The vanishing of these average values is associated with the fact that for an ensemble of osciallators in an energy eigenstate the phase is indeterminate.

† A is the amplitude (2.7).

Consider a superposition of energy eigenkets (2.78)

$$|\psi\rangle = \sum_{n=0}^{\infty} c_n \exp\left(-\frac{\mathrm{i}}{\hbar} E_n t\right)|n\rangle. \tag{2.96}$$

Then

$$\langle\psi|B|\psi\rangle = \sum_{n'=0}^{\infty} \sum_{n=0}^{\infty} c_{n'}^* c_n \exp[\mathrm{i}(n'-n)\omega t]\langle n'|B|n\rangle. \tag{2.97}$$

We see that if two or more of the c_n are non-zero, the average value of B oscillates at frequencies which are integral multiples of ω. There exist superpositions of energy eigenkets for which the averages $\langle\psi|x|\psi\rangle$ and $\langle\psi|p|\psi\rangle$ have the classical values (2.5) and (2.6): these superpositions (the so-called coherent states) are discussed in the next section.

2.5 Coherent states

In this section we construct states* for which the average value of $x(t)$ is equal to the classical value (2.5). Assume† that a possesses eigenkets $|z\rangle$

$$a|z\rangle = z|z\rangle. \tag{2.98}$$

The adjoint equation is

$$\langle z|a^\dagger = z^*\langle z|. \tag{2.99}$$

From (2.84)–(2.86)

$$\begin{aligned}\langle z|x(t)|z\rangle &= \left(\frac{\hbar}{2M\omega}\right)^{1/2}\langle z|a(t)+a^\dagger(t)|z\rangle \\ &= \left(\frac{\hbar}{2M\omega}\right)^{1/2}[\langle z|a|z\rangle\exp(-\mathrm{i}\omega t)+\langle z|a^\dagger|z\rangle\exp(\mathrm{i}\omega t)].\end{aligned} \tag{2.100}$$

If $|z\rangle$ is normalized, eqns (2.98)–(2.100) show that

$$\langle z|x(t)|z\rangle = \left(\frac{\hbar}{2M\omega}\right)^{1/2}[z\exp(-\mathrm{i}\omega t)+z^*\exp(\mathrm{i}\omega t)]. \tag{2.101}$$

If we write

$$z = |z|e^{\mathrm{i}\phi}, \tag{2.102}$$

the choice

$$|z| = \left(\frac{M\omega}{2\hbar}\right)^{1/2} A \tag{2.103}$$

---

* We work in the Heisenberg picture: the ket in the Schrödinger picture is given by (2.134).

† At this point it is not obvious that $|z\rangle$ exist: for example, one can show that eigenkets of $a^\dagger$ do not exist [10]. The kets $|z\rangle$ are given in terms of energy eigenkets by the superposition (2.111), and wavefunctions calculated in Section 2.6.

in (2.101) yields the desired classical form

$$\langle z|x(t)|z\rangle = A\cos(\omega t - \phi). \tag{2.104}$$

Similarly,

$$\langle z|p(t)|z\rangle = \mathrm{i}(\tfrac{1}{2}M\hbar\omega)^{1/2}[z^*\exp(\mathrm{i}\omega t) - z\exp(-\mathrm{i}\omega t)] \tag{2.105}$$

$$= -M\omega A\sin(\omega t - \phi). \tag{2.106}$$

Equation (2.98) is one definition of a coherent state [11]. Note that (2.98) cannot be derived from the condition (2.104) alone; to do so requires also (2.117), and the latter two equations provide an alternative definition of a coherent state [12]. Note also that it is not correct to suppose the coherent states are orthogonal (see (2.133)).

From (2.102), (2.103), (2.5), and (2.6) we have

$$z = \left(\frac{M\omega}{2\hbar}\right)^{1/2}[\mathrm{i}(M\omega)^{-1}p_c(0) + x_c(0)]. \tag{2.107}$$

Corresponding to the continuum of initial classical values, $x_c(0)$ and $p_c(0)$, the eigenvalue z takes on a continuum of values in the complex plane. Unlike the countable number of basis states $\{|n\rangle\}$, the set $\{|z\rangle\}$ contains an uncountable number of states. Coherent states have several interesting properties, and we discuss some of these below.

(i) We first express $|z\rangle$ as a superposition of energy eigenkets. From (2.36),

$$|z\rangle = \sum_{n=0}^{\infty}|n\rangle\langle n|z\rangle. \tag{2.108}$$

Multiplying both sides of (2.98) on the left by $\langle n-1|$, and using (2.30) and (1.19), we obtain

$$n^{1/2}\langle n|z\rangle = z\langle n-1|z\rangle.$$

Hence

$$\langle n|z\rangle = \frac{z^n}{(n!)^{1/2}}\langle 0|z\rangle,$$

and therefore

$$|z\rangle = \langle 0|z\rangle\sum_{n=0}^{\infty}\frac{z^n}{(n!)^{1/2}}|n\rangle. \tag{2.109}$$

The normalization $\langle z|z\rangle = 1$ requires

$$|\langle 0|z\rangle|^2\sum_{n=0}^{\infty}\frac{|z|^{2n}}{n!} = 1,$$

so that

$$\langle 0|z\rangle = \exp(-\tfrac{1}{2}|z|^2), \tag{2.110}$$

where a phase factor has been set equal to unity. Thus

$$|z\rangle = \exp(-\tfrac{1}{2}|z|^2)\sum_{n=0}^{\infty}\frac{z^n}{(n!)^{1/2}}|n\rangle. \tag{2.111}$$

According to (2.111), for an oscillator in the state $|z\rangle$ the probability that a measurement of the energy will yield the eigenvalue (2.23) is

$$P_n(z) = \frac{|z|^{2n}}{n!}e^{-|z|^2}, \tag{2.112}$$

which is a Poisson distribution with mean value $|z|^2$.

(ii) The average energy is

$$\langle E\rangle = \langle z|H|z\rangle = \langle z|a^{\dagger}a + \tfrac{1}{2}|z\rangle\hbar\omega \tag{2.113}$$

$$= (|z|^2 + \tfrac{1}{2})\hbar\omega. \tag{2.114}$$

(See (2.98) and (2.99).) From (2.114) and (2.103),

$$\langle E\rangle = \tfrac{1}{2}M\omega^2A^2 + \tfrac{1}{2}\hbar\omega, \tag{2.115}$$

which is equal to the classical energy (2.91) plus the zero-point energy (2.33). Note that it is the normal ordered Hamiltonian

$$:H: = a^{\dagger}a\hbar\omega, \tag{2.116}$$

that has the property

$$\langle z|:H:|z\rangle = E_c. \tag{2.117}$$

By a similar calculation to the above one finds

$$\langle z|H^2|z\rangle = (|z|^4 + 2|z|^2 + \tfrac{1}{4})\hbar^2\omega^2. \tag{2.118}$$

From (1.50), (2.114), and (2.118), the uncertainty in energy is

$$\Delta E = |z|\hbar\omega. \tag{2.119}$$

Therefore

$$\frac{\Delta E}{\langle E\rangle} = \frac{|z|}{|z|^2 + \tfrac{1}{2}}. \tag{2.120}$$

For a macroscopic oscillator $|z| \gg 1$,* and the fractional uncertainty in energy is small. The average value of N is equal to $|z|^2$ (see (2.114) and (2.20)). Hence a coherent state (2.111) that describes an approximately classical oscillator must involve a large number of energy eigenkets.

(iii) For an oscillator in the state $|z\rangle$ the average value of x^2 is

* Because $A \gg [\hbar/(M\omega)]^{1/2}$ in (2.103).

$$\langle z|x^2(t)|z\rangle = \frac{\hbar}{2M\omega}\langle z|a^2(t) + 2a^\dagger(t)a(t) + [a^\dagger(t)]^2 + 1|z\rangle$$

$$= A^2\cos^2(\omega t - \phi) + \frac{\hbar}{2M\omega}. \tag{2.121}$$

From (1.50), (2.104), and (2.121) the uncertainty in position is

$$\Delta x = \left(\frac{\hbar}{2M\omega}\right)^{1/2}. \tag{2.122}$$

Similarly,

$$\Delta p = (\tfrac{1}{2}M\hbar\omega)^{1/2}, \tag{2.123}$$

and therefore

$$\Delta x\Delta p = \tfrac{1}{2}\hbar. \tag{2.124}$$

We see that for all coherent states the uncertainty product $\Delta x\Delta p$ has the minimum value allowed by the uncertainty principle. Note that coherent states are not simply minimum uncertainty states: according to (2.122) and (2.123), they have the additional properties*

$$\Delta p = M\omega\Delta x \tag{2.125}$$

and

$$\frac{1}{2m}(\Delta p)^2 + \tfrac{1}{2}M\omega^2(\Delta x)^2 = \tfrac{1}{2}\hbar\omega. \tag{2.126}$$

The relations (2.124) and (2.125) provide an alternative definition of coherent states of an oscillator, as do (2.124) and (2.126).

(iv) By combining eqns (2.111) and (2.34) we have

$$|z\rangle = \exp(-\tfrac{1}{2}|z|^2)\sum_{n=0}^{\infty}\frac{(za^\dagger)^n}{n!}|0\rangle = \exp(-\tfrac{1}{2}|z|^2)\exp(za^\dagger)|0\rangle$$

$$= \exp(-\tfrac{1}{2}|z|^2)\exp(za^\dagger)\exp(-z^*a)|0\rangle, \tag{2.127}$$

because

$$\exp(-z^*a)|0\rangle = \sum_{k=0}^{\infty}\frac{(-z^*)^k}{k!}a^k|0\rangle = |0\rangle.$$

(See (2.31).) According to the identity (2.53),

$$\exp(za^\dagger - z^*a) = \exp(-\tfrac{1}{2}|z|^2)\exp(za^\dagger)\exp(-z^*a). \tag{2.128}$$

* States for which (2.124), but not (2.125), is valid are known as squeezed states: these states are generated from $|0\rangle$ by applying first a 'squeezing' operator and then the displacement operator (2.130) [13].

Therefore

$$|z\rangle = D(z)|0\rangle, \tag{2.129}$$

where $D(z)$ is the unitary operator

$$D(z) = \exp(za^\dagger - z^*a). \tag{2.130}$$

One can show that

$$D^\dagger(z)aD(z) = a + z. \tag{2.131}$$

Consequently, $D(z)$ is referred to as the displacement operator. All coherent states are generated by applying this operator to the ground state, and (2.129) is an alternative definition of coherent states.

(v) The scalar product of two coherent states is

$$\begin{aligned}\langle z_1|z_2\rangle &= \exp(-\tfrac{1}{2}|z_1|^2 - \tfrac{1}{2}|z_2|^2) \sum_{n_1=0}^{\infty} \sum_{n_2=0}^{\infty} \frac{(z_1^*)^{n_1} z_2^{n_2}}{(n_1!n_2!)^{1/2}} \langle n_1|n_2\rangle \\ &= \exp(-\tfrac{1}{2}|z_1|^2 - \tfrac{1}{2}|z_2|^2 + z_1^* z_2),\end{aligned} \tag{2.132}$$

where we have used (2.111) and (2.35). Thus

$$|\langle z_1|z_2\rangle|^2 = \exp(-|z_1 - z_2|^2). \tag{2.133}$$

We see that coherent states are not orthogonal. However, the set of coherent states is complete, and it provides a continuous, non-orthogonal basis for the Hilbert space of the oscillator. (The coherent states are actually overcomplete: any state of the oscillator can be expanded in terms of a subset of $\{|z\rangle\}$ [14].)

(vi) In the Schrödinger picture the ket of a coherent state is (Section 1.5)

$$\begin{aligned}|z\rangle_s &= \exp\left(-\frac{i}{\hbar}Ht\right)|z\rangle = \exp(-\tfrac{1}{2}|z|^2) \sum_{n=0}^{\infty} \frac{z^n}{(n!)^{1/2}} \exp[-i(n+\tfrac{1}{2})\omega t]|n\rangle \\ &= \exp(-i\tfrac{1}{2}\omega t) \exp(-\tfrac{1}{2}|z|^2) \sum_{n=0}^{\infty} \frac{[z\exp(-i\omega t)]^n}{(n!)^{1/2}} |n\rangle.\end{aligned} \tag{2.134}$$

The factor $\exp(-i\frac{1}{2}\omega t)$ in (2.134) is due to the zero-point energy. It is customary to omit this factor; doing so means that one uses the Hamiltonian (2.116) rather than H. Then (2.134) yields

$$|z\rangle_s = |z(t)\rangle, \tag{2.135}$$

where

$$z(t) = z\exp(-i\omega t). \tag{2.136}$$

Equations (2.101)–(2.106) and (2.136) show that (2.107) generalizes to

$$z(t) = \left(\frac{M\omega}{2\hbar}\right)^{1/2} [i(M\omega)^{-1}p_c(t) + x_c(t)]. \tag{2.137}$$

The above discussion shows that coherent states are 'quasi-classical' states of a quantum-mechanical oscillator: the average energy, displacement, and momentum in a coherent state are equal to the corresponding classical values, and the uncertainty product $\Delta x \Delta p$ has its minimum value. Coherent states can also be constructed for the two- and three-dimensional oscillators (see Sections 2.10 and 8.7), for the Coulomb problem (see Section 9.11), and other systems [15]. Further discussion of these states and their physical applications can be found in the extensive literature on this subject [16].

2.6 Wavefunctions

In the coordinate representation (2.30) and (2.31) are

$$\left(\frac{M\omega}{2\hbar}\right)^{1/2}\left(x - \frac{\hbar}{M\omega}\frac{\mathrm{d}}{\mathrm{d}x}\right)\psi_n(x) = (n+1)^{1/2}\psi_{n+1}(x) \tag{2.138}$$

and

$$\left(x + \frac{\hbar}{M\omega}\frac{\mathrm{d}}{\mathrm{d}x}\right)\psi_0(x) = 0, \tag{2.139}$$

where $\psi_n(x) = \langle x|n\rangle$ is the coordinate-space wavefunction (see Section 1.4). Equations (2.138), (2.139), and the normalization condition

$$\int_{-\infty}^{\infty} |\psi_n(x)|^2\, \mathrm{d}x = 1$$

determine $\psi_n(x)$.

The normalized solution to (2.139) is

$$\psi_0(x) = c_0 \exp\left(-\frac{M\omega}{2\hbar}x^2\right), \tag{2.140}$$

where $c_0 = (M\omega/\pi\hbar)^{1/4}$. Substituting (2.140) in (2.138) we find

$$\psi_1(x) = c_1\left(\frac{M\omega}{\hbar}\right)^{1/2} x \exp\left(-\frac{M\omega}{2\hbar}x^2\right), \tag{2.141}$$

where $c_1 = \sqrt{2}c_0$. The ground state ψ_0 and the first excited state ψ_1 are, respectively, even and odd functions of x: they are eigenfunctions of parity with eigenvalues $+1$ and -1. In general, (2.138) transforms between eigenfunctions of opposite parity. It is convenient to consider transformations between states of the same parity; by applying (2.138) twice and using (2.9) and (2.22) we have

$$\psi_{n+2}(u) = -\left(\frac{n+1}{n+2}\right)^{1/2}\left(1 - \frac{u}{n+1} + \frac{2u}{n+1}\frac{\mathrm{d}}{\mathrm{d}u}\right)\psi_n(u), \tag{2.142}$$

where

$$u = (M\omega/\hbar)x^2. \tag{2.143}$$

By repeated application of (2.142) we generate all the even eigenfunctions from ψ_0. Let

$$\psi_n(u) = c_0 e^{-(1/2)u}\phi_n(u) \qquad (n = 0, 2, 4, \ldots). \tag{2.144}$$

Then (2.142) yields the recurrence relation

$$\phi_{n+2}(u) = \gamma_n D_n \phi_n(u), \tag{2.145}$$

where

$$\gamma_n = -\left(\frac{n+1}{n+2}\right)^{1/2} \tag{2.146}$$

and

$$D_n = 1 - \frac{2u}{n+1} + \frac{2u}{n+1}\frac{\mathrm{d}}{\mathrm{d}u}. \tag{2.147}$$

Now $\phi_0 = 1$ and hence

$$\phi_{2n}(u) = \gamma_{2n-2} \cdots \gamma_2 \gamma_0 P_{2n}(u), \tag{2.148}$$

where

$$P_{2n}(u) = D_{2n-2} \ldots D_2 D_0 1. \tag{2.149}$$

By inspection

$$\gamma_{2n-2} \cdots \gamma_2 \gamma_0 = (-1)^n \left(\frac{(2n-1)\ldots 3.1}{(2n)\ldots 4.2}\right)^{1/2} = (-1)^n \left(\frac{(2n)!}{2^{2n}(n!)^2}\right)^{1/2}. \tag{2.150}$$

The function $P_{2n}(u)$ is a polynomial of order $2n$ in u: comparing (2.149) and (2.147) with (B.1) (in Appendix B) we have

$$P_{2n}(u) = {}_1F_1(-n; \tfrac{1}{2}; u), \tag{2.151}$$

where ${}_1F_1$ denotes the confluent hypergeometric function

$${}_1F_1(a; b; u) = 1 + \frac{a}{b}\frac{u}{1!} + \frac{a(a+1)}{b(b+1)}\frac{u^2}{2!} + \cdots. \tag{2.152}$$

From (2.144), (2.148), (2.150), and (2.151) we obtain the even wavefunctions

$$\psi_{2n}(x) = (-1)^n \left(\frac{M\omega}{\pi\hbar}\right)^{1/4} \left(\frac{(2n)!}{2^{2n}(n!)^2}\right)^{1/2} {}_1F_1\left(-n; \tfrac{1}{2}; \frac{M\omega}{\hbar}x^2\right) \exp\left(-\frac{M\omega}{2\hbar}x^2\right), \tag{2.153}$$

where $n = 0, 1, 2, \ldots$.

Starting from ψ_1, repeated application of (2.142) generates all the odd wavefunctions. By calculations that are similar to the above we find

$$\psi_{2n+1}(x) = (-1)^n \left(\frac{M\omega}{\pi\hbar}\right)^{1/4} \left(\frac{(2n+1)!}{2^{2n-1}(n!)^2}\right)^{1/2} \left(\frac{M\omega}{\hbar}\right)^{1/2} x\, {}_1F_1\left(-n; \tfrac{3}{2}; \frac{M\omega}{\hbar}x^2\right)$$

$$\times \exp\left(-\frac{M\omega}{2\hbar}x^2\right), \tag{2.154}$$

where $n = 0, 1, 2, \ldots$. This method of determining normalized wavefunctions can be applied to other problems, and additional examples are discussed in later sections.

We can obtain the wavefunctions (2.153) and (2.154) in a different form by applying (2.138) to (2.140). Then

$$\psi_n(v) = (2^n n!)^{-1/2}\left(\frac{M\omega}{\pi\hbar}\right)^{1/4}\left(v - \frac{d}{dv}\right)^n \exp\left(-\frac{1}{2}v^2\right), \tag{2.155}$$

where $v = (M\omega/\hbar)^{1/2}x$. The identity

$$\left(v - \frac{d}{dv}\right)f(v) = -\exp\left(\frac{1}{2}v^2\right)\frac{d}{dv}\exp\left(-\frac{1}{2}v^2\right)f(v)$$

shows that

$$\left(v - \frac{d}{dv}\right)^n f(v) = (-1)^n \exp\left(\frac{1}{2}v^2\right)\frac{d^n}{dv^n}\exp\left(-\frac{1}{2}v^2\right)f(v). \tag{2.156}$$

From (2.155) and (2.156),

$$\psi_n(x) = (2^n n!)^{-1/2}\left(\frac{M\omega}{\pi\hbar}\right)^{1/4} H_n\left[\left(\frac{M\omega}{\hbar}\right)^{1/2} x\right]\exp\left(-\frac{M\omega}{2\hbar}x^2\right), \tag{2.157}$$

where

$$H_n(v) = (-1)^n \exp(v^2)\frac{d^n}{dv^n}\exp(-v^2) \tag{2.158}$$

is a Hermite polynomial. Comparing (2.157) with (2.153) and (2.154) we obtain the formulae

$$H_{2n}(v) = (-1)^n\frac{(2n)!}{n!}\,{}_1F_1(-n; \tfrac{1}{2}; v^2) \tag{2.159}$$

and

$$H_{2n+1}(v) = (-1)^n\frac{(2n+1)!}{n!}2v\,{}_1F_1(-n; \tfrac{3}{2}; v^2). \tag{2.160}$$

In the momentum representation (2.30) and (2.31) are

$$-\frac{i}{(2M\hbar\omega)^{1/2}}\left(p - M\hbar\omega\frac{d}{dp}\right)\bar{\psi}_n(p) = (n+1)^{1/2}\bar{\psi}_{n+1}(p) \tag{2.161}$$

and

$$\left(p + M\hbar\omega\frac{d}{dp}\right)\bar{\psi}_0(p) = 0, \tag{2.162}$$

where $\bar{\psi}_n(p) = \langle p|n\rangle$ is the momentum-space wavefunction (see Section 1.4). Equations (2.161), (2.162), and the normalization

$$\int_{-\infty}^{\infty} |\bar{\psi}_n(p)|^2 \, \mathrm{d}p = 1$$

determine $\bar{\psi}_n(p)$. These calculations are similar to those for the coordinate-space wavefunction, and in terms of the Hermite polynomial the result is

$$\bar{\psi}_n(p) = \frac{(-\mathrm{i})^n}{(2^n n!)^{1/2}} (\pi M\hbar\omega)^{-1/4} H_n\left(\frac{p}{(M\hbar\omega)^{1/2}}\right) \exp\left(-\frac{p^2}{2M\hbar\omega}\right). \tag{2.163}$$

It can be verified that, as required by (1.95), the wavefunctions (2.157) and (2.163) are Fourier transforms of each other.

Consider next coherent states of the oscillator. In the coordinate representation (2.98), (2.11), and (2.107) yield

$$\left(\frac{\mathrm{d}}{\mathrm{d}x} + \frac{M\omega}{\hbar}[x - x_c(0)]\right)\psi_z(x) = \frac{\mathrm{i}}{\hbar} p_c(0)\psi_z(x), \tag{2.164}$$

where $\psi_z(x) = \langle x|z\rangle$. The normalized solution to (2.164) is

$$\psi_z(x) = \left(\frac{M\omega}{\pi\hbar}\right)^{1/4} \exp\left(\frac{\mathrm{i}}{\hbar} p_c(0)x - \frac{M\omega}{2\hbar}[x - x_c(0)]^2\right). \tag{2.165}$$

Equations (2.135)–(2.137) show that in the Schrödinger picture we must replace $x_c(0)$ and $p_c(0)$ with $x_c(t)$ and $p_c(t)$. Thus

$$\psi_z(x, t) = \left(\frac{M\omega}{\pi\hbar}\right)^{1/4} \exp\left(\frac{\mathrm{i}}{\hbar} p_c(t)x - \frac{M\omega}{2\hbar}[x - x_c(t)]^2\right), \tag{2.166}$$

where $x_c(t)$ and $p_c(t)$ are given by (2.5) and (2.6). The probability distribution $P_z(x, t) = |\psi_z(x, t)|^2$ is the Gaussian wave packet

$$P_z(x, t) = \left(\frac{M\omega}{\pi\hbar}\right)^{1/2} \exp\left(-\frac{M\omega}{\hbar}[x - x_c(t)]^2\right). \tag{2.167}$$

This wave packet has the properties: (i) it performs, without distortion, an oscillation of period $T = 2\pi/\omega$ along the x axis; (ii) the centre of the wave packet follows the classical path $x_c(t)$. (This second property holds for any wave packet of a one-dimensional oscillator [17]. Thus, as mentioned above, (2.104) is not sufficient to define a coherent state.) In a similar manner one can calculate the momentum-space wavefunction for a coherent state. The result is

$$\bar{\psi}_z(p, t) = (\pi M\hbar\omega)^{-1/4} \exp\left(-\frac{\mathrm{i}}{\hbar} x_c(t)p - \frac{1}{2M\hbar\omega}[p - p_c(t)]^2\right). \tag{2.168}$$

For further discussion of the wavefunctions derived above the reader is referred to the literature [18].

2.7 A charged oscillator in a uniform electric field

In Sections 2.7 and 2.8 we discuss two simple physical systems to which the theory of the one-dimensional harmonic oscillator applies, namely a charged oscillator in a uniform electric field, and a charged particle in a uniform magnetic field.

Suppose the particle considered in Section 2.1 has charge q and is acted on by a uniform electric field $\mathscr{E} = \mathscr{E}\hat{\mathbf{x}}$. Then (2.9) becomes

$$H = \frac{1}{2M}p^2 + \frac{1}{2}M\omega^2x^2 - q\mathscr{E}x. \tag{2.169}$$

In terms of

$$X = x - \frac{q\mathscr{E}}{M\omega^2}, \tag{2.170}$$

$$H = \frac{1}{2M}p^2 + \frac{1}{2}M\omega^2X^2 - \frac{q^2\mathscr{E}^2}{2M\omega^2}, \tag{2.171}$$

and the energy eigenvalue equation (1.108) becomes

$$\left(\frac{1}{2M}p^2 + \frac{1}{2}M\omega^2X^2\right)|n\rangle = \left(E_n + \frac{q^2\mathscr{E}^2}{2M\omega^2}\right)|n\rangle. \tag{2.172}$$

Because $[X, p] = i\hbar$, eqn (2.172) is equivalent to the problem considered in Sections 2.1 and 2.2. Hence

$$E_n = (n + \tfrac{1}{2})\hbar\omega - \frac{q^2\mathscr{E}^2}{2M\omega^2} \qquad (n = 0, 1, 2, \ldots). \tag{2.173}$$

In a similar manner we can extend the results derived in Sections 2.1–2.6 to the present problem. For example, from (2.157) and (2.170) the coordinate-space wavefunctions of the Hamiltonian (2.169) are

$$\psi_n(x) = (2^n n!)^{-1/2}\left(\frac{M\omega}{\pi\hbar}\right)^{1/4} H_n\left[\left(\frac{M\omega}{\hbar}\right)^{1/2}\left(x - \frac{q\mathscr{E}}{M\omega^2}\right)\right]$$
$$\times \exp\left[-\frac{M\omega}{2\hbar}\left(x - \frac{q\mathscr{E}}{M\omega^2}\right)^2\right]. \tag{2.174}$$

2.8 A charged particle in a uniform magnetic field [19]

The Hamiltonian for a spinless non-relativistic particle* of mass M and charge q in a magnetic field

$$\mathbf{B} = \nabla \times \mathbf{A} \tag{2.175}$$

* Relativistic particles with spin 0 and $\frac{1}{2}\hbar$ are treated in Section 11.5.

is

$$H = \frac{1}{2M}\mathbf{D}^2, \tag{2.176}$$

where [20]

$$\mathbf{D} = \mathbf{p} - q\mathbf{A}. \tag{2.177}$$

Now

$$[D_i, D_j] = -q([p_i, A_j] - [p_j, A_i]) = i\hbar q\left(\frac{\partial A_j}{\partial r_i} - \frac{\partial A_i}{\partial r_j}\right). \tag{2.178}$$

From (2.175), (1.154), and (1.157) we have

$$\varepsilon_{ijk} B_k = \frac{\partial A_j}{\partial r_i} - \frac{\partial A_i}{\partial r_j} \tag{2.179}$$

and hence we obtain the useful commutation relations

$$[D_i, D_j] = i\hbar q \varepsilon_{ijk} B_k. \tag{2.180}$$

In the following we consider the uniform field

$$\mathbf{B} = (0, 0, B), \tag{2.181}$$

where B is a positive constant. From (2.180) and (2.181)

$$[D_i, D_z] = 0, \tag{2.182}$$

and hence

$$[H, D_z] = 0. \tag{2.183}$$

Let $|E\gamma\rangle$ be an eigenket of the c.s.c.o. H and $\Gamma = \{D_z, \ldots\}$. Thus γ denotes the eigenvalues* $\{P_z, \ldots\}$. Then

$$H|E\gamma\rangle = E|E\gamma\rangle \tag{2.184}$$

and

$$D_z|E\gamma\rangle = P_z|E\gamma\rangle. \tag{2.185}$$

From (2.176), (2.184), and (2.185)

$$\frac{1}{2M}(D_x^2 + D_y^2)|E\gamma\rangle = \left(E - \frac{P_z^2}{2M}\right)|E\gamma\rangle. \tag{2.186}$$

We wish to determine the eigenvalues and eigenkets in (2.186).

In terms of the dimensionless operators

* We use the symbol P_z for the eigenvalue of D_z because one often chooses a vector potential such that $A_z = 0$, that is $D_z = p_z$ (see below).

$$\tilde{D}_i = (\hbar|q|B)^{-1/2} D_i, \tag{2.187}$$

(2.186) is

$$\frac{1}{2}(\tilde{D}_x^2 + \tilde{D}_y^2)|E\gamma\rangle = \frac{M}{\hbar|q|B}\left(E - \frac{P_z^2}{2M}\right)|E\gamma\rangle. \tag{2.188}$$

Also, (2.180) yields $[\tilde{D}_x, \tilde{D}_y] = i\hat{q}$, where $\hat{q} = \text{sign}\, q$. Thus

$$[\tilde{D}_-, \tilde{D}_+] = 1, \tag{2.189}$$

$$\tfrac{1}{2}(\tilde{D}_x^2 + \tilde{D}_y^2) = \tilde{D}_+\tilde{D}_- + \tfrac{1}{2}, \tag{2.190}$$

where

$$\tilde{D}_\pm = \frac{1}{\sqrt{2}}(\tilde{D}_x \mp i\hat{q}\tilde{D}_y).$$

Let $N = \tilde{D}_+\tilde{D}_-$ and denote $|E\gamma\rangle$ by $|n\gamma\rangle$, where n is the eigenvalue in

$$N|n\gamma\rangle = n|n\gamma\rangle. \tag{2.191}$$

Then, as in the theory of the one-dimensional harmonic oscillator (see Sections 2.1 and 2.2), it follows from (2.189), (2.191), and (2.182) that for normalized eigenkets*

$$\tilde{D}_-|n\gamma\rangle = n^{1/2}|n-1, \gamma\rangle, \tag{2.192}$$

$$\tilde{D}_+|n\gamma\rangle = (n+1)^{1/2}|n+1, \gamma\rangle, \tag{2.193}$$

where $n = 0, 1, 2, \ldots$. From (2.188), (2.190), (2.192), and (2.193) we obtain the energy eigenvalues

$$E = \frac{1}{2M}P_z^2 + (n + \tfrac{1}{2})\hbar\omega, \tag{2.194}$$

where

$$\omega = |q|B/M. \tag{2.195}$$

The second term on the right-hand side of (2.194) gives the Landau levels.

According to (2.192), $\tilde{D}_-$ annihilates the ground state

$$\tilde{D}_-|0\gamma\rangle = 0. \tag{2.196}$$

Equation (2.193) shows that all eigenkets can be obtained from the ground state by repeated application of $\tilde{D}_+$

$$|n\gamma\rangle = \frac{1}{(n!)^{1/2}}(\tilde{D}_+)^n|0\gamma\rangle. \tag{2.197}$$

---

* We assume that $\tilde{D}_\pm$ commute with the additional constants of the motion in Γ: this can readily be verified for the examples I–III given at the end of this section.

There are, apart from D_z, additional constants of the motion. Before determining them and the corresponding c.s.c.o.s and eigenvalues, it is helpful to review the classical motion. The classical equation of motion

$$M\ddot{\mathbf{r}}_c = q\dot{\mathbf{r}}_c \times \mathbf{B},$$

with the field (2.181), has the solution

$$\mathbf{r}_c = (R_c \cos \omega t + X_c, \ -R_c \sin \omega t + Y_c, v_z t), \tag{2.198}$$

where ω is the cyclotron frequency (2.195), and X_c, Y_c, R_c, and v_z are constants. The trajectory (2.198) is a helix with constant pitch and with axis parallel to $\mathbf{B}$: the projection of the motion on to the xy plane satisfies

$$(x_c - X_c)^2 + (y_c - Y_c)^2 = R_c^2,$$

which is a circle centred at (X_c, Y_c) and with radius R_c. The distance of the centre from the origin is

$$d_c = (X_c^2 + Y_c^2)^{1/2}. \tag{2.199}$$

Now let us determine the additional constants of the quantum-mechanical problem. Using (2.180) we can show that

$$[H, D_i] = -\mathrm{i}(\hbar qB/M)\varepsilon_{ij3} D_j. \tag{2.200}$$

It is also straightforward to prove that

$$[H, \mathbf{r}] = -\mathrm{i}(\hbar/M)\mathbf{D}. \tag{2.201}$$

Therefore

$$[H, X] = [H, Y] = 0, \tag{2.202}$$

where

$$X = x + (qB)^{-1}D_y, \tag{2.203}$$

$$Y = y - (qB)^{-1}D_x. \tag{2.204}$$

To interpret these constants of the motion, we note that the hypervirial theorem (Section 1.10) applied to (2.200) yields $\langle D_x \rangle = \langle D_y \rangle = 0$. Hence $\langle x \rangle = \langle X \rangle$ and $\langle y \rangle = \langle Y \rangle$. We therefore associate the operators X and Y with the coordinates X_c and Y_c of the axis of the helical trajectory.

Using (2.180) we find

$$[X, Y] = -\mathrm{i}\hat{q}\lambda^2 \tag{2.205}$$

where the constant

$$\lambda = (\hbar/|q|B)^{1/2} \tag{2.206}$$

is the magnetic length. The eigenvalues X_0 and Y_0 of X and Y are any real numbers in $(-\infty, \infty)$. The energy (2.194) is infinitely degenerate because it does not depend on X_0 or Y_0; for each E there are an infinite number of kets $|EP_zX_0\rangle$ or $|EP_zY_0\rangle$. According to (2.205), X and Y cannot have simultaneous eigenvalues and (see (1.49))

$$\Delta X \Delta Y \geq \tfrac{1}{2}\lambda^2. \tag{2.207}$$

Another constant of the motion is the operator

$$d^2 = X^2 + Y^2 \tag{2.208}$$

associated with the classical value (2.199). Because (2.205) and (2.208) have the same form as (2.9) and (2.10) for the oscillator, the eigenvalues of d^2 are

$$d_0^2 = (2n' + 1)\lambda^2 \qquad (n' = 0, 1, 2, \ldots). \tag{2.209}$$

Note that d^2 also commutes with D_z. The operator R associated with the radius of the classical orbit R_c is defined by

$$\begin{aligned} R^2 &= (x - X)^2 + (y - Y)^2 = (qB)^{-2}(D_x^2 + D_y^2) \\ &= (qB)^{-2}(2MH - D_z^2). \end{aligned} \tag{2.210}$$

Hence R^2 is a constant of the motion and

$$R^2|E\gamma\rangle = (2n + 1)\lambda^2|E\gamma\rangle. \tag{2.211}$$

So far we have not specified the vector potential $\mathbf{A}$ in (2.177), and this is a suitable point at which to discuss choices for $\mathbf{A}$. An important property of this vector is that it is not uniquely defined by $\mathbf{B}$ because $\mathbf{A}$ has the same curl as [21]

$$\mathbf{A}' = \mathbf{A} + \nabla\chi(\mathbf{r}, t), \tag{2.212}$$

where $\chi(\mathbf{r}, t)$ is an arbitrary single-valued function. For the uniform magnetic field (2.181) one choice of gauge is

$$\mathbf{A} = B(0, x, 0). \tag{2.213}$$

An alternative gauge that is often used is

$$\mathbf{A} = \tfrac{1}{2}\mathbf{B} \times \mathbf{r}, \tag{2.214}$$

which, for the field (2.181), is

$$\mathbf{A} = \tfrac{1}{2}B(-y, x, 0). \tag{2.215}$$

With the choice (2.215), eqns (2.203) and (2.204) become

$$X = \tfrac{1}{2}x + (qB)^{-1}p_y, \tag{2.216}$$

$$Y = \tfrac{1}{2}y - (qB)^{-1}p_x. \tag{2.217}$$

Consider the operator

$$L_z = xp_y - yp_x, \tag{2.218}$$

which is the z component of the orbital angular momentum operator (see

Section 4.1). From (2.216)–(2.218),

$$L_z = qB[-\tfrac{1}{2}(x^2 + y^2) + xX + yY] = \tfrac{1}{2}qB(d^2 - R^2). \tag{2.219}$$

Hence for the gauge (2.215), L_z is also a constant of the motion and its eigenvalues are (see (2.209) and (2.211))

$$\hbar(n' - n)\hat{q}. \tag{2.220}$$

Corresponding to the choices of gauge and the constants of the motion discussed above, there are various sets of commuting observables. For example:

I. With the vector potential (2.213), H, p_y, and p_z is a set of commuting observables with eigenkets $|EP_yP_z\rangle$.

II. With the vector potential* (2.215), H, p_z, and X is a set of commuting observables with eigenkets $|EP_zX_0\rangle$.

III. With the vector potential* (2.215), H, p_z, and d^2 is a set of commuting observables with eigenkets $|EP_zd_0^2\rangle$.

Each of these three sets of commuting operators is, in fact, a c.s.c.o. and in the next section we determine the corresponding wavefunctions.

2.9 Wavefunctions for a charged particle in a uniform magnetic field [19]

We consider the coordinate-space wavefunction; momentum-space wavefunctions can be obtained by similar calculations to those performed below.

I. In the coordinate representation and for the gauge (2.213), eqns (2.196) and (2.197) are

$$\left(\frac{\partial}{\partial x} + \mathrm{i}\hat{q}\frac{\partial}{\partial y} + \lambda^{-2}x\right)\psi_{0\gamma}(\mathbf{r}) = 0 \tag{2.221}$$

and

$$\psi_{n\gamma}(\mathbf{r}) = \frac{1}{(n!)^{1/2}}\left(-\frac{\mathrm{i}\lambda}{\sqrt{2}}\right)^n\left(\frac{\partial}{\partial x} - \mathrm{i}\hat{q}\frac{\partial}{\partial y} - \lambda^{-2}x\right)^n\psi_{0\gamma}(\mathbf{r}), \tag{2.222}$$

where $\psi_{n\gamma}(\mathbf{r}) = \langle\mathbf{r}|n\gamma\rangle$ is the coordinate-space wavefunction for energy eigenstates. The eigenvalue equations for p_y and p_z are

$$-\mathrm{i}\hbar\frac{\partial}{\partial y}\psi_{n\gamma} = P_y\psi_{n\gamma} \tag{2.223}$$

and

$$-\mathrm{i}\hbar\frac{\partial}{\partial z}\psi_{n\gamma} = P_z\psi_{n\gamma}. \tag{2.224}$$

---

* H, D_z, X, and H, D_z, d^2 are two sets of commuting operators in any gauge: the choice (2.215) means that $D_z = p_z$.

(Thus γ denotes $P_y P_z$.) Note that $X = \hat{q}\hbar^{-1}\lambda^2 p_y$ (see (2.203) and (2.213)) and thus $\psi_{n\gamma}$ is also an eigenfunction of X with eigenvalue

$$X_0 = \hat{q}\hbar^{-1}\lambda^2 P_y. \tag{2.225}$$

Equations (2.221)–(2.224) and the normalization condition determine $\psi_{n\gamma}(\mathbf{r})$. The solution that is normalized with respect to x is

$$\psi_{0\gamma}(\mathbf{r}) = \frac{1}{(\pi^{1/2}\lambda)^{1/2}} \exp\left[\frac{\mathrm{i}}{\hbar}(P_y y + P_z z) - \frac{1}{2}\left(\frac{x - X_0}{\lambda}\right)^2\right].$$

Substituting this in (2.222) we obtain

$$\psi_{n\gamma}(\mathbf{r}) = \frac{\mathrm{i}^n}{(\pi^{1/2}2^n n!\lambda)^{1/2}} \exp\left[\frac{\mathrm{i}}{\hbar}(P_y y + P_z z)\right]\left(u - \frac{d}{du}\right)^n \exp\left(-\frac{1}{2}u^2\right), \tag{2.226}$$

where $u = (x - X_0)/\lambda$. From (2.226) and (2.156),

$$\psi_{n\gamma}(\mathbf{r}) = \frac{\mathrm{i}^n}{(\pi^{1/2}2^n n!\lambda)^{1/2}} H_n\left(\frac{x - X_0}{\lambda}\right) \exp\left[\frac{\mathrm{i}}{\hbar}(P_y y + P_z z) - \frac{1}{2}\left(\frac{x - X_0}{\lambda}\right)^2\right], \tag{2.227}$$

where H_n is the Hermite polynomial (2.158). Equation (2.227) gives the wavefunctions for the c.s.c.o. H, p_y, p_z in the gauge (2.213).

II. With the gauge (2.215), eqns (2.196) and (2.197) become

$$\left(\frac{\partial}{\partial x} + \mathrm{i}\hat{q}\frac{\partial}{\partial y} + \frac{1}{2}\lambda^{-2}(x + \mathrm{i}\hat{q}y)\right)\psi_{0\gamma}(\mathbf{r}) = 0 \tag{2.228}$$

and

$$\psi_{n\gamma}(\mathbf{r}) = \frac{1}{(n!)^{1/2}}\left(-\frac{\mathrm{i}\lambda}{\sqrt{2}}\right)^n \left[\frac{\partial}{\partial x} - \mathrm{i}\hat{q}\frac{\partial}{\partial y} - \frac{1}{2}\lambda^{-2}(x - \mathrm{i}\hat{q}y)\right]^n \psi_{0\gamma}(\mathbf{r}). \tag{2.229}$$

In addition, we require that $\psi_{n\gamma}$ satisfy (2.224). This property and the fact that

$$\frac{\partial}{\partial x} + \mathrm{i}\hat{q}\frac{\partial}{\partial y}$$

commutes with any function $f(x + \mathrm{i}\hat{q}y)$ means that the solution to (2.228) has the form

$$\psi_{0\gamma}(\mathbf{r}) = \exp\left(\frac{\mathrm{i}}{\hbar}P_z z\right) f(x + \mathrm{i}\hat{q}y) \exp\left(-\frac{x^2 + y^2}{4\lambda^2}\right). \tag{2.230}$$

The appearance of an arbitrary function in (2.230) is a consequence of the fact that H and p_z are not a c.s.c.o.

Suppose $\psi_{n\gamma}$ is also an eigenfunction of the operator X (eqn. (2.203)) with eigenvalues X_0. (Thus γ denotes $P_z X_0$.) Then

$$\left(\frac{1}{2}x - \mathrm{i}\hat{q}\lambda^2\frac{\partial}{\partial y}\right)\psi_{n\gamma} = X_0\psi_{n\gamma}. \tag{2.231}$$

From (2.230) and (2.231)

$$\frac{\partial f}{\partial y} = \tfrac{1}{2}\lambda^{-2}(y - \mathrm{i}\hat{q}x + 2\mathrm{i}\hat{q}X_0)f. \tag{2.232}$$

Hence

$$f = c\exp\left(-\frac{1}{4\lambda^2}(x + \mathrm{i}\hat{q}y - 2X_0)^2\right), \tag{2.233}$$

where c is a constant. Substituting (2.233) in (2.230) and normalizing with respect to x, we find

$$\psi_{0\gamma}(\mathbf{r}) = \frac{1}{(\pi^{1/2}\lambda)^{1/2}}\exp\left(\frac{\mathrm{i}}{\hbar}P_z z - \frac{1}{4\lambda^2}[x^2 + y^2 - 2X_0^2 + (x + \mathrm{i}\hat{q}y - 2X_0)^2]\right). \tag{2.234}$$

The operator in square brackets in (2.229) is equal to

$$\exp\left(\frac{x^2 + y^2}{4\lambda^2}\right)\left(\frac{\partial}{\partial x} - \mathrm{i}\hat{q}\frac{\partial}{\partial y}\right)\exp\left(-\frac{x^2 + y^2}{4\lambda^2}\right). \tag{2.235}$$

Hence (2.229) and (2.234) yield

$$\psi_{n\gamma}(\mathbf{r}) = \frac{1}{(n!)^{1/2}}\left(-\frac{\mathrm{i}\lambda}{\sqrt{2}}\right)^n\frac{1}{(\pi^{1/2}\lambda)^{1/2}}\exp\left(\frac{\mathrm{i}}{\hbar}P_z z + \frac{1}{4\lambda^2}(x^2 + y^2 + 2X_0^2)\right)$$
$$\times\left[\frac{\partial}{\partial x} - \mathrm{i}\hat{q}\frac{\partial}{\partial y}\right]^n\exp\left(-\frac{1}{4\lambda^2}[2x^2 + 2y^2 + (x + \mathrm{i}\hat{q}y - 2X_0)^2]\right). \tag{2.236}$$

The differentiations in (2.236) can be simplified if we multiply both sides of this equation on the left by

$$\exp\left(-\frac{1}{4\lambda^2}(x - \mathrm{i}\hat{q}y - 2X_0)^2\right), \tag{2.237}$$

and use the fact that the operator in square brackets commutes with (2.237). Hence

$$\psi_{n\gamma}(\mathbf{r}) = \frac{1}{(n!)^{1/2}}\left(-\frac{\mathrm{i}\lambda}{\sqrt{2}}\right)^n\frac{1}{(\pi^{1/2}\lambda)^{1/2}}\exp\left(\frac{\mathrm{i}}{\hbar}P_z z + \frac{1}{2\lambda^2}[(x - X_0)^2\right.$$
$$\left. - \mathrm{i}\hat{q}y(x - 2X_0)]\right)\frac{\mathrm{d}^n}{\mathrm{d}x^n}\exp\left[-\left(\frac{x - X_0}{\lambda}\right)^2\right]$$
$$= \frac{\mathrm{i}^n}{(\pi^{1/2}n!2\lambda)^{1/2}}\exp\left(\frac{\mathrm{i}}{\hbar}P_z z - \frac{1}{2\lambda^2}[(x - X_0)^2 + \mathrm{i}\hat{q}y(x - 2X_0)]\right)$$

$$\times H_n\left(\frac{x - X_0}{\lambda}\right), \tag{2.238}$$

where H_n is the Hermite polynomial (2.158). Equation (2.238) gives the wavefunctions for the c.s.c.o. H, p_z, and X in the gauge (2.215).

III. Suppose that, instead of (2.231), $\psi_{n\gamma}$ is an eigenfunction of the operator d^2 with eigenvalues $(2n' + 1)\lambda^2$ (see (2.208) and (2.209)). Then it is also an eigenfunction of L_z

$$-\mathrm{i}\hbar\left(x\frac{\partial}{\partial y} - y\frac{\partial}{\partial x}\right)\psi_{n\gamma} = \hbar(n' - n)\hat{q}\psi_{n\gamma}. \tag{2.239}$$

(See (2.219) and (2.220).) Substituting (2.230) in (2.239) we obtain

$$(x + \mathrm{i}\hat{q}y)\frac{\partial f}{\partial x} = n'f. \tag{2.240}$$

Therefore

$$f = c(x + \mathrm{i}\hat{q}y)^{n'}, \tag{2.241}$$

where c is a constant. Substituting (2.241) in (2.230) and normalizing with respect to x and y, we find

$$\psi_{0\gamma}(\mathbf{r}) = \lambda^{-1}(\pi 2^{n'+1}n'!)^{-1/2}\left(\frac{x + \mathrm{i}\hat{q}y}{\lambda}\right)^{n'} \exp\left(\frac{\mathrm{i}}{\hbar}P_z z - \frac{x^2 + y^2}{4\lambda^2}\right). \tag{2.242}$$

From (2.229), (2.235), and (2.242)

$$\psi_{n\gamma}(\mathbf{r}) = \frac{1}{(n!)^{1/2}}\left(-\frac{\mathrm{i}\lambda}{\sqrt{2}}\right)^n \lambda^{-1}(\pi 2^{n'+1}n'!)^{-1/2} \exp\left(\frac{\mathrm{i}}{\hbar}P_z z + \frac{x^2 + y^2}{4\lambda^2}\right)$$
$$\times\left[\frac{\partial}{\partial x} - \mathrm{i}\hat{q}\frac{\partial}{\partial y}\right]^n \left(\frac{x + \mathrm{i}\hat{q}y}{\lambda}\right)^{n'} \exp\left(-\frac{x^2 + y^2}{2\lambda^2}\right). \tag{2.243}$$

The operator in square brackets commutes with

$$[(x - \mathrm{i}\hat{q}y)/\lambda]^{n'}. \tag{2.244}$$

Thus, multiplying both sides of (2.243) on the left by (2.244), we have

$$\psi_{n\gamma}(\mathbf{r}) = \frac{1}{(n!)^{1/2}}\left(-\frac{\mathrm{i}\lambda}{\sqrt{2}}\right)^n \lambda^{-1}(\pi 2^{n'+1}n'!)^{-1/2}\left(\frac{x - \mathrm{i}\hat{q}y}{\lambda}\right)^{-n'} \exp\left(\frac{\mathrm{i}}{\hbar}P_z z + \frac{x^2 + y^2}{4\lambda^2}\right)$$
$$\times\left[\frac{\partial}{\partial x} - \mathrm{i}\hat{q}\frac{\partial}{\partial y}\right]^n \left(\frac{x^2 + y^2}{\lambda^2}\right)^{n'} \exp\left(-\frac{x^2 + y^2}{2\lambda^2}\right). \tag{2.245}$$

This result can be simplified by expressing it in terms of cylindrical polar coordinates. Let $x = \rho\cos\phi$ and $y = \rho\sin\phi$. Then $\rho = (x^2 + y^2)^{1/2}$, and hence

$$\frac{\partial f(\rho)}{\partial v} = \frac{v}{\rho}\frac{\mathrm{d}f(\rho)}{\mathrm{d}\rho} \qquad (v = x \text{ or } y).$$

Thus (2.245) can be written

$$\psi_{n\gamma}(\mathbf{r}) = (-\mathrm{i})^n \lambda^{-1}(\pi 2^{n-n'+1} n! n'!)^{-1/2} \left(\frac{\rho}{\lambda}\right)^{n-n'} \exp\left(\frac{\mathrm{i}}{\hbar}P_z z - \frac{\rho^2}{4\lambda^2} - \mathrm{i}(n-n')\hat{q}\phi\right)$$
$$\times L_{n'n}\left(\frac{\rho^2}{2\lambda^2}\right). \tag{2.246}$$

Here

$$L_{n'n}(u) = \mathrm{e}^u \frac{\mathrm{d}^n}{\mathrm{d}u^n} u^{n'} \mathrm{e}^{-u} \tag{2.247}$$

is a polynomial of degree n'. Using induction, or proceeding as in Ref. 22, one can show that

$$\begin{aligned} L_{n'n}(u) &= (-1)^{n-n'} \frac{n!}{(n-n')!}\, {}_1F_1(-n'; n-n'+1; u) \quad \text{if } n > n' \\ &= \frac{n'!}{(n'-n)!} u^{n'-n}\, {}_1F_1(-n; n'-n+1; u) \qquad \text{if } n' > n, \end{aligned} \tag{2.248}$$

where ${}_1F_1$ is the confluent hypergeometric function (2.152).

Let* $m = (n' - n)\hat{q}$ and define the radial quantum number

$$N = \min(n, n') \qquad (=0, 1, 2, \ldots). \tag{2.249}$$

Then $\max(n, n') = N + |m|$. If we use these expressions to eliminate n and n' in favour of N and m in (2.246) and (2.248) we obtain

$$\psi_{NP_z m}(\mathbf{r}) = \frac{(-1)^N \mathrm{i}^{N+(1/2)|m|-(1/2)m\hat{q}}}{\lambda |m|!} \left(\frac{(N+|m|)!}{\pi 2^{|m|+1} N!}\right)^{1/2} \left(\frac{\rho}{\lambda}\right)^{|m|}$$
$$\times\, {}_1F_1\left(-N; |m|+1; \frac{\rho^2}{2\lambda^2}\right) \exp\left(\frac{\mathrm{i}}{\hbar}P_z z - \frac{\rho^2}{4\lambda^2} + \mathrm{i}m\phi\right). \tag{2.250}$$

These are the wavefunctions for the c.s.c.o. H, p_z, and L_z in the gauge (2.215). In terms of the quantum numbers N and m, the energy eigenvalues (2.194) are

$$E = \frac{1}{2M}P_z^2 + (N + \tfrac{1}{2}|m| - \tfrac{1}{2}m\hat{q} + \tfrac{1}{2})\hbar\omega. \tag{2.251}$$

These eigenvalues and wavefunctions can also be derived by using separation of variables to solve Schrödinger's equation expressed in terms of cylindrical polar coordinates [23].

* Thus the eigenvalues of L_z are equal to $\hbar m$; see (2.220).

2.10 The isotropic multi-dimensional harmonic oscillator

It is straightforward to generalize the theory for a one-dimensional oscillator, presented in Sections 2.1–2.6, to an isotropic D-dimensional oscillator. Here we give a brief outline of the procedure. The Hamiltonian is

$$H = \sum_{k=1}^{D} \left(\frac{1}{2M} p_k^2 + \tfrac{1}{2} M\omega^2 r_k^2 \right), \tag{2.252}$$

where r_k and p_k are the components of the position and momentum operators. (The oscillator is isotropic because in (2.252) each term in r_k^2 has the same coefficient.) We define

$$a_k = \frac{1}{(2M\hbar\omega)^{1/2}} (\mathrm{i}p_k + M\omega r_k) \qquad (k = 1, \ldots, D). \tag{2.253}$$

Then

$$a_k^\dagger = \frac{1}{(2M\hbar\omega)^{1/2}} (-\mathrm{i}p_k + M\omega r_k). \tag{2.254}$$

Using the canonical commutation relations (1.65)–(1.67) we can show that

$$[a_k, a_{k'}^\dagger] = \delta_{kk'}, \quad [a_k, a_{k'}] = 0, \quad [a_k^\dagger, a_{k'}^\dagger] = 0. \tag{2.255}$$

For the Hermitian operator

$$N_k = a_k^\dagger a_k, \tag{2.256}$$

(2.255) yield

$$[N_k, a_{k'}] = -\delta_{kk'} a_k \tag{2.257}$$

and

$$[N_k, a_{k'}^\dagger] = \delta_{kk'} a_k^\dagger. \tag{2.258}$$

(No summation is implied in (2.256)–(2.258).) Also

$$H = \sum_{k=1}^{D} (N_k + \tfrac{1}{2})\hbar\omega. \tag{2.259}$$

Next we determine the eigenvalues and eigenkets of H. Let $|n_k\rangle$ denote an eigenket of N_k with eigenvalue n_k. Using (2.257) and (2.258) we find

$$a_k |n_k\rangle = (n_k)^{1/2} |n_k - 1\rangle \tag{2.260}$$

and

$$a_k^\dagger |n_k\rangle = (n_k + 1)^{1/2} |n_k + 1\rangle, \tag{2.261}$$

where $n_k = 0, 1, 2, \ldots$. Consider the direct product (see Section 1.6)

$$|\ldots n_k \ldots n_{k'} \ldots\rangle = |\ldots\rangle \ldots |n_k\rangle \ldots |n_{k'}\rangle \ldots. \tag{2.262}$$

Because a_k and $a_{k'}$ commute when $k \neq k'$, eqns (2.260) and (2.262) yield

$$a_k |\ldots n_k \ldots\rangle = n_k^{1/2} |\ldots n_k - 1 \ldots\rangle. \tag{2.263}$$

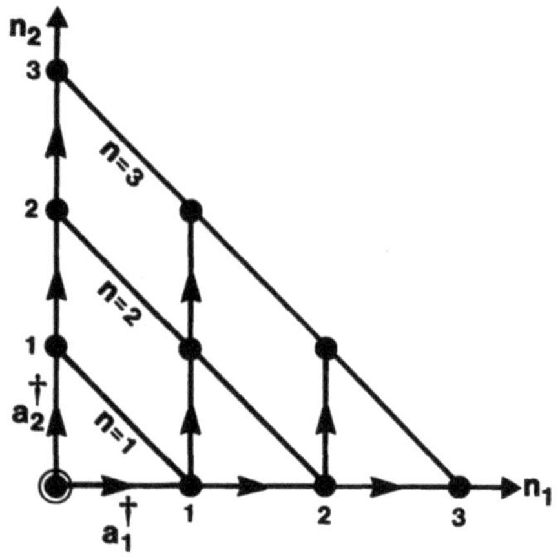

FIG. 2.1 Shift operations for a two-dimensional oscillator. The dots indicate the basis kets $|n_1 n_2\rangle$. Horizontal arrows depict the transformations effected by $a_1^\dagger$ and vertical arrows those effected by $a_2^\dagger$ (see (2.264)).

Similarly,

$$a_k^\dagger|\ldots n_k \ldots\rangle = (n_k + 1)^{1/2}|\ldots n_k + 1 \ldots\rangle. \tag{2.264}$$

Hence (2.262) is an eigenket of N_k with eigenvalue n_k: it is therefore also an eigenket of H with eigenvalue

$$E = \sum_{k=1}^{D} (n_k + \tfrac{1}{2})\hbar\omega = (n + \tfrac{1}{2}D)\hbar\omega, \tag{2.265}$$

where $n = n_1 + n_2 + \cdots + n_D$.

If $D > 1$, these energy levels are degenerate. For the nth level of a two-dimensional oscillator there are $n + 1$ possible values of n_1, and for each of these $n_2 = n - n_1$ is fixed. Thus the degeneracy of this level is

$$g_n = n + 1. \tag{2.266}$$

Similarly, for the nth level of a three-dimensional oscillator, corresponding to each value of n_1 there are $n - n_1 + 1$ possible values of n_2, and for each of these $n_3 = n - n_1 - n_2$ is fixed. The possible values of n_1 are $0, 1, \ldots, n$. Hence

$$g_n = \sum_{n_1=0}^{n} (n - n_1 + 1) = \tfrac{1}{2}(n + 1)(n + 2). \tag{2.267}$$

The operators

$$H_k = (N_k + \tfrac{1}{2})\hbar\omega \qquad (k = 1, \ldots, D) \tag{2.268}$$

constitute a c.s.c.o. for the oscillator with basis kets (2.262). Starting with any one of these kets we can generate all kets of the Hilbert space of the oscillator by applying the shift operators $a_k^\dagger$ and a_k. This is illustrated for a two-dimensional oscillator in Fig. 2.1, where kets are generated from the ground state $|00\rangle$ by applying $a_k^\dagger$ ($k = 1, 2$). In Chapter 5 we discuss an alternative c.s.c.o. that provides the so-called angular momentum basis.

The coordinate-space wavefunction is

$$\psi_{n_1 n_2 \ldots n_D}(\mathbf{r}) = \langle \mathbf{r}|n_1 n_2 \ldots n_D\rangle = \langle r_1|n_1\rangle\langle r_2|n_2\rangle \ldots \langle r_D|n_D\rangle,$$

where $r_1, r_2, \ldots, r_D$ are the components of $\mathbf{r}$. Thus

$$\psi_{n_1 n_2 \ldots n_D}(\mathbf{r}) = \psi_{n_1}(r_1)\psi_{n_2}(r_2)\ldots\psi_{n_D}(r_D), \tag{2.269}$$

where $\psi_{n_k}(r_k)$ is the wavefunction of the one-dimensional oscillator (2.157) with n replaced by n_k and x by r_k. Similarly, the momentum-space wavefunction is

$$\bar{\psi}_{n_1 n_2 \ldots n_D}(\mathbf{p}) = \bar{\psi}_{n_1}(p_1)\bar{\psi}_{n_2}(p_2)\ldots\bar{\psi}_{n_D}(p_D), \tag{2.270}$$

where $\bar{\psi}_{n_k}(p_k)$ is given by (2.163).

The multi-dimensional oscillator also possesses coherent states. These are eigenstates of the lowering operators a_k ($k = 1, \ldots, D$). For example, if $D = 2$ the coherent state is the direct product

$$|z_1 z_2\rangle = |z_1\rangle|z_2\rangle, \tag{2.271}$$

where $|z_k\rangle$ are the eigenkets in

$$a_k|z_k\rangle = z_k|z_k\rangle. \tag{2.272}$$

By similar calculations to those leading to (2.111) we can prove that (2.271) is given in terms of energy eigenkets by

$$|z_1 z_2\rangle = \exp\left[-\frac{1}{2}(|z_1|^2 + |z_2|^2)\right]\sum_{n_1=0}^{\infty}\sum_{n_2=0}^{\infty}\frac{z_1^{n_1} z_2^{n_2}}{(n_1!n_2!)^{1/2}}|n_1 n_2\rangle. \tag{2.273}$$

The states (2.271) have the property that for suitable values of the complex numbers z_1 and z_2, the average values of x and y are equal to the classical values; that is,

$$\langle z_1 z_2|x|z_1 z_2\rangle = A\cos(\omega t - \phi), \quad \langle z_1 z_2|y|z_1 z_2\rangle = B\sin(\omega t - \phi), \tag{2.274}$$

where A, B, and ϕ are constants (see Section 2.5).

It is straightforward to generalize the discussion to $D \geq 3$ and to derive properties for coherent states that are similar to those presented for the one-dimensional oscillator in Section 2.5. These states are sometimes referred to as Cartesian coherent states, to distinguish them from an alternative set of states known as coherent angular momentum states. The latter exist if $D \geq 2$: they are a superposition of the kets of an angular momentum basis (see Section 8.7).

Cartesian coherent states of the isotropic oscillator were discovered by Schrödinger in 1926 [24]. Working in the coordinate representation, Schrödinger constructed a wave packet which 'revolves in precisely the harmonic ellipses described by classical mechanics without dispersing'. He suggested that 'it is only a question of computational skill to accomplish the same thing for the electron in a hydrogen atom' [25]. As it turned out, the solution to the latter problem is not trivial (see Chapter 9).

2.11 Quantization of the free electromagnetic field

The theory of the oscillator has important applications to the electromagnetic field and to vibrations of a crystal lattice. In this section we discuss the former application.

The scalar and vector potential $\phi(\mathbf{r}, t)$ and $\mathbf{A}(\mathbf{r}, t)$ of the classical free electromagnetic field satisfy the equations

$$\left(\nabla^2 - \frac{1}{c^2}\frac{\partial^2}{\partial t^2}\right)\phi + \frac{\partial}{\partial t}\left(\nabla.\mathbf{A} + \frac{1}{c^2}\frac{\partial \phi}{\partial t}\right) = 0 \tag{2.275}$$

and

$$\left(\nabla^2 - \frac{1}{c^2}\frac{\partial^2}{\partial t^2}\right)\mathbf{A} - \nabla\left(\nabla.\mathbf{A} + \frac{1}{c^2}\frac{\partial \phi}{\partial t}\right) = 0. \tag{2.276}$$

The quantization of the electromagnetic field is complicated by the gauge dependence of these potentials [26]: this topic is beyond the scope of this book and we work in the particular gauge

$$\phi = 0, \qquad \nabla.\mathbf{A} = 0. \tag{2.277}$$

Then the classical field equations (2.275) and (2.276) become

$$\left(\nabla^2 - \frac{1}{c^2}\frac{\partial^2}{\partial t^2}\right)\mathbf{A}(\mathbf{r}, t) = 0. \tag{2.278}$$

The magnetic and electric fields are given in terms of the vector potential by (2.175) and

$$\mathbf{E} = -\frac{\partial \mathbf{A}}{\partial t}. \tag{2.279}$$

We suppose that the field is confined to a cube of volume $V = L^3$ and write the general solution to (2.278) as the expansion

$$\mathbf{A}(\mathbf{r}, t) = \sum_{\mathbf{k}} \mathbf{q}_{\mathbf{k}}(t)\exp(i\mathbf{k}.\mathbf{r}), \tag{2.280}$$

where*

$$\mathbf{k} = \frac{2\pi}{L}(n_x, n_y, n_z) \qquad (n_i = 0, \pm 1, \pm 2, \ldots), \tag{2.281}$$

and the coefficients $\mathbf{q}_{\mathbf{k}}$ satisfy

* Thus $\sum_{\mathbf{k}}$ denotes $\sum_{n_x=-\infty}^{\infty}\sum_{n_y=-\infty}^{\infty}\sum_{n_z=-\infty}^{\infty}$. In the limit $L \to \infty$ these sums become integrals [27].

$$\ddot{\mathbf{q}}_{\mathbf{k}} + c^2\mathbf{k}^2\mathbf{q}_{\mathbf{k}} = 0. \tag{2.282}$$

From (2.277) and (2.280) we obtain the transversality condition

$$\mathbf{k}.\mathbf{q}_{\mathbf{k}} = 0. \tag{2.283}$$

For the values of **k** given by (2.281) we have the orthonormality relation

$$\int_V \exp[\mathrm{i}(\mathbf{k}' - \mathbf{k}).\mathbf{r}]\, \mathrm{d}^3x = V\delta_{\mathbf{k}'\mathbf{k}}. \tag{2.284}$$

Equations (2.279), (2.175), and (2.280) yield the fields

$$\mathbf{E} = -\sum_{\mathbf{k}} \dot{\mathbf{q}}_{\mathbf{k}} \exp(\mathrm{i}\mathbf{k}.\mathbf{r}), \tag{2.285}$$

$$\mathbf{B} = \mathrm{i}\sum_{\mathbf{k}} \mathbf{k} \times \mathbf{q}_{\mathbf{k}} \exp(\mathrm{i}\mathbf{k}.\mathbf{r}). \tag{2.286}$$

The energy and momentum of the electromagnetic field are given by

$$H = \frac{1}{2}\varepsilon_0 \int_V (c^2\mathbf{B}^2 + \mathbf{E}^2)\, \mathrm{d}^3x,$$

$$= \frac{1}{2}\varepsilon_0 V \sum_{\mathbf{k}} (c^2\mathbf{k}^2\mathbf{q}_{-\mathbf{k}}.\mathbf{q}_{\mathbf{k}} + \dot{\mathbf{q}}_{-\mathbf{k}}\cdot\dot{\mathbf{q}}_{\mathbf{k}}) \tag{2.287}$$

and

$$\mathbf{P} = \varepsilon_0 \int_V \mathbf{E} \times \mathbf{B}\, \mathrm{d}^3x.$$

$$= -\mathrm{i}\varepsilon_0 V \sum_{\mathbf{k}} \mathbf{k}(\dot{\mathbf{q}}_{-\mathbf{k}}\cdot\mathbf{q}_{\mathbf{k}}), \tag{2.288}$$

where we have used (2.284)–(2.286).

Next we write the solution to (2.282) in the form

$$\mathbf{q}_{\mathbf{k}} = \beta_{\mathbf{k}}[\mathbf{a}_{\mathbf{k}} \exp(-\mathrm{i}\omega t) + \mathbf{a}^*_{-\mathbf{k}} \exp(\mathrm{i}\omega t)], \tag{2.289}$$

where $\omega = c|\mathbf{k}|$, $\mathbf{a}_{\mathbf{k}}$ is a constant, and $\beta_{\mathbf{k}}$ is a real constant such that $\beta_{-\mathbf{k}} = \beta_{\mathbf{k}}$ (see (2.297)). Then

$$\mathbf{A} = \sum_{\mathbf{k}} \beta_{\mathbf{k}}\{\mathbf{a}_{\mathbf{k}} \exp[\mathrm{i}(\mathbf{k}.\mathbf{r} - \omega t)] + \mathbf{a}^*_{-\mathbf{k}} \exp[\mathrm{i}(\mathbf{k}.\mathbf{r} + \omega t)]\}$$

$$= \sum_{\mathbf{k}} \beta_{\mathbf{k}}\{\mathbf{a}_{\mathbf{k}} \exp[\mathrm{i}(\mathbf{k}.\mathbf{r} - \omega t)] + \mathbf{a}^*_{\mathbf{k}} \exp[-\mathrm{i}(\mathbf{k}.\mathbf{r} - \omega t)]\}. \tag{2.290}$$

Hence the form (2.289) ensures that **A** is real.

According to (2.283) and (2.289)

$$\mathbf{a}_{\mathbf{k}} = \sum_{s=1}^{2} \boldsymbol{\varepsilon}_{\mathbf{k}s} a_{\mathbf{k}s}, \tag{2.291}$$

where $\boldsymbol{\varepsilon}_{\mathbf{k}1}$ and $\boldsymbol{\varepsilon}_{\mathbf{k}2}$ are orthogonal unit vectors perpendicular to $\mathbf{k}$. Substituting (2.289) and (2.291) in (2.287) and (2.288) we find

$$H = \varepsilon_0 V \sum_{\mathbf{k}} \sum_{s=1}^{2} \beta_{\mathbf{k}}^2 \omega^2 (a_{\mathbf{k}s}^* a_{\mathbf{k}s} + a_{\mathbf{k}s} a_{\mathbf{k}s}^*) \tag{2.292}$$

and*

$$\mathbf{P} = \varepsilon_0 V \sum_{\mathbf{k}} \sum_{s=1}^{2} \beta_{\mathbf{k}}^2 \omega \mathbf{k} (a_{\mathbf{k}s}^* a_{\mathbf{k}s} + a_{\mathbf{k}s} a_{\mathbf{k}s}^*). \tag{2.293}$$

For a classical field $a_{\mathbf{k}s}$ and $a_{\mathbf{k}s}^*$ commute and the quantity in parentheses in (2.292) and (2.293) is equal to $2a_{\mathbf{k}s}^* a_{\mathbf{k}s}$: we have preserved their order because for a quantized field $a_{\mathbf{k}s}$ and $a_{\mathbf{k}s}^*$ are non-commuting operators (see below).

In terms of the real variables

$$Q_{\mathbf{k}s} = (\varepsilon_0 V \beta_{\mathbf{k}}^2)^{1/2} [a_{\mathbf{k}s} \exp(-\mathrm{i}\omega t) + a_{\mathbf{k}s}^* \exp(\mathrm{i}\omega t)] \tag{2.294}$$

and

$$P_{\mathbf{k}s} = \dot{Q}_{\mathbf{k}s}, \tag{2.295}$$

the Hamiltonian is

$$H = \frac{1}{2} \sum_{\mathbf{k}} \sum_{s=1}^{2} (P_{\mathbf{k}s}^2 + \omega^2 Q_{\mathbf{k}s}^2). \tag{2.296}$$

We see that the dynamics of the free electromagnetic field is formally equivalent to an infinite collection of independent harmonic oscillators.

This equivalence is exploited in making the transition to a quantum theory of this field. First note that the choice

$$\beta_{\mathbf{k}} = \left(\frac{\hbar}{2\varepsilon_0 V \omega} \right)^{1/2} \tag{2.297}$$

makes the coefficients $a_{\mathbf{k}s}$ dimensionless (the factor 2 in (2.297) has been included for later convenience). By analogy with the oscillator, the transition to a quantum theory is achieved by replacing the complex numbers $a_{\mathbf{k}s}$ and $a_{\mathbf{k}s}^*$ with operators $a_{\mathbf{k}s}$ and $a_{\mathbf{k}s}^\dagger$ that satisfy the commutation relations (cf. (2.255))

$$[a_{\mathbf{k}s}, a_{\mathbf{k}'s'}^\dagger] = \delta_{\mathbf{k}\mathbf{k}'} \delta_{ss'}, \quad [a_{\mathbf{k}s}, a_{\mathbf{k}'s'}] = 0, \quad [a_{\mathbf{k}s}^\dagger, a_{\mathbf{k}'s'}^\dagger] = 0. \tag{2.298}$$

In terms of the Hermitian operators

$$N_{\mathbf{k}s} = a_{\mathbf{k}s}^\dagger a_{\mathbf{k}s}, \tag{2.299}$$

the operators corresponding to (2.292), (2.293), and the choice (2.297) are

* In arriving at (2.293) we use the fact that quantities such as

$$\sum_{\mathbf{k}} \beta_{\mathbf{k}}^2 \omega \mathbf{k} (\mathbf{a}_{-\mathbf{k}} \cdot \mathbf{a}_{\mathbf{k}}) \exp(-2\mathrm{i}\omega t)$$

are zero because the summand is an odd function of $\mathbf{k}$.

$$H = \sum_{\mathbf{k}} \sum_{s=1}^{2} \left(N_{\mathbf{k}s} + \frac{1}{2} \right) \hbar\omega, \tag{2.300}$$

$$\mathbf{P} = \sum_{\mathbf{k}} \sum_{s=1}^{2} N_{\mathbf{k}s} \hbar \mathbf{k}. \tag{2.301}$$

From (2.298) and (2.299)

$$[N_{\mathbf{k}s}, a_{\mathbf{k}'s'}] = -\delta_{\mathbf{k}\mathbf{k}'}\delta_{ss'}a_{\mathbf{k}s} \tag{2.302}$$

and

$$[N_{\mathbf{k}s}, a^{\dagger}_{\mathbf{k}'s'}] = \delta_{\mathbf{k}\mathbf{k}'}\delta_{ss'}a^{\dagger}_{\mathbf{k}s}. \tag{2.303}$$

Hence the Hamiltonian operator (2.300) is equivalent to a denumerably infinite collection of independent one-dimensional harmonic oscillators (see Section 2.1).

This equivalence enables us to obtain the eigenvalues and eigenkets of H and $\mathbf{P}$. Let $|n_{\mathbf{k}s}\rangle$ be an eigenket of $N_{\mathbf{k}s}$ with eigenvalue $n_{\mathbf{k}s}$. Then, as in Section 2.2, we use (2.302) and (2.303) to show that

$$a_{\mathbf{k}s}|n_{\mathbf{k}s}\rangle = n_{\mathbf{k}s}^{1/2}|n_{\mathbf{k}s} - 1\rangle \tag{2.304}$$

and

$$a^{\dagger}_{\mathbf{k}s}|n_{\mathbf{k}s}\rangle = (n_{\mathbf{k}s} + 1)^{1/2}|n_{\mathbf{k}s} + 1\rangle, \tag{2.305}$$

where $n_{\mathbf{k}s} = 0, 1, 2, \ldots$. Consider the direct product

$$|\ldots n_{\mathbf{k}s} \ldots n_{\mathbf{k}'s'} \ldots\rangle = |\ldots\rangle \ldots |n_{\mathbf{k}s}\rangle \ldots |n_{\mathbf{k}'s'}\rangle \ldots. \tag{2.306}$$

Because $a_{\mathbf{k}s}$ and $a_{\mathbf{k}'s'}$ commute if $\mathbf{k}'s' \neq \mathbf{k}s$, eqns (2.304) and (2.305) yield

$$a_{\mathbf{k}s}|\ldots n_{\mathbf{k}s} \ldots\rangle = n_{\mathbf{k}s}^{1/2}|\ldots n_{\mathbf{k}s} - 1 \ldots\rangle. \tag{2.307}$$

Similarly,

$$a^{\dagger}_{\mathbf{k}s}|\ldots n_{\mathbf{k}s} \ldots\rangle = (n_{\mathbf{k}s} + 1)^{1/2}|\ldots n_{\mathbf{k}s} + 1 \ldots\rangle. \tag{2.308}$$

Equations (2.307), (2.308), (2.300), and (2.301) show that (2.306) is an eigenket of H and P with eigenvalues

$$E = \sum_{\mathbf{k}} \sum_{s=1}^{2} \left(n_{\mathbf{k}s} + \frac{1}{2} \right) \hbar\omega, \tag{2.309}$$

$$\mathbf{P} = \sum_{\mathbf{k}} \sum_{s=1}^{2} n_{\mathbf{k}s} \hbar \mathbf{k}. \tag{2.310}$$

These results have the following interpretation. The quantized radiation field consists of photons, each of which has energy $\hbar\omega$ and momentum $\hbar\mathbf{k}$. The number of photons with energy $\hbar\omega$, momentum $\hbar\mathbf{k}$, and polarization $\boldsymbol{\varepsilon}_{\mathbf{k}s}$ is equal to $n_{\mathbf{k}s}$. The shift operator $a_{\mathbf{k}s}$ decreases the number of photons by one, and it is therefore referred to as the annihilation operator. Similarly $a^{\dagger}_{\mathbf{k}s}$, which increases the number of photons by one, is the creation operator. The space of the kets (2.306) is known as occupation number space and $N_{\mathbf{k}s}$ is the number operator. In (2.309) the sum over $\frac{1}{2}\hbar\omega$ is the zero-point energy of the field. Although this energy is infinite (because the number of modes is infinite), it can give rise to observable effects such as those associated with finite changes in the zero-point energy [28].

The operators corresponding to the electric and magnetic fields (2.279) and (2.175) are (see (2.290) and (2.297))

$$\mathbf{E} = \mathrm{i}\sum_{\mathbf{k}}\sum_{s=1}^{2}\left(\frac{\hbar}{2\varepsilon_0 V\omega}\right)^{1/2}\omega\boldsymbol{\varepsilon}_{\mathbf{k}s}\{a_{\mathbf{k}s}\exp[\mathrm{i}(\mathbf{k}.\mathbf{r}-\omega t)] - a^{\dagger}_{\mathbf{k}s}\exp[-\mathrm{i}(\mathbf{k}.\mathbf{r}-\omega t)]\} \tag{2.311}$$

and

$$\mathbf{B} = \mathrm{i}\sum_{\mathbf{k}}\sum_{s=1}^{2}\left(\frac{\hbar}{2\varepsilon_0 V\omega}\right)^{1/2}\mathbf{k}\boldsymbol{\varepsilon}_{\mathbf{k}s}\{a_{\mathbf{k}s}\exp[\mathrm{i}(\mathbf{k}.\mathbf{r}-\omega t)] - a^{\dagger}_{\mathbf{k}s}\exp[-\mathrm{i}(\mathbf{k}.\mathbf{r}-\omega t)]\}. \tag{2.312}$$

Because the kets (2.306) are orthogonal, it follows from (2.311), (2.312), (2.307), and (2.308) that the expectation values of **E** and **B** with respect to any state in occupation number space are zero.

Quasi-classical states of the free electromagnetic field can be defined as in Section 2.5. Let $|z_{\mathbf{k}s}\rangle$ be an eigenket of $a_{\mathbf{k}s}$:

$$a_{\mathbf{k}s}|z_{\mathbf{k}s}\rangle = z_{\mathbf{k}s}|z_{\mathbf{k}s}\rangle. \tag{2.313}$$

The kets $|z_{\mathbf{k}s}\rangle$ are given in terms of $|n_{\mathbf{k}s}\rangle$ by an obvious modification of (2.111). The direct product

$$|\ldots z_{\mathbf{k}s}\ldots z_{\mathbf{k}'s'}\ldots\rangle = |\ldots\rangle\ldots|z_{\mathbf{k}s}\rangle\ldots|z_{\mathbf{k}'s'}\rangle\ldots \tag{2.314}$$

is a coherent state of the free electromagnetic field. Equations (2.311)–(2.314) show that for a coherent state the expectation values of **E** and **B** are

$$\langle\mathbf{E}\rangle = \mathrm{i}\sum_{\mathbf{k}}\sum_{s=1}^{2}\left(\frac{\hbar}{2\varepsilon_0 V\omega}\right)^{1/2}\omega\boldsymbol{\varepsilon}_{\mathbf{k}s}\{z_{\mathbf{k}s}\exp[\mathrm{i}(\mathbf{k}.\mathbf{r}-\omega t)] - z^{*}_{\mathbf{k}s}\exp[-\mathrm{i}(\mathbf{k}.\mathbf{r}-\omega t)]\} \tag{2.315}$$

and

$$\langle\mathbf{B}\rangle = \mathrm{i}\sum_{\mathbf{k}}\sum_{s=1}^{2}\left(\frac{\hbar}{2\varepsilon_0 V\omega}\right)^{1/2}\mathbf{k}\times\boldsymbol{\varepsilon}_{\mathbf{k}s}\{z_{\mathbf{k}s}\exp[\mathrm{i}(\mathbf{k}.\mathbf{r}-\omega t)] - z^{*}_{\mathbf{k}s}\exp[-\mathrm{i}(\mathbf{k}.\mathbf{r}-\omega t)]\}. \tag{2.316}$$

If

$$z_{\mathbf{k}s} = -\mathrm{i}\left(\frac{2\varepsilon_0 V}{\hbar\omega}\right)^{1/2}E_{\mathbf{k}s}, \tag{2.317}$$

eqns (2.315) and (2.316) are Fourier expansions for classical electromagnetic fields, with the complex numbers $E_{\mathbf{k}s}$ being the Fourier coefficients of the electric field.

The harmonic oscillator discussed in Section 2.1 possesses a position operator x, eigenkets $|x\rangle$, and coordinate-space wavefunctions $\psi(x)$ (see Section 2.6). We note that in the above treatment of photons we have not introduced a position operator $\mathbf{r}$ for a photon, or the corresponding eigenkets $|\mathbf{r}\rangle$ and coordinate-space wavefunctions $\psi(\mathbf{r})$. It turns out that the concept of such a position operator requires careful formulation (see Ref. 29).

References

1. Pippard, A.B. (1978 and 1983). *The physics of vibration*, Vols. 1 and 2. Cambridge University Press.
2. Moshinsky, M. (1969). *The harmonic oscillator in modern physics: From atoms to quarks*. Gordon and Breach, New York.
3. Marcuse, D. (1980). *Principles of quantum electronics*. Academic Press, New York.
4. Avery, J. (1976). *Creation and annihilation operators*. McGraw-Hill, New York.
5. The zero-point energy is an intriguing property that has been studied by numerous researchers. See the review article by Boyer, T.H. (1980). *Foundations of radiation theory and quantum electrodynamics*, (ed. A.O. Barut), pp. 49–63. Plenum, New York.
6. See the discussion in Böhm, A. (1979). *Quantum mechanics*, pp. 24–6. Springer, New York.
7. Morales, J., Lopez-Bonilla, J., and Palma, A. (1987). Hypervirial theorem and parameter differentiation: Closed formulation for harmonic oscillator integrals. *Journal of Mathematical Physics*, **28**, 1032–5.
8. Morales, J. and Flores-Riveros, A. (1989). The generalization of the binomial theorem. *Journal of Mathematical Physics*, **30**, 393–7.
9. See, for example, Whittaker, E.T. and Watson, G.N. (1969). *A course in modern analysis*, 4th edn. Cambridge University Press, London.
10. See, for example, Miller, M.M. and Mishkin, E.A. (1966). Characteristic states of the electromagnetic radiation field. *Physical Review*, **152**, 1110–14.
11. Glauber, R.J. (1963). Coherent and incoherent states of the radiation field. *Physical Review*, **131**, 2766–88.
12. See, for example, Howard, S. and Roy, S.K. (1987). Coherent states of a harmonic oscillator. *American Journal of Physics*, **55**, 1109–17.
13. See Walls, D.F. (1983). Squeezed states of light. *Nature*, **306**, 141–6; (1986). Growing expectations from squeezed states of light. *Nature*, **324**, 210–11.
14. See, for example, Refs 11 and 12.
15. Perelomov, A. (1986). *Generalized coherent states and their applications*. Springer, Berlin.
16. See Refs 12 and 15, and references therein.
17. See Messiah, A. (1961). *Quantum mechanics*, Vol. I, p. 261. North-Holland, Amsterdam.
18. See Leubner, C., Alber, M., and Schupfer, N. (1988). Critique and correction of the textbook comparison between classical and quantum harmonic oscillator probability densities. *American Journal of Physics*, **56**, 1123–9.

19. Our presentation is based on that of Johnson, M.H. and Lippmann, B.A. (1949). Motion in a constant magnetic field. *Physical Review*, **76**, 828–32.
20. See Ref. 5, Chapter 11.
21. This non-uniqueness of **A** is part of an important general principle known as the gauge principle: see Ref. 5, Chapter 11.
22. Landau, L.D. and Lifschitz, Ė.M. (1977) *Quantum mechanics*, 3rd edn, p. 657. Pergamon, Oxford.
23. See Ref. 22, p. 458.
24. Schrödinger, E. (1926). Der stetige Übergang von der Mikro-zur Makromechanik. *Naturwissenschaften*, **14**, 664–6.
25. Pizibram, K. (ed.) (1967). *Letters on wave mechanics*, p. 10. Vision, London.
26. See, for example, Itzykson, C. and Zuber, J. (1980). *Quantum field theory*, Section 3.2. McGraw-Hill, New York.
27. See Messiah, A. (1962). *Quantum mechanics*, Vol. II, p. 974. North-Holland, Amsterdam.
28. See Refs 5 and 26.
29. Kraus, K. (1977). Position observables of the photon. In *The uncertainty principle and foundations of quantum mechanics*, (ed. W.C. Price and S.S. Chissick), pp. 293–320. Wiley, London.

3

OTHER ONE-DIMENSIONAL SYSTEMS

We discuss the application of operator methods to one-dimensional systems other than the harmonic oscillator. In Section 3.1 the method of 'one-way factorization' is used to obtain generalized quantum harmonic potentials. In Section 3.2 factorization and shift operators for an arbitrary one-dimensional potential are considered, and a classification is presented of the problems that can be treated by this method [1]. The use of the hypervirial theorem in the calculation of expectation values for one-dimensional systems is discussed in Section 3.3. In Sections 3.4–3.7 four examples are used to illustrate the applications of the preceding material. In these examples we consider in some detail how shift operators for energy can be derived and we also use these shift operators to calculate wavefunctions. Supersymmetric quantum mechanics and shape invariance are discussed in Sections 3.8 and 3.9.

3.1 Other Hamiltonians having the oscillator spectrum [2]

For a classical particle in a bound state of a one-dimensional potential $V(x_c)$, the angular frequency of oscillations is given by

$$\omega_c = 2\pi\left[\int_{x_1}^{x_2}\left(\frac{2M}{E - V(x_c)}\right)^{1/2} dx_c\right]^{-1}. \tag{3.1}$$

Here x_1 and x_2 are the classical turning points defined by

$$V(x_1) = V(x_2) = E. \tag{3.2}$$

For the harmonic oscillator potential

$$V(x_c) = \tfrac{1}{2}M\omega^2 x_c^2, \tag{3.3}$$

one has

$$\omega_c = \omega. \tag{3.4}$$

Thus ω_c is independent of energy. It can be shown that there are an uncountable number of potentials for which ω_c is independent of energy [3]. These potentials are known as generalized classical harmonic potentials.

The quantum-mechanical analogue of the property (3.4) is equal spacing of energy levels,* as in (2.23). There are an uncountable number of potentials $V(x)$ for which the energy eigenvalues are given by (2.23): these are known as generalized quantum harmonic potentials. In this section we use the method

* Equal spacing of energy levels means that any wave packet will return to its initial shape after a time $2\pi/\omega$ has elapsed [4].

of 'one-way factorization' to construct these generalized quantum potentials. The usefulness of such a factorization was first noted by Schrödinger, who remarked [5] 'We are likely to enlarge our technique if we keep our eye on this possibility. A one-way-round factorization may be efficient, even though on commutation it produces terms of entirely new form of which there is no trace in the original equation.'

Define the dimensionless operators

$$X = (M\omega/\hbar)^{1/2}x \tag{3.5}$$

and

$$p_X = (M\hbar\omega)^{-1/2}p. \tag{3.6}$$

In terms of these the commutation relation (2.10) is

$$[p_X, X] = -\mathrm{i}, \tag{3.7}$$

and the shift operator (2.11) is

$$a = \frac{1}{\sqrt{2}}(\mathrm{i}p_X + X). \tag{3.8}$$

This operator and its adjoint provide the 'two-way' factorization

$$aa^\dagger = H + \tfrac{1}{2} \tag{3.9}$$

and

$$a^\dagger a = H - \tfrac{1}{2}, \tag{3.10}$$

where

$$H = \tfrac{1}{2}p_X^2 + \tfrac{1}{2}X^2 \tag{3.11}$$

is the Hamiltonian (2.20) in units of $\hbar\omega$ (see Section 2.1).

Now consider an operator b which, together with its adjoint, provides the 'one-way factorization'*

$$bb^\dagger = H + \tfrac{1}{2}. \tag{3.12}$$

We look for such an operator in the form

$$b = \frac{1}{\sqrt{2}}[\mathrm{i}p_X + f(X)], \tag{3.13}$$

where f is a real function. Then

$$b^\dagger = \frac{1}{\sqrt{2}}[-\mathrm{i}p_X + f(X)]. \tag{3.14}$$

From (3.11)–(3.14) and (3.7) we find that f must satisfy the Ricatti equation

* Hence we do not require the additional relation $b^\dagger b = H - \frac{1}{2}$.

$$\frac{\mathrm{d}f}{\mathrm{d}X} + f^2 = X^2 + 1. \tag{3.15}$$

One solution is $f = X$, which means $b = a$. There are, however, additional solutions to (3.15). With

$$f(X) = X + g(X), \tag{3.16}$$

eqn (3.15) can be written

$$\frac{1}{g^2}\frac{\mathrm{d}g}{\mathrm{d}X} + \frac{2X}{g} + 1 = 0. \tag{3.17}$$

The substitution $g = G^{-1}$ transforms (3.17) into the first-order linear equation

$$\frac{\mathrm{d}G}{\mathrm{d}X} - 2XG - 1 = 0, \tag{3.18}$$

which has the general solution*

$$G = \left(\gamma + \int_0^X \exp(-u^2)\,\mathrm{d}u\right)\exp(X^2), \tag{3.19}$$

where γ is an arbitrary constant. Hence

$$g(X) = \frac{\exp(-X^2)}{\gamma + \int_0^X \exp(-u^2)\,\mathrm{d}u}. \tag{3.20}$$

Next, we evaluate the product $b^\dagger b$. Using (3.13), (3.14), (3.7), and (3.15), we find

$$b^\dagger b = \tilde{H} - \tfrac{1}{2}, \tag{3.21}$$

where

$$\tilde{H} = \tfrac{1}{2}p_X^2 + \tilde{V}(X), \tag{3.22}$$

and

$$\tilde{V}(X) = \frac{1}{2}X^2 - \frac{\mathrm{d}}{\mathrm{d}X}\frac{\exp(-X^2)}{\gamma + \int_0^X \exp(-u^2)\,\mathrm{d}u}. \tag{3.23}$$

Only if $\gamma = \infty$ is $\tilde{H} = H$: otherwise the product $b^\dagger b$ defines a new Hamiltonian. For $\tilde{H}$ to be Hermitian it is necessary that γ be real.

To determine the eigenkets and eigenvalues of $\tilde{H}$ we note that

$$\begin{aligned}\tilde{H}b^\dagger &= (b^\dagger b + \tfrac{1}{2})b^\dagger \\ &= b^\dagger(bb^\dagger + \tfrac{1}{2}) \\ &= b^\dagger(H + 1).\end{aligned} \tag{3.24}$$

* Note that

$$\frac{\mathrm{d}}{\mathrm{d}X}\int_0^X \exp(-u^2)\,\mathrm{d}u = \exp(-X^2).$$

Hence for

$$|\phi_n\rangle = c_n b^\dagger |n-1\rangle \qquad (n = 1, 2, \ldots), \tag{3.25}$$

where $|n\rangle$ is an eigenket of H and c_n is a constant,

$$\tilde{H}|\phi_n\rangle = (n + \tfrac{1}{2})|\phi_n\rangle. \tag{3.26}$$

Also

$$\begin{aligned}\langle\phi_{n'}|\phi_n\rangle &= c_{n'}^* c_n \langle n'-1|bb^\dagger|n-1\rangle \\ &= c_{n'}^* c_n n \delta_{n'n},\end{aligned} \tag{3.27}$$

and therefore $|\phi_n\rangle$ is normalized if

$$c_n = n^{-1/2} \qquad (n = 1, 2, \ldots). \tag{3.28}$$

Equation (3.25) can be written

$$\begin{aligned}|\phi_n\rangle &= c_n(a^\dagger + 2^{-1/2}g)|n-1\rangle \\ &= |n\rangle + (2n)^{-1/2} g|n-1\rangle.\end{aligned} \tag{3.29}$$

Because

$$\langle\phi_n|\chi\rangle = c_n^* \langle n-1|b|\chi\rangle, \tag{3.30}$$

if there exists a ket $|\phi_0\rangle$ such that

$$b|\phi_0\rangle = 0, \tag{3.31}$$

then

$$\langle\phi_n|\phi_0\rangle = 0 \qquad (n = 1, 2, \ldots). \tag{3.32}$$

According to (3.21) and (3.31)

$$\tilde{H}|\phi_0\rangle = \tfrac{1}{2}|\phi_0\rangle, \tag{3.33}$$

and $|\phi_0\rangle$ is the ground state.

If we operate on both sides of (3.25) with $b^\dagger a^\dagger b$ and use (3.12), (3.28), and (2.30), we find that

$$b^\dagger a^\dagger b|\phi_n\rangle = n(n+1)^{1/2}|\phi_{n+1}\rangle. \tag{3.34}$$

This equation is a raising operation for $|\phi_n\rangle$. We see that the raising operators for $|\phi_n\rangle$ differ from those for $|n\rangle$: in particular, the former are third order in the momentum whereas the latter are linear in the momentum. Also, unlike the transformations (2.34) where the kets are generated from the ground state, the kets in (3.34) are generated from the first excited state $|\phi_1\rangle$.

Finally, we determine the values of γ for which the states $|\phi_n\rangle$ $(n = 0, 1, \ldots)$ exist. The task of doing this for a potential as complicated as (3.23) is simplified by working in the coordinate representation. Because

$$\left|\int_0^X \exp(-u^2)\,du\right| \leq \frac{1}{2}\pi^{1/2}, \tag{3.35}$$

we see that for

$$|\gamma| > \tfrac{1}{2}\pi^{1/2} \tag{3.36}$$

the potential $\tilde{V}$ has no singularities and $\tilde{V} \to \frac{1}{2}X^2$ as $X \to \pm\infty$. Hence if γ is real and satisfies (3.36), the Hermitian operators (3.22) are a one-parameter family of Hamiltonians.

From (3.29)

$$\phi_n(X) = \psi_n(X) + (2n)^{-1/2}g(X)\psi_{n-1}(X) \qquad (n = 1, 2, \ldots), \tag{3.37}$$

where g is given by (3.20) and ψ_n is the wavefunction (2.157). For each γ that satisfies (3.36) the wavefunctions (3.37) are normalized. Similarly, from (3.31)

$$\left(\frac{\mathrm{d}}{\mathrm{d}X} + f\right)\phi_0(X) = 0, \tag{3.38}$$

and hence

$$\phi_0(X) = c_0\psi_0 \exp\left(-\int g \, \mathrm{d}X\right), \tag{3.39}$$

where c_0 is a constant. If γ satisfies (3.36), ϕ_0 is square-integrable. The functions (3.37) and (3.39) are a complete set of wavefunctions for the Hamiltonians (3.22). For finite γ the potentials of these Hamiltonians are different from the oscillator potential; nevertheless, from (3.26) and (3.33) the energy eigenvalues are identical to those of the oscillator. Because of these properties the parameter-dependent potentials (3.23) are also referred to as isospectral deformations of the oscillator potential.

We will see in Section 3.8 that $H + \frac{1}{2}$ and $\tilde{H} - \frac{1}{2}$ are supersymmetric partner Hamiltonians; the condition (3.31) corresponds to unbroken supersymmetry [6]. The one-way factorization method can also be used to obtain generalized potentials for other systems (see Section 5.8).

3.2 Factorization and shift operators for one-dimensional systems

In Section 2.2 we discussed shift operators for the one-dimensional oscillator. These operators have the property that they factorize the Hamiltonian (see (3.9) and (3.10)). Here we discuss the generalization of this method to other one-dimensional systems. Suppose the Hamiltonian is

$$H_k = \frac{1}{2M}p^2 + V_k(x), \tag{3.40}$$

where k is a real parameter. The energy eigenvalue equation is

$$H_k|Ek\rangle = E|Ek\rangle. \tag{3.41}$$

Let

$$A_k = \mathrm{i}p + f_k(x) \tag{3.42}$$

where f_k is a real function. Then

$$A_k^\dagger = -ip + f_k(x). \tag{3.43}$$

We wish to determine the potentials V_k for which H_k can be factorized by A_k and $A_k^\dagger$; that is, for which

$$A_k^\dagger A_k = 2MH_k + F_k, \tag{3.44}$$

$$A_k A_k^\dagger = 2MH_{k+1} + G_k, \tag{3.45}$$

where F_k and G_k are independent of x and p.

Let us first note the significance of these factorizations. If we multiply (3.41) on the left by A_k and use (3.44) and (3.45) we find

$$H_{k+1}(A_k|Ek\rangle) = \left(E + \frac{1}{2M}(F_k - G_k)\right)(A_k|Ek\rangle). \tag{3.46}$$

Thus $A_k|Ek\rangle$ is an eigenket of H_{k+1} corresponding to energy $E + (F_k - G_k)/2M$. Similarly, by multiplying (3.41) on the left by $A_{k-1}^\dagger$ and using (3.44) and (3.45), with k replaced by $k - 1$, we obtain

$$H_{k-1}(A_{k-1}^\dagger|Ek\rangle) = \left(E - \frac{1}{2M}(F_{k-1} - G_{k-1})\right)(A_{k-1}^\dagger|Ek\rangle). \tag{3.47}$$

Hence $A_{k-1}^\dagger|Ek\rangle$ is an eigenket of H_{k-1} corresponding to energy $E - (F_{k-1} - G_{k-1})/2M$.

It is convenient to introduce the notation

$$A_k^+ = A_k, \tag{3.48}$$

$$A_k^- = A_{k-1}^\dagger. \tag{3.49}$$

Then

$$A_k^+|Ek\rangle = \alpha^+(E, k)|E + \frac{1}{2M}(F_k - G_k), k + 1\rangle, \tag{3.50}$$

$$A_k^-|Ek\rangle = \alpha^-(E, k)|E - \frac{1}{2M}(F_{k-1} - G_{k-1}), k - 1\rangle. \tag{3.51}$$

Suppose $|Ek\rangle$ is normalizable and p is Hermitian with respect to $|Ek\rangle$ and $A_k^\pm|Ek\rangle$ (for bound states of the potentials considered below, a sufficient condition for the latter is finiteness of the average potential [17]). Then the norm of the right-hand side of (3.50) is equal to

$$\langle Ek|(A_k^+)^\dagger A_k^+|Ek\rangle = \langle A_k^\dagger A_k\rangle = \langle 2MH_k + F_k\rangle,$$

which is finite. Similarly for (3.51). Thus the transformations (3.50) and (3.51) preserve the normalizability of kets. The conditions for the validity of this important property can also be expressed in terms of F_k and G_k [1]; in most of the applications considered in this book these conditions are satisfied. For normalized kets in (3.50) and (3.51), and choosing a phase factor equal to unity,

$$\alpha^+ = (2ME + F_k)^{1/2}, \tag{3.52}$$

$$\alpha^- = (2ME + G_{k-1})^{1/2}. \tag{3.53}$$

Note that non-negativity of the norm in (3.50) and (3.51) requires $2ME + F_k \geq 0$ and $2ME + G_{k-1} \geq 0$.

Next we determine the potentials V_k for which the Hamiltonian can be factorized as in (3.44) and (3.45). Substituting (3.42) and (3.43) in (3.44) and (3.45), and using (1.128) we obtain the non-linear equations*

$$-\hbar\frac{\mathrm{d}f_k}{\mathrm{d}x} + f_k^2 = 2MV_k + F_k, \tag{3.54}$$

$$\hbar\frac{\mathrm{d}f_k}{\mathrm{d}x} + f_k^2 = 2MV_{k+1} + G_k. \tag{3.55}$$

Subtracting (3.55) and (3.54), with k replaced by $k + 1$ in the latter, we have

$$\hbar\frac{\mathrm{d}f_{k+1}}{\mathrm{d}x} + \hbar\frac{\mathrm{d}f_k}{\mathrm{d}x} - f_{k+1}^2 + f_k^2 = G_k - F_{k+1}. \tag{3.56}$$

This equation is a self-consistency condition for the factorization method of constructing shift operators, and its solutions lead to a classification of all factorizable one-dimensional problems [1, 7]. The following discussion is based on the classic paper by Infeld and Hull [1]. They showed that for the function $f_k(x)$ the general forms

$$f_k(x) = -\hbar k u(x) + v(x) \tag{3.57}$$

and

$$f_k(x) = -\hbar k u(x) + k^{-1}v(x), \tag{3.58}$$

where u and v are independent of k, are sufficient to determine all factorizations (3.44) and (3.45). (See also Section 3.9.)

Consider first (3.57). Substituting this in (3.56) we find

$$\hbar^2(2k+1)\left(\frac{\mathrm{d}u}{\mathrm{d}x} + u^2\right) - 2\hbar\left(\frac{\mathrm{d}v}{\mathrm{d}x} + uv\right) = F_{k+1} - G_k. \tag{3.59}$$

Hence

$$\frac{\mathrm{d}u}{\mathrm{d}x} + u^2 = \text{constant} \tag{3.60}$$

and

* Note that if V is independent of k then $\mathrm{d}f/\mathrm{d}x$ is constant: the oscillator and the free particle are the only possibilities.

$$\frac{dv}{dx} + uv = \text{constant}. \tag{3.61}$$

We proceed as follows. Equations (3.60) and (3.61) are solved for $u(x)$ and $v(x)$, and then $V_k(x)$ is obtained in terms of F_k from (3.54) and (3.57). We choose F_k to give the potential a simple form. Finally, (3.59) is used to calculate G_k.

There are four solutions to (3.60) and (3.61), and we discuss each of these in turn.

1.
$$u = 0, \qquad v = \beta x + \gamma = f_k, \tag{3.62}$$

where β and γ are arbitrary real constants. Then

$$V_k = \frac{1}{2M}[(\beta x + \gamma)^2 - \hbar\beta - F_k]. \tag{3.63}$$

With $F_k = -\hbar\beta$,*

$$V_k = \frac{1}{2M}(\beta x + \gamma)^2. \tag{3.64}$$

From (3.59) and (3.62) we have

$$F_{k+1} - G_k = -2\hbar\beta. \tag{3.65}$$

Thus $G_k = \hbar\beta$. The shift operations (3.50) and (3.51) become

$$A^{\pm}|E\rangle = (2ME \mp \hbar\beta)^{1/2}|E \mp (\hbar\beta/M)\rangle, \tag{3.66}$$

where

$$A^{\pm} = \pm ip + \beta x + \gamma. \tag{3.67}$$

The following three cases are performed in a similar manner and we simply state the results.

2.
$$u = x^{-1}, \qquad v = \beta x - \hbar\gamma x^{-1}. \tag{3.68}$$

$$V_k = \frac{1}{2M}[\beta^2 x^2 + \hbar^2(k+\gamma)(k+\gamma-1)x^{-2} - 2\hbar\beta\gamma], \tag{3.69}$$

if $F_k = -(2k+1)\hbar\beta$. Then $G_k = -(2k-1)\hbar\beta$ and

$$A_k^{\pm}|Ek\rangle = [2ME - (2k - 1 \pm 2)\hbar\beta]^{1/2}|E \mp (\hbar\beta/M), k \pm 1\rangle, \tag{3.70}$$

where

$$A_k^{\pm} = \pm ip + \beta x - \hbar(k - \tfrac{1}{2} \pm \tfrac{1}{2} + \gamma)x^{-1}. \tag{3.71}$$

3.
$$u = \gamma, \qquad v = \beta \exp(-\gamma x) + \hbar\delta, \tag{3.72}$$

where δ is an arbitrary real constant.

* An alternative choice is $F_k = G_k = -2\hbar\beta k$. With this choice the shift operations (3.66) do not change the energy.

$$V_k = \frac{1}{2M}[\beta^2 \exp(-2\gamma x) - \hbar\beta(2\gamma k - \gamma - 2\delta) \exp(-\gamma x)], \quad (3.73)$$

if $F_k = \hbar^2(\gamma k - \delta)^2$. Then $G_k = F_k$ and

$$A_k^\pm |Ek\rangle = \{2ME + \hbar^2[(k - \tfrac{1}{2} \pm \tfrac{1}{2})\gamma - \delta]^2\}^{1/2} |E, k \pm 1\rangle, \quad (3.74)$$

where

$$A_k^\pm = \pm ip + \beta \exp(-\gamma x) - \hbar(k - \tfrac{1}{2} \pm \tfrac{1}{2})\gamma + \hbar\delta. \quad (3.75)$$

4.
$$\begin{aligned} u &= \beta \cot \beta(x + d), \\ v &= \hbar\beta\gamma \cot \beta(x + d) + \varepsilon \operatorname{cosec} \beta(x + d), \end{aligned} \quad (3.76)$$

where ε and d are arbitrary constants.

$$\begin{aligned} V_k = \frac{1}{2M}\{&[\hbar^2\beta^2(k - \gamma)(k - \gamma - 1) + \varepsilon^2] \operatorname{cosec}^2\beta(x + d) \\ &- 2\hbar\beta\varepsilon(k - \gamma - \tfrac{1}{2}) \operatorname{cosec} \beta(x + d) \cot \beta(x + d)\}, \end{aligned} \quad (3.77)$$

if $F_k = -\hbar^2\beta^2(k - \gamma)^2$. Then $G_k = F_k$ and

$$A_k^\pm |Ek\rangle = [2ME - \hbar^2\beta^2(k - \tfrac{1}{2} \pm \tfrac{1}{2} - \gamma)^2]^{1/2} |E, k \pm 1\rangle, \quad (3.78)$$

where

$$A_k^\pm = \pm ip - \hbar\beta(k - \tfrac{1}{2} \pm \tfrac{1}{2} - \gamma) \cot \beta(x + d) + \varepsilon \operatorname{cosec} \beta(x + d). \quad (3.79)$$

Consider next (3.58). Substituting this in (3.56) we obtain

$$\hbar^2(2k + 1)\left(\frac{du}{dx} + u^2\right) - \frac{2k + 1}{k(k + 1)}\left(\hbar\frac{dv}{dx} + \frac{v^2}{k(k + 1)}\right) = F_{k+1} - G_k. \quad (3.80)$$

Hence u must satisfy (3.60), while

$$\frac{dv}{dx} + \frac{v^2}{k(k + 1)} = \text{constant}. \quad (3.81)$$

Because v is independent of k the solution to (3.81) is $v = q$, an arbitrary real constant. Corresponding to the solutions for u we obtain two more factorizations:

5.
$$u = x^{-1}. \quad (3.82)$$

$$V_k = \frac{1}{2M}[-2\hbar q x^{-1} + \hbar^2 k(k - 1)x^{-2}], \quad (3.83)$$

if $F_k = q^2k^{-2}$. Then $G_k = F_k$ and

$$A_k^\pm |Ek\rangle = [2ME + q^2(k - \tfrac{1}{2} \pm \tfrac{1}{2})^{-2}]^{1/2} |E, k \pm 1\rangle, \quad (3.84)$$

where

$$A_k^{\pm} = \pm ip - \hbar(k - \tfrac{1}{2} \pm \tfrac{1}{2})x^{-1} + q(k - \tfrac{1}{2} \pm \tfrac{1}{2})^{-1}. \tag{3.85}$$

6.
$$u = \beta \cot \beta(x + d). \tag{3.86}$$

$$V_k = \frac{1}{2M}[-2\hbar\beta q \cot \beta(x + d) + \hbar^2\beta^2 k(k - 1) \operatorname{cosec}^2\beta(x + d)], \tag{3.87}$$

if $F_k = q^2k^{-2} - \hbar^2\beta^2k^2$. Then $G_k = F_k$ and

$$A_k^{\pm}|Ek\rangle = [2ME + q^2(k - \tfrac{1}{2} \pm \tfrac{1}{2})^{-2} - \hbar^2\beta^2(k - \tfrac{1}{2} \pm \tfrac{1}{2})^2]^{1/2}|E, k \pm 1\rangle, \tag{3.88}$$

where

$$A_k^{\pm} = \pm ip - \hbar\beta(k - \tfrac{1}{2} \pm \tfrac{1}{2}) \cot \beta(x + d) + q(k - \tfrac{1}{2} \pm \tfrac{1}{2})^{-1}. \tag{3.89}$$

In the above, certain terms that depend on k (but not on x) were eliminated from the potentials by our choices for F_k. This has the advantage that the resulting potentials correspond in special cases to potentials that are of physical interest (see Sections 3.4–3.7). For the factorization types 3–6 a consequence of our choices for F_k is that the shift operations (3.50) and (3.51) do not change the energy (because $F_k = G_k$ in each case). Shift operations for energy are important and in the examples considered below we will show how these can be obtained.

3.3 Expectation values: use of the hypervirial theorem

In the case of eigenkets $|Ek\rangle$ of the Hamiltonian H_k, the hypervirial theorem (eqn (1.193)) becomes

$$\langle Ek|[H_k, W]|Ek\rangle = 0. \tag{3.90}$$

Here it is assumed that H_k is Hermitian with respect to $W|Ek\rangle$; otherwise the operator W is arbitrary. By considering various forms for W in (3.90) we obtain expectation values for certain one-dimensional potentials.

Using (3.40) and the commutation relation (1.128), we can show that

$$[g(x), H_k] = \frac{i\hbar}{M}\frac{dg}{dx}p + \frac{\hbar^2}{2M}\frac{d^2g}{dx^2} \tag{3.91}$$

and

$$[f(x)p, H_k] = -i\hbar\frac{dV_k}{dx}f + 2i\hbar\frac{df}{dx}(H_k - V_k) + \frac{\hbar^2}{2M}\frac{d^2f}{dx^2}p. \tag{3.92}$$

In the examples discussed below the Hamiltonian contains a parameter a having the dimension of length.* It is convenient to define the dimensionless operators

$$X = x/a \tag{3.93}$$

* From the context in which it is used, no confusion between this parameter and the annihilation operator (3.8) should occur.

and

$$p_X = \frac{a}{\hbar} p. \tag{3.94}$$

These satisfy

$$[p_X, f(X)] = -\mathrm{i}(\mathrm{d}f/\mathrm{d}X). \tag{3.95}$$

Equations (3.90) and (3.91) with $W = g(X)$ and $\mathrm{d}g/\mathrm{d}X = F(X)$ yield

$$\langle F(X)p_X \rangle = \tfrac{1}{2}\mathrm{i}\langle \mathrm{d}F/\mathrm{d}X \rangle, \tag{3.96}$$

where the expectation values are with respect to $|Ek\rangle$. Similarly, with

$$W = F(X)p_X + \tfrac{1}{2}\mathrm{i}(\mathrm{d}F/\mathrm{d}X), \tag{3.97}$$

eqns (3.90)–(3.92) yield

$$\left\langle \frac{\mathrm{d}^3 F}{\mathrm{d}X^3} + \frac{8Ma^2}{\hbar^2}[E - V_k(aX)]\frac{\mathrm{d}F}{\mathrm{d}X} - \frac{4Ma^2}{\hbar^2}\frac{\mathrm{d}V_k(aX)}{\mathrm{d}X}F \right\rangle = 0. \tag{3.98}$$

Equations (3.96) and (3.98) are useful results that enable us to calculate expectation values for certain one-dimensional problems (see below).

3.4 The one-dimensional Morse potential [8]

In the remainder of this chapter we illustrate the use of operator methods to solve one-dimensional problems. We do not discuss the physical applications of these problems; for this the reader should consult Refs 8–15.

The one-dimensional Morse potential is

$$V(x) = D[\exp(-2x/a) - 2\exp(-x/a)], \tag{3.99}$$

where D and a are positive constants. The classical potential has a minimum value $-D$ at $x = 0$; it tends to zero as $x \to \infty$, and to $+\infty$ as $x \to -\infty$ (see Fig. 3.1).

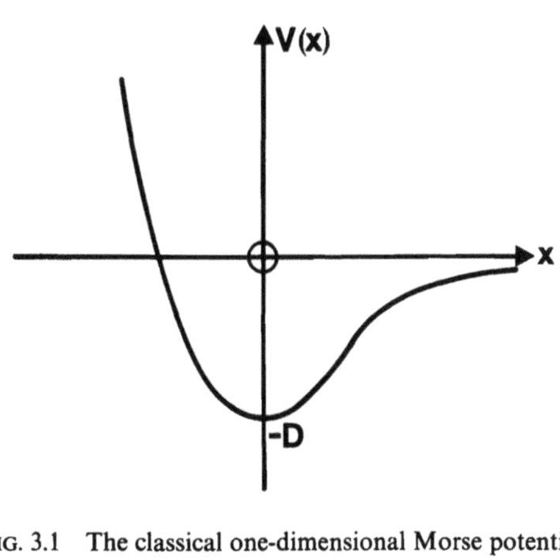

FIG. 3.1 The classical one-dimensional Morse potential.

Let

$$k = (2Ma^2\hbar^{-2}D)^{1/2} - \tfrac{1}{2}. \tag{3.100}$$

The eigenvalue equation (3.41) is

$$\left(\frac{1}{2M}p^2 + D[\exp(-2x/a) - 2\exp(-x/a)]\right)|Ek\rangle = E|Ek\rangle. \tag{3.101}$$

For bound states ($E < 0$) put

$$E = -\frac{\hbar^2}{2Ma^2}n^2, \tag{3.102}$$

where the dimensionless variable n is real. Define p_X by (3.94) and let

$$X = \frac{x}{a} - \ln(2k + 1), \tag{3.103}$$

which differs from (3.93) by a constant.* In terms of these variables (3.101) becomes

$$\left(\frac{1}{2M}p_X^2 + \frac{1}{8M}[e^{-2X} - 2(2k + 1)e^{-X}]\right)|nk\rangle = -\frac{n^2}{2M}|nk\rangle, \tag{3.104}$$

where $|nk\rangle$ denotes $|Ek\rangle$. The orthonormality condition is

$$\langle n'k|nk\rangle = \delta_{n'n}. \tag{3.105}$$

3.4.1 *Shift operators for k*

We first discuss shift operators for the Morse potential that change the parameter k. Consider factorization 3 of Section 3.2 with $\beta = \frac{1}{2}\hbar a^{-1}$, $\gamma = a^{-1}$, and $\delta = -a^{-1}$. Then (3.73) is equal to $\hbar^2/a^2$ times the potential in (3.104), and hence from (3.74) and (3.75) we obtain the shift operations†

$$(\mathrm{i}p_X + \tfrac{1}{2}e^{-X} - k - 1)|nk\rangle = [(k + 1)^2 - n^2]^{1/2}|n, k + 1\rangle \tag{3.106}$$

and

$$(-\mathrm{i}p_X + \tfrac{1}{2}e^{-X} - k)|nk\rangle = [k^2 - n^2]^{1/2}|n, k - 1\rangle. \tag{3.107}$$

Consider bound states ($n^2 > 0$). The coefficients in (3.106) and (3.107) violate the condition (1.164) in an interval of width $\Delta k = |2n|$, and if $|n| < \frac{1}{2}$ this interval is too small to conclude that either (3.106) or (3.107) should terminate. However, k has a lower bound equal to $-\frac{1}{2}$ (see (3.100)) and hence a sequence of lowering operations (3.107) must terminate at a minimum value $k_{\min}^2 = n^2$. We can, without loss of generality, take $k_{\min} = n$. From this result and (3.106)

---

* This constant does not affect previous results: thus (3.95), (3.96), and (3.98) remain valid.

† These shift operations can also be obtained *ab initio* by factorizing (3.104).

we conclude that the possible values of k are (see also (3.144))

$$k = n, n + 1, \ldots. \tag{3.108}$$

The shift operations (3.106) and (3.107) are between eigenkets of the same energy (same n) belonging to different Morse potentials (different k). These transformations are depicted in Fig. 3.2. For a given Morse oscillator (k fixed)

$$n = k, k - 1, \ldots, n_{\min} \tag{3.109}$$

(see Fig. 3.2). The value of $n_{\min}$ is discussed below. The vibrational quantum number

$$N = k - n \tag{3.110}$$

has the values

$$N = 0, 1, \ldots, k - n_{\min}. \tag{3.111}$$

From (3.102), (3.110), and (3.100) we obtain the energy eigenvalues for bound states of the Morse potential

$$E = -D\left[1 - 2\left(\frac{\hbar^2}{2Ma^2D}\right)^{1/2}\left(N + \frac{1}{2}\right) + \frac{\hbar}{2Ma^2D}\left(N + \frac{1}{2}\right)^2\right]. \tag{3.112}$$

If we treat D as a continuous variable in (3.101) and apply the Hellmann–Feynman theorem (Section 1.11)

$$\left\langle \frac{\partial H_k}{\partial D} \right\rangle = \frac{\partial E}{\partial D},$$

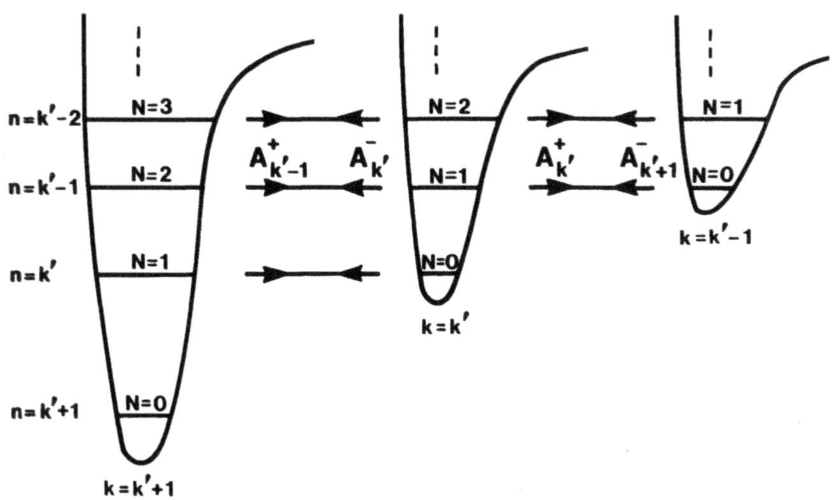

FIG. 3.2 Shift operations for Morse potentials: A_k^+ and A_k^-, respectively, are the shift operators in (3.106) and (3.107), k is the parameter (3.100), and N is the vibrational quantum number (3.110).

we obtain

$$\langle \exp(-2x/a) - 2\exp(-x/a) \rangle = -\frac{2n}{2k+1}. \tag{3.113}$$

Hence

$$\langle V \rangle = -\frac{2nD}{2k+1}. \tag{3.114}$$

For bound states $\langle V \rangle < 0$: because D and $2k + 1$ are positive, this requires $n > 0$. Thus the minimum value of n in (3.109) must satisfy

$$0 < n_{\min} \leq 1. \tag{3.115}$$

From (3.111) and (3.115) the number of bound states of a Morse oscillator is the greatest integer that is less than $k + 1$. Thus if $k < 0$, which means $D < \hbar^2/(8Ma^2)$, the potential cannot bind the particle.

3.4.2 *Shift operators for energy*

Next we discuss shift operators that change the energy eigenvalue. These operators transform between eigenkets of a given Morse potential: that is, they change n but not k. We present two methods of deducing such operators. The first is an *ab initio* calculation in which one rewrites the energy eigenvalue equation and factorizes it to obtain shift operators. The coefficients in the shift operations are evaluated by algebraic methods. The second method involves transformation of the energy eigenvalue equation into one of the factorization types of Section 3.2 in which the energy eigenvalue plays the role of the parameter k. We consider each of these methods in turn.

By multiplying both sides of (3.104) on the left with e^{2X}, rearranging terms and using (3.110), we obtain

$$O_N|N\rangle = -\tfrac{1}{4}|N\rangle. \tag{3.116}$$

Here*

$$O_N = e^{2X}p_X^2 - (k + \tfrac{1}{2})e^X + (N - k)^2 e^{2X} \tag{3.117}$$

and $|N\rangle$ denotes $|nk\rangle$. Inspection of O_N suggests that it can be factorized by the operators†

$$B_N^{\pm} = \pm ie^X p_X + (N - k)e^X + K_N^{\pm}, \tag{3.118}$$

---

* In the following we will not be concerned with changes in k and therefore, to simplify the notation, we have omitted the k in O_{Nk} and $|Nk\rangle$.

† Strictly speaking we should replace the quantum number N in (3.118) by an operator $\mathcal{N}$ such that $\mathcal{N}|N\rangle = N|N\rangle$; that is, by (see (3.112))

$$\mathcal{N} = -(-2Ma^2\hbar^{-2}H)^{1/2} + (2Ma^2\hbar^{-2}D)^{1/2} - \tfrac{1}{2}.$$

where $K_N^{\mp}$ are independent of X and p_X. A short calculation yields

$$B_{N+1}^- B_N^+ = e^{2X} p_X^2 + (N-k)^2 e^{2X} + [(N-k)K_{N+1}^- + (N-k+1)K_N^+]e^X + i(K_{N+1}^- - K_N^+)e^X p_X + K_N^+ K_{N+1}^- .$$

Comparing this with (3.117) we require

$$K_{N+1}^- = K_N^+ \tag{3.119}$$

and

$$(N-k)K_{N+1}^- + (N-k+1)K_N^+ = -(k+\tfrac{1}{2}). \tag{3.120}$$

Hence (3.118) with

$$K_N^{\mp} = \frac{k+\frac{1}{2}}{2k-2N\mp 1} \tag{3.121}$$

provide the desired factorization

$$B_{N+1}^- B_N^+ = O_N + \left(\frac{k+\frac{1}{2}}{2k-2N-1}\right)^2. \tag{3.122}$$

Similarly, one can readily show that

$$B_{N-1}^+ B_N^- = O_N + \left(\frac{k+\frac{1}{2}}{2k-2N+1}\right)^2. \tag{3.123}$$

From (3.116), (3.122), and (3.123)

$$B_N^{\pm}|N\rangle = \beta_N^{\pm}|N\pm 1\rangle, \tag{3.124}$$

where the coefficients $\beta_N^{\pm}$ are to be determined. For normalized kets in (3.124) we have

$$|\beta_N^{\pm}|^2 = \langle (B_N^{\pm})^{\dagger} B_N^{\pm}\rangle, \tag{3.125}$$

where the expectation values are with respect to $|N\rangle$. The adjoints of (3.118) are

$$(B_N^+)^{\dagger} = B_{N+1}^- - 2e^X, \qquad (B_N^-)^{\dagger} = B_{N-1}^+ + 2e^X. \tag{3.126}$$

Thus in (3.125) we require the expectation values of $e^{2X}p_X$, e^{2X}, and e^X. With $F = e^{2X}$, eqn (3.96) yields

$$\langle e^{2X} p_X\rangle = i\langle e^{2X}\rangle. \tag{3.127}$$

Similarly, with $F = e^{2X}$, eqns (3.98)–(3.100), (3.102), (3.103), and (3.110) yield

$$\langle e^{2X}\rangle = \tfrac{1}{8}[(k-N)^2 - 1]^{-1}[3(2k+1)\langle e^X\rangle - 1]. \tag{3.128}$$

The expectation value of e^X can be written down by adding eqns (3.124), multiplying on the left by $\langle N|$, and using the orthonormality property

$$\langle N'|N\rangle = \delta_{N'N}.$$

Hence

$$\langle e^X \rangle = \frac{2k+1}{(2k-2N-1)(2k-2N+1)}. \tag{3.129}$$

Substituting (3.126)–(3.129) in (3.125) and using (3.116), (3.122), and (3.123), we obtain after some straightforward though tedious algebra

$$\beta_N^{\pm} = \left(\frac{(N + \frac{1}{2} \pm \frac{1}{2})(k - N)(2k - N + \frac{1}{2} \mp \frac{1}{2})}{(2k - 2N \mp 1)^2 (k - N \mp 1)} \right)^{1/2}. \tag{3.130}$$

In (3.130) a phase factor has been set equal to one. Note that $\beta_0^- = 0$ and $(\beta_{N-1}^+)^2 > 0$ for N given by (3.111) and (3.115): thus the spectrum generated by the energy shift operators is the same as that generated by the shift operators for k. Note also that the raising operation in (3.124) does not terminate at the ket with the largest N; instead, this ket is transformed out of the space of bound states (see also Sections 3.5.2 and 3.7.2).

The energy shift operations can also be derived by transforming (3.104) into one of the factorization types discussed in Section 3.2. Let

$$Y = (k + \tfrac{1}{2})e^{-X}, \tag{3.131}$$

$$p_Y = -(k + \tfrac{1}{2})^{-1} e^X p_X. \tag{3.132}$$

Then

$$[p_Y, Y] = -i, \tag{3.133}$$

$$p_X = -Y p_Y, \tag{3.134}$$

$$p_X^2 = Y^2 p_Y^2 - iY p_Y. \tag{3.135}$$

Using (3.131) and (3.135) we can express (3.104) in terms of Y and p_Y. The term that is linear in p_Y can be removed by multiplying both sides of (3.104) on the left by $Y^{1/2}$ and using the identity

$$Y^{1/2}(Y^2 p_Y^2 - iY p_Y) = (Y^2 p_Y^2 - \tfrac{1}{4}) Y^{1/2}.$$

Hence we obtain the transformed equation

$$\left(\frac{1}{2M} p_Y^2 - \frac{1}{2MY} + \frac{(n - \frac{1}{2})(n + \frac{1}{2})}{2MY^2} \right) \{Y^{1/2} |nk\rangle\}$$
$$= -[2M(2k+1)^2]^{-1} \{Y^{1/2} |nk\rangle\}. \tag{3.136}$$

Comparing (3.136) with factorization 5 in Section 3.2, we can write down the shift operations

$$\left(\pm i p_Y - \frac{n \pm \frac{1}{2}}{Y} + \frac{1}{2n \pm 1} \right) \{a_n Y^{1/2} |nk\rangle\}$$
$$= \left(\frac{1}{(2n \pm 1)^2} - \frac{1}{(2k+1)^2} \right)^{1/2} \{a_{n\pm 1} Y^{1/2} |n \pm 1, k\rangle\}. \tag{3.137}$$

Here the constants a_n are determined by normalization of the quantities in braces; that is,

$$\begin{aligned} 1 &= |a_n|^2 \int_0^\infty |\langle Y| Y^{1/2} |nk\rangle|^2 \, \mathrm{d}Y \\ &= |a_n|^2 (k + \tfrac{1}{2}) \int_\infty^{-\infty} |\langle X| \mathrm{e}^{-(1/2)X} |nk\rangle|^2 \frac{\mathrm{d}Y}{\mathrm{d}X} \mathrm{d}X \\ &= |a_n|^2 (k + \tfrac{1}{2})^2 \langle nk| \mathrm{e}^{-2X} |nk\rangle. \end{aligned} \tag{3.138}$$

To evaluate the expectation value in (3.138) we note that (3.113) and (3.103) provide one equation involving this quantity, namely,

$$\langle \mathrm{e}^{-2X} - 2(2k+1)\mathrm{e}^{-X} \rangle = -2n(2k+1). \tag{3.139}$$

Another such equation can be obtained by treating k in the Hamiltonian in (3.104) as a continuous variable and applying the Hellmann–Feynman theorem. Thus

$$\left\langle -\frac{1}{2M} \mathrm{e}^{-X} \right\rangle = -\frac{1}{2M} \frac{\partial n^2}{\partial k} = -\frac{n}{M}.$$

Hence

$$\langle \mathrm{e}^{-X} \rangle = 2n, \tag{3.140}$$

so that

$$\langle \mathrm{e}^{-2X} \rangle = 2n(2k+1). \tag{3.141}$$

From (3.138) and (3.141) we have (choosing a phase factor equal to one)

$$a_n = \left(\frac{2}{n(2k+1)^3} \right)^{1/2}. \tag{3.142}$$

Substituting (3.142) in (3.137) and expressing the shift operators in terms of X and p_X, we find

$$\left(\mathrm{i}\mathrm{e}^X p_X - n\mathrm{e}^X + \frac{k+\frac{1}{2}}{2n-1} \right) |nk\rangle = \left(\frac{n(k+n)(k+1-n)}{(2n-1)^2(n-1)} \right)^{1/2} |n-1, k\rangle \tag{3.143}$$

and

$$\left(-\mathrm{i}\mathrm{e}^X p_X - n\mathrm{e}^X + \frac{k+\frac{1}{2}}{2n+1} \right) |nk\rangle = \left(\frac{n(k+1+n)(k-n)}{(2n+1)^2(n+1)} \right)^{1/2} |n+1, k\rangle. \tag{3.144}$$

When expressed in terms of N (eqn (3.110)) and with $|nk\rangle$ denoted by $|N\rangle$, the shift operations (3.143) and (3.144) are the same as those obtained previously (see (3.124), (3.118), (3.121), and (3.130)).

The energy shift operations for the Morse oscillator resemble those for the harmonic oscillator in that there is a ground state or 'vacuum' $|0\rangle$, which is annihilated by a lowering operator, and from which one can generate all other bound states by repeated application of raising operators. We consider two applications of the energy shift operators, namely the calculation of matrix elements and of normalized coordinate-space wavefunctions.

3.4.3 *Matrix elements*

It is possible to calculate matrix elements of any operator that can be expressed as a power series of the shift operators $B_N^{\pm}$. For example,

$$e^X = \tfrac{1}{2}(k-N)^{-1}[K_N^+ + K_N^- - (B_N^+ + B_N^-)]$$

and hence

$$e^X|N\rangle = \tfrac{1}{2}(k-N)^{-1}[(K_N^+ + K_N^-)|N\rangle - \beta_N^+|N+1\rangle - \beta_N^-|N-1\rangle]. \quad (3.145)$$

Using (3.145) and (3.121) we can calculate the matrix elements of e^{MX} $(M = 1, 2, \ldots)$. For example,

$$\begin{aligned}\langle N'|e^X|N\rangle &= \frac{2k+1}{(2k-2N-1)(2k-2N+1)}\delta_{N'N} \\ &\quad - \frac{1}{2}\left(\frac{(N+1)(2k-N)}{(k-N)(2k-2N-1)^2(k-N-1)}\right)^{1/2}\delta_{N',N+1} \\ &\quad - \frac{1}{2}\left(\frac{N(2k-N+1)}{(k-N)(2k-2N+1)^2(k-N+1)}\right)^{1/2}\delta_{N',N-1}. \quad (3.146)\end{aligned}$$

Knowledge of these matrix elements enables one to treat a perturbed Morse oscillator [11, 12].

3.4.4 *Coordinate-space wavefunctions*

In the coordinate representation the shift operations (3.124) yield

$$\left(e^X\frac{d}{dX} + ke^X - \frac{1}{2}\right)\psi_0(X) = 0 \quad (3.147)$$

and

$$\left(e^X\frac{d}{dX} + (N-k)e^X + \frac{k+\frac{1}{2}}{2k-2N-1}\right)\psi_N(X) = \beta_N^+\psi_{N+1}(X). \quad (3.148)$$

Here $\psi_N(X) = \langle N|X\rangle$ is the coordinate-space wavefunction for bound states: it is normalized according to

$$\int_{-\infty}^{\infty} |\psi_N(X)|^2\, dX = 1. \quad (3.149)$$

Equations (3.147)–(3.149) determine $\psi_N(X)$. The solution to (3.147) is

$$\psi_0(X) = c_0 \exp(-kX - \tfrac{1}{2}e^{-X}), \quad (3.150)$$

where c_0 is a constant. In terms of $u = e^{-X}$ we have

$$\psi_0 = c_0 u^k e^{-(1/2)u}. \quad (3.151)$$

Using the integral

$$\int_0^\infty u^{2k-1} e^{-u}\, du = \Gamma(2k), \quad (3.152)$$

we see that (3.149) requires

$$c_0 = [\Gamma(2k)]^{-1/2}. \quad (3.153)$$

If we write

$$\psi_N(u) = c_0 u^{k-N} e^{-(1/2)u} \phi_N(u), \quad (3.154)$$

then (3.148) and (3.130) yield the recurrence relation

$$\phi_{N+1} = \gamma_N D_N \phi_N \qquad (N = 0, 1, 2, \ldots), \quad (3.155)$$

where

$$\gamma_N = -(2k-2N-1)(2k-2N)\left(\frac{k-N-1}{(N+1)(k-N)(2k-N)}\right)^{1/2} \quad (3.156)$$

and

$$D_N = 1 - \frac{2k-N}{(2k-2N-1)(2k-2N)}u + \frac{u}{2k-2N}\frac{d}{du}. \quad (3.157)$$

From (3.151) and (3.154), $\phi_0 = 1$. Thus by repeated application of (3.155) we have

$$\phi_N(u) = \gamma_{N-1} \cdots \gamma_1 \gamma_0 P_N(u) \qquad (N = 1, 2, \ldots), \quad (3.158)$$

where

$$P_N(u) = D_{N-1} \ldots D_1 D_0 1. \quad (3.159)$$

We evaluate (3.158) in two parts. Consider first $P_N(u)$. In general $P_N(u)$ is a polynomial of order N, and the first few of these are given in Table 3.1 where we have defined $P_0 = \phi_0 = 1$. In terms of the confluent hypergeometric func-

TABLE 3.1 The polynomials P_N for $N = 0, 1, 2$. For $N = 1$ and 2 the expressions for P_N are calculated from (3.159) and (3.157)

N	$P_N(u)$
0	1
1	$1 - \frac{1}{2k-1}u$
2	$1 - \frac{2}{2k-3}u + \frac{2}{(2k-3)(2k-2)}\frac{u^2}{2!}$

tion (2.152) the entries in Table 3.1 are given by

$$P_N(u) = {}_1F_1(-N; 2k - 2N + 1; u). \tag{3.160}$$

The extension of (3.160) to all non-negative integer values of N is established by putting $B = 2k - 2N + 1$ in (B.2).

Consider next the product of the γ in (3.158):

$$\begin{aligned}\gamma_{N-1}\ldots\gamma_1\gamma_0 &= (-1)^N \prod_{j=0}^{N-1} (2k-2j-1)(2k-2j)\left(\frac{k-j-1}{(j+1)(k-j)(2k-j)}\right)^{1/2} \\ &= (-1)^N(2k-2N+1)\ldots(2k-1)(2k) \\ &\quad\times\left(\frac{1}{N!}\,\frac{k-N}{k}\,\frac{1}{(2k-N+1)\ldots(2k-1)(2k)}\right)^{1/2} \\ &= \frac{(-1)^N}{\Gamma(2k-2N+1)}\left(\frac{2(k-N)}{N!}\Gamma(2k-N+1)\Gamma(2k)\right)^{1/2}, \end{aligned} \tag{3.161}$$

where we have used the result

$$\alpha(\alpha+1)\ldots(\alpha+n) = \frac{\Gamma(\alpha+n+1)}{\Gamma(\alpha)}. \tag{3.162}$$

Equations (3.153), (3.154), (3.158), (3.160), and (3.161) yield the coordinate-space wavefunctions for bound states of the Morse potential

$$\begin{aligned}\psi_N(X) &= (-1)^N\left(\frac{2(k-N)\Gamma(2k-N+1)}{N![\Gamma(2k-2N+1)]^2}\right)^{1/2} {}_1F_1(-N; 2k-2N+1; e^{-X}) \\ &\quad\times \exp[-(k-N)X]\exp(-\tfrac{1}{2}e^{-X}),\end{aligned} \tag{3.163}$$

where X is given by (3.103).

3.5 The Rosen–Morse potential [13]

We consider the potential

$$V(x) = -U_0 \operatorname{sech}^2\frac{x}{a} + U_1 \tanh\frac{x}{a}, \tag{3.164}$$

where U_0, U_1, and a are positive constants. The classical potential shown in Fig. 3.3 tends to $\pm U_1$ as $x \to \pm\infty$ and if $U_1 < 2U_0$ there is a minimum value $-U_0[1 + \frac{1}{4}(U_1/U_0)^2]$ at $x = -a\tanh^{-1}(U_1/2U_0)$.

The eigenvalue equation (3.41) is

$$\left(\frac{1}{2M}p^2 - U_0 \operatorname{sech}^2\frac{x}{a} + U_1 \tanh\frac{x}{a}\right)|Ek\rangle = E|Ek\rangle, \tag{3.165}$$

where the parameter k is defined in (3.170).

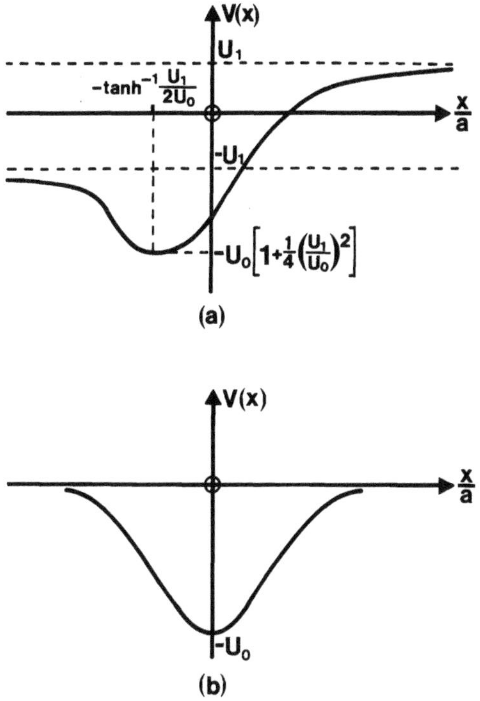

3.3 The classical Rosen–Morse potential: (a) if $U_1 \neq 0$, (b) the symmetric case ($U_1 = 0$).

Shift operators for k

ɪider factorizaton 6 of Section 3.2 with $d = -\pi/(2\beta)$ and afterwards β ɪced by $\mathrm{i}\beta$. Then

$$\cot \beta(x + d) \to -\mathrm{i} \tanh \beta x, \tag{3.166}$$

$$\operatorname{cosec} \beta(x + d) \to -\operatorname{sech} \beta x, \tag{3.167}$$

(3.87) becomes

$$V_k(x) = \frac{1}{2M}[-2\hbar\beta q \tanh \beta x - \hbar^2 k(k-1)\beta^2 \operatorname{sech}^2 \beta x]. \tag{3.168}$$

convenient to define

$$H_k = \frac{1}{2M}p^2 + V_{k+1} \tag{3.169}$$

ad of (3.40): then in the operators and coefficients for factorization 6 (but n $|Ek\rangle$) we must replace k by $k + 1$. From (3.168), V_{k+1} is the potential in 5) if $\beta = a^{-1}$, $q = Ma\hbar^{-1}U_1$, and

$$k(k+1) = 2Ma^2\hbar^{-2}U_0. \tag{3.170}$$

n (3.88), (3.89), and (3.166) we obtain the shift operations

$$[ip_X - (k+1)\tanh X - Ma^2\hbar^{-2}U_1(k+1)^{-1}]|Ek\rangle$$
$$= [2Ma^2\hbar^{-2}E + (Ma^2\hbar^{-2}U_1)^2(k+1)^{-2} + (k+1)^2]^{1/2}|E, k+1\rangle \quad (3.171)$$

and

$$(-ip_X - k\tanh X - Ma^2\hbar^{-2}U_1k^{-1})|Ek\rangle$$
$$= [2Ma^2\hbar^{-2}E + (Ma^2\hbar^{-2}U_1)^2k^{-2} + k^2]^{1/2}|E, k-1\rangle, \quad (3.172)$$

where X and p_X are defined in (3.93) and (3.94).

Non-negativity of the norm requires that the coefficients in (3.171) and (3.172) satisfy (1.164). Now the function

$$g(k) = k^2 + (Ma^2\hbar^{-2}U_1)^2k^{-2}$$

has a minimum value $g_{\min} = 2Ma^2\hbar^{-2}U_1$ when $k^2 = Ma^2\hbar^{-2}U_1$. Hence if $E > -U_1$, the coefficients in (3.171) and (3.172) satisfy (1.164) for all k: the energy spectrum is continuous. If $E < -U_1$ these coefficients violate (1.164) in intervals of width

$$\Delta k = (-Ma^2\hbar^{-2}E)^{1/2}\{[1 + (1 - U_1^2/E^2)^{1/2}]^{1/2} - [1 - (1 - U_1^2/E^2)^{1/2}]^{1/2}\},$$

which is not necessarily greater than or equal to one. Thus, by itself, non-negativity of the norm is not sufficient to require that either (3.171) or (3.172) should terminate. There are two possibilities corresponding to the solutions

$$k = \tfrac{1}{2}[-1 \pm (1 + 8Ma^2\hbar^{-2}U_0)^{1/2}] \quad (3.173)$$

of (3.170). Consider the positive solution in (3.173). Because k has a lower bound equal to zero, the sequence of lowering operations (3.172) must terminate at some minimum value $k_{\min} = n$ where $n > 0$. (See also (3.190) and (3.196).) This requires

$$E = -\frac{\hbar^2}{2Ma^2}n^2 - \frac{Ma^2U_1^2}{2\hbar^2}\frac{1}{n^2}. \quad (3.174)$$

The raising operation (3.171) produces kets $|Ek\rangle$ with $k = n+1, n+2, \ldots$; thus the values of k are $k = n, n+1, \ldots$. For a given potential (k fixed) the values of n are

$$n = k, k-1, \ldots, n_{\min}, \quad (3.175)$$

where $n_{\min}$ satisfies (3.179). The vibrational quantum number

$$N = k - n \quad (3.176)$$

has the values (see (3.179))

$$N = 0, 1, \ldots, k - n_{\min} < k - (Ma^2\hbar^{-2}U_1)^{1/2}. \quad (3.177)$$

We can determine the condition for bound states to exist by treating U_0 as a continuous variable and applying the Hellmann–Feynman theorem (Section 1.11)

$$\left\langle \frac{\partial H}{\partial U_0} \right\rangle = \frac{\partial E}{\partial U_0} = \frac{\partial k}{\partial U_0} \frac{\partial E}{\partial k}.$$

These partial derivatives can be evaluated using (3.165), (3.173), (3.174), and (3.176). Hence

$$\langle \operatorname{sech}^2 X \rangle = \frac{2n}{2k+1}\left[1 - \left(\frac{Ma^2 U_1}{\hbar^2 n^2}\right)^2\right]. \tag{3.178}$$

The left-hand side of (3.178) is positive and therefore

$$n > (Ma^2 \hbar^{-2} U_1)^{1/2}. \tag{3.179}$$

Because $k \geq n$, (3.179) and (3.173) show that for bound states to exist we must have

$$(Ma^2 \hbar^{-2} U_1)^{1/2} < \tfrac{1}{2}[(1 + 8Ma^2 \hbar^{-2} U_0)^{1/2} - 1]. \tag{3.180}$$

Thus for given M, a, and U_0 there is a maximum value of U_1 above which there are no bound states.

When $k = n$, (3.172) and (3.174) yield

$$[-\mathrm{i}p_X - n \tanh X - Ma^2 \hbar^{-2} U_1 n^{-1}]|nn\rangle = 0. \tag{3.181}$$

In the coordinate representation p_X is replaced by $-\mathrm{i}\,\mathrm{d}/\mathrm{d}X$ and the solution to (3.181) is

$$\psi_{nn}(X) = c \cosh^{-n} X \exp\left(-\frac{Ma^2 U_1}{\hbar^2 n} X\right), \tag{3.182}$$

where c is a constant. To normalize (3.182) with respect to X we change the variable to $u = \mathrm{e}^{-2X}$ and use the integral

$$\int_0^\infty \frac{u^{z-1}}{(1+u)^{z+\omega}}\,\mathrm{d}u = B(z, \omega) = \frac{\Gamma(z)\Gamma(\omega)}{\Gamma(z+\omega)}. \tag{3.183}$$

Thus provided (3.179) is satisfied

$$c = \left[\Gamma(2n) \Big/ 2^{2n-1} \Gamma\left(n + \frac{Ma^2 U_1}{\hbar^2 n}\right) \Gamma\left(n - \frac{Ma^2 U_1}{\hbar^2 n}\right)\right]^{1/2}. \tag{3.184}$$

If we consider the negative solution in (3.173) and repeat the above analysis, we find it is the raising operation (3.171) that must terminate if $E < -U_1$. The resulting eigenvalues and eigenkets are the same as those obtained above.

3.5.2 *Shift operators for energy*

Here we consider the symmetric Rosen–Morse potential (eqn (3.164) with $U_1 = 0$); we return to eqn (3.164) in Section 3.7. From (3.174)

$$E = -\frac{\hbar^2}{2Ma^2} n^2. \tag{3.185}$$

Using (3.185), (3.176), and (3.170), the energy eigenvalue equation (3.165) can be rewritten

$$O_N|N\rangle = k(k+1)|N\rangle, \tag{3.186}$$

where $|N\rangle$ denotes $|nk\rangle$ and

$$O_N = \cosh^2 X p_X^2 + (k-N)^2 \cosh^2 X. \tag{3.187}$$

Shift operators for energy can be obtained by factorizing O_N. These calculations are similar to those in Section 3.4 and the results are

$$\begin{aligned} B_{N+1}^- B_N^+ &= O_N - (k-N)(k-N-1), \\ B_{N-1}^+ B_N^- &= O_N - (k-N)(k-N+1), \end{aligned} \tag{3.188}$$

where

$$B_N^\pm = \pm \mathrm{i} \cosh X p_X - (k-N) \sinh X. \tag{3.189}$$

It follows that

$$B_N^\pm |N\rangle = \beta_N^\pm |N \pm 1\rangle. \tag{3.190}$$

The coefficients $\beta_N^\pm$ are determined below (see (3.196)).

These shift operations can also be derived by transforming (3.165) into one of the factorization types of Section 3.2. Let

$$\cos Y = \tanh X, \qquad p_Y = -\cosh X p_X.$$

The operators Y and p_Y satisfy (3.133). In terms of Y and p_Y, eqn (3.165) with $U_1 = 0$ becomes

$$\left[\frac{1}{2M} p_Y^2 + \frac{(n-\frac{1}{2})(n+\frac{1}{2})}{2M \sin^2 Y}\right]\{\sin^{1/2} Y|nk\rangle\} = \frac{(2k+1)^2}{8M}\{\sin^{1/2} Y|nk\rangle\}. \tag{3.191}$$

Here the purpose of the factor $\sin^{1/2} Y$ is the same as that of the factor $Y^{1/2}$ in (3.136), namely to eliminate a term that is linear in p_Y.

Comparing (3.191) with factorization 4 in Section 3.2 we obtain the shift operations

$$\begin{aligned} &[\pm \mathrm{i} p_Y - (n \pm \tfrac{1}{2}) \cot Y]\{a_n \sin^{1/2} Y|nk\rangle\} \\ &\quad = [(k+\tfrac{1}{2})^2 - (n \pm \tfrac{1}{2})^2]^{1/2}\{a_{n\pm 1} \sin^{1/2} Y|n \pm 1, k\rangle\}, \end{aligned} \tag{3.192}$$

where a_n are constants such that the quantities in braces are normalized. Thus

$$\begin{aligned} 1 &= |a_n|^2 \int_0^\pi |\langle Y| \sin^{1/2} Y|nk\rangle|^2 \, \mathrm{d}Y \\ &= |a_n|^2 \int_\infty^{-\infty} |\langle X| \cosh^{-1/2} X|nk\rangle|^2 \frac{\mathrm{d}Y}{\mathrm{d}X} \mathrm{d}X \\ &= |a_n|^2 \langle nk| \operatorname{sech}^2 X|nk\rangle. \end{aligned} \tag{3.193}$$

Using (3.178) and choosing a phase factor equal to unity we have

$$a_n = \left(\frac{2k+1}{2n}\right)^{1/2}. \tag{3.194}$$

Substituting (3.194) in (3.192) and expressing the results in terms of X and p_X, we find

$$(\pm \mathrm{i} \cosh X p_X - n \sinh X)|nk\rangle = \left(\frac{n}{n \mp 1}(k \mp n + 1)(k \pm n)\right)^{1/2} |n \mp 1, k\rangle. \tag{3.195}$$

Equations (3.195) are the shift operations (3.190) with

$$\beta_N^{\pm} = \left(\frac{(N + \frac{1}{2} \pm \frac{1}{2})(k - N)(2k - N + \frac{1}{2} \mp \frac{1}{2})}{k \mp 1 - N}\right)^{1/2}. \tag{3.196}$$

Equation (3.196) shows that the spectrum generated by the energy shift operators is consistent with that generated by the shift operators for k.

It is also useful to construct operators that change N by two. Consider raising N by two. Starting from $B_{N+1}^{+} B_N^{+} |N\rangle$ and using (3.186) and (3.187) to eliminate terms that are quadratic in p_X, we find

$$\left(-\mathrm{i} \cosh X \sinh X p_X + (k - N)\sinh^2 X - \frac{(N+1)(2k-N)}{2(k-N-1)}\right)|N\rangle$$
$$= \frac{\beta_{N+1}^{+} \beta_N^{+}}{2(k-N-1)}|N+2\rangle. \tag{3.197}$$

With the aid of the shift operations (3.190) one can determine matrix elements of any operator that can be expressed as a power series of $B_N^{\pm}$. These calculations are similar to those in Section 3.4.3.

3.5.3 *Coordinate-space wavefunctions*

The ground state wavefunction for the symmetric Rosen–Morse potential is

$$\psi_0(X) = c_0 \cosh^{-k} X, \tag{3.198}$$

where

$$c_0 = \left(\frac{\Gamma(k + \frac{1}{2})}{\Gamma(\frac{1}{2})\Gamma(k)}\right)^{1/2}. \tag{3.199}$$

(See (3.182) and (3.184).) By applying B_0^{+} in differential form to the ground state, we generate the wavefunction for the first excited state (see (3.189), (3.190), and (3.196))

$$\psi_1(X) = c_1 \cosh^{-k} X \sinh X, \tag{3.200}$$

where

$$c_1 = -\left(\frac{2\Gamma(k + \frac{1}{2})}{\Gamma(\frac{1}{2})\Gamma(k-1)}\right)^{1/2}. \tag{3.201}$$

We see that ψ_0 and ψ_1 are, respectively, even and odd functions of X: they are eigenfunctions of parity with eigenvalues $+1$ and -1.

By repeated application of (3.197) in differential form to ψ_0 we can generate all the even wavefunctions $\psi_{2N}^{(e)}$. Let

$$\psi_N^{(e)}(X) = c_0 \cosh^{-k} X \phi_N^{(e)}(X) \tag{3.202}$$

and $u = -\sinh^2 X$. Then (3.197) and (3.196) provide the recurrence relation

$$\phi_{N+2}^{(e)}(u) = \gamma_N^{(e)} D_N^{(e)} \phi_N^{(e)}(u) \qquad (N = 0, 2, 4, \ldots), \tag{3.203}$$

where

$$\gamma_N^{(e)} = -\left(\frac{(N+1)(k-N-2)(2k-N)}{(N+2)(k-N)(2k-N-1)}\right)^{1/2} \tag{3.204}$$

and

$$D_N^{(e)} = 1 + \frac{2(k-N-1)}{N+1}u + \frac{4(k-N-1)}{(N+1)(2k-N)}u(1-u)\frac{\mathrm{d}}{\mathrm{d}u}. \tag{3.205}$$

From (3.198) and (3.202), $\phi_0^{(e)} = 1$. Hence (3.203) yields

$$\phi_{2N}^{(e)} = \gamma_{2N-2}^{(e)} \ldots \gamma_2^{(e)} \gamma_0^{(e)} P_{2N}^{(e)} \qquad (N = 1, 2, \ldots). \tag{3.206}$$

Here

$$P_{2N}^{(e)} = D_{2N-2}^{(e)} \ldots D_2^{(e)} D_0^{(e)} 1 \tag{3.207}$$

is a polynomial of order $2N$ in u. The first few of these are given in Table 3.2 where we have defined $P_0^{(e)} = \phi_0^{(e)} = 1$. In terms of the hypergeometric function

$${}_2F_1(a, b; c; x) = 1 + \frac{ab}{c}x + \frac{a(a+1)b(b+1)}{c(c+1)}\frac{x^2}{2!} + \cdots, \tag{3.208}$$

the entries in Table 3.2 are given by

$$P_{2N}^{(e)} = {}_2F_1(-N, N-k; \tfrac{1}{2}; u). \tag{3.209}$$

Equation (B.3) with $B = N - k$ and $C = \frac{1}{2}$ shows that (3.209) is valid for all non-negative integer values of N.

The product of the coefficients in (3.206) is

TABLE 3.2 The polynomials $P_{2N}^{(e)}$ for $N = 0, 1, 2$. For $N = 1$ and 2 the expressions for $P_{2N}^{(e)}$ are calculated from (3.207) and (3.205)

N	$P_{2N}^{(e)}(u)$
0	1
1	$1 - 2(1-k)u$
2	$1 - 4(2-k)u + \frac{4}{3}(2-k)(3-k)u^2$

$$\gamma_{2N-2}^{(e)} \cdots \gamma_2^{(e)}\gamma_0^{(e)} = (-1)^N \left[\prod_{j=0}^{N-1} \left(\frac{j+\frac{1}{2}}{j+1} \right) \left(\frac{k-2j-2}{k-2j} \right) \left(\frac{k-j}{k-j-\frac{1}{2}} \right) \right]^{1/2}$$

$$= (-1)^N \left[\left(\frac{\Gamma(N+\frac{1}{2})}{\Gamma(\frac{1}{2})N!} \right) \left(\frac{k-2N}{k} \right) \right.$$

$$\left. \times \left(\frac{k\Gamma(k)}{\Gamma(k-N+1)} \frac{\Gamma(k-N+\frac{1}{2})}{\Gamma(k+\frac{1}{2})} \right) \right]^{1/2}. \qquad (3.210)$$

In (3.210) each factor in large parentheses is equal to the product of the corresponding factor in large parentheses in the preceding line. In writing down these products we have used (3.162).

Collecting the above results we obtain the even wavefunctions

$$\psi_{2N}^{(e)}(X) = (-1)^N \left(\frac{(k-2N)\Gamma(N+\frac{1}{2})\Gamma(k-N+\frac{1}{2})}{N![\Gamma(\frac{1}{2})]^2\Gamma(k-N+1)} \right)^{1/2}$$

$$\times\ {}_2F_1(-N, N-k; \tfrac{1}{2}; -\sinh^2 X) \cosh^{-k} X, \qquad (3.211)$$

for $N = 0, 1, 2, \ldots$.

In a similar manner, from $\psi_1(X)$ we can generate all the odd wavefunctions $\psi_{2N+1}^{(o)}$. Defining

$$\psi_N^{(o)}(X) = c_1 \cosh^{-k} X \sinh X \phi_N^{(o)}(X), \qquad (3.212)$$

we obtain from (3.197) and (3.196) the recurrence relation

$$\phi_{N+2}^{(o)}(u) = \gamma_N^{(o)} D_N^{(o)} \phi_N^{(o)}(u) \qquad (N = 1, 3, 5, \ldots), \qquad (3.213)$$

where

$$\gamma_N^{(o)} = -\left(\frac{(N+2)(k-N-2)(2k-N-1)}{(N+1)(k-N)(2k-N)} \right)^{1/2} \qquad (3.214)$$

and

$$D_N^{(o)} = 1 + \frac{2(k-N-1)}{N+2} u + \frac{4(k-N-1)}{(N+2)(2k-N-1)} u(1-u) \frac{\mathrm{d}}{\mathrm{d}u}. \qquad (3.215)$$

Proceeding as before we find for $N = 0, 1, 2, \ldots$

$$\psi_{2N+1}^{(o)}(X) = (-1)^{N+1} \left(\frac{4(k-2N-1)\Gamma(N+\frac{3}{2})\Gamma(k-N+\frac{1}{2})}{N![\Gamma(\frac{1}{2})]^2\Gamma(k-N)} \right)^{1/2}$$

$$\times\ {}_2F_1(-N, N+1-k; \tfrac{3}{2}; -\sinh^2 X) \cosh^{-k} X \sinh X. \qquad (3.216)$$

3.6 The first Pöschl–Teller potential [15]

The first Pöschl–Teller potential is

$$V(x) = \frac{\hbar^2}{2Ma^2} \left(\frac{\mu(\mu-1)}{\sin^2 x/a} + \frac{\lambda(\lambda-1)}{\cos^2 x/a} \right), \qquad (3.217)$$

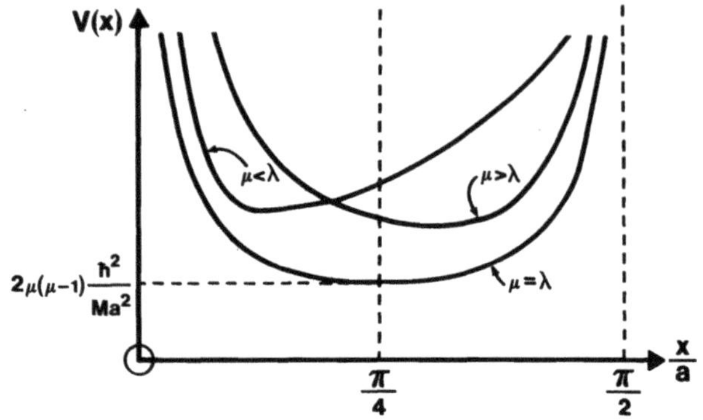

FIG. 3.4 The classical Pöschl–Teller potential (3.217).

where a is a positive constant and μ and λ are constants greater than one. The classical potential is periodic with period $\frac{1}{2}\pi a$. Consider this potential in the interval $[0, \frac{1}{2}\pi a]$. It is infinite at $x = 0$ and $\frac{1}{2}\pi a$. If $\mu = \lambda$, V is symmetric about $x/a = \frac{1}{4}\pi$; if $\mu > \lambda$ it is asymmetric with a minimum at $x/a > \frac{1}{4}\pi$; and if $\mu < \lambda$ it is asymmetric with a minimum at $x/a < \frac{1}{4}\pi$ (see Fig. 3.4).

If we put

$$E = \frac{\hbar^2}{2Ma^2}(2n)^2 \tag{3.218}$$

and use the dimensionless operators (3.93) and (3.94), the eigenvalue equation (3.41) can be written

$$\left(p_X^2 + \frac{\mu(\mu-1)}{\sin^2 X} + \frac{\lambda(\lambda-1)}{\cos^2 X}\right)|nk\rangle = (2n)^2|nk\rangle. \tag{3.219}$$

The parameter k is defined in (3.220).

3.6.1 *Shift operators for k*

Consider factorization 4 of Section 3.2 with $\beta = 2/a$, $\gamma = 0$, $\varepsilon = -2\hbar v/a$, and $d = 0$. Then (3.77) is the potential (3.217) with $k + v = \mu$ and $k - v = \lambda$.* In terms of μ and λ,

$$k = \tfrac{1}{2}(\mu + \lambda), \qquad v = \tfrac{1}{2}(\mu - \lambda). \tag{3.220}$$

Equations (3.78) and (3.79) yield the shift operations

$$(ip_X - \mu \cot X + \lambda \tan X)|nk\rangle = 2(n^2 - k^2)^{1/2}|n, k+1\rangle \tag{3.221}$$

and

* In denoting the eigenkets by $|nk\rangle$ we have omitted the parameter v defined in (3.220), which is irrelevant for our purposes.

$$[-ip_X - (\mu - 1)\cot X + (\lambda - 1)\tan X]|nk\rangle$$
$$= 2[n^2 - (k-1)^2]^{1/2}|n, k-1\rangle. \quad (3.222)$$

These transformations change k but not v; thus $k \to k \pm 1$ means $\mu \to \mu \pm 1$ and $\lambda \to \lambda \pm 1$.

Non-negativity of the norm requires that the raising operation (3.221) terminates at a maximum value $k_{max} = n$. This result and the lowering operation (3.222) show that the values of k are $k = n, n-1, \ldots$. For a given potential (k fixed)

$$n = k, k+1, \ldots. \quad (3.223)$$

The vibrational quantum number

$$N = n - k \quad (3.224)$$

has the values $N = 0, 1, 2, \ldots$. From (3.220), (3.224), and (3.218) we obtain the energy eigenvalues

$$E = \frac{\hbar^2}{2Ma^2}(\mu + \lambda + 2N)^2. \quad (3.225)$$

3.6.2 *Shift operators for energy*

Equation (3.219) can be rewritten

$$O_N|N\rangle = -\mu(\mu - 1)|N\rangle, \quad (3.226)$$

where $|N\rangle$ denotes $|nk\rangle$ and

$$O_N = \sin^2 X \cos^2 X p_X^2 + (\mu + \lambda + 2N)^2 \sin^4 X$$
$$+ [\lambda(\lambda - 1) - \mu(\mu - 1) - (\mu + \lambda + 2N)^2]\sin^2 X. \quad (3.227)$$

The factorization of O_N can be performed as in Section 3.4 and the results are

$$B_{N+1}^- B_N^+ = O_N + K_{N+1}^- K_N^+, \qquad B_{N-1}^+ B_N^- = O_N + K_{N-1}^+ K_N^-. \quad (3.228)$$

Here

$$B_N^\pm = \pm i \sin X \cos X p_X - (\mu + \lambda + 2N)\sin^2 X + K_N^\pm, \quad (3.229)$$

$$K_N^+ = \frac{\mu(\mu - 1) - \lambda(\lambda - 1)}{2(\mu + \lambda + 1 + 2N)} + \frac{1}{2}(\mu + \lambda + 2N), \quad (3.230)$$

and

$$K_N^- = K_{N-1}^+ + 1. \quad (3.231)$$

It follows that

$$B_N^\pm|N\rangle = \beta_N^\pm|N \pm 1\rangle. \quad (3.232)$$

The coefficients $\beta_N^\pm$ are calculated below.

The same shift operations can be deduced by transforming (3.219) into factorization 6 of Section 3.2. Let

$$Y = \ln \tan X, \qquad p_Y = \sin X \cos X p_X.$$

These operators satisfy (3.133). In terms of Y and p_Y, eqn (3.219) becomes

$$\left(\frac{1}{2M}p_Y^2 - \frac{1}{2M}\left(n-\frac{1}{2}\right)\left(n+\frac{1}{2}\right)\operatorname{sech}^2 Y + \frac{1}{4M}[\lambda(\lambda-1) - \mu(\mu-1)]\tanh Y\right)$$
$$\times \{\cosh^{1/2} Y|N\rangle\}$$
$$= -\frac{1}{4M}\left(\lambda(\lambda-1) + \mu(\mu-1) + \frac{1}{2}\right)\{\cosh^{1/2} Y|N\rangle\}. \tag{3.233}$$

From (3.168), V_{k+1} is the 'potential' in (3.233) if $\beta = \hbar^{-1}$, $k = n - \frac{1}{2}$, $q = \frac{1}{4}[\mu(\mu-1) - \lambda(\lambda-1)]$, and βx is replaced by Y. Thus proceeding as in the derivation of (3.171) and (3.172) we find

$$\left(\pm ip_Y - \left(n \pm \frac{1}{2}\right)\tanh Y + \frac{\mu(\mu-1) - \lambda(\lambda-1)}{4(n\pm\frac{1}{2})}\right)\{a_N \cosh^{1/2} Y|N\rangle\}$$
$$= \alpha_N^{\pm}\{a_{N\pm1}\cosh^{1/2} Y|N\pm1\rangle\}, \tag{3.234}$$

where a_N are constants such that the quantities in braces are normalized,

$$\alpha_N^{\pm} = \left[\left(n\pm\frac{1}{2}\right)^2 + \left(\frac{\mu(\mu-1) - \lambda(\lambda-1)}{4(n\pm\frac{1}{2})}\right)^2\right.$$
$$\left. - \frac{1}{2}\left(\mu(\mu-1) + \lambda(\lambda-1) + \frac{1}{2}\right)\right]^{1/2}, \tag{3.235}$$

and $n = \frac{1}{2}(\mu + \lambda + 2N)$. With the help of the identities

$$\mu(\mu-1) - \lambda(\lambda-1) = (\mu-\lambda)(\mu+\lambda-1)$$

and

$$\tfrac{1}{2}[\mu(\mu-1) + \lambda(\lambda-1) + \tfrac{1}{2}] = \tfrac{1}{4}(\mu+\lambda-1)^2 + \tfrac{1}{4}(\mu-\lambda)^2,$$

(3.235) simplifies to

$$\alpha_N^+ = \left(\frac{(N+1)(\mu+\lambda+N)(2\lambda+1+2N)(2\mu+1+2N)}{(\mu+\lambda+1+2N)^2}\right)^{1/2}$$

and $\alpha_N^- = \alpha_{N-1}^+$.

To determine a_N we have

$$1 = |a_N|^2 \int_{-\infty}^{\infty} |\langle Y|\cosh^{1/2} Y|N\rangle|^2\,\mathrm{d}Y = |a_N|^2 \int_0^{\pi/2} |\langle X|N\rangle|^2 \cosh Y \frac{\mathrm{d}Y}{\mathrm{d}X}\mathrm{d}X$$
$$= \tfrac{1}{2}|a_N|^2[\langle \sin^{-2} X\rangle + \langle \cos^{-2} X\rangle].$$

These expectation values can be calculated by applying the Hellmann–Feynman theorem to (3.219). Treating μ as a continuous variable yields

$$\langle \sin^{-2}X \rangle = \frac{2(\mu + \lambda + 2N)}{2\mu - 1}.$$

Similarly, treating λ as a continuous variable gives

$$\langle \cos^{-2}X \rangle = \frac{2(\mu + \lambda + 2N)}{2\lambda - 1}.$$

Thus

$$a_N = \left(\frac{(2\mu - 1)(2\lambda - 1)}{2(\mu + \lambda + 2N)(\mu + \lambda - 1)}\right)^{1/2},$$

where we have set a phase factor equal to unity.

Using the above results and expressing (3.234) in terms of X and p_X, we obtain the shift operations (3.232) with $B_N^{\pm}$ given by (3.229)–(3.231) and

$$\beta_N^{\pm} = \left(\frac{(N+\frac{1}{2}\pm\frac{1}{2})(\mu+\lambda+N-\frac{1}{2}\pm\frac{1}{2})(\mu+\lambda+2N)(2\lambda\pm1+2N)(2\mu\pm1+2N)}{(\mu+\lambda\pm2+2N)(\mu+\lambda\pm1+2N)^2}\right)^{1/2}. \tag{3.236}$$

These shift operations enable one to calculate matrix elements of any operator that can be written as a power series in $B_N^{\pm}$. The calculations can be performed as in Section 3.4.3.

3.6.3 *Coordinate-space wavefunctions*

In the coordinate representation the shift operations (3.232) yield the first-order equations

$$\left(\frac{d}{dX} + \lambda \tan X - \mu \cot X\right)\psi_0(X) = 0 \tag{3.237}$$

and

$$\left(\sin X \cos X \frac{d}{dX} - (\mu + \lambda + 2N)\sin^2 X + K_N^+\right)\psi_N(X) = \beta_N^+ \psi_{N+1}(X). \tag{3.238}$$

For normalization with respect to X

$$\int_0^{\pi/2} |\psi_N(X)|^2 \, dX = 1. \tag{3.239}$$

Equations (3.237)–(3.239) determine $\psi_N(X)$. The solution to (3.237) is

$$\psi_0(X) = c_0 \sin^{\mu} X \cos^{\lambda} X, \tag{3.240}$$

where c_0 is a constant. Substituting (3.240) in (3.239), changing the variable to $z = \sin^2 X$, and using the integral

$$\int_0^1 z^{\mu-(1/2)}(1 - z)^{\lambda-(1/2)} \, dz = B(\mu + \tfrac{1}{2}, \lambda + \tfrac{1}{2}) = \frac{\Gamma(\mu + \frac{1}{2})\Gamma(\lambda + \frac{1}{2})}{\Gamma(\mu + \lambda + 1)}, \tag{3.241}$$

we obtain

$$c_0 = \left(\frac{2\Gamma(\mu+\lambda+1)}{\Gamma(\mu+\frac{1}{2})\Gamma(\lambda+\frac{1}{2})}\right)^{1/2}. \tag{3.242}$$

In (3.238) let

$$\psi_N(X) = c_0 \sin^\mu X \cos^\lambda X \phi_N(X). \tag{3.243}$$

Then $\phi_0 = 1$. The rest of the calculation proceeds as in Section 3.4: substituting (3.230) and (3.236) in (3.238), and changing the variable to $u = \sin^2 X$, we find that

$$\phi_N(u) = \gamma_{N-1} \cdots \gamma_1 \gamma_0 P_N(u) \qquad (N = 1, 2, \ldots). \tag{3.244}$$

Here

$$P_N(u) = D_{N-1} \ldots D_1 D_0 1, \tag{3.245}$$

$$D_N = 1 - \frac{\mu+\lambda+1+2N}{\mu+\frac{1}{2}+N} u + \frac{\mu+\lambda+1+2N}{(\mu+\lambda+N)(\mu+\frac{1}{2}+N)} u(1-u)\frac{\mathrm{d}}{\mathrm{d}u}, \tag{3.246}$$

and

$$\gamma_N = \left(\frac{(\mu+\lambda+2+2N)(\mu+\lambda+N)(\mu+\frac{1}{2}+N)}{(N+1)(\mu+\lambda+2N)(\lambda+\frac{1}{2}+N)}\right)^{1/2}. \tag{3.247}$$

Equation (B.3) shows that (3.245) and (3.246) generate the polynomials

$$P_N(u) = {}_2F_1(-N, \mu+\lambda+N; \mu+\tfrac{1}{2}; u), \tag{3.248}$$

where ${}_2F_1$ is the hypergeometric function (3.208). The product of the γ in (3.244) can be written down with the aid of (3.162): the result is

$$\gamma_{N-1} \cdots \gamma_1 \gamma_0 = \left(\frac{(\mu+\lambda+2N)\Gamma(\mu+\lambda+N)\Gamma(\mu+\frac{1}{2}+N)\Gamma(\lambda+\frac{1}{2})}{N!\Gamma(\mu+\lambda+1)\Gamma(\mu+\frac{1}{2})\Gamma(\lambda+\frac{1}{2}+N)}\right)^{1/2}. \tag{3.249}$$

Collecting the above results we have the normalized coordinate-space wavefunctions

$$\psi_N(X) = \left(\frac{2(\mu+\lambda+2N)\Gamma(\mu+\lambda+N)\Gamma(\mu+\frac{1}{2}+N)}{N![\Gamma(\mu+\frac{1}{2})]^2\Gamma(\lambda+\frac{1}{2}+N)}\right)^{1/2} \times {}_2F_1(-N, \mu+\lambda+N; \mu+\tfrac{1}{2}; \sin^2 X) \sin^\mu X \cos^\lambda X. \tag{3.250}$$

It is interesting to consider the potential (3.217) with $\mu = \lambda$. In this symmetric case it is convenient to use the operators $u = 2x - \frac{1}{2}\pi a$ and $p_u = \frac{1}{2}p$. In terms of these the eigenvalue equation is

$$\left(\frac{1}{2M}p_u^2 + \frac{\hbar^2}{2Ma^2}\frac{\lambda(\lambda-1)}{\cos^2 u/a}\right)|N\rangle = E|N\rangle,$$

where (see (3.225))

$$E = \frac{\hbar^2}{2Ma^2}(\lambda+N)^2. \tag{3.251}$$

The coordinate-space wavefunctions are given by (3.250) with $\mu = \lambda$. In the limit $\lambda \to 1$ these wavefunctions and the eigenvalues (3.251) become those for an infinite square-well potential with barriers at $u = \pm\frac{1}{2}\pi a$.

3.7 The second Pöschl–Teller potential [15]

Pöschl and Teller also discussed the solution to the Schrödinger equation with the potential

$$V(x) = \frac{\hbar^2}{2Ma^2}\left(\frac{\mu(\mu-1)}{\sinh^2 xa} - \frac{\lambda(\lambda+1)}{\cosh^2 x/a}\right), \tag{3.252}$$

where μ and λ are constants greater than 1. A similar problem had been studied previously by Weyl [16]. The classical potential is shown in Fig. 3.5 for $\lambda > \mu - 1 > 0$. Unlike the potential (3.217) which has only bound states, the second Pöschl–Teller potential possesses both bound and unbound states.

We consider bound states and put

$$E = -\frac{\hbar^2}{2Ma^2}(2n)^2. \tag{3.253}$$

The energy eigenvalue equation can be written

$$\left(p_X^2 + \frac{\mu(\mu-1)}{\sinh^2 X} - \frac{\lambda(\lambda+1)}{\cosh^2 X}\right)|nk\rangle = -(2n)^2|nk\rangle, \tag{3.254}$$

where k is defined by (3.257). The analysis of this problem is similar to that of the previous section and in the following we give an outline of the calculations.

3.7.1 *Shift operators for k*

In factorization 4 of Section 3.2 let $\beta = 2i/a$, $\gamma = 0$, $\varepsilon = -2i\hbar\nu/a$, and $d = 0$. Then (3.77) becomes

$$V_k(x) = \frac{\hbar^2}{2Ma^2}\left(\frac{(k+\nu)(k+\nu-1)}{\sinh^2 x/a} - \frac{(k-\nu)(k-\nu-1)}{\cosh^2 x/a}\right). \tag{3.255}$$

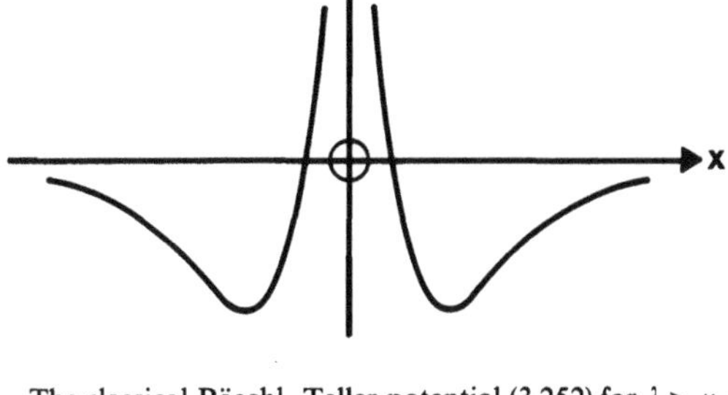

FIG. 3.5 The classical Pöschl–Teller potential (3.252) for $\lambda > \mu - 1 > 0$.

This is the potential (3.252) if

$$k + \nu = -\mu + 1, \qquad k - \nu = \lambda + 1; \tag{3.256}$$

that is,

$$k = \tfrac{1}{2}(\lambda - \mu + 2) \tag{3.257}$$

and

$$\nu = -\tfrac{1}{2}(\mu + \lambda). \tag{3.258}$$

From (3.78) and (3.79) we obtain the shift operations

$$[ip_X + (\mu - 1)\coth X - (\lambda + 1)\tanh X]|nk\rangle = 2(k^2 - n^2)^{1/2}|n, k + 1\rangle \tag{3.259}$$

and

$$(-ip_X + \mu\coth X - \lambda\tanh X)|nk\rangle = 2[(k - 1)^2 - n^2]^{1/2}|n, k - 1\rangle. \tag{3.260}$$

These transformations change k but not ν; consequently $k \to k \pm 1$ means $\mu \to \mu \mp 1$ and $\lambda \to \lambda \pm 1$.

The coefficients in (3.259) and (3.260) violate (1.164) in an interval of width $\Delta k = |2n|$. Thus if $|n| \geq \frac{1}{2}$, non-negativity of the norm requires that the lowering operation (3.260) must terminate at $k = n + 1$.* This result and (3.259) show that the possible values of k are $k = n + 1, n + 2, \ldots$. Thus for a given potential

$$n = k - 1 - N \qquad (N = 0, 1, 2, \ldots). \tag{3.261}$$

Equations (3.253), (3.257), and (3.261) yield the energy eigenvalues

$$E = -\frac{\hbar^2}{2Ma^2}(\lambda - \mu - 2N)^2. \tag{3.262}$$

If $|n| < \frac{1}{2}$ it does not follow from non-negativity of the norm alone that (3.260) must terminate. By constructing shift operators for energy we can avoid this difficulty (see below).

The condition for the existence of states with eigenvalue (3.262) can be obtained using the Hellmann–Feynman theorem. Treating λ as a continuous variable in (3.252) and (3.262) we find

$$\langle\cosh^{-2}X\rangle = \frac{2(\lambda - \mu - 2N)}{2\lambda + 1}. \tag{3.263}$$

The left-hand side of (3.263) is positive and therefore

$$\lambda - \mu > 2N. \tag{3.264}$$

* We can, without loss of generality, suppose $n > 0$.

Hence the number of bound states is the largest integer that is less than

$$1 + \tfrac{1}{2}(\lambda - \mu). \tag{3.265}$$

By treating μ as a continuous variable and applying the Hellman–Feynman theorem we obtain

$$\langle \sinh^{-2} X \rangle = \frac{2(\lambda - \mu - 2N)}{2\mu - 1}. \tag{3.266}$$

From (3.252), (3.263), and (3.266) the average value of the potential is

$$\langle V \rangle = \frac{\hbar^2}{Ma^2}(\lambda - \mu - 2N)\left(\frac{\mu(\mu - 1)}{2\mu - 1} - \frac{\lambda(\lambda + 1)}{2\lambda + 1}\right). \tag{3.267}$$

3.7.2 *Shift operators for energy*

By means of (3.262) the eigenvalue equation (3.254) can also be written in the form (3.226) where

$$\begin{aligned} O_N = {} & \sinh^2 X \cosh^2 X p_X^2 + (\lambda - \mu - 2N)^2 \sinh^4 X \\ & + [\mu(\mu - 1) - \lambda(\lambda + 1) + (\lambda - \mu - 2N)^2] \sinh^2 X. \end{aligned} \tag{3.268}$$

For O_N one can readily derive the factorizations (3.228) where

$$B_N^{\pm} = \pm \mathrm{i} \sinh X \cosh X p_X - (\lambda - \mu - 2N) \sinh^2 X + K_N^{\pm}, \tag{3.269}$$

$$K_N^{+} = -\frac{\mu(\mu - 1) - \lambda(\lambda + 1)}{2(\lambda - \mu - 1 - 2N)} - \frac{1}{2}(\lambda - \mu - 2N), \tag{3.270}$$

and

$$K_N^{-} = K_{N-1}^{+} + 1. \tag{3.271}$$

The operators $B_N^{\pm}$ perform the shift operations (3.232); the coefficients $\beta_N^{\pm}$ in these formulae can be deduced by calculations that are similar to those leading to (3.236) and the results are

$$\beta_N^{\pm} = \left(\frac{(N + \frac{1}{2} \pm \frac{1}{2})(\lambda - \mu - N + \frac{1}{2} \mp \frac{1}{2})(\lambda - \mu - 2N)(2\lambda \mp 1 - 2N)(2\mu \pm 1 + 2N)}{(\lambda - \mu \mp 2 - 2N)(\lambda - \mu \mp 1 - 2N)^2}\right)^{1/2}. \tag{3.272}$$

From (3.232), (3.272), and (3.269)–(3.271) we have

$$(-\mathrm{i}p_X + \mu \coth X - \lambda \tanh X)|0\rangle = 0, \tag{3.273}$$

which is just (3.260) with $k = n + 1$. According to (3.272) and (3.264), $|\beta_0^+|^2$, $|\beta_1^+|^2, \ldots, |\beta_{N-1}^+|^2$ are all positive. Hence the kets $|1\rangle, |2\rangle, \ldots, |N\rangle$ obtained by applying B_0^+, $B_1^+ B_0^+$, $\ldots$, $B_{N-1}^+ \ldots B_1^+ B_0^+$ to a normalizable ket $|0\rangle$ are also normalizable. According to (3.272) and (3.264), for any non-integral value of

N the sequence of lowering operations for N will produce kets which violate non-negativity of the norm.

3.7.3 *Coordinate-space wavefunctions*

In the coordinate representation the solution to (3.273) is

$$\psi_0(X) = c_0 \sinh^\mu X \cosh^{-\lambda} X. \tag{3.274}$$

The normalization condition is

$$\int_0^\infty |\psi_N|^2 \, dX = 1. \tag{3.275}$$

Substituting (3.274) in (3.275), changing the variable to $u = \tanh^2 X$, and using the integral (3.241) we find

$$c_0 = \left(\frac{2\Gamma(\lambda + \frac{1}{2})}{\Gamma(\mu + \frac{1}{2})\Gamma(\lambda - \mu)} \right)^{1/2}. \tag{3.276}$$

By applying $B_0^+, B_1^+ B_0^+, \ldots$ to the ground state $\psi_0(X)$, we generate the excited states. The calculations are similar to those in the previous section. With

$$\psi_N(X) = c_0 \sinh^\mu X \cosh^{-\lambda} X \phi_N(X) \tag{3.277}$$

and $u = -\sinh^2 X$, $\phi_N(u)$ satisfies (3.244) with

$$D_N = 1 - \frac{\mu - \lambda + 1 + 2N}{\mu + \frac{1}{2} + N} u + \frac{\mu - \lambda + 1 + 2N}{(\mu - \lambda + N)(\mu + \frac{1}{2} + N)} u(1 - u) \frac{d}{du} \tag{3.278}$$

and

$$\gamma_N = \left(\frac{(\lambda - \mu - 2 - 2N)(\lambda - \mu - N)(\mu + \frac{1}{2} + N)}{(N + 1)(\lambda - \mu - 2N)(\lambda - \frac{1}{2} - N)} \right)^{1/2}. \tag{3.279}$$

Hence we obtain the normalized coordinate-space wavefunctions for bound states of the potential (3.252)

$$\psi_N(X) = \left(\frac{2(\lambda - \mu - 2N)\Gamma(\mu + \frac{1}{2} + N)\Gamma(\lambda + \frac{1}{2} - N)}{N![\Gamma(\mu + \frac{1}{2})]^2 \Gamma(\lambda - \mu + 1 - N)} \right)^{1/2}$$
$$\times\, {}_2F_1(-N, N + \mu - \lambda; \mu + \tfrac{1}{2}; -\sinh^2 X) \sinh^\mu X \cosh^{-\lambda} X. \tag{3.280}$$

The conditions for these wavefunctions to be square-integrable are (3.264) and $\mu > -\frac{1}{2}$. Thus the assumption $\mu > 1$ made above can be relaxed. For example, if $0 < \mu < 1$ the classical potential (3.252) is attractive. The above analysis still applies and the energy eigenvalues are given by (3.262). The only change is to the normalization (3.275) where the range of integration is from $-\infty$ to ∞: this requires a factor $2^{-1/2}$ on the right-hand sides of (3.276) and (3.280).

Note that for each value of $\mu(\mu - 1)$ between $-\frac{1}{4}$ and $\frac{3}{4}$ there are two normalizable solutions, namely those corresponding to $-\frac{1}{2} < \mu < \frac{1}{2}$ and $\frac{1}{2} < \mu < \frac{3}{2}$. A property that distinguishes these two solutions is finiteness of the average potential (or the average kinetic energy) [17]; according to (3.280), (3.264), and (3.252), $\langle V \rangle$ is finite if $\mu > \frac{1}{2}$. This restriction is also apparent from the positivity of the left-hand side of (3.266).

Because (3.252) is unchanged if $\mu \to -\mu + 1$ and/or $\lambda \to -\lambda - 1$, if we make these substitutions in (3.280) we obtain solutions that are valid for other ranges of μ and λ. For example, replacing μ with $-\mu + 1$ yields the normalized wavefunctions

$$\psi_N(X) = \left(\frac{2(\lambda + \mu - 1 - 2N)\Gamma(-\mu + \frac{3}{2} + N)\Gamma(\lambda + \frac{1}{2} - N)}{N![\Gamma(-\mu + \frac{3}{2})]^2\Gamma(\lambda + \mu - N)}\right)^{1/2}$$
$$\times {}_2F_1(-N, N + 1 - \mu - \lambda; -\mu + \tfrac{3}{2}; -\sinh^2 X) \sinh^{-\mu+1} X \cosh^{-\lambda} X, \tag{3.281}$$

where $\lambda > -\mu + 1 + 2N$ and $\mu < 0$. In the limit $\mu \to 0$ and including the factor $2^{-1/2}$ mentioned above, we see that (3.280) and (3.281) are (to within a phase factor) the even and odd solutions (3.211) and (3.216) for the symmetric Rosen–Morse potential.

3.7.4 *Relationship to the Rosen–Morse potential*

The Rosen–Morse and second Pöschl–Teller equations are closely related [18]. Let

$$2Ma^2\hbar^{-2}E = -A^2 - B^2, \tag{3.282}$$

$$2Ma^2\hbar^{-2}U_0 = -C - \tfrac{1}{4}, \tag{3.283}$$

$$2Ma^2\hbar^{-2}U_1 = -2AB, \tag{3.284}$$

and

$$X = -\mathrm{i}\frac{\pi}{2} + \ln \tanh Y. \tag{3.285}$$

From (3.285)

$$\mathrm{e}^X = -\mathrm{i} \tanh Y, \tag{3.286}$$

and hence

$$\cosh X = \mathrm{i}(\sinh 2Y)^{-1}, \qquad \tanh X = -\cosh 2Y. \tag{3.287}$$

Define $p_Y = -2\mathrm{i} \cosh X p_X$. Then $[p_Y, Y] = -\mathrm{i}$.

In terms of p_Y, Y, and the parameters A, B, and C the Rosen–Morse equation (3.165) becomes

$$\left(\frac{1}{4}p_Y^2 + \frac{A^2 + B^2 - \frac{1}{4} + 2AB \cosh 2Y}{\sinh^2 2Y}\right)\{\sinh^{1/2} 2Y|E\rangle\} = C\{\sinh^{1/2} 2Y|E\rangle\};$$

that is,

$$\left(p_Y^2 + \frac{(A+B+\frac{1}{2})(A+B-\frac{1}{2})}{\sinh^2 Y} - \frac{(A-B-\frac{1}{2})(A-B+\frac{1}{2})}{\cosh^2 Y}\right)\{\sinh^{1/2}2Y|E\rangle\}$$
$$= 4C\{\sinh^{1/2}2Y|E\rangle\}. \tag{3.288}$$

This has the same form as the second Pöschl–Teller equation (3.254) and therefore we identify

$$A + B + \tfrac{1}{2} = \mu, \qquad A - B - \tfrac{1}{2} = \lambda. \tag{3.289}$$

From (3.280) and (3.289) we obtain the coordinate-space wavefunction

$$\psi_N(X) = \sinh^{-1/2}2Y c_N\, {}_2F_1(-N, N+2B+1; A+B+1; -\sinh^2 Y)$$
$$\times (\sinh Y)^{A+B+(1/2)}(\cosh Y)^{-A+B+(1/2)} \tag{3.290}$$
$$= \mathrm{i}^{A+B}2^{-B-(1/2)}c_N\, {}_2F_1\left(-N, N+2B+1; A+B+1; \frac{\mathrm{e}^X}{\mathrm{e}^X+\mathrm{e}^{-X}}\right)$$
$$\times\, \mathrm{e}^{AX}\cosh^{-B}X. \tag{3.291}$$

In the last step we have used (3.286) and (3.287). The wavefunction (3.291) is square-integrable if

$$-B < A < B. \tag{3.292}$$

Thus

$$B > 0. \tag{3.293}$$

According to (3.262) and (3.289) the eigenvalue in (3.288) is given by

$$4C = -(2B + 1 + 2N)^2,$$

and hence (3.283) and (3.293) yield

$$B = -N - \tfrac{1}{2} + \tfrac{1}{2}(1 + 8Ma^2\hbar^{-2}U_0)^{1/2}. \tag{3.294}$$

Equations (3.282), (3.284), and (3.294) yield the energy eigenvalues (3.174). Also, according to (3.292) and (3.284),

$$B > (Ma^2\hbar^{-2}U_1)^{1/2}, \tag{3.295}$$

which with $N = 0$ is the condition (3.180) for bound states to exist.

Finally we determine the constant c_N in (3.291). If we normalize

$$\phi_N(Y) = \sinh^{1/2}2Y\psi_N(X)$$

to unity, then

$$c_N = \gamma_{N-1}\cdots\gamma_1\gamma_0 c_0,$$

where*

* Equation (3.296) is (3.279) expressed in terms of the positive quantity $\mu - \lambda = 2B + 1$.

$$\gamma_N = \left(\frac{(\mu - \lambda + 2 + 2N)(\mu - \lambda + N)(\mu + \frac{1}{2} + N)}{(N+1)(\mu - \lambda + 2N)(N + \frac{1}{2} - \lambda)}\right)^{1/2}. \tag{3.296}$$

Hence

$$\gamma_{N-1} \cdots \gamma_1 \gamma_0$$
$$= \left(\frac{(2B + 1 + 2N)\Gamma(2B + 1 + N)\Gamma(B - A + 1)\Gamma(A + B + 1 + N)}{N!(2B + 1)\Gamma(2B + 1)\Gamma(B - A + 1 + N)\Gamma(A + B + 1)}\right)^{1/2}.$$

The normalization of $\phi_N(Y)$ to unity means

$$\int_{-\infty}^{\infty} |\sinh^{1/2} 2Y \, \psi_N(X)|^2 \left|\frac{dY}{dX}\right| dX = 1,$$

and hence

$$\frac{1}{2}\int_{-\infty}^{\infty} |\psi_N(X)|^2 \cosh^{-2} X \, dX = 1. \tag{3.297}$$

Substituting $\psi_0(X)$ given by (3.291) in (3.297), changing the variable to $u = \exp(2X)$, and using the integral (3.183), we find

$$c_0 = \left(\frac{2\Gamma(2B + 2)}{\Gamma(A + B + 1)\Gamma(B - A + 1)}\right)^{1/2},$$

where a phase factor has been set equal to unity.

We wish $\psi_N(X)$ to be normalized to unity: this requires instead of (3.297)

$$\frac{1}{2}\int_{-\infty}^{\infty} |\psi_N(X)|^2 \cosh^{-2} X \, dX = (2B + 1 + 2N)^{-1} \frac{B^2 - A^2}{B}. \tag{3.298}$$

(See (3.178), (3.173), (3.284), and (3.294).) Thus we must include in c_N a factor equal to the square root of the right-hand side of (3.298). Collecting the above results we obtain

$$c_N = \left(\frac{2(B^2 - A^2)\Gamma(2B + 1 + N)\Gamma(A + B + 1 + N)}{N! B[\Gamma(A + B + 1)]^2 \Gamma(B - A + 1 + N)}\right)^{1/2}, \tag{3.299}$$

where B and A are given by (3.294) and (3.284). Equation (3.291), with c_N given by (3.299), is the coordinate-space wavefunction for the Rosen–Morse potential.

3.8 Supersymmetric quantum mechanics of one-dimensional systems

We conclude this chapter by considering the factorization method from another point of view, namely shape invariance. In this section we give a brief introduction to supersymmetric quantum mechanics (SUSYQM) and in the next section we show how shape invariance of supersymmetric potentials enables one to solve certain quantum-mechanical problems.

Let

$$A_{\pm} = \pm\frac{\mathrm{i}}{(2M)^{1/2}}p + W(x), \tag{3.300}$$

where W is a real function, and consider the one-dimensional operators

$$H_{\pm} = A_{\pm}A_{\mp} \tag{3.301}$$

$$= \frac{1}{2M}p^2 + V_{\pm}(x), \tag{3.302}$$

where

$$V_{\pm}(x) = W^2 \pm \frac{\hbar}{(2M)^{1/2}}\frac{\mathrm{d}W}{\mathrm{d}x}. \tag{3.303}$$

Suppose H_- possesses a discrete spectrum

$$H_-|\psi_N^-\rangle = E_N^-|\psi_N^-\rangle \qquad (N = 0, 1, 2, \ldots). \tag{3.304}$$

(Our usage of the label N conforms to the notation in Sections 3.4–3.7; note that for the oscillator N corresponds to the quantum number n in (2.22).) Because $(A_+)^\dagger = A_-$ we see that $E_N^- \geq 0$. We suppose initially that A_+ annihilates the ground state

$$A_+|\psi_0^-\rangle = 0; \tag{3.305}$$

then it follows from (3.301) that the ground state eigenvalue is zero

$$E_0^- = 0. \tag{3.306}$$

Note that in the coordinate representation (3.305) is the first-order differential equation

$$\left(\frac{\hbar}{(2M)^{1/2}}\frac{\mathrm{d}}{\mathrm{d}x} + W\right)\psi_0^-(x) = 0, \tag{3.307}$$

and hence the ground state wavefunction is

$$\psi_0^-(x) = c_0 \exp\left(-\frac{(2M)^{1/2}}{\hbar}\int^x W(x)\,\mathrm{d}x\right), \tag{3.308}$$

where c_0 is a constant. We are assuming that $|\psi_0^-\rangle$ in (3.305) is normalizable, and hence that (3.308) is square-integrable. It is clear from (3.308) that ψ_0^- has no nodes, as one expects for the ground state of a one-dimensional system (see Section 5.7).

The eigenvalue equation for H_+ is

$$H_+|\psi_N^+\rangle = E_N^+|\psi_N^+\rangle \qquad (N = 0, 1, 2, \ldots). \tag{3.309}$$

It is straightforward to show that there exist special relationships between the eigenkets of H_+ and H_-: the reasoning is similar to that leading to (3.50) and

(3.51), and we find

$$A_+ \, |\psi_{N+1}^-\rangle = (E_{N+1}^-)^{1/2} \, |\psi_N^+\rangle, \tag{3.310}$$

$$A_- \, |\psi_N^+\rangle = (E_N^+)^{1/2} \, |\psi_{N+1}^-\rangle, \tag{3.311}$$

and

$$E_N^+ = E_{N+1}^-, \tag{3.312}$$

for $N = 0, 1, \ldots$. In (3.310) and (3.311) we have assumed normalized kets and we have set a phase factor equal to unity. We see that H_+ and H_- have the same spectrum except for the ground state (only H_- has a normalized ground state with eigenvalue zero) and that $A_\pm$ transform between the eigenkets of $H_\pm$. These results are depicted in Fig. 3.6.

Before proceeding we note that in addition to (3.305) there are two other possibilities. The first is $A_- \, |\psi_0^+\rangle = 0$. Then $E_0^+ = 0$ and the spectra of $H_\pm$ are identical except that H_- has no normalizable eigenket with zero eigenvalue: the relationships between the eigenvalues and eigenkets are those shown in Fig. 3.6 but with the labels + and − interchanged everywhere. The final possibility is that there are no normalizable eigenkets of H_- or H_+ such that either $A_- |\psi_0^-\rangle = 0$ or $A_+ |\psi_0^+\rangle = 0$. Then the eigenvalues of both H_- and H_+ begin at positive values and every eigenvalue of H_- is an eigenvalue of H_+ and vice versa. This possibility corresponds to broken supersymmetry and we will not discuss it any further here [6]. In what follows we assume (3.305).

Consider now the Hamiltonian

$$H_{ss} = \begin{pmatrix} H_- & 0 \\ 0 & H_+ \end{pmatrix}, \tag{3.313}$$

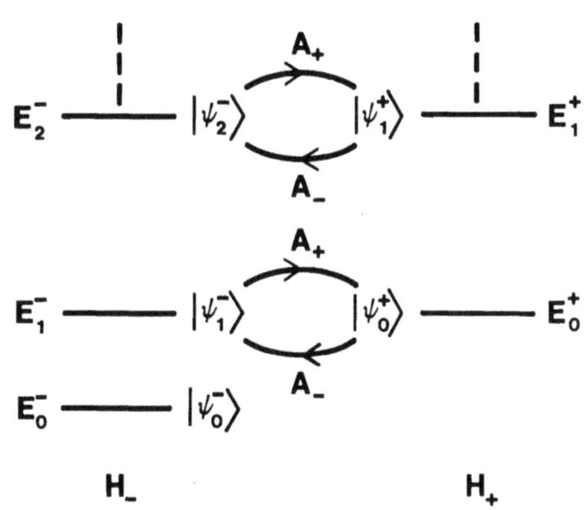

FIG. 3.6 Schematic illustration of the eigenvalues of the Hamiltonians (3.301) and the transformations (3.310) and (3.311). The spectra of $H_\pm$ are identical except for the ground state of H_-.

which has the linearly independent eigenkets

$$|N\rangle_- = \begin{pmatrix} |\psi_N^-\rangle \\ 0 \end{pmatrix} \quad \text{and} \quad |N\rangle_+ = \begin{pmatrix} 0 \\ |\psi_N^+\rangle \end{pmatrix}, \tag{3.314}$$

for $N = 0, 1, 2, \ldots$, with eigenvalues E_N^- and $E_N^+ = E_{N+1}^-$. One expects that the two-fold degeneracy of each excited state of H_{ss} is a manifestation of an underlying symmetry. The nature of this symmetry can be studied as follows. Let

$$Q = \begin{pmatrix} 0 & 0 \\ A_+ & 0 \end{pmatrix}, \tag{3.315}$$

where A_+ is defined by (3.300). Then

$$Q^\dagger = \begin{pmatrix} 0 & A_- \\ 0 & 0 \end{pmatrix}. \tag{3.316}$$

These operators transform between the degenerate eigenkets of H_{ss} (see (3.313)–(3.315), (3.310), and (3.311))

$$Q|N\rangle_- = (E_N^-)^{1/2}|N-1\rangle_+ \tag{3.317}$$

and

$$Q^\dagger|N\rangle_+ = (E_N^+)^{1/2}|N+1\rangle_- \tag{3.318}$$

for $N = 0, 1, \ldots$. In Fig. 3.7 we have depicted the spectrum of H_{ss} and the transformations (3.317) and (3.318).

It is simple to show that

$$Q^2 = (Q^\dagger)^2 = 0, \tag{3.319}$$

$$H_{ss} = QQ^\dagger + Q^\dagger Q, \tag{3.320}$$

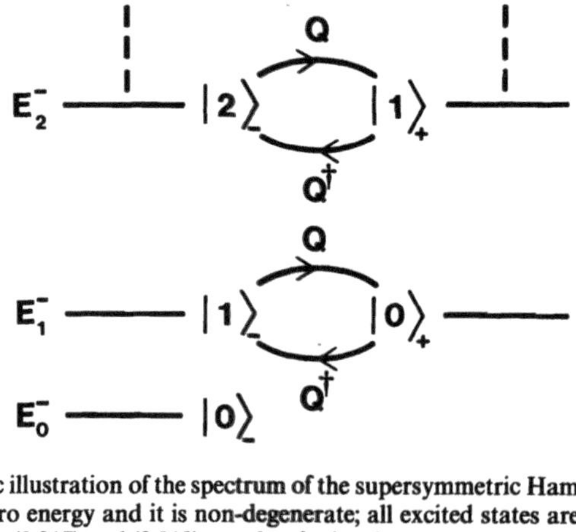

FIG. 3.7 Schematic illustration of the spectrum of the supersymmetric Hamiltonian (3.313). The ground state has zero energy and it is non-degenerate; all excited states are doubly degenerate. The transformations (3.317) and (3.318) are also depicted.

and

$$[Q, H_{ss}] = [Q^\dagger, H_{ss}] = 0. \tag{3.321}$$

Thus Q and $Q^\dagger$ are nilpotent operators whose anticommutator is H_{ss}, and which commute with H_{ss}. Equations (3.319)–(3.321) show that the operators H_{ss}, Q, and $Q^\dagger$ close under commutation and anticommutation;* they form a graded Lie algebra.

The realization (3.315) and (3.316) of this algebra is a simple example of SUSYQM [20]. In the terminology of SUSYQM, H_{ss} is referred to as the *supersymmetric Hamiltonian*, H_+ and H_- are *supersymmetric partners*, as are V_+ and V_-. The operator W is the *superpotential* and the generators Q and $Q^\dagger$ of supersymmetry are the *charge operators*.

It is useful to note that Q and H_{ss} can also be written $Q = \frac{1}{2}A_+\sigma_-$ and

$$H_{ss} = \frac{1}{2M}p^2 + W^2 - \frac{\hbar}{(2M)^{1/2}}\frac{dW}{dx}\sigma_z, \tag{3.322}$$

where $\sigma_- = \sigma_1 - i\sigma_2$ and σ_i are Pauli matrices (Section 4.14).

The discussion so far has been rather formal and we now consider a simple example [21, 22]. With

$$W = \left(\frac{1}{2}M\right)^{1/2}\omega x \tag{3.323}$$

(ω a positive constant) we have $Q = \frac{1}{2}(\hbar\omega)^{1/2}a\sigma_-$, $Q^\dagger = \frac{1}{2}(\hbar\omega)^{1/2}a^\dagger\sigma_+$, and

$$H_{ss} = \begin{pmatrix} H - \frac{1}{2}\hbar\omega & 0 \\ 0 & H + \frac{1}{2}\hbar\omega \end{pmatrix}, \tag{3.324}$$

where H is the oscillator Hamiltonian (2.9) and a and $a^\dagger$ are the annihilation and creation operators (2.11) and (2.12). The eigenkets of H_{ss} are

$$|N\rangle_- = \begin{pmatrix} |N\rangle \\ 0 \end{pmatrix} \quad \text{and} \quad |N\rangle_+ = \begin{pmatrix} 0 \\ |N\rangle \end{pmatrix}, \tag{3.325}$$

where $|N\rangle$ is an eigenket of H (Section 2.2). The spectra of H_{ss} and H are compared in Fig. 3.8.

To identify a physical system described by the supersymmetric Hamiltonian (3.324) we first use (3.322) and (3.323) to write this as

$$H_{ss} = \frac{1}{2M}p^2 + \frac{1}{2}M\omega^2x^2 - \omega S_z, \tag{3.326}$$

where $S_z = \frac{1}{2}\hbar\sigma_z$ is the z component of the spin angular momentum operator for a non-relativistic spin $\frac{1}{2}$ particle (Section 4.14). Consider a non-relativistic

* In the supersymmetric algebra of four-space the Q have indices and the nilpotent condition (3.319) is replaced by anticommutation relations [19].

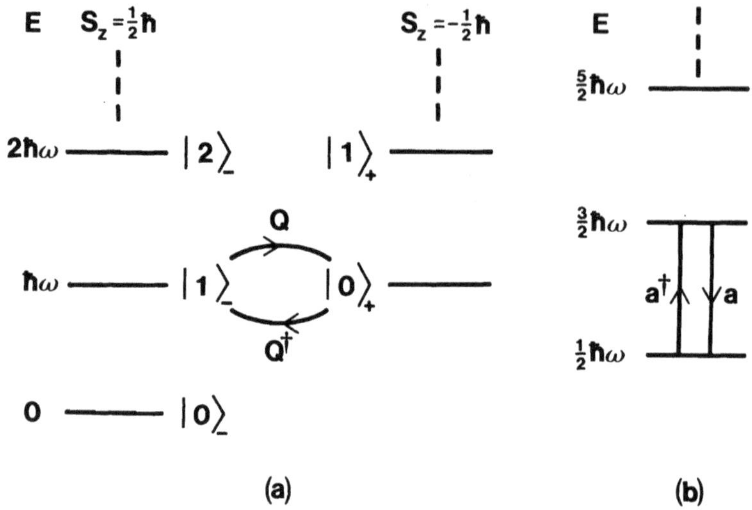

FIG. 3.8 (a) The spectrum of the supersymmetric Hamiltonian (3.324). The eigenkets $|N\rangle_\pm$ are given by (3.325). The transformations (3.317) and (3.318) are depicted. The labels $S_z = \pm\frac{1}{2}\hbar$ refer to the forms (3.326) and (3.329). (b) The spectrum of the oscillator Hamiltonian. The transformations (2.29) and (2.30) are depicted.

spin $\frac{1}{2}$ particle of charge q in a uniform magnetic field $\mathbf{B} = (0, 0, B)$: if we choose the gauge (2.213) the Hamiltonian is

$$H = \frac{1}{2M}(p_x^2 + p_z^2) + \frac{1}{2M}(p_y - qBx)^2 - \frac{qB}{M}S_z. \tag{3.327}$$

By a simple extension of the discussion in Section 2.9 one can show that H, p_y, p_z, and S_z are a c.s.c.o. The effect of H on the eigenkets $|EP_yP_zS_z\rangle$ is the same as that of the one-dimensional Hamiltonian

$$\hat{H} = \frac{1}{2M}p_x^2 + \frac{1}{2}M\omega^2[x - (qB)^{-1}P_y]^2 + \frac{1}{2M}P_z^2 - \hat{q}\omega S_z, \tag{3.328}$$

where ω is the cyclotron frequency (2.195) and $\hat{q} = \text{sign}\, q$. We see that for $q > 0$,

$$\hat{H} - \frac{1}{2M}P_z^2 \tag{3.329}$$

corresponds to (3.326) with p replaced by p_x and x by $x - (qB)^{-1}P_y$. Thus (3.326) and (3.329) have the same spectrum. This is indicated in Fig. 3.8(a) where the labels $S_z = \pm\frac{1}{2}\hbar$ refer to the eigenvalues of (3.329) for $q > 0$; if $q < 0$ these labels are interchanged in Fig. 3.8(a).

The restriction to non-relativistic motion that we made in the above can be removed without spoiling the supersymmetry of the problem. Thus the Dirac Hamiltonian for a charged particle in a uniform magnetic field is supersymmetric [22]. If we write the last term in (3.327) as

$$-g\frac{qB}{2M}S_z,$$

where according to the Dirac theory $g = 2$, it is clear that the supersymmetry of (3.329) is restricted to $g = 2$. Quantum electrodynamic corrections [23] make $g \neq 2$ and they therefore destroy the supersymmetry for states of a Dirac particle in a uniform magnetic field.

To construct a supersymmetric Hamiltonian H_{ss} and the generators Q and $Q^\dagger$ of supersymmetry, we need to know the superpotential W that enters via (3.300). In the above example we started by choosing a simple linear expression for W. This is one method of generating a supersymmetric system; alternatively one can start with a known potential $V_-(x)$ and solve the Ricatti equation (3.303) for W. Another approach is to determine the ground state wavefunction $\psi_0^-(x)$ of H_- and then use the relation (see (3.308))

$$W(x) = -\frac{\hbar}{(2M)^{1/2}}\frac{\mathrm{d}}{\mathrm{d}x}\ln\psi_0^-(x). \tag{3.330}$$

There are many examples of one-dimensional Hamiltonians H_- for which a partner Hamiltonian H_+ can be constructed such that the combined Hamiltonian (3.313) is supersymmetric, and some of these are discussed in the next section. Often it is not clear whether the combined system exists in nature, and it has been remarked that in this sense supersymmetry is a 'solution looking for a problem' [22]. Nevertheless there has been considerable interest in the applications of SUSYQM to atomic, nuclear, and solid state physics [24], and particularly to particle physics [25]. SUSYQM is also of considerable theoretical interest in its own right, for example for the insights it provides into the factorization method (see Section 3.9).

The term 'supersymmetry' was originally used to denote a symmetry built into certain quantum field theories. In supersymmetric field theories the generators of supersymmetry have the remarkable property that they transform between component fields whose intrinsic spins differ by $\frac{1}{2}\hbar$; they change fermion states into boson states and vice versa.* Thus supersymmetric multiplets contain both bosons and fermions, and it has been suggested that this property may provide a unifying principle in particle physics. Supersymmetric field theories possess other attractive features. For example, they are not as divergent and they are apparently less sensitive to adjustment of parameters than their less symmetric counterparts. A supersymmetric theory (superstring theory) may provide a unified theory of quantum gravity and the other

---

* As we have mentioned above, the terminology 'supersymmetric system' refers also to any system governed by the algebra (3.319)–(3.321) which is a limiting case of the algebra of supersymmetry in quantum field theory; included among such systems are cases which do not involve transformations of spin angular momentum.

fundamental interactions. These topics are beyond the scope of this book and the interested reader is referred to the extensive literature [19].

3.9 Shape invariance

We have seen in previous sections that for a variety of physical problems shift operators can be obtained by the factorization method. One way of classifying these problems is via the solutions (3.57) and (3.58) to the factorization condition (3.56) (see Section 3.2). There are other (equivalent) ways of expressing the condition for the factorization method to apply to an eigenvalue problem and in this section we discuss one of them, namely shape invariance in SUSYQM [26].

First we explain what is meant by shape invariance. Usually the supersymmetric partner potentials $V_{\pm}(x)$ in (3.303) also depend on a set of parameters a_0, and so we denote them by $V_{\pm}(x, a_0)$. The concept of shape invariance relates supersymmetric partner potentials through the parameters appearing in them; in particular, $V_+(x, a_0)$ and $V_-(x, a_0)$ are said to be *shape invariant* if [26]

$$V_+(x, a_0) = V_-(x, a_1) + R(a_1), \tag{3.331}$$

where a_1 is a function of a_0

$$a_1 = f(a_0),$$

and the remainder $R(a_1)$ is independent of x. According to (3.331) the potentials $V_{\pm}$ have similar shapes and they differ only in the parameters of the potential. By using (3.303) in (3.331) we see that in terms of the superpotential the shape invariance condition is

$$W^2(x, a_0) + \frac{\hbar}{(2M)^{1/2}} \frac{\mathrm{d}W(x, a_0)}{\mathrm{d}x} = W^2(x, a_1) - \frac{\hbar}{(2M)^{1/2}} \frac{\mathrm{d}W(x, a_1)}{\mathrm{d}x} + R(a_1). \tag{3.332}$$

It can be shown that (3.332) is equivalent to the factorization condition (3.56) [27].

We now show that shape invariance is a sufficient condition* for the eigenvalue equation (3.304) to be analytically soluble and, furthermore, that the eigenvalues and eigenkets can be obtained by a simple algebraic procedure. First we construct a hierarchy of Hamiltonians $H^{(s)}$ ($s = 0, 1, \ldots$), where $H^{(0)} = H_-$ and

$$H^{(s)} = \frac{1}{2M} p^2 + V_-(x, a_s) + \sum_{k=1}^{s} R(a_k) \qquad (s = 1, 2, \ldots). \tag{3.333}$$

* It is apparently not a necessary condition [28].

Here $a_s = f(a_{s-1})$. Note that $H^{(0)}$ and $H^{(1)} = H_+$ are supersymmetric partners.
In terms of a_s and a_{s+1} the shape invariance condition (3.331) is

$$V_+(x, a_s) = V_-(x, a_{s+1}) + R(a_{s+1}) \qquad (s = 0, 1, 2, \ldots), \tag{3.334}$$

and therefore

$$\begin{aligned} H^{(s+1)} &= \frac{1}{2M}p^2 + V_-(x, a_{s+1}) + \sum_{k=1}^{s+1} R(a_k) \\ &= \frac{1}{2M}p^2 + V_+(x, a_s) + \sum_{k=1}^{s} R(a_k). \end{aligned} \tag{3.335}$$

Equations (3.333) and (3.335) show that $H^{(s)}$ and $H^{(s+1)}$ are supersymmetric partner Hamiltonians, and hence the results of Section 3.8 apply to these pairs of Hamiltonians. In particular, the relationship between their eigenvalues is

$$E_N^{(s)} = E_{N-1}^{(s+1)} \qquad (N = 1, 2, \ldots) \tag{3.336}$$

(see Fig. 3.9). Also, the ground state energy $E_0^{(N)}$ of $H^{(N)}$ is equal to the eigenvalue E_N^- of $H^{(0)}$; because the ground state eigenvalue corresponding to the first two terms in (3.333) is zero (see (3.306)) we have

$$E_N^- = \sum_{k=1}^{N} R(a_k). \tag{3.337}$$

We can also determine the eigenkets of H_-. If we denote the eigenkets of (3.333) and (3.335), respectively, by $|\psi_N^{(s)}(a_s)\rangle$ and $|\psi_N^{(s+1)}(a_s)\rangle$ then the trans-

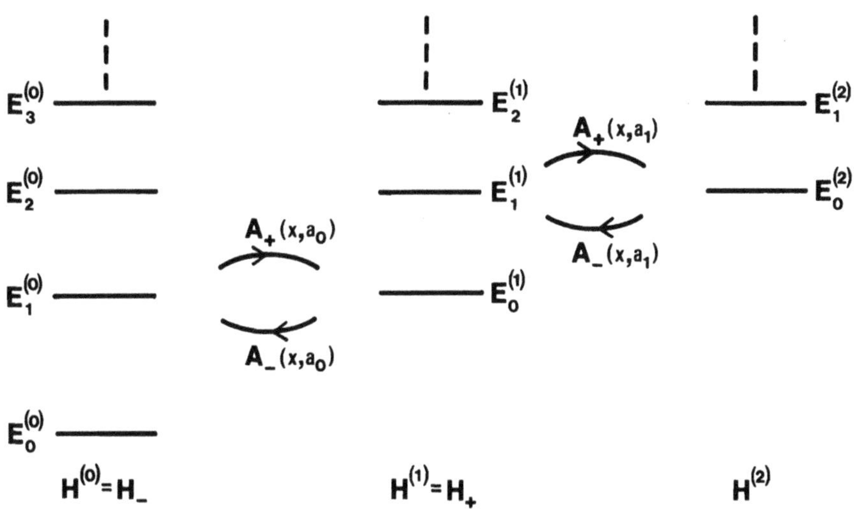

FIG. 3.9 Energy eigenvalues of the hierarchy of Hamiltonians $H^{(s)}$ for $s = 0, 1, 2$. Also depicted are examples of the transformations (3.338) and (3.339).

formations (3.310) and (3.311) show that

$$A_+(x, a_s)|\psi_{N+1}^{(s)}(a_s)\rangle = (E_{N+1}^{(s)} - E_s^{(0)})^{1/2}|\psi_N^{(s+1)}(a_s)\rangle \qquad (3.338)$$

and

$$A_-(x, a_s)|\psi_N^{(s+1)}(a_s)\rangle = (E_{N+1}^{(s)} - E_s^{(0)})^{1/2}|\psi_{N+1}^{(s)}(a_s)\rangle, \qquad (3.339)$$

where

$$A_\pm(x, a_s) = \pm\frac{\mathrm{i}}{(2M)^{1/2}}p + W(x, a_s). \qquad (3.340)$$

Examples of the transformations (3.338) and (3.339) are shown in Fig. 3.9.

Let $|\psi_N^{(s)}(a_{s'})\rangle$ denote an eigenket of a Hamiltonian given by (3.333) with a_s replaced by $a_{s'}$. Then (see (3.334))

$$|\psi_N^{(1)}(a_s)\rangle = |\psi_N^{(0)}(a_{s+1})\rangle. \qquad (3.341)$$

From (3.339), (3.341), and (3.336) we obtain

$$A_-(x, a_s)|\psi_N^{(0)}(a_{s+1})\rangle = (E_{N+1}^{(0)})^{1/2}|\psi_{N+1}^{(0)}(a_s)\rangle \qquad (N = 0, 1, 2, \ldots). \quad (3.342)$$

By repeating this procedure we can generate all the excited states of H_- from $|\psi_0^{(0)}(a_N)\rangle$

$$|\psi_N^{(0)}(a_0)\rangle = \alpha_N A_-(x, a_0)A_-(x, a_1)\ldots A_-(x, a_{N-1})|\psi_0^{(0)}(a_N)\rangle, \quad (3.343)$$

where $N = 1, 2, \ldots$ and (see (3.339) and Fig. 3.9)

$$\alpha_N = \prod_{k=1}^{N} (E_N^{(0)} - E_{k-1}^{(0)})^{-1/2}. \qquad (3.344)$$

Equations (3.337) and (3.343) are the formal algebraic solutions for the eigenvalues and eigenkets of shape invariant potentials. We now consider some applications of these results.

(i) The simplest example is the one-dimensional oscillator, for which the superpotential is given by (3.323). Substituting this in (3.303) we obtain the supersymmetric partners

$$V_\pm(x, a_0) = \tfrac{1}{2}M\omega^2x^2 \pm \tfrac{1}{2}\hbar\omega. \qquad (3.345)$$

If we identify $a_s = \omega$ $(s = 0, 1, \ldots)$ we see from (3.345) that $V_\pm(x, a_s)$ satisfy the shape invariance condition (3.334) with remainder $R(a_s) = \hbar\omega$. Hence (3.337) and (3.306) yield the eigenvalues $E_N^- = N\hbar\omega$ $(N = 0, 1, \ldots)$. The operators (3.300) are given in terms of the annihilation and creation operators (2.11) and (2.12) by $A_+ = (\hbar\omega)^{1/2}a$ and $A_- = (\hbar\omega)^{1/2}a^\dagger$. Thus (3.342) is the raising operation (2.30) while (3.305) is (2.31). From (3.305), (3.343), and (3.344) we can evaluate the coordinate- and momentum-space wavefunctions (see Section 2.6).

(ii) Consider the superpotential

$$W = A - Be^{-\alpha x} \tag{3.346}$$

where A, B, and α are constants. Then

$$V_{\pm}(x, a_0) = A^2 - 2ABe^{-\alpha x} + B^2 e^{-2\alpha x} \pm \frac{\hbar \alpha B}{(2M)^{1/2}} e^{-\alpha x}. \tag{3.347}$$

If we identify

$$a_s = A - \frac{\hbar\alpha}{(2M)^{1/2}} s \qquad (s = 0, 1, \ldots), \tag{3.348}$$

then $V_{\pm}(x, a_s)$ are given by (3.347) with A replaced by a_s. It is easy to show that these potentials satisfy (3.334) with

$$\begin{aligned} R(a_s) &= 2A\frac{\hbar\alpha}{(2M)^{1/2}} - \frac{\hbar^2\alpha^2}{2M}(2s-1) \\ &= a_{s-1}^2 - a_s^2. \end{aligned}$$

Hence (3.337) and (3.306) yield the eigenvalues

$$\begin{aligned} E_N^- &= a_0^2 - a_N^2 \\ &= A^2 - \left(A - \frac{\hbar\alpha}{(2M)^{1/2}} N\right)^2. \end{aligned} \tag{3.349}$$

For the particular values $B^2 = D$, $A = B - \hbar\alpha/2(2M)^{1/2}$, and $\alpha = a^{-1}$, $V_-(x, a_0)$ is equal to the Morse potential (3.99) plus A^2 and E_N^- is equal to the eigenvalue (3.112) plus A^2. The normalized solution (3.308) is (3.150) with c_0 given by (3.153). The wavefunctions obtained from (3.343), (3.344), (3.349), and (3.150) are consistent with (3.163).

(iii) If

$$W = A \tanh \alpha x + \frac{B}{A}, \tag{3.350}$$

then

$$V_{\pm}(x, a_0) = A^2 + \frac{B^2}{A^2} + 2B \tanh \alpha x - A\left(A \mp \frac{\hbar\alpha}{(2M)^{1/2}}\right)\operatorname{sech}^2 \alpha x. \tag{3.351}$$

With a_s given by (3.348), $V_{\pm}(x, a_s)$ satisfy (3.334) with

$$R(a_s) = a_{s-1}^2 - a_s^2 + B^2(a_{s-1}^{-2} - a_s^{-2}).$$

Hence (3.337) and (3.306) yield the eigenvalues

$$\begin{aligned} E_N^- &= a_0^2 - a_N^2 + B^2(a_0^{-2} - a_N^{-2}) \\ &= A^2 + B^2 A^{-2} - \left(A - \frac{\hbar\alpha}{(2M)^{1/2}} N\right)^2 - B^2\left(A - \frac{\hbar\alpha}{(2M)^{1/2}} N\right)^{-2}. \end{aligned} \tag{3.352}$$

For the particular values $B = \frac{1}{2}U$, $\alpha = a^{-1}$, and

$$A\left(A + \frac{\hbar\alpha}{(2M)^{1/2}}\right) = U_0, \tag{3.353}$$

$V_{-}(x, a_0)$ in (3.351) is the Rosen–Morse potential (3.164) plus $A^2 + B^2A^{-2}$. The solution to (3.353) is

$$A = \frac{\hbar k}{(2M)^{1/2}a},$$

where k is given by (3.173). Taking the positive solution in (3.173) we find that (3.352) is (3.174) plus $A^2 + B^2A^{-2}$.

The normalized solution (3.308) is (3.182) with c given by (3.184). The wavefunctions obtained from (3.343), (3.344), (3.352), (3.182) are consistent with those obtained in Section 3.7.4. (Note that the number of bound states of the Morse and Rosen–Morse potentials is finite: see Sections 3.4 and 3.5.)

The reader may find it instructive to perform the above calculations with

$$W = A\tanh\alpha x + B\,\mathrm{sech}\,\alpha x$$

$$W = A\cot\alpha x - B\,\mathrm{cosec}\,\alpha x \qquad (A > B)$$

$$W = A\tan\alpha x - B\cot\alpha x \qquad (A, B > 0)$$

$$W = A\tanh\alpha x - B\coth\alpha x \qquad (A > B > 0).$$

These superpotentials are derived in Ref. 28, and solutions are given in Refs. 28 and 29.

There are several extensions that can be made to the methods discussed in this chapter:

(i) We have considered shift operators which change the value of a parameter or a quantum number by unity. This is sufficient for bound state problems. It is not difficult to remove the restriction to integer shifts [30], and this enables one to apply the operator method, including the use of shape invariance, to scattering problems [31].

(ii) The use of SUSYQM and shape invariance can be extended to n-dimensional systems [32].

(iii) In general, the supersymmetric partner H_+ of a given Hamiltonian H_- is not unique: there exists a one-parameter family of Hamiltonians $H_+(\gamma)$ and this allows the construction of generalized potentials [2, 6]. Simple examples are the generalized harmonic potentials (Section 3.1) and generalized Morse potentials [33].

(iv) The factorization method can be extended to treat non-factorizable problems within the framework of perturbation theory, by a procedure that is well adapted for computer algebra [34].

References

1. Infeld, L. and Hull, T.E. (1951). The factorization method. *Reviews of Modern Physics*, **23**, 21–68. The history of the factorization method in the theory of differential equations has been surveyed by Stahlhofen, A. (1988). The Ricatti equation as common basis for supersymmetric quantum mechanics and the factorization method. *Duke University Preprint*. See also Stahlhofen, A. and Bleuler, K. (1989). An algebraic form of the factorization method. *Il Nuovo Cimento*, **104** B, 447–65.
2. Mielnik, Bogden (1984). Factorization method and new potentials with the oscillator spectrum. *Journal of Mathematical Physics*, **25**, 3387–9. The derivation of generalized potentials via the factorization method is related to the Darboux transformation: see Pöschel, J. and Trubowitz, E. (1987). *Inverse spectral theory*, Ch. 5. Academic Press, Boston.
3. Landau, L.D. and Lifschitz, E.M. (1960). *Mechanics*, p. 25. Pergamon, Oxford.
4. Nieto, M. M. and Gutschick, V.P. (1981). Inequivalence of the classes of classical and quantum harmonic potentials: Proof by example. *Physical Review*, D **23**, 922–6.
5. Schrödinger, E. (1941). Further studies on solving eigenvalue problems by factorization. *Proceedings of the Royal Irish Academy* (Section A), **46**, 183–206.
6. See Sukumar, C.V. (1985). Supersymmetric quantum mechanics of one-dimensional systems. *Journal of Physics A: Math. Gen.*, **18**, 2917–36. The connection with broken supersymmetry is discussed in Khare, A. and Maharana, J. (1984). Supersymmetric quantum mechanics in one, two and three dimensions. *Nuclear Physics*, **B 244**, 409–20, and Fubini, S. and Rabinovici, E. (1984). Superconformal quantum mechanics. *Nuclear Physics*, **B 245**, 17–44.
7. The factorization method is related to the representation theory of certain Lie groups. This has been discussed in detail by Miller, W. Jr (1968). *Lie theory and special functions*. Academic Press, New York.
8. Morse, P.M. (1929). Diatomic molecules according to wave mechanics. II. Vibrational levels. *Physical Review*, **34**, 57–64.
9. King, G.W. (1964). *Spectroscopy and molecular structure*. Holt, New York.
10. Morse, P.M., Fisk, J.B. and Schiff, L.I. (1936). Collision of neutron and proton. *Physical Review*, **50**, 748–54.
11. Huffaker, J.N. and Dwivedi, P.H. (1975). Factorization—method treatment of the perturbed Morse oscillator. *Journal of Mathematical Physics*, **16**, 862–7.
12. Huffaker, J.N. (1976). Diatomic molecules as perturbed Morse oscillators. *Journal of Chemical Physics*, **64**, 3175–84.
13. Rosen, N. and Morse, P.M. (1932). On the vibrations of polyatomic molecules. *Physical Review*, **42**, 210–17.
14. Englefield, M.J. (1987). Algebra representations on eigenfunctions of the Rosen–Morse potential. *Journal of Mathematical Physics*, **28**, 827–32.
15. Pöschl, G. and Teller, E. (1933). Bemerkungen zur Quantenmechanik des anharmonischen Oszillators. *Zeitschrift für Physik*, **83**, 143–51. For more recent work on the Pöschl–Teller potentials, see Grosche, C. (1989). Path integral solution of a class of potentials related to the Pöschl–Teller potential. *Journal of Physics A: Math. Gen.*, **22** 5073–87 and references therein.
16. Weyl, H. (1909). *Königlichen Gesellschaft der Wissenschaften zu Göttingen, Nachrichten*, 37.
17. Finiteness of the average potential is related to the Hermitian property of the

Hamiltonian and other results. See Prugovečki, E. (1981). *Quantum mechanics in Hilbert space*, 2nd edn, pp. 354–68. Academic Press, New York.

18. Barut, A.O., Inomata, I., and Wilson, R. (1987). Algebraic treatment of second Pöschl–Teller, Morse–Rosen, and Eckart equations. *Journal of Physics A: Math. Gen.*, **30**, 4083–96.
19. See, for example, West, P. (1986). *Introduction to supersymmetry and supergravity*. World Scientific, Singapore.
20. Witten, E. (1981). Dynamical breaking of supersymmetry. *Nuclear Physics*, B **188**, 513–54.
21. Blockley, C.A. and Stedman, G.E. (1985). Simple supersymmetry: I. Basic examples. *European Journal of Physics*, **6**, 218–24.
22. Haymaker, R.W. and Rau, A.R.P. (1986). Supersymmetry in quantum mechanics. *American Journal of Physics*, **54**, 928–36.
23. Itzykson, C. and Zuber, J. (1980). *Quantum field theory*. McGraw-Hill, New York.
24. See References in Sukumar, C.V. (1986). Supersymmetry, potentials with bound states at arbitrary energies and multi-soliton configurations. *Journal of Physics A: Math. Gen.*, **19**, 2297–316.
25. See, for example, Dragon, N., Ellwanger, U., and Schmidt, M.G. (1987). Supersymmetry and supergravity. In *Progress in particle and nuclear physics* (ed. A. Faessler), Vol. 18, pp. 1–91. Pergamon, Oxford.
26. The concept of shape invariance was introduced by Gendenshtein, L. E. (1983). Derivation of exact spectra of the Schrödinger equation by means of supersymmetry. *Journal of Experimental and Theoretical Physics Letters*, **38**, 356–9.
27. Stahlhofen, A. (1989). Remarks on the equivalence between the shape-invariance condition and the factorization condition. *Journal of Physics A: Math. Gen.*, **22**, 1053–8.
28. Cooper, F., Ginocchio, J.N., and Khare, A. (1987). Relationship between supersymmetry and solvable potentials. *Physical Review*, D **36**, 2458–73.
29. Dutt, R., Khare, A., and Sukhatme, U.P. (1988). Supersymmetry, shape invariance, and exactly solvable potentials. *American Journal of Physics*, **56**, 163–8.
30. Humi, M. (1968). Extension of the factorization method. *Journal of Mathematical Physics*, **9**, 1258–65.
31. Khare, A. and Sukhatme, U.P. (1988). Scattering amplitudes for supersymmetric shape-invariant potentials by operator methods. *Journal of Physics A: Math. Gen.*, **21**, L501–8.
32. See Ref. 6 and Stedman, G.E. (1985). Simple supersymmetry: II. Factorization method in quantum mechanics. *European Journal of Physics*, **6**, 225–31.
33. Drigo Filho, E. (1988). The Morse oscillator generalized from supersymmetry. *Journal of Physics A: Math. Gen.*, **21**, L1025–8.
34. Bessis, N. and Bessis, G. (1990). Perturbed ladder operator method: An algebraic recursive solution of perturbed wave equations. *Physical Review*, A **42**, 1096–122.

4
ANGULAR MOMENTUM

The main purpose of this chapter is to present shift operators for the quantum numbers of angular momentum. These operators have useful applications, some of which are discussed in this and subsequent chapters. We also give a brief discussion of the non-relativistic treatment of spin angular momentum.

4.1 Introduction

An angular momentum operator **J** is a vector whose Cartesian components J_i ($i = 1, 2, 3$ or x, y, z) are linear Hermitian operators that satisfy the commutation relations

$$[J_i, J_j] = i\hbar\varepsilon_{ijk}J_k, \tag{4.1}$$

where ε_{ijk} is defined by (1.155).

Two types of angular momentum operator are given special symbols: the orbital angular momentum operator **L** and the spin angular momentum operator **S**. The former operator is obtained from the definition of classical angular momentum by replacing the classical position and momentum vectors by linear Hermitian operators that satisfy the canonical commutation relations (1.65)–(1.67). Thus for a particle

$$\mathbf{L} = \mathbf{r} \times \mathbf{p}, \tag{4.2}$$

that is

$$L_i = \varepsilon_{ijk}r_jp_k. \tag{4.3}$$

From this definition **L** is linear and

$$[L_i, L_j] = i\hbar\varepsilon_{ijk}L_k \tag{4.4}$$

(see Appendix C). Also

$$L_i^\dagger = \varepsilon_{ijk}p_k^\dagger r_j^\dagger = \varepsilon_{ijk}r_jp_k = L_i. \tag{4.5}$$

Thus **L** is an angular momentum operator. For the spin angular momentum operator **S** one postulates that S_i satisfy the defining commutation relations (4.1) [1, 2].

For a system of N particles the total orbital angular momentum operator

$$\mathbf{L} = \sum_{\mu=1}^{N} \mathbf{r}^{(\mu)} \times \mathbf{p}^{(\mu)} \tag{4.6}$$

(μ labels the particle) also satisfies (4.4) because the position and momentum operators of different particles commute.

It follows from (4.1) that

$$[\mathbf{J}^2, J_i] = 0, \tag{4.7}$$

where

$$\mathbf{J}^2 = J_x^2 + J_y^2 + J_z^2. \tag{4.8}$$

Therefore $\mathbf{J}^2$ and one of J_i (traditionally taken to be J_z) can have simultaneous eigenkets:*

$$J_z|km\rangle = \hbar m|km\rangle, \tag{4.9}$$

$$\mathbf{J}^2|km\rangle = \hbar^2 k^2|km\rangle. \tag{4.10}$$

From (4.10) and (4.8) we have

$$\hbar^2 k^2\langle km|km\rangle = \langle km|J_x^2 + J_y^2 + J_z^2|km\rangle = \langle km|J_i^\dagger J_i|km\rangle \geq 0.$$

Hence $k^2 \geq 0$. The real numbers m and k^2 are determined below.

4.2 Shift operators for *m*

The operators

$$J_\pm = J_x \pm iJ_y \tag{4.11}$$

satisfy

$$[J_z, J_\pm] = \pm\hbar J_\pm, \qquad [\mathbf{J}^2, J_\pm] = 0. \tag{4.12}$$

Thus

$$J_z\{J_\pm|km\rangle\} = \hbar(m \pm 1)\{J_\pm|km\rangle\} \tag{4.13}$$

and

$$\mathbf{J}^2\{J_\pm|km\rangle\} = \hbar^2 k^2\{J_\pm|km\rangle\}. \tag{4.14}$$

It follows that $J_\pm$ are shift operators for m. We write

$$J_\pm|km\rangle = \alpha_\pm(k, m)|k, m \pm 1\rangle, \tag{4.15}$$

where the coefficients $\alpha_\pm(k, m)$ are to be determined. In (4.15) we have assumed that any additional operators to $\mathbf{J}^2$ and J_z in the c.s.c.o. commute with $J_\pm$ (see, for example, Section 5.1).

4.3 Eigenvalues

Taking the norm of both sides of (4.15), we obtain

$$\begin{aligned}|\alpha_\pm(k, m)|^2\langle k, m \pm 1|k, m \pm 1\rangle &= \langle km|J_\mp J_\pm|km\rangle \\ &= \hbar^2[k^2 - m(m \pm 1)]\langle km|km\rangle, \end{aligned} \tag{4.16}$$

* Eigenkets of a c.s.c.o. that includes additional operators to $\mathbf{J}^2$ and J_z should be denoted $|\gamma_1\gamma_2\ldots km\rangle$; it is convenient to suppose that the additional quantum numbers are implicit in $|km\rangle$.

where we have used $J_\pm^\dagger = J_\mp$ and

$$J_\mp J_\pm = \mathbf{J}^2 - J_z(J_z \pm \hbar). \tag{4.17}$$

Non-negativity of the norm in (4.16) requires that a sequence of raising operations must terminate at a maximum value m' of m that satisfies $(m')^2 + m' - k^2 = 0$; similarly, a sequence of lowering operations must terminate at a minimum value m'' that satisfies $(m'')^2 - m'' - k^2 = 0$. The solutions for which $m' > m''$ are

$$m' = -m'' = (k^2 + \tfrac{1}{4})^{1/2} - \tfrac{1}{2}.$$

But $m' - m'' = N\,(= 0, 1, 2, \ldots)$, the number of times that J_+ must be applied to transform $|km''\rangle$ into $|km'\rangle$. Thus

$$k^2 = j(j+1), \tag{4.18}$$

$$m' = j, \tag{4.19}$$

where

$$j = \tfrac{1}{2}N = 0, \tfrac{1}{2}, 1, \tfrac{3}{2}, \ldots. \tag{4.20}$$

The possible values of the quantum number m are

$$m = -j, -j+1, \ldots, j. \tag{4.21}$$

It is customary to use the quantum number j instead of k to label the eigenkets in (4.9) and (4.10). Then

$$J_z|jm\rangle = \hbar m|jm\rangle, \tag{4.22}$$

$$\mathbf{J}^2|jm\rangle = \hbar^2 j(j+1)|jm\rangle. \tag{4.23}$$

If the kets in (4.15) and (4.16) are normalized,

$$J_\pm|jm\rangle = \hbar \exp(i\theta_\pm)[j(j+1) - m(m \pm 1)]^{1/2}|j, m \pm 1\rangle, \tag{4.24}$$

where $\theta_\pm = \theta_\pm(j, m)$ are real. The identity (4.17) requires

$$\theta_-(j, m) = -\theta_+(j, m-1), \tag{4.25}$$

otherwise the phases are arbitrary. We adopt the conventional choice, $\theta_\pm = 0$.

The kets $|jm\rangle$ form a basis of the state space [3]. The orthonormality and closure relations are

$$\langle j'm'|jm\rangle = \delta_{j'j}\delta_{m'm} \tag{4.26}$$

and

$$\sum_j \sum_{m=-j}^{j} |jm\rangle\langle jm| = 1. \tag{4.27}$$

4.4 Orbital angular momentum

In the case of orbital angular momentum (4.22)–(4.24) are written

$$L_z|lm\rangle = \hbar m|lm\rangle, \tag{4.28}$$

$$\mathbf{L}^2|lm\rangle = \hbar^2 l(l+1)|lm\rangle, \tag{4.29}$$

and*

$$L_\pm|lm\rangle = \hbar[l(l+1) - m(m \pm 1)]^{1/2}|l, m \pm 1\rangle. \tag{4.30}$$

Here $m = -l, -l+1, \ldots, l$ and $l = 0, \frac{1}{2}, 1, \frac{3}{2}, \ldots$. It is shown below that the half-integral values of l are inadmissible (see Section 4.6).

The operators $L_\pm$ cannot generate all kets $|lm\rangle$ from a given ket; to do so requires also shift operators that change l, and we now give a derivation of such operators. In this derivation the scalar and vector properties of operators play an important role. By definition, K is a scalar operator with respect to $\mathbf{L}$ if

$$[L_i, K] = 0, \tag{4.31}$$

and $\mathbf{v}$ is a vector operator with respect to $\mathbf{L}$ if

$$[L_i, v_j] = i\hbar\varepsilon_{ijk}v_k. \tag{4.32}$$

From (4.32) it follows that (see Appendix C)

$$[\mathbf{L}^2, \mathbf{v}] = 2\hbar^2\mathbf{v} + 2i\hbar\mathbf{v} \times \mathbf{L}, \tag{4.33}$$

$$[L_z, v_\pm] = \pm\hbar v_\pm, \tag{4.34}$$

and

$$[\mathbf{L}^2, v_\pm] = 2\hbar^2 v_\pm + 2\hbar(v_\pm L_z - v_z L_\pm), \tag{4.35}$$

where $v_\pm = v_x \pm i v_y$. Let both sides of (4.34) and (4.35) operate on $|l, \pm l\rangle$. Then

$$L_z\{v_\pm|l, \pm l\rangle\} = \pm\hbar(l+1)\{v_\pm|l, \pm l\rangle\} \tag{4.36}$$

and

$$\mathbf{L}^2\{v_\pm|l, \pm l\rangle\} = \hbar^2(l+1)(l+2)\{v_\pm|l, \pm l\rangle\}. \tag{4.37}$$

If $\mathbf{L}^2$ and L_z are a c.s.c.o., it follows from (4.36) and (4.37) that

$$v_\pm|l, \pm l\rangle = \beta_\pm|l+1, \pm l \pm 1\rangle, \tag{4.38}$$

where $\beta_\pm$ are scalars. With an appropriate choice of $\mathbf{v}$, $\beta_\pm$ are functions only of l (see (4.63)).

* Note that if the c.s.c.o. does not include $\mathbf{L}^2$ then $L_\pm$ are not necessarily shift operators (see Section 10.1).

To extend the above results and obtain shift operators for l and m in the kets $|lm\rangle$, we need an additional vector operator to $\mathbf{v}$. Now $\mathbf{v} \times \mathbf{L}$ is a vector operator because (see Appendix C)

$$[L_i, (\mathbf{v} \times \mathbf{L})_j] = i\hbar\varepsilon_{ijk}(\mathbf{v} \times \mathbf{L})_k. \tag{4.39}$$

Replacing $\mathbf{v}$ with $\mathbf{v} \times \mathbf{L}$ in (4.33) gives

$$[\mathbf{L}^2, \mathbf{v} \times \mathbf{L}] = 2\hbar^2\mathbf{v} \times \mathbf{L} + 2i\hbar(\mathbf{v} \times \mathbf{L}) \times \mathbf{L}. \tag{4.40}$$

If

$$\mathbf{v}.\mathbf{L} = 0, \tag{4.41}$$

then

$$(\mathbf{v} \times \mathbf{L}) \times \mathbf{L} = -\mathbf{v}\mathbf{L}^2 + i\hbar(\mathbf{v} \times \mathbf{L}) \tag{4.42}$$

(see Appendix C) and (4.40) becomes

$$[\mathbf{L}^2, \mathbf{v} \times \mathbf{L}] = -2i\hbar\mathbf{v}\mathbf{L}^2. \tag{4.43}$$

Now consider the operators

$$\mathbf{U}^{\pm} = \pm i\mathbf{v} \times \mathbf{L} \pm \mathbf{v}K, \tag{4.44}$$

where $\mathbf{v}$ and K are to be determined. (We have placed K to the right of $\mathbf{v}$ to allow it to operate directly on eigenkets $|lm\rangle$. This simplifies the calculations.) We suppose that K is a scalar operator. Then $\mathbf{U}^{\pm}$ are vector operators and

$$[L_i, U_j^{\pm}] = i\hbar\varepsilon_{ijk}U_k^{\pm}. \tag{4.45}$$

If $\mathbf{v}$ satisfies (4.41), eqns (4.43) and (4.33) show that

$$[\mathbf{L}^2, \mathbf{U}^{\pm}] = 2\hbar\mathbf{U}^{\pm}K \mp 2\hbar\mathbf{v}\Omega, \tag{4.46}$$

where

$$\Omega = K^2 - \hbar K - \mathbf{L}^2. \tag{4.47}$$

If $\mathbf{U}^{\pm}$ are to be shift operators, we must have $\Omega = 0$. The solutions to $\Omega = 0$ are $K = \hbar(S + \frac{1}{2})$ and $K = -\hbar(S - \frac{1}{2})$ where*

$$S = (\hbar^{-2}\mathbf{L}^2 + \tfrac{1}{4})^{1/2}. \tag{4.48}$$

Define

$$K^{\pm} = \hbar(S \pm \tfrac{1}{2}) \tag{4.49}$$

and in (4.44) make the choices $K = \pm K^{\pm}$. Then

$$\mathbf{U}^{\pm} = \pm i\mathbf{v} \times \mathbf{L} + \mathbf{v}K^{\pm} \tag{4.50}$$

* The distinction between the operator S, defined by (4.48), and the eigenvalues of the spin angular momentum operator $\mathbf{S}$ should be clear from the context in which they are used.

and (4.46) becomes

$$[\mathbf{L}^2, \mathbf{U}^\pm] = \pm 2\hbar \mathbf{U}^\pm K^\pm. \tag{4.51}$$

The kets $|lm\rangle$ are eigenkets of the Hermitian operators S and $K^\pm$ (see (1.45) and (4.29)):

$$S|lm\rangle = (l + \tfrac{1}{2})|lm\rangle \tag{4.52}$$

and

$$K^\pm|lm\rangle = \hbar(l + \tfrac{1}{2} \pm \tfrac{1}{2})|lm\rangle. \tag{4.53}$$

Note that $K^\pm$ are scalar operators which satisfy the operator equations

$$(K^\pm)^2 \mp \hbar K^\pm - \mathbf{L}^2 = 0. \tag{4.54}$$

From (4.45), (4.51), and (4.53) it follows that

$$L_z\{U_\mu^\pm|lm\rangle\} = \hbar(m + \mu)\{U_\mu^\pm|lm\rangle\} \tag{4.55}$$

and

$$\mathbf{L}^2\{U_\mu^\pm|lm\rangle\} = \hbar^2(l + \tfrac{1}{2} \pm \tfrac{1}{2})(l + \tfrac{1}{2} \pm \tfrac{3}{2})\{U_\mu^\pm|lm\rangle\}, \tag{4.56}$$

where $\mu = \pm 1$ or 0, and

$$U_{+1}^\pm = U_x^\pm + iU_y^\pm, \qquad U_{-1}^\pm = U_x^\pm - iU_y^\pm, \qquad U_0^\pm = U_z^\pm. \tag{4.57}$$

If $\mathbf{L}^2$ and L_z are a c.s.c.o., it follows from (4.55) and (4.56) that*

$$U_\mu^\pm|lm\rangle = \beta_\mu^\pm|l \pm 1, m + \mu\rangle, \tag{4.58}$$

where $\beta_\mu^\pm$ are scalars. With an appropriate choice of $\mathbf{v}$, $\beta_\mu^\pm$ are functions only of l and m (see (4.70)–(4.72)). In (4.58) the superscript $\pm$ indicates the shift in l, whereas the subscript μ denotes the shift in m.

It is useful to note that there are four identities relating the six operators $U_\mu^\pm$ (see Appendix D):

$$U_{+1}^\pm(L_z + \hbar \mp K^\pm) = U_0^\pm L_+ \tag{4.59}$$

and

$$U_{-1}^\pm(L_z - \hbar \pm K^\pm) = U_0^\pm L_-. \tag{4.60}$$

4.5 Shift operators for *l*

Using $\mathbf{r}$ and $\mathbf{p}$, one can readily construct operators that are scalars and vectors with respect to $\mathbf{L}$. For example, $\mathbf{r}.\mathbf{r}$, $\mathbf{r}.\mathbf{p}$, and $\mathbf{p}.\mathbf{p}$ and any differentiable function of these are scalar operators. Examples of vector operators are $\mathbf{r}$, $\mathbf{p}$,

* We emphasize that if $\mathbf{L}^2$ and L_z are not a c.s.c.o., $U_\mu^\pm$ will be shift operators only if $U_\mu^\pm|\gamma_1\gamma_2\ldots lm\rangle$ are eigenkets of the additional operators in the c.s.c.o. (see Section 4.5).

and any (space) vector constructed from **r** and **p**, such as $\mathbf{r} \times \mathbf{L}$ and $\mathbf{p} \times \mathbf{L}$. (See Appendix C.)

We can now give explicit expressions for the shift operators discussed above. The simplest vector operators are **r** and **p**. We take $\mathbf{v} = \hat{\mathbf{r}}\ (= r^{-1}\mathbf{r})$ and $\mathbf{v} = \hat{\mathbf{p}}\ (= p^{-1}\mathbf{p})$, where scalars r^{-1} and p^{-1} are included to make the coefficients $\beta_\pm$ in (4.38) and $\beta_\mu^\pm$ in (4.58) independent of r and p (see (4.63) and (4.70)–(4.72)).

With $\mathbf{v} = \hat{\mathbf{r}}$, (4.38) becomes

$$\hat{r}_\pm |l, \pm l\rangle = \beta_\pm |l+1, \pm l \pm 1\rangle, \tag{4.61}$$

where $\hat{r}_\pm = \hat{x} \pm \mathrm{i}\hat{y}$, and $\hat{x} = r^{-1}x$, etc. If the kets in (4.61) are normalized,

$$|\beta_\pm|^2 = \langle l, \pm l|(\hat{r}_\pm)^\dagger \hat{r}_\pm |l, \pm l\rangle = \langle l, \pm l|1 - \hat{z}^2|l, \pm l\rangle. \tag{4.62}$$

Using the expectation value of $\hat{z}^2$ calculated in Appendix E we obtain

$$\beta_\pm = \mp\left(\frac{2(l+1)}{2l+3}\right)^{1/2}. \tag{4.63}$$

Here the phase factors have been set equal to ∓ 1 to be consistent with the choice $+1$ for the phase factors in (4.30) and (4.70) (see below). Similarly, taking $\mathbf{v} = \hat{\mathbf{p}}$ gives

$$\hat{p}_\pm |l, \pm l\rangle = \beta_\pm |l+1, \pm l \pm 1\rangle, \tag{4.64}$$

where $\beta_\pm$ is given by (4.63) (see Appendix E).

Because

$$\hat{\mathbf{r}}\,.\,\mathbf{L} = 0, \tag{4.65}$$

(see Appendix C), $\mathbf{v} = \hat{\mathbf{r}}$ is an acceptable choice in the operators (4.50). Then

$$\mathbf{U}^\pm = \pm\mathrm{i}\hat{\mathbf{r}} \times \mathbf{L} + \hat{\mathbf{r}}K^\pm, \tag{4.66}$$

where $K^\pm$ are given by (4.49).

To calculate the six coefficients $\beta_\mu^\pm$ in the shift operation (4.58), it is sufficient to calculate $\beta_0^\pm$. The coefficients $\beta_{+1}^\pm$ and $\beta_{-1}^\pm$ can be obtained from these by using the identities (4.59) and (4.60). If the kets in (4.58) are normalized,

$$|\beta_0^\pm|^2 = \langle lm|(U_0^\pm)^\dagger U_0^\pm|lm\rangle. \tag{4.67}$$

The adjoint of (4.66) is

$$(\mathbf{U}^\pm)^\dagger = \mp\mathrm{i}\hat{\mathbf{r}} \times \mathbf{L} + (K^\pm \mp 2\hbar)\hat{\mathbf{r}}. \tag{4.68}$$

From (4.66)–(4.68) and using (4.32), (4.54), and the Hermitian property of $K^\pm$ we have

$$|\beta_0^\pm|^2 = \langle lm|(\hat{\mathbf{r}} \times \mathbf{L})_z^2 - 2\mathrm{i}\hbar\hat{z}(\hat{\mathbf{r}} \times \mathbf{L})_z + \hat{z}^2(\mathbf{L}^2 \mp 2\hbar K^\pm) \pm \hbar K^\pm|lm\rangle. \tag{4.69}$$

The three expectation values in (4.69) are calculated in Appendix E. Hence we find

$$\beta_0^\pm = \hbar\left(\frac{(2l+1)(2l+1\mp 2)}{(2l+3)(2l-1)}(l+m+\tfrac{1}{2}\pm\tfrac{1}{2})(l-m+\tfrac{1}{2}\pm\tfrac{1}{2})\right)^{1/2}, \tag{4.70}$$

$$\beta_{+1}^\pm = \mp\hbar\left(\frac{(2l+1)(2l+1\mp 2)}{(2l+3)(2l-1)}(l\pm m+\tfrac{1}{2}\pm\tfrac{1}{2})(l\pm m+\tfrac{1}{2}\pm\tfrac{3}{2})\right)^{1/2}, \tag{4.71}$$

$$\beta_{-1}^\pm = \pm\hbar\left(\frac{(2l+1)(2l+1\mp 2)}{(2l+3)(2l-1)}(l\mp m+\tfrac{1}{2}\pm\tfrac{1}{2})(l\mp m+\tfrac{1}{2}\pm\tfrac{3}{2})\right)^{1/2}. \tag{4.72}$$

In (4.70) the phase factors have been set equal to one; the phase factors in (4.71) and (4.72) are given in terms of these and the phase factor in (4.30) by the identities (4.59) and (4.60). An alternative derivation of shift operators for l is given in Section 4.13.

We can readily verify that (4.58), with $\mathbf{U}^\pm$ given by (4.66), contains (4.61) as a special case. For example, using $U_{+1}^-|ll\rangle = 0$ we can eliminate $(\hat{\mathbf{r}}\times\mathbf{L})_+$ from $U_{+1}^+|ll\rangle = \beta_{+1}^+|l+1, l+1\rangle$. Hence we obtain the raising operation in (4.61) with β_+ given by (4.63).

For the purpose of operating on a ket $|lm\rangle$, the operators $K^\pm$ in $\mathbf{U}^\pm$ can be replaced by their eigenvalues $\hbar(l+\frac{1}{2}\pm\frac{1}{2})$ (see (4.53)). Thus the effect of $\mathbf{U}^\pm$ on $|lm\rangle$ is identical to that of

$$\mathbf{U}_l^\pm = \pm i\hat{\mathbf{r}}\times\mathbf{L} + \hbar(l+\tfrac{1}{2}\pm\tfrac{1}{2})\hat{\mathbf{r}}. \tag{4.73}$$

The latter operators are linear in $\mathbf{p}$ and they are also non-linear functions of $\mathbf{r}$.

The vector operators $\mathbf{U}^\pm$ in (4.66) correspond to the choice $\mathbf{v} = \hat{\mathbf{r}}$ in (4.50). An equally simple choice satisfying (4.41), namely $\mathbf{v} = \hat{\mathbf{p}}$, produces the shift operators

$$\mathbf{V}^\pm = \pm i\hat{\mathbf{p}}\times\mathbf{L} + \hat{\mathbf{p}}K^\pm. \tag{4.74}$$

All the above results involving $\mathbf{U}^\pm$ remain valid if $\mathbf{U}^\pm$ is replaced by $\mathbf{V}^\pm$. In particular

$$V_\mu^\pm|lm\rangle = \beta_\mu^\pm|l\pm 1, m+\mu\rangle \tag{4.75}$$

where $\beta_\mu^\pm$ are given by (4.70)–(4.72). The effect of $\mathbf{V}^\pm$ on $|lm\rangle$ is identical to that of the operators

$$\mathbf{V}_l^\pm = \pm i\hat{\mathbf{p}}\times\mathbf{L} + \hbar(l+\tfrac{1}{2}\pm\tfrac{1}{2})\hat{\mathbf{p}}. \tag{4.76}$$

$\mathbf{V}_l^\pm$ are linear in $\mathbf{r}$, and they are also non-linear functions of $\mathbf{p}$.

At this point it is natural to enquire whether the derivation of the shift operators $U_\mu^\pm$ in (4.58) is more general than that given above; that is, whether there are additional possibilities for $\mathbf{v}$ in (4.50) which provide shift operators for a c.s.c.o. that is larger than $\mathbf{L}^2$ and L_z. We shall see later that there are choices which yield shift operators for the eigenkets (in an angular momentum basis) of a particle in a constant potential (Section 8.1), an isotropic harmonic oscillator (Section 8.1), and the Coulomb problem (Section 9.1). There is also

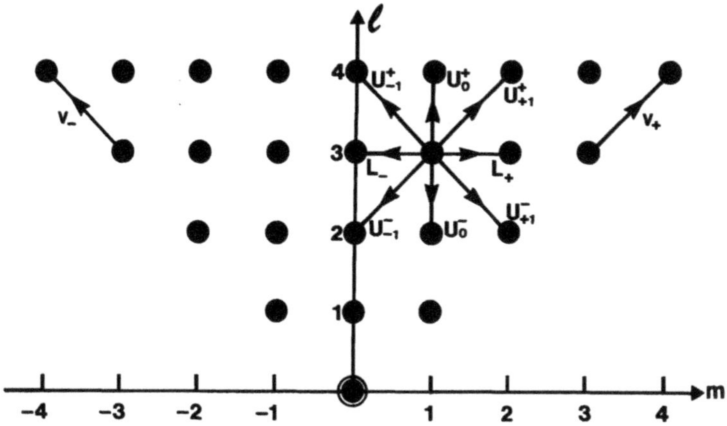

FIG. 4.1 Schematic illustration of the shift operations (4.30), (4.38), and (4.58) for integral values of l. The simplest expressions for $\mathbf{v}$ are $\hat{\mathbf{r}}$ and $\hat{\mathbf{p}}$ (see text).

an alternative choice to (4.49) for $K^{\pm}$ which yields shift operators for a relativistic oscillator (Section 12.9).

The effect on $|lm\rangle$ of the various shift operators discussed above is depicted in Fig. 4.1 for integral values of l. Here all lowering sequences terminate at $|00\rangle$ or $|l, \pm l\rangle$ (see (4.30) and (4.70)–(4.72)). A similar diagram applies if l is half-integral: however, in this case, not all lowering sequences terminate at $|\frac{1}{2}, \pm\frac{1}{2}\rangle$ or $|l, \pm l\rangle$ (see below). The shift operators have a number of applications, and we consider a few of these.

4.6 Restriction of l to integral values*

We prove that the orbital angular momentum operator $\mathbf{L}$ is not Hermitian with respect to a Hilbert space spanned by eigenkets $|lm\rangle$ if l is a half-integer.

Suppose $\mathbf{L}$ is Hermitian. From a single ket one can generate a set of basis kets by repeated application of shift operators that change l and m: the operators $L_{\pm}$, U^+_{+1}, and U^-_{-1}, say. Repeated application of U^-_{-1} to $|ll\rangle$, where l is half-integral, yields the sequence

$$|l-1, l-1\rangle, \ldots, |\tfrac{1}{2}\tfrac{1}{2}\rangle, |-\tfrac{1}{2}, -\tfrac{1}{2}\rangle, \ldots. \tag{4.77}$$

In this sequence

$$|\tfrac{1}{2}\tfrac{1}{2}\rangle \sim U^-_{-1}|\tfrac{3}{2}\tfrac{3}{2}\rangle \neq 0 \tag{4.78}$$

* This topic has attracted considerable attention since the early days of quantum mechanics and a review is given in Ref. 4.

(see (4.58) and (4.72)), and

$$|-\tfrac{1}{2}, -\tfrac{1}{2}\rangle \sim U^{-}_{-1}|\tfrac{1}{2}\tfrac{1}{2}\rangle \neq 0. \tag{4.79}$$

(Expression (4.79) can be proved by contradiction: suppose $U^{-}_{-1}|\tfrac{1}{2}\tfrac{1}{2}\rangle = 0$; that is, from (4.66)

$$[\hbar(\hat{x} - \mathrm{i}\hat{y}) - \hat{z}L_{-}]|\tfrac{1}{2}\tfrac{1}{2}\rangle = 0. \tag{4.80}$$

Multiply (4.80) on the left by $\hat{x} + \mathrm{i}\hat{y}$, and use

$$0 = \hat{\mathbf{r}}.\mathbf{L} = \tfrac{1}{2}(\hat{x} + \mathrm{i}\hat{y})L_{-} + \tfrac{1}{2}(\hat{x} - \mathrm{i}\hat{y})L_{+} + \hat{z}L_{z}$$

and $L_{+}|\tfrac{1}{2}\tfrac{1}{2}\rangle = 0$. Then (4.80) becomes $\hbar|\tfrac{1}{2}\tfrac{1}{2}\rangle = 0$, which contradicts (4.78).) But

$$\mathbf{L}^2|-\tfrac{1}{2}, -\tfrac{1}{2}\rangle = -\tfrac{1}{4}\hbar^2|-\tfrac{1}{2}, -\tfrac{1}{2}\rangle,$$

which contradicts the result proved in Section 4.1, that the eigenvalues of $\mathbf{L}^2$ cannot be negative if $\mathbf{L}$ is Hermitian.

The above result and (4.20) show that for orbital angular momentum the possible values of the quantum number l are*

$$l = 0, 1, 2, \ldots. \tag{4.81}$$

4.7 Matrix elements

Addition of the two equations in (4.66) gives $\hat{\mathbf{r}}(K^{+} + K^{-}) = \mathbf{U}^{+} + \mathbf{U}^{-}$. Operating with this on $|lm\rangle$ yields

$$\hat{r}_{\mu}|lm\rangle = [\hbar(2l+1)]^{-1}(\beta^{+}_{\mu}|l+1, m+\mu\rangle + \beta^{-}_{\mu}|l-1, m+\mu\rangle). \tag{4.82}$$

Using (4.82) one can calculate any matrix element of the form $\langle l'm'|\hat{x}^{n_1}\hat{y}^{n_2}\hat{z}^{n_3}|lm\rangle$, where n_i are non-negative integers. Thus, multiplying (4.82) by $\langle l'm'|$, and using (4.70)–(4.72) and the orthonormality property of $|lm\rangle$, we find

$$\langle l'm'|\hat{z}|lm\rangle = \left(\frac{(l+1)^2 - m^2}{(2l+1)(2l+3)}\right)^{1/2} \delta_{l',l+1}\delta_{m'm} + \left(\frac{l^2 - m^2}{(2l-1)(2l+1)}\right)^{1/2} \delta_{l',l-1}\delta_{m'm} \tag{4.83}$$

and

$$\langle l'm'|\hat{x} \pm \mathrm{i}\hat{y}|lm\rangle = \mp\left(\frac{(l \pm m+1)(l \pm m+2)}{(2l+1)(2l+3)}\right)^{1/2} \delta_{l',l+1}\delta_{m',m\pm1} \pm \left(\frac{(l \mp m)(l \mp m-1)}{(2l+1)(2l-1)}\right)^{1/2} \delta_{l',l-1}\delta_{m',m\pm1}. \tag{4.84}$$

* See also Sections 8.2, 9.2, and 9.5.

These matrix elements occur in the theory of electric dipole transitions of an atom described by determinantal wavefunctions [5], and as (4.83) and (4.84) indicate they account for the selection rules $\Delta l = \pm 1$, $\Delta m = 0, \pm 1$.

As a second application, consider the expectation values of a traceless quadrupole moment tensor $Q_{ij} = 3r_i r_j - r^2\delta_{ij}$ for a spherically symmetric potential:

$$\langle \tilde{Q}_{ij} \rangle = \langle r^2 \rangle \langle lm | 3\hat{r}_i\hat{r}_j - \delta_{ij} | lm \rangle. \tag{4.85}$$

Here $\langle r^2 \rangle = \langle El | r^2 | El \rangle$, where $|El\rangle$ is an eigenket of the radial Hamiltonian (Chapter 5). Consider first the diagonal elements in (4.85). Write $\hat{x} = \frac{1}{2}(\hat{r}_{+1} + \hat{r}_{-1})$ and $\hat{y} = -\frac{1}{2}\mathrm{i}(\hat{r}_{+1} - \hat{r}_{-1})$. Then

$$\hat{x}^2 = \tfrac{1}{4}(\hat{r}_{+1}\hat{r}_{+1} + \hat{r}_{-1}\hat{r}_{-1} + 2\hat{r}_{+1}\hat{r}_{-1})$$

and

$$\hat{y}^2 = \tfrac{1}{4}(-\hat{r}_{+1}\hat{r}_{+1} - \hat{r}_{-1}\hat{r}_{-1} + 2\hat{r}_{+1}\hat{r}_{-1}).$$

From (4.82) and (4.26), $\langle \hat{r}_{+1}\hat{r}_{+1} \rangle = \langle \hat{r}_{-1}\hat{r}_{-1} \rangle = 0$. Thus $\langle \hat{x}^2 \rangle = \langle \hat{y}^2 \rangle = \frac{1}{2}\langle \hat{r}_{+1}\hat{r}_{-1} \rangle = \frac{1}{2}\langle 1 - \hat{z}^2 \rangle$, and (4.85) yields

$$\langle Q_{xx} \rangle = \langle Q_{yy} \rangle = -\tfrac{1}{2}\langle Q_{zz} \rangle. \tag{4.86}$$

The non-diagonal elements of $\langle Q_{ij} \rangle$ are zero: for example, $\langle \hat{x}\hat{y} \rangle = -\frac{1}{4}\mathrm{i}\langle \hat{r}_{+1}\hat{r}_{+1} - \hat{r}_{-1}\hat{r}_{-1} \rangle = 0$. Therefore we need calculate only $\langle Q_{zz} \rangle$. From (4.85) and (E4),

$$\langle Q_{zz} \rangle = \langle r^2 \rangle \frac{2l(l+1) - 6m^2}{(2l-1)(2l+3)}. \tag{4.87}$$

4.8 The rigid rotator (top)

The rigid rotator is a useful model which provides an approximate treatment of the rotational energy levels of polyatomic molecules. The classical Hamiltonian of a rigid rotating body is

$$\begin{aligned} H &= \tfrac{1}{2}(I_1\omega_1^2 + I_2\omega_2^2 + I_3\omega_3^2) \\ &= \frac{1}{2}\left(\frac{L_1^2}{I_1} + \frac{L_2^2}{I_2} + \frac{L_3^2}{I_3}\right), \end{aligned} \tag{4.88}$$

where I_i are the principal moments of inertia, and ω_i and L_i are respectively the components of angular velocity and angular momentum relative to the principal axes 1, 2, 3.

The quantum-mechanical Hamiltonian is given by (4.88) with L_i replaced by Hermitian operators $\tilde{L}_i$. In general $\tilde{L}_i$ do not satisfy the same commutation relations as the orbital angular momentum operators referred to 'space-fixed' axes (the x, y, and z axes of an inertial frame). The commutation relations of the latter operators are given by (4.4), whereas it can be shown that [6]

$$[\tilde{L}_i, \tilde{L}_j] = -i\hbar\varepsilon_{ijk}\tilde{L}_k, \tag{4.89}$$

which differs from (4.4) by a sign on the right-hand side.

The consequences of (4.89) can be deduced by the same method that was used in Sections 4.1–4.3. From (4.89)

$$[\tilde{\mathbf{L}}^2, \tilde{L}_i] = 0. \tag{4.90}$$

Therefore $\tilde{\mathbf{L}}^2$ and $\tilde{L}_3$ (say) can have simultaneous eigenkets. Also

$$[\tilde{L}_3, \tilde{L}_\pm] = \mp\hbar\tilde{L}_\pm, \tag{4.91}$$

where

$$\tilde{L}_\pm = \tilde{L}_1 \pm i\tilde{L}_2. \tag{4.92}$$

Thus if the eigenkets of $\tilde{\mathbf{L}}^2$ and $\tilde{L}_3$ are denoted by $|lm\rangle$, we have

$$\tilde{\mathbf{L}}^2|lm\rangle = \hbar^2 l(l+1)|lm\rangle, \tag{4.93}$$

$$\tilde{L}_3|lm\rangle = \hbar m|lm\rangle, \tag{4.94}$$

and

$$\tilde{L}_\pm|lm\rangle = \hbar[l(l+1) - m(m \mp 1)]^{1/2}|l, m \mp 1\rangle, \tag{4.95}$$

where $m = -l, -l+1, \ldots, l$ and* $l = 0, 1, 2, \ldots$.

We can now write down the energy eigenvalues of the Hamiltonian (4.88) for certain types of top. For the spherical top, $I_1 = I_2 = I_3 \equiv I$ and

$$H = \frac{1}{2I}\tilde{\mathbf{L}}^2. \tag{4.96}$$

Hence

$$E = \frac{1}{2I}\hbar^2 l(l+1). \tag{4.97}$$

For a top which is symmetric about the 3 axis, $I_1 = I_2 \neq I_3$ and

$$H = \frac{1}{2I_1}\tilde{\mathbf{L}}^2 + \tfrac{1}{2}\left(\frac{1}{I_3} - \frac{1}{I_1}\right)\tilde{L}_3^2. \tag{4.98}$$

Hence

$$E = \frac{1}{2I_1}\hbar^2 l(l+1) + \tfrac{1}{2}\left(\frac{1}{I_3} - \frac{1}{I_1}\right)m^2. \tag{4.99}$$

4.9 Lie algebras

The angular momentum operators J_i are generators of a Lie group [8]. If $\mathbf{J} = \mathbf{L}$, then $2l + 1$ (the dimension of the irreducible representations of the

---

* It can be shown that half-integral values of l are excluded. In the case of molecules containing identical nuclei there may be additional restrictions on l [7].

group) is necessarily odd because $l = 0, 1, \ldots$; the group is denoted by SO(3), the three-dimensional rotation group. In general, $2j + 1$ can be even or odd because $j = 0, \frac{1}{2}, 1, \ldots$; the group is the covering group of SO(3), denoted by SU(2). (The notation is explained in Section 4.14.) The commutation relations satisfied by the generators of a Lie group (eqn (4.1) in this case) define the Lie algebra of the group. The Lie groups SO(3) and SU(2) are locally isomorphic in that they share a common Lie algebra, so(3) [9].

The theory of Lie algebras and Lie groups has many useful applications in quantum mechanics [9], and in this book we will occasionally discuss the Lie algebra and subalgebras associated with a particular problem. Here we consider an example, namely the Lie algebra associated with orbital angular momentum. The shift operators $L_\pm$ that can be constructed from the generators of SO(3) affect only the quantum number m in $|lm\rangle$. We have also constructucted shift operators for l and it is natural to ask whether these operators, together with L_i, can form a Lie algebra.

Let

$$\mathbf{U} = \hbar^{-1} S^{1/2} \mathbf{U}^- S^{-1/2} \tag{4.100}$$

and

$$\tilde{\mathbf{U}} = \hbar^{-1} S^{1/2} \mathbf{U}^+ S^{-1/2}, \tag{4.101}$$

where $\mathbf{U}^\pm$ and S are given in (4.66) and (4.48). Then (see (4.58), (4.70)–(4.72), and (4.52))

$$U_z|lm\rangle = [(l+m)(l-m)]^{1/2}|l-1, m\rangle, \tag{4.102}$$

$$U_{\pm 1}|lm\rangle = \pm[(l \mp m)(l \mp m - 1)]^{1/2}|l-1, m \pm 1\rangle, \tag{4.103}$$

$$\tilde{U}_z|lm\rangle = [(l+m+1)(l-m+1)]^{1/2}|l+1, m\rangle, \tag{4.104}$$

$$\tilde{U}_{\pm 1}|lm\rangle = \mp[(l \pm m + 1)(l \pm m + 2)]^{1/2}|l+1, m \pm 1\rangle. \tag{4.105}$$

Using (4.102)–(4.105) one can readily show that $\tilde{\mathbf{U}}$ is the adjoint of $\mathbf{U}$. For example,

$$\langle l'm'|\tilde{U}_z|lm\rangle = [(l+m+1)(l-m+1)]^{1/2}\delta_{l',l+1}\delta_{m'm}, \tag{4.106}$$

and

$$\langle lm|U_z|l'm'\rangle = [(l'+m')(l'-m')]^{1/2}\delta_{l'-1,l}\delta_{m'm}; \tag{4.107}$$

the complex conjugate of the right-hand side of (4.107) is equal to (4.106) and hence (see (1.24) and (1.25)) $U_z^\dagger = \tilde{U}_z$.

Define the Hermitian operators

$$\mathbf{X} = \tfrac{1}{2}(\mathbf{U} + \mathbf{U}^\dagger), \qquad \mathbf{Y} = \tfrac{1}{2}\mathrm{i}(\mathbf{U} - \mathbf{U}^\dagger). \tag{4.108}$$

It follows from (4.28), (4.30), (4.52), and (4.102)–(4.105) that

$$[L_i, L_j] = \mathrm{i}\hbar\varepsilon_{ijk}L_k, \qquad [L_i, X_j] = \mathrm{i}\hbar\varepsilon_{ijk}X_k,$$

$$\begin{aligned} &[L_i, Y_j] = \mathrm{i}\hbar\varepsilon_{ijk}Y_k, && [L_i, S] = 0, \\ &[X_i, X_j] = -\mathrm{i}\hbar^{-1}\varepsilon_{ijk}L_k, && [X_i, Y_j] = -\mathrm{i}\delta_{ij}S, \\ &[X_i, S] = -\mathrm{i}Y_i, && [Y_i, Y_j] = -\mathrm{i}\hbar^{-1}\varepsilon_{ijk}L_k, \\ &[Y_i, S] = \mathrm{i}X_i. \end{aligned} \tag{4.109}$$

The 45 commutation relations (4.109) can be expressed in the concise form

$$[L_{ab}, L_{cd}] = \mathrm{i}(g_{ac}L_{bd} + g_{bd}L_{ac} - g_{bc}L_{ad} - g_{ad}L_{bc}), \tag{4.110}$$

where

$$(L_{ab}) = \begin{bmatrix} 0 & \hbar^{-1}L_3 & -\hbar^{-1}L_2 & X_1 & Y_1 \\ & 0 & \hbar^{-1}L_1 & X_2 & Y_2 \\ & & 0 & X_3 & Y_3 \\ & & & 0 & S \\ & & & & 0 \end{bmatrix}, \tag{4.111}$$

$L_{ba} = -L_{ab}$, and $g_{ab} = 0$ if $a \neq b$, $g_{ab} = 1$ if $a = b = 1, 2, 3$, and $g_{ab} = -1$ if $a = b = 4, 5$. We see that the 10 operators L_i, X_i, Y_i ($i = 1, 2, 3$) and S close under commutation; the Lie algebra corresponding to (4.110) is called* so(3, 2).

4.10 Spherical harmonics

In the coordinate representation eqns (4.28)–(4.30) become

$$L_z Y_{lm}(\hat{\mathbf{r}}') = \hbar m Y_{lm}(\hat{\mathbf{r}}'), \tag{4.112}$$

$$\mathbf{L}^2 Y_{lm}(\hat{\mathbf{r}}') = \hbar^2 l(l+1) Y_{lm}(\hat{\mathbf{r}}'), \tag{4.113}$$

and

$$L_\pm Y_{lm}(\hat{\mathbf{r}}') = \hbar[l(l+1) - m(m \pm 1)]^{1/2} Y_{l,m\pm 1}(\hat{\mathbf{r}}'). \tag{4.114}$$

Here

$$Y_{lm}(\hat{\mathbf{r}}') = \langle \hat{\mathbf{r}}' | lm \rangle, \tag{4.115}$$

where $|\hat{\mathbf{r}}'\rangle$ is an eigenket of $\hat{\mathbf{r}}$ with eigenvalue $\hat{\mathbf{r}}'$, and (see Section 1.4)

$$\mathbf{L} = -\mathrm{i}\hbar\mathbf{r}' \times \nabla_{\mathbf{r}'}. \tag{4.116}$$

In the following it is convenient to omit the prime on $\mathbf{r}$. From (4.116)

$$L_\pm = \pm\hbar z\left(\frac{\partial}{\partial x} \pm \mathrm{i}\frac{\partial}{\partial y}\right) \mp \hbar(x \pm \mathrm{i}y)\frac{\partial}{\partial z}. \tag{4.117}$$

---

* so(M, $N - M$) is the Lie algebra associated with the pseudorotation group SO(M, $N - M$), the set of coordinate transformations that leaves $x_1^2 + x_2^2 + \cdots + x_M^2 - x_{M+1}^2 - \cdots - x_N^2$ invariant. SO(3, 2) is the de Sitter group.

The functions $Y_{lm}(\hat{\mathbf{r}})$ are the coefficients in the expansion

$$|lm\rangle = \int \mathrm{d}\Omega |\hat{\mathbf{r}}\rangle\langle\hat{\mathbf{r}}|lm\rangle, \tag{4.118}$$

where the solid angle $\mathrm{d}\Omega$ denotes a surface element on the unit sphere at $\hat{\mathbf{r}}$, and the integration extends over this sphere. If (4.118) is multiplied on both sides by $\langle l'm'|$ it yields the orthonormality condition

$$\int \mathrm{d}\Omega \; Y^*_{l'm'}(\hat{\mathbf{r}})\, Y_{lm}(\hat{\mathbf{r}}) = \delta_{l'l}\delta_{m'm}. \tag{4.119}$$

The normalized solutions to (4.112) and (4.113) are called spherical harmonics. (In the above we are anticipating (4.129) which shows that these solutions are functions only of $\hat{\mathbf{r}}$.)

We use the c number $\hat{r}_+ = \hat{x} + \mathrm{i}\hat{y}$ and the differential operator L_- to calculate these spherical harmonics: this method involves the solution to a simple differential equation, an elementary integration, and some differentiation [10]. We start with

$$L_i Y_{00} = 0 \qquad (i = 1, 2, 3). \tag{4.120}$$

(See (4.112) and (4.114).) From (4.116) and (4.120)

$$\frac{\partial Y_{00}}{\partial x^2} = \frac{\partial Y_{00}}{\partial y^2} = \frac{\partial Y_{00}}{\partial z^2}. \tag{4.121}$$

Therefore $Y_{00} = Y_{00}(r)$ and (4.119) requires

$$Y_{00} = 1/(4\pi)^{1/2}. \tag{4.122}$$

Equations (4.61) and (4.63) show that repeated application of $\hat{r}_+$ to Y_{00} yields

$$Y_{ll} = c_l(\hat{x} + \mathrm{i}\hat{y})^l \tag{4.123}$$

where

$$c_l = \frac{(-1)^l}{2^l l!}\left(\frac{(2l+1)!}{4\pi}\right)^{1/2}. \tag{4.124}$$

From Y_{ll} and by repeated application of L_- we can generate Y_{lm}. First multiply both sides of (4.123) by $\hat{r}_-^l$. Then

$$\hat{r}_-^l\, Y_{ll} = c_l(1 - \hat{z}^2)^l. \tag{4.125}$$

Now L_- given by (4.117) commutes with $\hat{r}_-$. Also

$$\frac{\partial}{\partial x_i} f(\hat{z}) = \frac{1}{r^3}(-x_i z + r^2\delta_{i3})\frac{\mathrm{d}f}{\mathrm{d}\hat{z}},$$

and hence

$$L_- f(\hat{z}) = \hbar \hat{r}_- \frac{\mathrm{d}f}{\mathrm{d}\hat{z}}. \tag{4.126}$$

Therefore, operating on both sides of (4.125) with L_-^N gives

$$\hat{r}_-^l L_-^N Y_{ll} = c_l(\hbar \hat{r}_-)^N \frac{\mathrm{d}^N}{\mathrm{d}\hat{z}^N}(1-\hat{z}^2)^l \tag{4.127}$$

for $N = 0, 1, 2, \ldots$. From (4.114)

$$\begin{aligned} L_-^N Y_{ll} &= \hbar^N[(2l)(1)(2l-1)(2)\ldots(2l-N+1)(N)]^{1/2} Y_{l,l-N} \\ &= \hbar^N \left(\frac{(2l)!N!}{(2l-N)!}\right)^{1/2} Y_{l,l-N}. \end{aligned} \tag{4.128}$$

Equations (4.127) and (4.128) with $N = l - m$ give the spherical harmonics in terms of Cartesian coordinates

$$Y_{lm}(\hat{\mathbf{r}}) = \frac{(-1)^l}{2^l l!}\left(\frac{2l+1}{4\pi}\frac{(l+m)!}{(l-m)!}\right)^{1/2}(\hat{x}-\mathrm{i}\hat{y})^{-m}\frac{\mathrm{d}^{l-m}}{\mathrm{d}\hat{z}^{l-m}}(1-\hat{z}^2)^l. \tag{4.129}$$

There are various definitions of the spherical harmonics that differ according to the phase conventions used. The phase in (4.129) is determined by our choice of $+1$ for the phase factors in (4.30) and (4.70). (This choice requires the phase factors ∓ 1 in (4.63).)

In a similar manner we can use the c number $\hat{r}_- = \hat{x} - \mathrm{i}\hat{y}$ and the operator L_+ to generate Y_{lm} from Y_{00} (see (4.61) and (4.114)). The result is

$$Y_{lm}(\hat{\mathbf{r}}) = \frac{(-1)^{l+m}}{2^l l!}\left(\frac{2l+1}{4\pi}\frac{(l-m)!}{(l+m)!}\right)^{1/2}(\hat{x}+\mathrm{i}\hat{y})^{m}\frac{\mathrm{d}^{l+m}}{\mathrm{d}\hat{z}^{l+m}}(1-\hat{z}^2)^l. \tag{4.130}$$

Inspection of (4.129) and (4.130) yields the important properties

$$Y_{lm}(-\hat{\mathbf{r}}) = (-1)^l Y_{lm}(\hat{\mathbf{r}}) \tag{4.131}$$

and

$$Y_{l,-m}(\hat{\mathbf{r}}) = (-1)^m Y_{lm}^*(\hat{\mathbf{r}}). \tag{4.132}$$

We can readily express the spherical harmonics in terms of spherical polar coordinates. In terms of the angles θ and ϕ shown in Fig. 4.2,

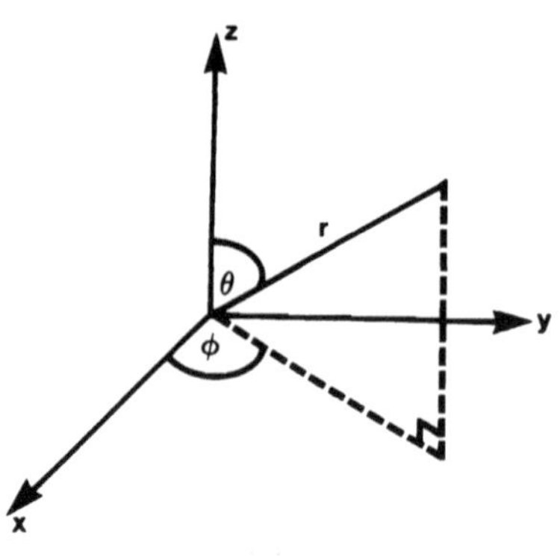

FIG. 4.2 The system of spherical polar coordinates (r, θ, ϕ).

$$\hat{x} \pm i\hat{y} = \exp(\pm i\phi) \sin \theta, \tag{4.133}$$

$$\hat{z} = \cos \theta. \tag{4.134}$$

Thus (4.129) and (4.130) become respectively

$$Y_{lm}(\theta, \phi) = \frac{(-1)^l}{2^l l!}\left(\frac{2l+1}{4\pi}\frac{(l+m)!}{(l-m)!}\right)^{1/2} \exp(im\phi)(\sin \theta)^{-m}\frac{d^{l-m}}{d(\cos \theta)^{l-m}}(\sin \theta)^{2l} \tag{4.135}$$

$$= \frac{(-1)^{l+m}}{2^l l!}\left(\frac{2l+1}{4\pi}\frac{(l-m)!}{(l+m)!}\right)^{1/2} \exp(im\phi)(\sin \theta)^{m}\frac{d^{l+m}}{d(\cos \theta)^{l+m}}(\sin \theta)^{2l}. \tag{4.136}$$

Thus Y_{lm} is equal to a function of ϕ times a function of θ. The latter function is proportional to $(1 - \cos^2\theta)^{|m|/2}$ times a polynomial of order $l - |m|$ in $\cos \theta$; it has $l - |m|$ zeros for different values of θ. Therefore Y_{lm} has $l - |m|$ nodal lines on the surface of the unit sphere.

Equations (4.135) and (4.136) are often written in the form

$$Y_{lm}(\theta, \phi) = (-1)^m\left(\frac{2l+1}{4\pi}\frac{(l-m)!}{(l+m)!}\right)^{1/2} \exp(im\phi)P_l^m(\cos \theta), \tag{4.137}$$

where P_l^m is an associated Legendre function. From (4.135)–(4.137)

$$P_l^m(\cos \theta) = \frac{(-1)^{l+m}}{2^l l!}\frac{(l+m)!}{(l-m)!}(\sin \theta)^{-m}\frac{d^{l-m}}{d(\cos \theta)^{l-m}}(\sin \theta)^{2l} \tag{4.138}$$

$$= \frac{(-1)^l}{2^l l!}(\sin \theta)^m\frac{d^{l+m}}{d(\cos \theta)^{l+m}}(\sin \theta)^{2l}. \tag{4.139}$$

The associated Legendre functions $P_l^m(\cos \theta)$ are a complete set of orthogonal functions (in the index l for each m) on the interval $-1 \leq \cos \theta \leq 1$; the spherical harmonics $Y_{lm}(\theta, \phi)$ are a complete set of orthonormal functions (in the indices l and m) on the surface of the unit sphere ($0 \leq \theta \leq \pi$ and $0 \leq \phi \leq 2\pi$) [11].

According to (4.139) the associated Legendre function is given in terms of derivatives of $P_l^0 = P_l$ by

$$P_l^m(\cos \theta) = (\sin \theta)^m\frac{d^m}{d(\cos \theta)^m}P_l(\cos \theta). \tag{4.140}$$

Here $m \geq 0$ and

$$P_l(\cos \theta) = \frac{(-1)^l}{2^l l!}\frac{d^l}{d(\cos \theta)^l}(1 - \cos^2\theta)^l \tag{4.141}$$

is the Legendre polynomial. These polynomials can be expressed in terms of the hypergeometric function (3.208) by using the binomial theorem to expand $(1 - \cos^2\theta)^l$ and then performing the differentiations in (4.141). The results are

$$P_{2l}(u) = \frac{(2l)!}{2^{2l}(l!)^2}\,{}_2F_1(-l, l+\tfrac{1}{2}; \tfrac{1}{2}; u^2) \tag{4.142}$$

and

$$P_{2l+1}(u) = \frac{(2l+1)!}{2^{2l}(l!)^2}u\,{}_2F_1(-l, l+\tfrac{3}{2}; \tfrac{3}{2}; u^2). \tag{4.143}$$

For the first few spherical harmonics (4.135) and (4.132) give

$$\begin{aligned} Y_{00} &= 1/(4\pi)^{1/2}, \\ Y_{10} &= (3/4\pi)^{1/2}\cos\theta, \\ Y_{1,\pm1} &= \mp(3/8\pi)^{1/2}\sin\theta\exp(\pm \mathrm{i}\phi), \\ Y_{20} &= (5/16\pi)^{1/2}(3\cos^2\theta - 1), \\ Y_{2,\pm1} &= \mp(15/8\pi)^{1/2}\cos\theta\sin\theta\exp(\pm \mathrm{i}\phi), \\ Y_{2,\pm2} &= (15/32\pi)^{1/2}\sin^2\theta\exp(\pm \mathrm{i}2\phi). \end{aligned} \tag{4.144}$$

In the momentum representation similar calculations yield the spherical harmonics

$$Y_{lm}(\hat{\mathbf{p}}) = \langle \hat{\mathbf{p}} | lm \rangle, \tag{4.145}$$

which can be obtained from $Y_{lm}(\hat{\mathbf{r}})$ by replacing $\hat{\mathbf{r}}$ with $\hat{\mathbf{p}}$. Thus in (4.135) and (4.136), θ and ϕ will be the angular coordinates of $\hat{\mathbf{p}}$ instead of $\hat{\mathbf{r}}$.

4.11 Spherical polar coordinates

It is useful to construct the conjugate momentum operators p_θ and p_ϕ of the angles

$$\theta = \cos^{-1}\hat{z} \quad \text{and} \quad \phi = \tan^{-1}(y/x). \tag{4.146}$$

Consider first p_ϕ. Because

$$\begin{aligned} [L_z, \phi] = [xp_y - yp_x, \tan^{-1}(y/x)] &= -\mathrm{i}\hbar x^2/(x^2+y^2) - \mathrm{i}\hbar y^2/(x^2+y^2) \\ &= -\mathrm{i}\hbar, \end{aligned} \tag{4.147}$$

and L_z is Hermitian with respect to $|lm\rangle$, we see that

$$p_\phi = L_z. \tag{4.148}$$

(There is a well-known 'fallacy' associated with (4.147): if the hypervirial theorem is applied to (4.147) it yields $0 = -\mathrm{i}\hbar\langle lm|lm\rangle$. The explanation of this contradiction is that ϕ takes $|lm\rangle$ out of the space on which L_z is Hermitian; for a detailed discussion see Ref. 12.)

To determine p_θ, consider the operator

$$(\hat{\mathbf{r}} \times \mathbf{L})_z = r^{-1}[xzp_x + yzp_y - (x^2+y^2)p_z]. \tag{4.149}$$

Because

$$\frac{\partial}{\partial r_i}\cos^{-1}u = -(1-u^2)^{-1/2}\frac{\partial u}{\partial r_i},$$

we have

$$[(\hat{\mathbf{r}}\times\mathbf{L})_z, \theta] = -i\hbar(1-\hat{z}^2)^{1/2}. \tag{4.150}$$

Also

$$(\hat{\mathbf{r}}\times\mathbf{L})_z^\dagger = (\hat{\mathbf{r}}\times\mathbf{L})_z - 2i\hbar\hat{z}. \tag{4.151}$$

From (4.150) and (4.151) we see that

$$p_\theta = (1-\hat{z}^2)^{-1/2}[(\hat{\mathbf{r}}\times\mathbf{L})_z - \tfrac{1}{2}i\hbar\hat{z}] \tag{4.152}$$

has the properties

$$[p_\theta, \theta] = -i\hbar, \tag{4.153}$$

$$(\sin\theta\, p_\theta)^\dagger = \sin\theta\, p_\theta - i\hbar\cos\theta = p_\theta \sin\theta. \tag{4.154}$$

(In the last step we have expanded $\sin\theta$ and used (4.153).)

From (4.154) and (1.23)

$$\langle lm|(p_\theta^\dagger - p_\theta)\sin\theta|l'm'\rangle = 0. \tag{4.155}$$

If

$$\langle l'm'|(\sin\theta)^{-1}|lm\rangle < \infty, \tag{4.156}$$

then (4.155) and the closure relation (4.27) yield

$$p_\theta^\dagger = p_\theta. \tag{4.157}$$

The condition (4.156) is valid for the kets $|lm\rangle$ (see Section 4.10), and (4.153) and (4.157) show that p_θ is the conjugate momentum operator of θ. For the functions $f(\theta)$ considered below we can generalize (4.153) to (see Section 1.8)

$$[p_\theta, f(\theta)] = -i\hbar\frac{\mathrm{d}f}{\mathrm{d}\theta}. \tag{4.158}$$

By multiplying out terms and using (E.3) and (4.32) we obtain the identity

$$[(\hat{\mathbf{r}}\times\mathbf{L})_z - \tfrac{1}{2}i\hbar\hat{z}]^2 = (1-\hat{z}^2)(\mathbf{L}^2 + \tfrac{1}{4}\hbar^2) - i\hbar\hat{z}[(\hat{\mathbf{r}}\times\mathbf{L})_z - \tfrac{1}{2}i\hbar\hat{z}] - L_z^2 + \tfrac{1}{4}\hbar^2. \tag{4.159}$$

The left-hand side of (4.159) is

$$(\sin\theta\, p_\theta)^2 = \sin^2\theta\, p_\theta^2 - i\hbar\sin\theta\cos\theta\, p_\theta. \tag{4.160}$$

Substituting (4.160) and (4.148) in (4.159) and solving for $\mathbf{L}^2$ yields the useful relation

$$\mathbf{L}^2 = p_\theta^2 + (\sin^{-2}\theta)(p_\phi^2 - \tfrac{1}{4}\hbar^2) - \tfrac{1}{4}\hbar^2. \tag{4.161}$$

Note that in the coordinate representation (4.148) and (4.152) yield the first-order differential operators

$$p_\phi = -i\hbar\frac{\partial}{\partial\phi} \tag{4.162}$$

and

$$p_\theta = -i\hbar\left(\frac{\partial}{\partial\theta} + \tfrac{1}{2}\cot\theta\right). \tag{4.163}$$

4.12 Applications of the hypervirial theorem

Applying the hypervirial theorem to (4.29) we have (see Section 1.10)

$$\langle lm|[\mathbf{L}^2, W]|lm\rangle = 0. \tag{4.164}$$

Here it is assumed that $\mathbf{L}^2$ is Hermitian with respect to $W|lm\rangle$; otherwise the operator W is arbitrary. Certain expectation values can be obtained from (4.164) by choosing various forms for W.

From (4.158) and (4.161) we can show that

$$[g(\theta), \mathbf{L}^2] = 2i\hbar\frac{dg}{d\theta}p_\theta + \hbar^2\frac{d^2g}{d\theta^2} \tag{4.165}$$

and

$$[f(\theta)p_\theta, \mathbf{L}^2] = i\hbar\frac{d^2f}{d\theta^2}p_\theta + 2i\hbar\frac{df}{d\theta}p_\theta^2 + i\hbar(p_\phi^2 - \tfrac{1}{4}\hbar^2)2\cos\theta\sin^{-3}\theta\, f. \tag{4.166}$$

Thus with $W = g(\theta)$ and $dg/d\theta = f(\theta)$, (4.164) becomes

$$\langle f(\theta)p_\theta\rangle = \tfrac{1}{2}i\hbar\langle df/d\theta\rangle. \tag{4.167}$$

Similarly with

$$W = f(\theta)p_\theta + \frac{1}{2}i\hbar\frac{df}{d\theta},$$

we have

$$\left\langle\frac{d^3f}{d\theta^3} + [(2l+1)^2 - (4m^2-1)\sin^{-2}\theta]\frac{df}{d\theta} + (4m^2-1)\cos\theta\sin^{-3}\theta\, f\right\rangle = 0. \tag{4.168}$$

Consider first (4.168). Setting $f = \sin^{N+1}\theta$ and $f = \cos\theta\sin^{N+1}\theta$ gives respectively

$$\langle (N+1)[(2l+1)^2 - (N+1)^2]\cos\theta\sin^N\theta$$
$$+ N(N^2 - 4m^2)\cos\theta\sin^{N-2}\theta\rangle = 0 \tag{4.169}$$

and

$$\langle (N+2)[(N+2)^2-(2l+1)^2]\sin^{N+2}\theta + (N+1)[(2l+1)^2+4m^2 - 2(N+1)^2-3]\sin^N\theta + N(N^2-4m^2)\sin^{N-2}\theta\rangle = 0. \tag{4.170}$$

With $N = 0, 2, 4, \ldots$ these recurrence relations yield $\langle\cos\theta\sin^n\theta\rangle$ and $\langle\sin^n\theta\rangle$ (and hence $\langle\cos^n\theta\rangle$) for $n = 0, 2, 4, \ldots$. For example, taking $N = 0$ we obtain

$$\langle\cos\theta\rangle = 0 \tag{4.171}$$

and

$$\langle\cos^2\theta\rangle = \frac{2l^2+2l-2m^2-1}{(2l+3)(2l-1)}. \tag{4.172}$$

Using these results and (4.167) we can write down expectation values involving p_θ

$$\langle\sin\theta\, p_\theta\rangle = 0, \tag{4.173}$$

$$\langle\cos\theta\sin\theta\, p_\theta\rangle = i\hbar\frac{1-4m^2}{2(2l+3)(2l-1)}, \tag{4.174}$$

etc.

4.13 Shift operators obtained by factorization

We use the factorization method of Chapter 3 to give an alternative derivation of shift operators for l. Using (4.161) and (4.28) we rewrite (4.29) as

$$O_l|lm\rangle = -(m^2-\tfrac{1}{4})|lm\rangle \tag{4.175}$$

where

$$O_l = \sin^2\theta[\hbar^{-2}p_\theta^2 - (l+\tfrac{1}{2})^2]. \tag{4.176}$$

The factorizations of O_l are

$$B_{l+1}^- B_l^+ = O_l + (l+\tfrac{1}{2})(l+\tfrac{3}{2}) \tag{4.177}$$

and

$$B_{l-1}^+ B_l^- = O_l + (l-\tfrac{1}{2})(l+\tfrac{1}{2}), \tag{4.178}$$

where

$$B_l^\pm = \pm i\hbar^{-1}\sin\theta\, p_\theta + (l+\tfrac{1}{2})\cos\theta. \tag{4.179}$$

From (4.178), (4.179), and (4.175)

$$B_l^\pm|lm\rangle = \beta_l^\pm|l\pm 1, m\rangle. \tag{4.180}$$

Comparing (4.179) and (4.73) we see that $B_l^\pm = \hbar^{-1}(\mathbf{U}_l^\pm)_z$.

Now

$$(B_l^+)^\dagger = B_{l+1}^- - 2\cos\theta, \qquad (B_l^-)^\dagger = B_{l-1}^+ + 2\cos\theta. \tag{4.181}$$

Thus for normalized kets in (4.180), eqns (4.181) and (4.175) yield

$$|\beta_l^\pm|^2 = -m^2 + \tfrac{1}{4} + (l + \tfrac{1}{2})(l + \tfrac{1}{2} \pm 1) \mp 2\langle\cos\theta\, B_l^\pm\rangle. \tag{4.182}$$

Substituting (4.179), (4.172), and (4.174) in (4.182) and choosing a phase factor equal to unity, we find that $\hbar\beta_l^\pm$ is given by (4.70).

4.14 Spin angular momentum

The description of the spin or intrinsic angular momentum possessed by certain particles is most naturally done in relativistic quantum mechanics (see Chapter 11). In this section we give a brief account of the non-relativistic treatment of particles with spin.

The eigenket of a particle with spin s must depend on the $2s + 1$ components of spin corresponding to $m_s = -s, \ -s + 1, \ \ldots, \ s$. In the non-relativistic limit the eigenket separates into the direct product

$$|\gamma\rangle \otimes |sm_s\rangle, \tag{4.183}$$

where γ denotes a set of quantum numbers such as energy, orbital angular momentum, etc. Here $|sm_s\rangle$ is the eigenket in

$$\mathbf{S}^2|sm_s\rangle = \hbar^2 s(s+1)|sm_s\rangle, \tag{4.184}$$

$$S_3|sm_s\rangle = \hbar m_s|sm_s\rangle, \tag{4.185}$$

and $s = 0, \frac{1}{2}, 1, \frac{3}{2}, \ldots$. These kets are completely specified for each s by a set of $2s + 1$ numbers. (In the following we simplify the notation by omitting the subscript on m_s.)

The spin operators and eigenkets can be represented in terms of matrices (see Section 1.2 and Ref. 13). The matrix elements of **S** follow from (4.22), (4.24), and (4.26):

$$\langle s'm'|S_\pm|sm\rangle = \hbar[(s \mp m)(s \pm m + 1)]^{1/2}\delta_{s's}\delta_{m',m\pm 1} \tag{4.186}$$

and

$$\langle s'm'|S_3|sm\rangle = \hbar m\delta_{s's}\delta_{m',m\pm 1}. \tag{4.187}$$

Using (4.186) and (4.187) we can write down matrix representations of **S**. We do not use a special symbol for these representations. In going from (4.186) and (4.187) to the spin matrices we adopt the convention: $m' = s, s-1, \ldots, -s$ correspond to columns 1 to $2s + 1$ and $m = s, s-1, \ldots -s$ correspond to rows 1 to $2s + 1$. The first few spin matrices are

(i) Spin 0:

$$\mathbf{S} = 0. \tag{4.188}$$

(ii) Spin $\frac{1}{2}\hbar$:

$$S_k = \tfrac{1}{2}\hbar\sigma_k, \tag{4.189}$$

where σ_k are the Pauli matrices

$$\sigma_1 = \begin{pmatrix} 0 & 1 \\ 1 & 0 \end{pmatrix}, \quad \sigma_2 = \begin{pmatrix} 0 & -\mathrm{i} \\ \mathrm{i} & 0 \end{pmatrix}, \quad \sigma_3 = \begin{pmatrix} 1 & 0 \\ 0 & -1 \end{pmatrix}. \tag{4.190}$$

The eigenvectors of $\mathbf{S}^2$ and S_3 in this representation are

$$v_1 = |\tfrac{1}{2}\tfrac{1}{2}\rangle = \begin{pmatrix} 1 \\ 0 \end{pmatrix} \tag{4.191}$$

and

$$v_2 = |\tfrac{1}{2}, -\tfrac{1}{2}\rangle = \begin{pmatrix} 0 \\ 1 \end{pmatrix}. \tag{4.192}$$

The shift operators for m are

$$S_+ = S_1 + \mathrm{i}S_2 = \hbar\begin{pmatrix} 0 & 1 \\ 0 & 0 \end{pmatrix} \tag{4.193}$$

and

$$S_- = S_1 - \mathrm{i}S_2 = \hbar\begin{pmatrix} 0 & 0 \\ 1 & 0 \end{pmatrix}. \tag{4.194}$$

(iii) Spin $\hbar$:

$$S_1 = \frac{\hbar}{\sqrt{2}}\begin{bmatrix} 0 & 1 & 0 \\ 1 & 0 & 1 \\ 0 & 1 & 0 \end{bmatrix}, \quad S_2 = \frac{\hbar}{\sqrt{2}}\begin{bmatrix} 0 & -\mathrm{i} & 0 \\ \mathrm{i} & 0 & -\mathrm{i} \\ 0 & \mathrm{i} & 0 \end{bmatrix},$$
$$S_3 = \hbar\begin{bmatrix} 1 & 0 & 0 \\ 0 & 0 & 0 \\ 0 & 0 & -1 \end{bmatrix}. \tag{4.195}$$

The eigenvectors of $\mathbf{S}^2$ and S_3 are

$$|1\,1\rangle = \begin{bmatrix} 1 \\ 0 \\ 0 \end{bmatrix}, \quad |1\,0\rangle = \begin{bmatrix} 0 \\ 1 \\ 0 \end{bmatrix}, \quad |1, -1\rangle = \begin{bmatrix} 0 \\ 0 \\ 1 \end{bmatrix}, \tag{4.196}$$

and the shift operators are

$$S_+ = \hbar\begin{bmatrix} 0 & \sqrt{2} & 0 \\ 0 & 0 & \sqrt{2} \\ 0 & 0 & 0 \end{bmatrix}, \quad S_- = \hbar\begin{bmatrix} 0 & 0 & 0 \\ \sqrt{2} & 0 & 0 \\ 0 & \sqrt{2} & 0 \end{bmatrix}. \tag{4.197}$$

One can readily verify that (4.195) satisfy the algebra su(2) (eqn (4.1)).

In a similar manner one can write down matrix representations for the orbital angular momentum operators. These representations are restricted to integral values of l; thus (4.195) also represent L_1, L_2, L_3 for $l = 1$.

If A is an $n \times n$ matrix then

$$\exp(\mathrm{i}A) = 1 + \frac{1}{1!}(\mathrm{i}A) + \frac{1}{2!}(\mathrm{i}A)^2 + \cdots \tag{4.198}$$

is also an $n \times n$ matrix. If A is Hermitian then $\exp(\mathrm{i}A)$ is unitary because

$$\exp(\mathrm{i}A)[\exp(\mathrm{i}A)]^\dagger = \exp(\mathrm{i}A)\exp(-\mathrm{i}A) = 1.$$

Also, if Trace $A = 0$ the determinant of $\exp(\mathrm{i}A)$ is unity because*

$$|\exp(\mathrm{i}A)| = \exp(\mathrm{i}\ \mathrm{Trace}\ A). \tag{4.199}$$

Thus from the traceless Hermitian 2×2 matrices σ_k we can construct unitary 2×2 matrices

$$U = \exp\left(\frac{\mathrm{i}}{2}\sum_{k=1}^{3} \varepsilon_k \sigma_k\right) \tag{4.200}$$

for which $|U| = 1$.

The state of a spin $\frac{1}{2}$ particle can be represented by

$$V = a_1 v_1 + a_2 v_2 = \begin{pmatrix} a_1 \\ a_2 \end{pmatrix}, \tag{4.201}$$

where $a_1 (a_2)$ is the amplitude that the particle has spin up (down) and $|a_1|^2 + |a_2|^2 = 1$. The matrix U performs the linear transformations

$$V' = UV = \begin{pmatrix} U_{11}a_1 + U_{12}a_2 \\ U_{21}a_1 + U_{22}a_2 \end{pmatrix}. \tag{4.202}$$

Because U is unitary the norm is unchanged:

$$(V')^\dagger V' = (UV)^\dagger UV = V^\dagger U^\dagger UV = V^\dagger V. \tag{4.203}$$

If $\boldsymbol{\varepsilon} = -\hat{\mathbf{n}}\phi$ where $\hat{\mathbf{n}}$ is a unit vector, the transformation (4.202) corresponds to a rotation of a spin $\frac{1}{2}$ system through an angle ϕ about the direction $\hat{\mathbf{n}}$. The matrices U provide a unitary matrix representation for the group of rotations for this system [13]. The matrices $\frac{1}{2}\sigma_k$ in (4.200) are a representation of the group generators.

In a similar manner one can use the matrices S_i corresponding to spin $\hbar$, $\frac{3}{2}\hbar, \ldots$ to generate 3×3, $4 \times 4, \ldots$ unitary matrices which are also representations of a rotation group. The matrices $\frac{1}{2}\sigma_k$ are the lowest-order matrix representation (a fundamental representation) of the group generators. The

---

* If A, and hence $\exp(\mathrm{i}A)$, are diagonal, (4.199) is obvious. Because a square matrix can be diagonalized by a similarity transformation without changing its trace or determinant, (4.199) follows.

group is denoted by SU(2), the group whose matrix representations are special (determinant = 1) unitary matrices, the lowest order of these matrices being 2 × 2.

We conclude this section with a derivation of several useful properties of the Pauli matrices (4.190). From (4.189) and (4.1)

$$\sigma_i\sigma_j - \sigma_j\sigma_i = 2\mathrm{i}\varepsilon_{ijk}\sigma_k. \tag{4.204}$$

From (4.190)

$$\sigma_i\sigma_j + \sigma_j\sigma_i = 2\delta_{ij}. \tag{4.205}$$

Hence

$$\sigma_i\sigma_j = \delta_{ij} + \mathrm{i}\varepsilon_{ijk}\sigma_k. \tag{4.206}$$

If $[\hat{a}_i, \hat{a}_j] = 0$ and $[\hat{a}_i, \sigma_j] = 0$, then

$$(\boldsymbol{\sigma}.\hat{\mathbf{a}})^2 = \hat{a}_i\hat{a}_j\sigma_i\sigma_j = \tfrac{1}{2}\hat{a}_i\hat{a}_j(\sigma_i\sigma_j + \sigma_j\sigma_i) = \hat{a}_i\hat{a}_j\delta_{ij} = 1. \tag{4.207}$$

In particular,

$$(\boldsymbol{\sigma}.\hat{\mathbf{r}})^2 = 1 \tag{4.208}$$

and

$$(\boldsymbol{\sigma}.\hat{\mathbf{p}})^2 = 1. \tag{4.209}$$

If $\mathbf{a}$ commutes with σ_i, then

$$\begin{aligned}(\boldsymbol{\sigma}.\mathbf{a})(\boldsymbol{\sigma}.\mathbf{b}) &= \sigma_i\sigma_j a_i b_j = (\delta_{ij} + \mathrm{i}\varepsilon_{ijk}\sigma_k)a_i b_j \\ &= \mathbf{a}.\mathbf{b} + \mathrm{i}\boldsymbol{\sigma}.(\mathbf{a} \times \mathbf{b}).\end{aligned} \tag{4.210}$$

In particular, with $\mathbf{a} = \mathbf{r}$ and $\mathbf{b} = \mathbf{p}$ we have

$$(\boldsymbol{\sigma}.\mathbf{r})(\boldsymbol{\sigma}.\mathbf{p}) = \mathbf{r}.\mathbf{p} + \mathrm{i}\boldsymbol{\sigma}.\mathbf{L}. \tag{4.211}$$

Similarly, with $\mathbf{a} = \mathbf{b} = \mathbf{L}$ and

$$\mathbf{L} \times \mathbf{L} = \mathrm{i}\hbar\mathbf{L} \tag{4.212}$$

(see (4.4)), we obtain the factorization

$$\mathbf{L}^2 = \hbar^2\mathscr{K}(\mathscr{K} - 1) \tag{4.213}$$

where

$$\mathscr{K} = \hbar^{-1}\boldsymbol{\sigma}.\mathbf{L} + 1. \tag{4.214}$$

These identities are used in Chapters 11 and 12.

References

1. It is usually asserted that spin has no classical counterpart. For a criticism of this point of view see Bulgac, A. and Kusnezov, D. (1990). Classical limit for Lie

algebras. *Annals of Physics*, **199**, 187–224. For an interpretation of spin as the zero-point angular momentum associated with the zero-point energy of the electron, see Hestenes, D. (1979). Spin and uncertainty in the interpretation of quantum mechanics. *American Journal of Physics*, **47**, 399–415.

2. In general J_i are defined as the generators of the rotation operator and (4.1) is derived from this definition. See, for example, Sakurai, J.J. (1985). *Modern quantum mechanics*, Ch. 3. Benjamin, Menlo Park, CA.
3. See Cohen-Tannoudji, C., Diu, B. and Laloë, F. (1977). *Quantum mechanics*, Vol. 1, Ch. 6. Wiley, New York.
4. Biedenharn, L.C. and Louck, J.D. (1981). *Angular momentum in quantum physics*, pp. 319–23. Addison-Wesley, Reading, MA. See also Jordan, T.F. (1976). Conditions on wave functions derived from operator domains. *American Journal of Physics*, **44**, 567–70. For work on half-integral l in the relativistic case, see Armstrong, B.H. (1963). Anomalous solutions to the Dirac and Schrödinger equations for the Coulomb potential. *Physical Review*, **130**, 2506–18 and references therein.
5. See, for example, Bethe, H.A. and Jackiw, R. (1969). *Intermediate quantum mechanics*, Ch. 11. Benjamin, New York.
6. See, for example, Greenberg, N.I. (1977). Discussion of the rigid rotator problem. *American Journal of Physics*, **45**, 199–204.
7. See, for example, Davydov, A.S. (1977). *Quantum mechanics*, Ch. 6. Pergamon, Oxford.
8. See Ref. 2, Chapter 3.
9. See, for example, Wybourne, B.G. (1974). *Classical groups for physicists*. Wiley, New York. Elliott, J.P. and Dawber, P.G. (1979). *Symmetry in physics*, Vols. 1 and 2. Macmillan, London.
10. Andrews, M. (1979). Vector operators and spherical harmonics in quantum mechanics. *American Journal of Physics*, **47**, 274–7.
11. See, for example, Prugovečki, E. (1981). *Quantum mechanics in Hilbert space*, 2nd edn, Ch. 7. Academic Press, New York.
12. Biedenharn, L.C. and Louck, J.D. (1981). *The Racah–Wigner algebra in quantum theory*, pp. 313–17. Addison-Wesley, Reading, MA. Loss, D. and Mullen, K. (1991). Effect of dissipation on phase periodicity and the quantum dynamics of Josephson junctions. *Physical Review* A, **43**, 2129–38.
13. See Ref. 2, Chapter 3, and Ref. 3, Chapter 6.

5

SPHERICALLY SYMMETRIC POTENTIALS

A number of quantum-mechanical problems involve a potential that is spherically symmetric, at least to a first approximation. For such potentials one can consider a 'one-dimensional' problem, namely the eigenvalue equation for the radial Hamiltonian (see (5.26)), and in this chapter we use operator methods to treat certain important cases for which exact solutions are possible.

5.1 The two-body problem

Consider two non-relativistic spinless particles that interact via a spherically symmetric potential represented by the operator $V(r)$, where $r = (\mathbf{r}\,.\,\mathbf{r})^{1/2}$ and

$$\mathbf{r} = \mathbf{r}_1 - \mathbf{r}_2. \tag{5.1}$$

The Hamiltonian is

$$H_\mathrm{T} = \frac{1}{2m_1}\mathbf{p}_1^2 + \frac{1}{2m_2}\mathbf{p}_2^2 + V(r), \tag{5.2}$$

where $\mathbf{r}_1$ and $\mathbf{p}_1$ are conjugate operators, as are $\mathbf{r}_2$ and $\mathbf{p}_2$. Also, $\mathbf{r}_1$ commutes with $\mathbf{p}_2$ and $\mathbf{r}_2$ commutes with $\mathbf{p}_1$. H_T can be rewritten as

$$H_\mathrm{T} = H_\mathrm{cm} + H. \tag{5.3}$$

Here

$$H_\mathrm{cm} = \frac{1}{2(m_1 + m_2)}\mathbf{P}^2, \tag{5.4}$$

$$H = \frac{1}{2M}\mathbf{p}^2 + V(r), \tag{5.5}$$

where

$$\mathbf{P} = \mathbf{p}_1 + \mathbf{p}_2, \tag{5.6}$$

$$\mathbf{p} = \frac{m_2\mathbf{p}_1 - m_1\mathbf{p}_2}{m_1 + m_2}, \tag{5.7}$$

$$M = \frac{m_1 m_2}{m_1 + m_2}. \tag{5.8}$$

M is the reduced mass and H_cm and H are, respectively, the Hamiltonians associated with motion of the centre of mass (translational motion) and the relative motion of the two particles (rotational and vibrational motion). It is straightforward to show that **r** and **p** are Hermitian and

$$[p_i, r_j] = -i\hbar\delta_{ij}. \tag{5.9}$$

Thus **r** and **p** are canonically conjugate.

We define

$$\mathbf{L} = \mathbf{r} \times \mathbf{p}, \tag{5.10}$$

which is the orbital angular momentum operator associated with motion about the centre of mass (internal angular momentum). Because **P** commutes with **p** and **r**, it follows that H_{cm}, H, $\mathbf{L}^2$, and L_z constitute a set of commuting operators; we assume that it is a c.s.c.o.

Let $|Elm\rangle$ be an eigenket of H, $\mathbf{L}^2$, and L_z

$$H|Elm\rangle = E|Elm\rangle, \tag{5.11}$$

$$\mathbf{L}^2|Elm\rangle = \hbar^2 l(l+1)|Elm\rangle, \tag{5.12}$$

$$L_z|Elm\rangle = \hbar m|Elm\rangle. \tag{5.13}$$

Let $|E_{\text{cm}}\rangle$ be an eigenket of H_{cm}

$$H_{\text{cm}}|E_{\text{cm}}\rangle = E_{\text{cm}}|E_{\text{cm}}\rangle. \tag{5.14}$$

The basis kets of H_{T} can be rewritten as the direct-product basis $|E_{\text{cm}}\rangle \otimes |Elm\rangle$ (see Section 1.6), where

$$H_{\text{T}}|E_{\text{cm}}\rangle \otimes |Elm\rangle = (E_{\text{cm}} + E)|E_{\text{cm}}\rangle \otimes |Elm\rangle. \tag{5.15}$$

The eigenkets $|Elm\rangle$ of (5.11)–(5.13) constitute an *angular momentum basis.*

The kets $|Elm\rangle$ are orthogonal because H, $\mathbf{L}^2$, and L_z are Hermitian (see Section 1.2). We suppose that $|Elm\rangle$ are normalized. Then

$$\langle E'l'm'|Elm\rangle = \delta_{E'E}\delta_{l'l}\delta_{m'm} \tag{5.16}$$

for discrete energy eigenvalues and

$$\langle E'l'm'|Elm\rangle = \delta(E' - E)\delta_{l'l}\delta_{m'm} \tag{5.17}$$

for continuous energy eigenvalues. The eigenvalues E_{cm} have a continuous spectrum (see Chapter 6) and therefore

$$\langle E'_{\text{cm}}|E_{\text{cm}}\rangle = \delta(E'_{\text{cm}} - E_{\text{cm}}). \tag{5.18}$$

Because H commutes with **L**,

$$H(L_\pm|Elm\rangle) = E(L_\pm|Elm\rangle). \tag{5.19}$$

Now $L_\pm|Elm\rangle \sim |E, l, m \pm 1\rangle$ (see Chapter 4) and thus it follows from (5.19) that E cannot depend on the quantum number m. This degeneracy is a general property of spherically symmetric potentials. The eigenvalues E are determined below for certain potentials.

If one of the masses m_1 and m_2 is infinite, then $H_{\text{cm}} = 0$ and the above reduces to the problem of a single particle in a spherically symmetric potential.

5.2 The radial Hamiltonian

It is useful to define the radial momentum operator

$$p_r = \tfrac{1}{2}(\hat{\mathbf{r}}\cdot\mathbf{p} + \mathbf{p}\cdot\hat{\mathbf{r}}). \tag{5.20}$$

This can also be written (see (1.138) and (1.150))

$$p_r = \frac{1}{r}(\mathbf{r}\cdot\mathbf{p} - \mathrm{i}\hbar). \tag{5.21}$$

It follows from (5.21) and (1.136) that*

$$[p_r, f(r)] = -\mathrm{i}\hbar\frac{\mathrm{d}f}{\mathrm{d}r}. \tag{5.22}$$

The radial momentum operator is related to the linear and the orbital angular momentum operators by the operator identity

$$\mathbf{L}^2 = \mathbf{r}^2(\mathbf{p}^2 - p_r^2) \tag{5.23}$$

(see Appendix G). Thus $\mathbf{p}^2$ can be eliminated from (5.5) to give

$$H = \frac{1}{2M}p_r^2 + \frac{1}{2Mr^2}\mathbf{L}^2 + V(r). \tag{5.24}$$

(These results have classical counterparts: the classical radial momentum is $p_r = \hat{\mathbf{r}}\cdot\mathbf{p}$, and (5.23) and (5.24) apply also in the classical limit, where (5.24) is the Hamiltonian for one-dimensional motion in a non-inertial (rotating) frame of reference.)

The action of H on a ket $|Elm\rangle$ is the same as that of the radial Hamiltonian

$$H_l = \frac{1}{2M}p_r^2 + \frac{\hbar^2 l(l+1)}{2Mr^2} + V(r), \tag{5.25}$$

which is obtained by replacing $\mathbf{L}^2$ in (5.24) with its eigenvalues. Let $|El\rangle$ denote an eigenket of H_l with eigenvalue E

$$H_l|El\rangle = E|El\rangle. \tag{5.26}$$

If H_l is Hermitian with respect to $|El\rangle$ and $|E'l\rangle$, these kets are orthogonal†. We suppose that they are normalized. Then

$$\langle E'l|El\rangle = \delta_{E'E} \tag{5.27}$$

for discrete energy eigenvalues and

$$\langle E'l|El\rangle = \delta(E' - E) \tag{5.28}$$

for continuous energy eigenvalues.

---

* Thus if p_r is Hermitian (see Appendix F), r and p_r are conjugate operators.

† However $|El\rangle$ and $|El'\rangle$ are not necessarily orthogonal.

The three-dimensional problem defined by (5.11)–(5.13) is considered for certain potentials in Chapters 8 and 9. In the present chapter we discuss properties of the radial Hamiltonian H_l and the kets $|El\rangle$.

It is straightforward to express any of the operator equations in this chapter in terms of a particular representation. For example, in the coordinate representation (5.11)–(5.13) become

$$H\psi_{Elm}(\mathbf{r}') = E\psi_{Elm}(\mathbf{r}'), \tag{5.29}$$

$$\mathbf{L}^2\psi_{Elm}(\mathbf{r}') = \hbar^2 l(l+1)\psi_{Elm}(\mathbf{r}'), \tag{5.30}$$

$$L_z\psi_{Elm}(\mathbf{r}') = \hbar m\psi_{Elm}(\mathbf{r}'), \tag{5.31}$$

Here the coordinate-space wavefunction is defined by

$$\psi_{Elm}(\mathbf{r}') = \langle \mathbf{r}'|Elm\rangle, \tag{5.32}$$

where $|\mathbf{r}'\rangle$ is an eigenket of $\mathbf{r}$ with eigenvalue $\mathbf{r}'$, and H, $\mathbf{L}^2$, and L_z are the differential forms obtained by making the substitutions (see Section 1.4).

$$\mathbf{r} \to \mathbf{r}', \tag{5.33}$$

$$\mathbf{p} \to -i\hbar\nabla_{\mathbf{r}'}, \tag{5.34}$$

in the corresponding abstract operators. From (5.21), (5.33), and (5.34)

$$p_r \to -i\hbar\left(\frac{\partial}{\partial r'} + \frac{1}{r'}\right). \tag{5.35}$$

It is convenient to omit the prime on $\mathbf{r}$ in the above formulae; thus in applications involving the coordinate representation it is understood that $\mathbf{r}$ is a c number.

Normalizable solutions to the partial differential equations (5.29)–(5.31) are given by the separation of variables

$$\psi_{Elm}(\mathbf{r}) = R_{El}(r)Y_{lm}(\hat{\mathbf{r}}) \tag{5.36}$$

(see also Chapters 8 and 9). Here $Y_{lm}(\hat{\mathbf{r}})$ is a spherical harmonic (see Section 4.10) and

$$R_{El}(r) = \langle r|El\rangle \tag{5.37}$$

is the radial coordinate-space wavefunction; it is the normalized solution to the differential form of (5.26), namely,

$$\left[-\frac{\hbar^2}{2m}\left(\frac{\mathrm{d}}{\mathrm{d}r} + \frac{1}{r}\right)^2 + \frac{\hbar^2 l(l+1)}{2Mr^2} + V(r)\right]R_{El}(r) = ER_{El}(r). \tag{5.38}$$

By a similar procedure the abstract operator equations can also be expressed in the momentum representation (see Chapter 7).

5.3 Factorization of the radial Hamiltonian [1]

For certain potentials the radial Hamiltonian can be factorized: that is, it can be expressed in terms of a product of two operators, each of which is a linear function* of p_r. We first determine the potentials for which this factorization is possible.

Let

$$R_l = \mathrm{i}p_r + f_l(r), \tag{5.39}$$

$$R_l' = -\mathrm{i}p_r + f_l(r), \tag{5.40}$$

and require the factorizations

$$R_l' R_l = 2MH_l + F_l \tag{5.41}$$

and

$$R_l R_l' = 2MH_{l+1} + G_l, \tag{5.42}$$

where F_l and G_l are independent of r and p_r. In the following we consider potentials that are independent of angular momentum. Substituting (5.25), (5.39), and (5.40) in (5.41) and (5.42), and using (5.22), we find

$$-\hbar\frac{\mathrm{d}f_l}{\mathrm{d}r} + f_l^2 = \frac{\hbar^2 l(l+1)}{r^2} + 2MV(r) + F_l \tag{5.43}$$

and

$$\hbar\frac{\mathrm{d}f_l}{\mathrm{d}r} + f_l^2 = \frac{\hbar^2 (l+1)(l+2)}{r^2} + 2MV(r) + G_l. \tag{5.44}$$

Addition of (5.43) and (5.44) gives

$$f_l^2 = \frac{\hbar^2(l+1)^2}{r^2} + 2MV(r) + \tfrac{1}{2}(F_l + G_l). \tag{5.45}$$

Subtracting (5.43) and (5.44), and integrating, we find

$$f_l = -\frac{\hbar(l+1)}{r} - \frac{F_l - G_l}{2\hbar}r + A_l, \tag{5.46}$$

where A_l is a constant. From (5.45) and (5.46),

$$V(r) = \frac{(F_l - G_l)^2}{8M\hbar^2}r^2 - A_l\frac{F_l - G_l}{2M\hbar}r - A_l\frac{\hbar(l+1)}{M}\frac{1}{r}, \tag{5.47}$$

where constant terms in $V(r)$ have been suppressed.

The coefficient of each of r^2, r, and r^{-1} in (5.47) must be independent of l.

* We also consider factorizations that yield operators non-linear in p_r: see (5.118)–(5.121) and (5.126).

Therefore: (i) $F_l - G_l$ must be independent of l; (ii) if $F_l = G_l$, either $A_l = 0$ or $A_l \sim (l+1)^{-1}$; (iii) if $F_l \neq G_l$, then $A_l = 0$. Thus (5.47) can be written

$$V(r) = \frac{(F_l - G_l)^2}{8M\hbar^2} r^2 - A_l \frac{\hbar(l+1)}{M} \frac{1}{r}, \tag{5.48}$$

and there are three cases.

5.3.1 *A constant potential*

If $F_l = G_l$ and $A_l = 0$, then $V(r) = 0$, $f_l = -\hbar(l+1)r^{-1}$, $F_l = 0$,

$$R_l = \mathrm{i}p_r - \hbar(l+1)r^{-1}, \tag{5.49}$$

$$R_l^\prime R_l = 2MH_l, \tag{5.50}$$

and

$$R_l R_l^\prime = 2MH_{l+1}. \tag{5.51}$$

Obviously, these results can be generalized to a non-zero constant potential: if a constant V is added and subtracted on the right-hand sides of (5.50) and (5.51), we have

$$R_l^\prime R_l = 2MH_l - 2MV \tag{5.52}$$

and

$$R_l R_l^\prime = 2MH_{l+1} - 2MV, \tag{5.53}$$

where H_l is given by (5.25) with $V(r) = V$.

5.3.2 *The Coulomb potential*

If $F_l = G_l$ and $A_l = (M\kappa/\hbar)(l+1)^{-1}$, where κ is a constant, then (5.48) is the Coulomb potential

$$V(r) = -\frac{\kappa}{r}. \tag{5.54}$$

It is convenient to define the length

$$a = \frac{\hbar^2}{M\kappa}. \tag{5.55}$$

(Note that a is positive for an attractive potential ($\kappa > 0$) and negative for a repulsive potential ($\kappa < 0$). In the case of the hydrogen-like atom, for example, $\kappa = Ze^2(4\pi\varepsilon_0)^{-1}$ and

$$a = \frac{4\pi\varepsilon_0\hbar^2}{ZMe^2} \tag{5.56}$$

is the Bohr radius divided by the atomic number Z.) Then

$$R_l = \mathrm{i}p_r - \frac{\hbar(l+1)}{r} + \frac{\hbar}{a(l+1)}, \tag{5.57}$$

$$F_l = \hbar^2 a^{-2}(l+1)^{-2}, \tag{5.58}$$

$$R_l' R_l = 2MH_l + \hbar^2 a^{-2}(l+1)^{-2}, \tag{5.59}$$

and

$$R_l R_l' = 2MH_{l+1} + \hbar^2 a^{-2}(l+1)^{-2}. \tag{5.60}$$

5.3.3 *The three-dimensional isotropic harmonic oscillator*

If $F_l \neq G_l$ then $A_l = 0$, and (5.48) becomes

$$V(r) = \tfrac{1}{2}M\omega^2 r^2, \tag{5.61}$$

where

$$\omega^2 = \frac{(F_l - G_l)^2}{4M^2\hbar^2}. \tag{5.62}$$

Then

$$R_l(\omega) = \mathrm{i}p_r - \hbar(l+1)r^{-1} - M\omega r, \tag{5.63}$$

$$F_l(\omega) = (2l+3)M\hbar\omega, \tag{5.64}$$

$$G_l(\omega) = (2l+1)M\hbar\omega, \tag{5.65}$$

$$R_l'(\omega)R_l(\omega) = 2MH_l + (2l+3)M\hbar\omega, \tag{5.66}$$

and

$$R_l(\omega)R_l'(\omega) = 2MH_{l+1} + (2l+1)M\hbar\omega. \tag{5.67}$$

In the following we assume that $\omega^2 > 0$. Then (5.61) is the potential of a three-dimensional isotropic harmonic oscillator (hereafter referred to as the oscillator).

This completes the factorization of the radial Hamiltonian (5.25) for potentials that do not depend on angular momentum. These factorizations are more restrictive than those for a one-dimensional potential (Section 3.2) because of the centrifugal term in (5.25); if this term is zero there are additional factorizations of (5.25) (see Section 5.10).

5.4 Shift operators

Let

$$R_l^+ = R_l \tag{5.68}$$

and

$$R_l^- = R'_{l-1}. \tag{5.69}$$

By calculations that are similar to those leading to (3.50) and (3.51), it follows from the factorizations (5.41) and (5.42) that

$$R_l^+ \,|El\rangle = \alpha^+(E, l)|E + \frac{1}{2M}(F_l - G_l), l + 1\rangle \tag{5.70}$$

and

$$R_l^- \,|El\rangle = \alpha^-(E, l)|E - \frac{1}{2M}(F_{l-1} - G_{l-1}), l - 1\rangle. \tag{5.71}$$

Thus $R_l^\pm$ are shift operators for energy and angular momentum in the eigenkets of the radial Hamiltonian; the superscripts $\pm$ indicate the shift in l. Where no ambiguity results, we shall omit the functional dependence in coefficients such as $\alpha^\pm(E, l)$.

If p_r is Hermitian with respect to the kets $|El\rangle$,

$$(R_l^+)^\dagger = R_{l+1}^- = R'_l. \tag{5.72}$$

Thus for Hermitian p_r and normalized kets*

$$|\alpha^+(E, l)|^2 = \langle El|(R_l^+)^\dagger R_l^+ \,|El\rangle = 2ME + F_l, \tag{5.73}$$

$$|\alpha^-(E, l)|^2 = \langle El|(R_l^-)^\dagger R_l^- \,|El\rangle = 2ME + G_{l-1}, \tag{5.74}$$

where we have used (5.41) and (5.42). According to (5.73) and (5.74), nonnegativity of the norm in the shift operations (5.70) and (5.71) requires $2ME + F_l \geq 0$ and $2ME + G_{l-1} \geq 0$.

Whether p_r is Hermitian depends on the kets $|El\rangle$: for the oscillator and bound states of an attractive Coulomb potential, p_r is Hermitian (see Appendix F). Hermitian p_r is a sufficient condition for the validity of (5.73) and (5.74); however, it is not a necessary condition and there are important cases for which (5.73) and (5.74) are suitable choices for $|\alpha^\pm|$ even though p_r is not Hermitian with respect to $|El\rangle$. These include a free particle, a repulsive Coulomb potential, and scattering states ($E \geq 0$) of an attractive Coulomb potential. In these cases the energy spectrum is continuous and the normalization condition is (5.28) (or, alternatively, (6.113)); we will see in Chapter 6 that the choices (5.73) and (5.74) preserve this normalization.

5.5 Summary

The above results can be summarized.

---

* We emphasize that (5.70) and (5.71) are valid even when $|El\rangle$ are not normalizable and p_r is not Hermitian. For example, when expressed in differential form they apply to both regular and irregular solutions of (5.38) (see Sections 6.4 and 6.5).

5.5.1 *A constant potential*

If $V(r)$ is constant,

$$R_l^{\pm}|El\rangle = \alpha^{\pm}|E, l \pm 1\rangle \tag{5.75}$$

where

$$R_l^{\pm} = \pm i p_r - \hbar(l + \tfrac{1}{2} \pm \tfrac{1}{2})r^{-1}. \tag{5.76}$$

Applications of these shift operators are discussed in Chapter 6.

5.5.2 *The Coulomb potential*

If $V(r) = -kr^{-1}$,

$$R_l^{\pm}|El\rangle = \alpha^{\pm}|E, l \pm 1\rangle, \tag{5.77}$$

where*

$$R_l^{\pm} = \pm i p_r - \hbar(l + \tfrac{1}{2} \pm \tfrac{1}{2})r^{-1} + \hbar a^{-1}(l + \tfrac{1}{2} \pm \tfrac{1}{2})^{-1}, \tag{5.78}$$

$$\alpha^{\pm} = [2ME + \hbar^2 a^{-2}(l + \tfrac{1}{2} \pm \tfrac{1}{2})^{-2}]^{1/2}, \tag{5.79}$$

and a is given by (5.55). In (5.79) the phase factors have been set equal to unity.

The energy eigenvalues for negative energy states (bound states) of an attractive Coulomb potential can be obtained from (5.77) and (5.79). For these states it follows from (5.79) and non-negativity of the norm that the sequence of kets

$$R_l^{+}|El\rangle, R_{l+1}^{+}R_l^{+}|El\rangle, \ldots . \tag{5.80}$$

must terminate at a maximum value $l_{\max} = n - 1$, where† $n = 1, 2, \ldots$ is the principal quantum number. Thus the possible values of l are $l = 0, 1, \ldots, n - 1$, and the energy eigenvalues are

$$E = -\frac{\hbar^2}{2Ma^2}\frac{1}{n^2}. \tag{5.81}$$

These eigenvalues are degenerate in l and therefore the operators $R_l^{\pm}$ transform between degenerate kets. If we adopt the convenient notation that $|nl\rangle$ denotes $|El\rangle$ for the bound states, the condition that R_l^{+} must annihilate $|El\rangle$ when $l = n - 1$ can be expressed

$$R_l^{+}|l + 1, l\rangle = 0. \tag{5.82}$$

For a repulsive potential and for scattering states of an attractive potential, $E \geq 0$ in (5.79); consequently the sequence (5.80) need not terminate, and the energy spectrum is continuous.

* In (5.78) R_0^{-} is not defined. However, this is not a serious difficulty: see Chapter 9.

† It is shown in Chapter 9 that half-integral values of l must be excluded.

5.5.3 *The oscillator*

If $V(r) = \frac{1}{2}M\omega^2 r^2$,

$$R_l^{\pm}(\omega)|El\rangle = \alpha^{\pm}(\omega)|E \pm \hbar\omega, l \pm 1\rangle \tag{5.83}$$

and

$$R_l^{\pm}(-\omega)|El\rangle = \alpha^{\pm}(-\omega)|E \mp \hbar\omega, l \pm 1\rangle, \tag{5.84}$$

where

$$R_l^{\pm}(\omega) = \pm ip_r - \hbar(l + \tfrac{1}{2} \pm \tfrac{1}{2})r^{-1} - M\omega r, \tag{5.85}$$

$$\alpha^{\pm}(\omega) = [2ME + 2M\hbar\omega(l + \tfrac{1}{2} \pm 1)]^{1/2}. \tag{5.86}$$

In (5.86) the phase factors have been taken to be unity.

The energy eigenvalues can be obtained from (5.84) and (5.86). Non-negativity of the norm requires that the sequence of kets

$$R_l^{+}(-\omega)|El\rangle, R_{l+1}^{+}(-\omega)R_l^{+}(-\omega)|El\rangle, \ldots, \tag{5.87}$$

must terminate at a maximum value $l_{max} = n$, where* $n = 0, 1, \ldots$ is the principal quantum number. From (5.86)

$$E = (n + \tfrac{3}{2})\hbar\omega, \tag{5.88}$$

which are also degenerate in l. If $|El\rangle$ is denoted by $|nl\rangle$, the requirement that the sequence (5.87) must terminate is

$$R_l^{+}(-\omega)|ll\rangle = 0. \tag{5.89}$$

5.6 Additional factorizations and shift operators

Operating with r^2 on the left of (5.26) and rearranging terms, we obtain

$$O_E|El\rangle = -l(l+1)|El\rangle, \tag{5.90}$$

where

$$O_E = \frac{1}{\hbar^2}r^2p_r^2 + \frac{2M}{\hbar^2}r^2[V(r) - E]. \tag{5.91}$$

For the oscillator and bound states of a Coulomb potential, O_E can be factorized to provide additional useful shift operators. It is convenient to define the dimensionless parameters

$$n = \frac{E}{\hbar\omega} - \frac{3}{2} \tag{5.92}$$

for the oscillator, and

* It is shown in Chapter 8 that half-integral values of l must be excluded.

$$n = \left(-\frac{\hbar^2}{2Ma^2E}\right)^{1/2} \tag{5.93}$$

for the bound states of a Coulomb potential. (Note that we are not making any assumptions about the values of n.)

5.6.1 *The oscillator*

For the oscillator (5.91) is

$$O_n = \frac{1}{\hbar^2}r^2p_r^2 + \left(\frac{r}{a}\right)^4 - 2\left(n + \frac{3}{2}\right)\frac{r^2}{a^2}, \tag{5.94}$$

where the constant

$$a = (\hbar/M\omega)^{1/2} \tag{5.95}$$

has the dimension of length. It is straightforward to derive the operators

$$Q_n^{\pm} = \pm\frac{\mathrm{i}}{\hbar}rp_r - \frac{r^2}{a^2} + n + \frac{3}{2} \pm \frac{1}{2} \tag{5.96}$$

that provide the factorizations

$$Q_{n+2}^{-}Q_n^{+} = O_n + (n+2)(n+3), \qquad Q_{n-2}^{+}Q_n^{-} = O_n + n(n+1). \tag{5.97}$$

It follows from (5.97) and (5.90) that

$$Q_n^{\pm}|nl\rangle = \beta^{\pm}(n, l)|n \pm 2, l\rangle. \tag{5.98}$$

Now

$$(Q_n^{+})^{\dagger} = Q_n^{-} = Q_{n+2}^{-} - 2, \tag{5.99}$$

and

$$(Q_n^{-})^{\dagger} = Q_n^{+} = Q_{n-2}^{+} + 2. \tag{5.100}$$

Using (5.97)–(5.100), and assuming normalized kets, we obtain

$$\begin{aligned}\beta^{\pm}(n, l) &= [(n + 1 \pm 1)(n + 2 \pm 1) - l(l + 1)]^{1/2}\\ &= [(n - l + 1 \pm 1)(n + l + 2 \pm 1)]^{1/2}.\end{aligned} \tag{5.101}$$

The value $+1$ for the phase factors in (5.101) is required by (5.104).

The energy eigenvalues can be obtained from (5.98) and (5.101). Non-negativity of the norm requires that the sequence of kets

$$Q_n^{-}|nl\rangle,\ Q_{n-2}^{-}Q_n^{-}|nl\rangle, \ldots$$

must terminate at a minimum value $n_{\min} = l$. From the ket $|ll\rangle$, and by application of Q_l^{+}, $Q_{l+2}^{+}Q_l^{+}, \ldots$, one obtains a sequence of kets $|nl\rangle$ with

$$n = 2N + l, \tag{5.102}$$

where the radial quantum number $N = 0, 1, 2, \ldots$ denotes the number of times raising operators have been applied to $|ll\rangle$. Equations (5.92) and (5.102) show that the energy eigenvalues can be written

$$E = (2N + l + \tfrac{3}{2})\hbar\omega. \tag{5.103}$$

From (5.83), (5.84), and (5.98) it is clear that there must be relationships between $Q_n^\pm$ and products of R_l^+ and R_l^-. These are given by the identities

$$R_{l\pm1}^\mp(-\omega)R_l^\pm(\omega) = 2M\hbar\omega Q_n^\pm + 2M(H_l - E). \tag{5.104}$$

We note that the operator $r^{-2}(H_l - E)$ can also be factorized. For

$$Q_{nl}^\pm = \pm ir^{-1}p_r - \hbar(l + \tfrac{1}{2} \pm \tfrac{1}{2})r^{-2} + M\omega(n + \tfrac{3}{2})(l + \tfrac{1}{2} \pm 1)^{-1}, \tag{5.105}$$

it can be shown that

$$Q_{n,l\pm2}^\mp Q_{nl}^\pm = 2Mr^{-2}(H_l - E) + M^2\omega^2\frac{(n - l + 1 \mp 1)(n + l + 2 \pm 1)}{(l + \tfrac{1}{2} \pm 1)^2} \tag{5.106}$$

and

$$Q_{nl}^\pm|nl\rangle = M\omega\left(\frac{(n - l + 1 \mp 1)(n + l + 2 \pm 1)}{(l + \tfrac{1}{2} \pm 1)^2}\right)^{1/2}|n, l \pm 2\rangle. \tag{5.107}$$

The operators $Q_{nl}^\pm$ transform between kets with energy eigenvalues that are degenerate in l.

5.6.2 *The Coulomb potential*

For bound states of an attractive Coulomb potential (5.91) is

$$O_n = \frac{1}{\hbar^2}r^2p_r^2 - 2\frac{r}{a} + \frac{r^2}{a^2n^2}. \tag{5.108}$$

We wish to derive operators $Q_n^\pm$ such that

$$Q_{n\pm1}^\mp Q_n^\pm = O_n + F^\pm, \tag{5.109}$$

where $F^\pm$ are independent of r and p_r. These factorizations are more complicated than those discussed above, and we perform the derivation of $Q_n^\pm$ in three parts:

(i) Consider first a simpler problem. Let

$$\tilde{O}_n = \frac{1}{\hbar^2}u^2p_u^2 - \frac{2nu}{a} + \frac{u^2}{a^2}, \tag{5.110}$$

where u and p_u satisfy $[u, p_u] = i\hbar$. It is straightforward to factorize $\tilde{O}_n$:

$$\tilde{Q}_{n\pm1}^\mp\tilde{Q}_n^\pm = \tilde{O}_n + n(n \pm 1), \tag{5.111}$$

where

$$\tilde{Q}_n^{\pm} = \pm \frac{\mathrm{i}}{\hbar} u p_u - \frac{u}{a} + n. \tag{5.112}$$

Let $u = r/n$ and $p_u = np_r$. Then (5.111) becomes

$$\left(\mp \frac{\mathrm{i}}{\hbar} rp_r - \frac{r}{an} + n \pm 1\right)\left(\pm \frac{i}{\hbar} rp_r - \frac{r}{an} + n\right) = O_n + n(n \pm 1). \tag{5.113}$$

(ii) Equation (5.113) would have the required form (5.109) if the term $-r/(an)$ in the first parentheses in (5.113) were $-r/[a(n \pm 1)]$. To obtain (5.109) from (5.113), we introduce operators $\Delta_n^{\pm}$ such that

$$\Delta_n^{\pm} \frac{r}{n} = \frac{r}{n \pm 1} \Delta_n^{\pm} \tag{5.114}$$

and

$$\Delta_n^{\pm} np_r = (n \pm 1)p_r \Delta_n^{\pm} \tag{5.115}$$

($n \neq 1$ in Δ_n^-). Thus $\Delta_n^{\pm}$ commute with rp_r. Equations (5.114) and (5.115) define $\Delta_n^{\pm}$ only to within an arbitrary constant factor. It is convenient to require that

$$(\Delta_n^{\pm})^{\dagger} \Delta_n^{\pm} = 1. \tag{5.116}$$

Equations (5.114)–(5.116) define $\Delta_n^{\pm}$ (see (5.126) and (5.127)). From (5.114)–(5.116)

$$(\Delta_n^{\pm})^{\dagger} = \Delta_{n \pm 1}^{\mp}. \tag{5.117}$$

Operating on the left of both sides of (5.113) with $\Delta_{n\pm1}^{\mp} \Delta_n^{\pm} = 1$ yields

$$Q_{n+1}^{-} Q_n^{+} = O_n + n(n+1) \tag{5.118}$$

and

$$Q_{n-1}^{+} Q_n^{-} = O_n + n(n-1) \qquad (n > 1), \tag{5.119}$$

where

$$Q_n^{\pm} = \Delta_n^{\pm} T_n^{\pm} \tag{5.120}$$

and

$$T_n^{\pm} = \pm \frac{\mathrm{i}}{\hbar} rp_r - \frac{r}{an} + n. \tag{5.121}$$

In the above Q_1^- is undefined: we define it in (5.133).

(iii) Equations (5.114) and (5.115) are scaling operations for the conjugate operators r and p_r. Next we derive expressions for the scaling operators

$\Delta_n^\pm$. Because $\Delta_n^\pm$ commute with rp_r, we expect that they are functions of rp_r. Consider the operators

$$\Delta^\pm(rp_r) = \sum_{m=0}^{\infty} b_m^\pm (rp_r)^m, \tag{5.122}$$

where $b_m^\pm$ are independent of r and p_r. Now $r(rp_r)^m = (rp_r + i\hbar)^m r$ and $p_r(rp_r)^m = (rp_r - i\hbar)^m p_r$. Therefore

$$r\Delta^\pm(rp_r) = \Delta^\pm(rp_r + i\hbar)r \tag{5.123}$$

and

$$p_r\Delta^\pm(rp_r) = \Delta^\pm(rp_r - i\hbar)p_r. \tag{5.124}$$

Equation (5.123) will be the scaling operation (5.114) if

$$\Delta^\pm(rp_r + i\hbar) = \frac{n \pm 1}{n}\Delta^\pm(rp_r). \tag{5.125}$$

The solutions to (5.125) are

$$\Delta^\pm(rp_r) = C_n^\pm \left(\frac{n}{n \pm 1}\right)^{i\hbar^{-1}rp_r} = C_n^\pm \exp\left(\frac{i}{\hbar} rp_r \ln \frac{n}{n \pm 1}\right) \tag{5.126}$$

where $C_n^\pm$ are arbitrary constants. Because $(\exp A)^\dagger = \exp(A^\dagger)$ and $(rp_r)^\dagger = p_r r = rp_r - i\hbar$, we see that (5.116) requires

$$C_n^\pm = \exp\left(\frac{1}{2} \ln \frac{n}{n \pm 1}\right). \tag{5.127}$$

From (5.124) and (5.126) one can readily verify (5.115). This completes the derivation of the operators $Q_n^\pm$.

The factorizations (5.118) and (5.119) show that

$$Q_n^\pm |nl\rangle = \beta^\pm(n, l)|n \pm 1, l\rangle. \tag{5.128}$$

For normalized kets in (5.128)

$$\begin{aligned} |\beta^\pm|^2 &= \langle nl|(Q_n^\pm)^\dagger Q_n^\pm |nl\rangle \\ &= \langle nl|(T_n^\pm)^\dagger (\Delta_n^\pm)^\dagger \Delta_n^\pm T_n^\pm |nl\rangle \\ &= \langle nl|(T_n^\pm)^\dagger T_n^\pm |nl\rangle. \end{aligned} \tag{5.129}$$

Now $(T_n^\pm)^\dagger = T_n^\mp \mp 1$. Substituting this and (5.121) in (5.129), and using (5.108) and (5.90), we find

$$|\beta^\pm|^2 = \langle nl| -2i\hbar^{-1} rp_r \pm 2n^{-1}a^{-1}r + n^2 \mp n - l(l+1)|nl\rangle. \tag{5.130}$$

The expectation values in (5.130) are evaluated in Chapter 6: using (6.14) and (6.51) we have

$$\beta^\pm(n, l) = \{n^{-1}(n \pm 1)[n(n \pm 1) - l(l+1)]\}^{1/2}. \tag{5.131}$$

The phase factors in (5.131) have been set equal to unity.

We have not yet defined Q_1^-. Because Q_n^- annihilates $|n, n-1\rangle$ if $n > 1$ (see below), it is desirable to define Q_1^- such that

$$Q_1^- \,|1\,0\rangle = 0, \tag{5.132}$$

which is satisfied if (see (5.82) and (5.121))

$$Q_1^- = T_1^-. \tag{5.133}$$

The energy eigenvalues can be deduced from (5.128) and (5.131). Non-negativity of the norm requires that the sequence of kets

$$Q_n^- \,|nl\rangle,\, Q_{n-1}^- Q_n^- \,|nl\rangle, \ldots$$

must terminate at a minimum value $n_{\min} = l + 1$. By application of Q_{l+1}^+, $Q_{l+2}^+ Q_{l+1}^+, \ldots$, to $|l+1, l\rangle$, one obtains a sequence of kets with

$$n = N + l + 1. \tag{5.134}$$

Here the radial quantum number $N = 0, 1, 2, \ldots$ denotes the number of times raising operators have been applied to $|l+1, l\rangle$. From (5.93) and (5.134) the energy eigenvalues can be written

$$E = -\frac{\hbar^2}{2Ma^2}(N + l + 1)^{-2}. \tag{5.135}$$

The shift operators $Q_n^\pm$ for a Coulomb potential are more complicated than those encountered previously. Nevertheless, $Q_n^\pm$ are useful operators (see Chapter 9). The transformations effected by some of the shift operators discussed above are depicted in Fig. 5.1.

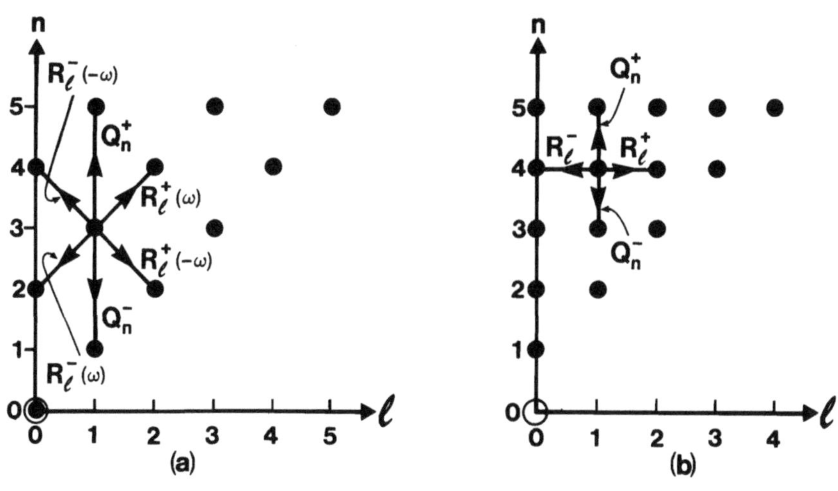

FIG. 5.1 Schematic illustration of the transformations effected by shift operators discussed in the text: (a) for the kets $|nl\rangle$ of the three-dimensional isotropic harmonic oscillator; (b) for the kets $|nl\rangle$ of bound states of an attractive Coulomb potential.

From (5.77) and (5.128) it is clear that products of $R_l^\pm$ and $Q_n^\pm$ shift both n and l in $|nl\rangle$. When $n = l + 1$ these shift operators have a simple form because we can use (5.82) and (5.78) to eliminate terms that are linear in p_r. For example,

$$R_l^+ Q_n^+ |nl\rangle = \alpha^+(n+1, l)\beta^+(n, l)|n+1, l+1\rangle \tag{5.136}$$

simplifies to

$$\Delta_{l+1}^+ \frac{r}{l+1}|l+1, l\rangle = -\frac{1}{2}a[(2l+3)(2l+4)]^{1/2}|l+2, l+1\rangle. \tag{5.137}$$

Equation (5.137) can be derived *ab initio* by using (5.90), (5.108), (5.114), and (5.115) to show that

$$O_{l+2}(\Delta_{l+1}^+ r|l+1, l\rangle) = -(l+1)(l+2)(\Delta_{l+1}^+ r|l+1, l\rangle), \tag{5.138}$$

and then using (5.116) and (6.52) to calculate the norm of $\Delta_{l+1}^+ r|l+1, l\rangle$.

5.7 Discussion

1. The shift operators derived in this chapter act on the eigenkets $|nl\rangle$ of a radial Hamiltonian H_l. In general, they are not shift operators for the eigenkets $|nlm\rangle$ of the Hamiltonian H (see Chapters 8 and 9). The exceptions are the energy shift operators $Q_n^\pm$ (Eqns (5.96) and (5.120)) and it is useful to consider the effect of these on $|nlm\rangle$.

   For the oscillator

$$H(Q_n^\pm |nlm\rangle) = E_{n\pm 2}(Q_n^\pm |nlm\rangle), \tag{5.139}$$

where E_n is given by (5.88). Because $Q_n^\pm$ are scalars they do not change l and m. Thus

$$Q_n^\pm |nlm\rangle = \beta^\pm(n, l)|n \pm 2, l, m\rangle. \tag{5.140}$$

The coefficients $\beta^\pm$ can be evaluated by similar calculations to those leading to (5.101): the result is (5.101) because of the relation (6.55). From (5.96) we construct operators

$$Q^\pm = \pm\frac{\mathrm{i}}{\hbar}rp_r - \frac{r^2}{a^2} + \frac{H}{\hbar\omega} \pm \frac{1}{2}, \tag{5.141}$$

that satisfy

$$Q^\pm |nlm\rangle = Q_n^\pm |nlm\rangle. \tag{5.142}$$

Hence

$$Q^\pm |nlm\rangle = \beta^\pm(n, l)|n \pm 2, l, m\rangle. \tag{5.143}$$

   Similarly, for bound states of a Coulomb potential

$$H(Q_n^\pm |nlm\rangle) = E_{n\pm 1}(Q_n^\pm |nlm\rangle), \tag{5.144}$$

where E_n is given by (5.81). Thus

$$Q_n^{\pm}|nlm\rangle = \beta^{\pm}(n, l)|n \pm 1, l, m\rangle, \tag{5.145}$$

where $\beta^{\pm}$ are given by (5.131).

From $Q_n^{\pm}$ we can construct shift operators $Q^{\pm}$ that do not depend explicitly on n. This is more difficult than for the oscillator because of the scaling operators in (5.120). First define the operator*

$$\mathcal{N} = (-2\hbar^{-2}Ma^2H)^{-1/2}. \tag{5.146}$$

Then

$$\mathcal{N}|Elm\rangle = (-2\hbar^{-2}Ma^2E)^{-1/2}|Elm\rangle, \tag{5.147}$$

which can be written in terms of n (Eqn (5.93)) as

$$\mathcal{N}|nlm\rangle = n|nlm\rangle. \tag{5.148}$$

The scaling operators can be expressed as (see (5.126) and (5.127))

$$\Delta_n^{\pm} = \sum_{k=0}^{\infty} \frac{1}{k!}\left(\frac{\mathrm{i}}{\hbar}rp_r + \frac{1}{2}\right)^k \left(\ln\frac{n}{n \pm 1}\right)^k. \tag{5.149}$$

The effect of these on $|nlm\rangle$ is the same as that of

$$\Delta^{\pm} = \sum_{k=0}^{\infty} \frac{1}{k!}\left(\frac{\mathrm{i}}{\hbar}rp_r + \frac{1}{2}\right)^k \left(\ln\frac{\mathcal{N}}{\mathcal{N} \pm 1}\right)^k. \tag{5.150}$$

Now use (5.114) and (5.115) to write $Q_n^{\pm}$ in the form

$$Q_n^{\pm} = \pm\frac{\mathrm{i}}{\hbar}rp_r\Delta_n^{\pm} - \frac{r}{a}\Delta_n^{\pm}(n \pm 1)^{-1} + \Delta_n^{\pm}n. \tag{5.151}$$

It is clear that

$$Q^{\pm} = \pm\frac{\mathrm{i}}{\hbar}rp_r\Delta^{\pm} - \frac{r}{a}\Delta^{\pm}(\mathcal{N} \pm 1)^{-1} + \Delta^{\pm}\mathcal{N} \tag{5.152}$$

have the same effect on $|nlm\rangle$ as do $Q_n^{\pm}$. (Note that the order of the operators $\Delta^{\pm}$ and $\mathcal{N}$ in (5.152) is important.) Thus

$$Q^{\pm}|nlm\rangle = \beta^{\pm}(n, l)|n \pm 1, l, m\rangle, \tag{5.153}$$

where $\beta^{\pm}$ are given by (5.131).

2. For the oscillator and the bound states of a Coulomb potential we have (see (5.103) and (5.135)):

   (i) For a given l the energy eigenvalues are non-degenerate; written as E_{Nl},

---

* Note that we are restricting ourselves to the negative energy subspace of H (the space of bound states). On this subspace $-H$ is a Hermitian, positive definite operator and $\mathcal{N}$ is well defined.

where $N = 0, 1, 2, \ldots$ is the radial quantum number, they are ordered according to

$$E_{0l} < E_{1l} < E_{2l} \ldots . \tag{5.154}$$

(In the coordinate representation N gives the number of nodes of the radial wavefunction. See Chapter 6.)

(ii) The ground state is a state with $l = 0$ (an s state).

These properties are general results for the discrete spectrum of a two-body spherically symmetric potential. The non-degeneracy of E_{Nl} (for a given l) is a consequence of the one-dimensional nature of the radial equation (5.38) [2]. The connection between the ordering (5.154) and the number of nodes of the radial wavefunction is the content of the oscillation theorem [2].

Property (ii) can be proved using (i) and the fact that the expectation value of the centrifugal term in (5.25) cannot be negative. Write (5.26) as

$$H_l|Nl\rangle = E_{Nl}|Nl\rangle,$$

where E_{Nl} satisfies (5.154) and $\langle N'l|Nl\rangle = \delta_{N'N}$. Then

$$\begin{aligned} E_{Nl} &= \langle Nl|H_l|Nl\rangle \\ &= \langle Nl|H_{l'}|Nl\rangle + \frac{\hbar^2}{2M}[l(l+1) - l'(l'+1)]\langle Nl|r^{-2}|Nl\rangle, \end{aligned} \tag{5.155}$$

where we have assumed that V does not depend on l. Now suppose that V is a real function. Then $|Nl\rangle$ form a complete set and we can write*

$$|Nl\rangle = \sum_{N'=0}^{\infty} |N'l'\rangle\langle N'l'|Nl\rangle. \tag{5.156}$$

Now $\langle r^{-2}\rangle > 0$. Thus if $l > l'$, it follows from (5.154)–(5.156) that

$$E_{Nl} > \sum_{N'=0}^{\infty} |\langle N'l'|Nl\rangle|^2 E_{N'l'} > E_{0l'} \sum_{N'=0}^{\infty} |\langle N'l'|Nl\rangle|^2.$$

According to the closure relation (1.39), the last sum is equal to unity. Therefore $E_{Nl} > E_{0l'}$ if $l > l'$, and the ground state is an s state.

3. The factorizations considered in Section 5.6 can be generalized to apply to an oscillator or attractive Coulomb potential plus a repulsive inverse-square term; that is, to the potentials

$$V(r) = \tfrac{1}{2}M\omega^2 r^2 + \lambda r^{-2} \tag{5.157}$$

and

$$V(r) = -\mathscr{k} r^{-1} + \lambda r^{-2}, \tag{5.158}$$

---

* If the energy spectrum is partly continuous we must include a contribution from the continuous spectrum (see (1.62)). This does not affect the result.

where λ is a positive constant. (These potentials have applications in molecular physics.) By substituting (5.157) and (5.158) in (5.91) and factorizing $O_E - 2\hbar^{-2}M\lambda$ one can deduce shift operations for energy. From the coefficients in these shift operations one obtains the energy eigenvalues. The calculations are similar to those in Section 5.6 and they are left as an exercise for the reader.* The results are

$$E = \{2N + 1 + [(l + \tfrac{1}{2})^2 + 2\hbar^{-2}M\lambda]^{1/2}\}\hbar\omega \tag{5.159}$$

for the potential (5.157), and

$$E = -\frac{\hbar^2}{2Ma^2}\{N + \tfrac{1}{2} + [(l + \tfrac{1}{2})^2 + 2\hbar^{-2}M\lambda]^{1/2}\}^{-2} \tag{5.160}$$

for bound states of the potential (5.158). Here $N = 0, 1, \ldots$ is the radial quantum number and a is given by (5.55). Note that in each case the degeneracy of E with respect to l is removed by the inverse-square term in (5.157) and (5.158).

5.8 One-way factorization: generalized potentials

In this section the method of one-way factorization (Sections 1.9 and 3.1) is used to obtain generalized Coulomb and oscillator potentials.

Consider the operators

$$D_l = \mathrm{i}p_r + f_{l-1}(r), \qquad D_l' = -\mathrm{i}p_r + f_{l-1}(r), \tag{5.161}$$

that provide the one-way factorization

$$D_l D_l' = 2MH_l + C_l. \tag{5.162}$$

Here H_l is the radial Hamiltonian (5.25) and C_l is independent of r and p_r. The function $f_{l-1}(r)$ must satisfy the Ricatti equation

$$\hbar\frac{\mathrm{d}f_{l-1}}{\mathrm{d}r} + f_{l-1}^2 = \frac{\hbar^2 l(l+1)}{r^2} + 2MV(r) + C_l. \tag{5.163}$$

(See (5.44).) With the aid of (5.22) and (5.163) we can show that

$$D_{l+1}' D_{l+1} = 2M\tilde{H}_l + \tilde{C}_l, \tag{5.164}$$

where

$$\tilde{H}_l = \frac{1}{2M}p_r^2 + \frac{\hbar^2 l(l+1)}{2Mr^2} + \tilde{V}(r), \tag{5.165}$$

and

$$\tilde{V}(r) = V(r) + \frac{\hbar^2(l+1)}{Mr^2} - \frac{\hbar}{M}\frac{\mathrm{d}f_l}{\mathrm{d}r} + \frac{1}{2M}(C_{l+1} - \tilde{C}_l). \tag{5.166}$$

---

* A similar problem to (5.158) is considered in Chapter 12.

Here $\tilde{C}_l$ is independent of r and p_r, and its value can be chosen to eliminate constants from the potential $\tilde{V}$ (see below).

In Section 5.3 we considered a 'two-way' factorization of H_l (see (5.41) and (5.42)) and obtained a solution to (5.163) for the oscillator and Coulomb potentials. For this solution (5.166) yields $\tilde{V} = V$. In this section we show that the less restrictive one-way factorization allows additional solutions to (5.163) and hence new radial Hamiltonian operators $\tilde{H}_l$ that have the same energy spectrum as H_l.

5.8.1 *The Coulomb potential* [3]

With

$$C_l = \left(\frac{\hbar}{al}\right)^2, \tag{5.167}$$

eqn (5.162) is the factorization (5.60) with l replaced by $l-1$. Hence one solution to (5.163) is

$$f_{l-1} = -\frac{\hbar l}{r} + \frac{\hbar}{al}, \tag{5.168}$$

which means

$$D_l = R_{l-1}. \tag{5.169}$$

Let

$$f_{l-1}(r) = -\frac{\hbar l}{r} + \frac{\hbar}{al} + g_l(r). \tag{5.170}$$

Then (5.163) becomes

$$\hbar\frac{\mathrm{d}g_l}{\mathrm{d}r} + 2\hbar\left(\frac{1}{al} - \frac{l}{r}\right)g_l + g_l^2 = 0. \tag{5.171}$$

The substitution $g_l = G_l^{-1}$ transforms (5.171) into the first-order linear equation

$$\hbar\frac{\mathrm{d}G_l}{\mathrm{d}r} - 2\hbar\left(\frac{1}{al} - \frac{l}{r}\right)G_l - 1 = 0. \tag{5.172}$$

The general solution is

$$G_l(u) = \frac{a}{\hbar}\left(\gamma_l + \int_0^u t^{2l}\exp(-2t/l)\,\mathrm{d}t\right)u^{-2l}\exp(2u/l), \tag{5.173}$$

where

$$u = r/a, \tag{5.174}$$

and γ_l is an arbitrary constant. Hence

$$g_l(u) = \frac{\hbar}{a} \frac{u^{2l} \exp(-2u/l)}{\gamma_l + \int_0^u t^{2l} \exp(-2t/l)\, \mathrm{d}t}. \tag{5.175}$$

Substituting (5.170), (5.175), and (5.167) in (5.166) and choosing

$$\tilde{C}_l = \frac{\hbar^2}{a^2(l+1)^2}, \tag{5.176}$$

we obtain the generalized Coulomb potential

$$\tilde{V}(u) = -\frac{\hbar^2}{Ma^2}\left(\frac{1}{u} + \frac{\mathrm{d}}{\mathrm{d}u} \frac{u^{2(l+1)} \exp[-2u/(l+1)]}{\gamma_{l+1} + \int_0^u t^{2(l+1)} \exp[-2t/(l+1)]\, \mathrm{d}t}\right). \tag{5.177}$$

Only if $\gamma_l = \infty$ is $\tilde{H}_l = H_l$: otherwise the product $D'_{l+1} D_{l+1}$ defines a new radial Hamiltonian. For $\tilde{H}_l$ to be Hermitian with respect to $|El\rangle$, γ_l must be real.

Next we determine the eigenvalues and eigenkets of $\tilde{H}_l$. Because p_r is Hermitian with respect to these kets (see below), $D'_l = D_l^\dagger$. Now

$$\tilde{H}_l D_{l+1}^\dagger = D_{l+1}^\dagger H_{l+1}. \tag{5.178}$$

(See (5.164), (5.162), (5.167), (5.176).) Define

$$|\phi_{nl}\rangle = c_{nl} D_{l+1}^\dagger |n, l+1\rangle \qquad (n = l+2, l+3, \ldots), \tag{5.179}$$

where c_{nl} is a constant and $|nl\rangle$ is an eigenket of H_l with eigenvalue E given by (5.81). Then

$$\tilde{H}_l |\phi_{nl}\rangle = E|\phi_{nl}\rangle \qquad (n = l+2, l+3, \ldots). \tag{5.180}$$

Also

$$\begin{aligned}\langle \phi_{n'l} | \phi_{nl}\rangle &= c_{n'l}^* c_{nl} \langle n', l+1 | D_{l+1} D_{l+1}^\dagger | n, l+1\rangle \\ &= c_{n'l}^* c_{nl}\left(2ME + \frac{\hbar^2}{a^2(l+1)^2}\right)\langle n', l+1 | n, l+1\rangle \\ &= c_{n'l}^* c_{nl} \frac{\hbar^2}{a^2}\left(\frac{1}{(l+1)^2} - \frac{1}{n^2}\right)\delta_{n'n},\end{aligned} \tag{5.181}$$

and therefore $|\phi_{nl}\rangle$ are normalized if

$$c_{nl} = \frac{a}{\hbar}\left(\frac{1}{(l+1)^2} - \frac{1}{n^2}\right)^{-1/2}. \tag{5.182}$$

Because

$$\langle \phi_{nl} | \chi\rangle = c_{nl}^* \langle n, l+1 | D_{l+1} | \chi\rangle, \tag{5.183}$$

if there exists a ket $|\phi_{l+1,l}\rangle$ such that

$$D_{l+1}|\phi_{l+1,l}\rangle = 0, \tag{5.184}$$

then

$$\langle\phi_{nl}|\phi_{l+1,l}\rangle = 0 \qquad (n = l+2, l+3, \ldots). \tag{5.185}$$

Also, (5.184), (5.164), and (5.176) show that

$$\tilde{H}_l|\phi_{l+1,l}\rangle = -\frac{\hbar^2}{2Ma^2(l+1)^2}|\phi_{l+1,l}\rangle. \tag{5.186}$$

It is useful to note that according to (5.161) and (5.168)–(5.170),

$$D_l = R_{l-1} + g_l(r). \tag{5.187}$$

Hence

$$D_{l+1}^\dagger = R_l^\dagger + g_{l+1}(r) = R_{l+1}^- + g_{l+1}(r). \tag{5.188}$$

(See (5.68) and (5.72).) Thus we can write (5.179) in the convenient form

$$|\phi_{nl}\rangle = |nl\rangle + \frac{a}{\hbar}\left(\frac{1}{(l+1)^2} - \frac{1}{n^2}\right)^{-1/2} g_{l+1}(r)|n, l+1\rangle. \tag{5.189}$$

(Here we have used (5.77), (5.79), and (5.182).)

Next we determine the value of the parameter γ_l for which the states $|\phi_{nl}\rangle$ exist. The potential (5.177) is complicated and therefore we consider the coordinate representation. Because

$$0 \le \int_0^u t^{2l}\exp(-2t/l)\,dt \le \left(\frac{l}{2}\right)^{2l+1}(2l)!, \tag{5.190}$$

we see that for

$$\gamma_l > 0 \quad \text{or} \quad \gamma_l < -\left(\frac{l}{2}\right)^{2l+1}(2l)!, \tag{5.191}$$

the potential $\tilde{V}$ has no singularities and $\tilde{V} \to -\hbar^2/(Mar)$ as $r \to \infty$. If γ_l satisfies (5.191), the operators (5.165) are a one-parameter family of radial Hamiltonians.

From (5.189) we have the radial coordinate-space wavefunctions of $\tilde{H}_l$ for $n = l+2, l+3, \ldots,$

$$\phi_{nl}(r) = R_{nl}(r) + \frac{a}{\hbar}\left(\frac{1}{(l+1)^2} - \frac{1}{n^2}\right)^{-1/2} g_{l+1}(r)R_{n,l+1}(r). \tag{5.192}$$

Here R_{nl} is the radial coordinate-space wavefunction of H_l (see (6.98)). For each γ_l that satisfies (5.191) the wavefunctions (5.192) are normalized. Similarly, from (5.184)

$$\left[-\hbar\left(\frac{d}{dr} + \frac{1}{r}\right) + f_l(r)\right]\phi_{l+1,l}(r) = 0, \tag{5.193}$$

and hence

$$\phi_{l+1,l}(r) = b_l r^{-1} \exp\left(-\hbar^{-1}\int f_l(r)\,\mathrm{d}r\right)$$

$$= b_l r^l \exp\left(-\frac{r}{a(l+1)} - \frac{1}{\hbar}\int g_{l+1}(r)\,\mathrm{d}r\right), \tag{5.194}$$

where b_l is a constant. If γ_l satisfies (5.191), $\phi_{l+1,l}$ is square-integrable. These wavefunctions have the property

$$r\phi_{nl} \to 0 \quad \text{as} \quad r \to 0 \quad (n = l+1, l+2, \ldots). \tag{5.195}$$

Hence p_r is Hermitian with respect to ϕ_{nl} (see Appendix F).

The functions (5.192) and (5.194) are a complete set of wavefunctions for the Hamiltonians (5.165). For finite γ_l the potentials of these Hamiltonians are different from the Coulomb potential; however, the energy eigenvalues are identical to those of the Coulomb Hamiltonian.

5.8.2 *The oscillator* [4]

A one-way factorization can also be performed for the oscillator, and we give an outline of the procedure. If

$$C_l = -(2l-1)M\hbar\omega, \tag{5.196}$$

(5.162) is the factorization (5.67) with $\omega \to -\omega$ and $l \to l-1$. Hence one solution to (5.163) is

$$f_{l-1}(r) = -\hbar l r^{-1} + M\omega r, \tag{5.197}$$

which means

$$D_l = R_{l-1}(-\omega). \tag{5.198}$$

With

$$f_{l-1}(r) = -\hbar l r^{-1} + M\omega r + g_l(r), \tag{5.199}$$

the solution to (5.163) gives

$$g_l(u) = (M\hbar\omega)^{1/2}\frac{u^{2l}\exp(-u^2)}{\gamma_l + \int_0^u t^{2l}\exp(-t^2)\,\mathrm{d}t}. \tag{5.200}$$

Here γ_l is an arbitrary constant and

$$u = r/a, \tag{5.201}$$

where a is given by (5.95). With the choice

$$\tilde{C}_l = -(2l + 3)M\hbar\omega, \tag{5.202}$$

(5.166) yields the generalized oscillator potential

$$\tilde{V}(u) = \hbar\omega\left(\frac{1}{2}u^2 - \frac{\mathrm{d}}{\mathrm{d}u}\frac{u^{2(l+1)}\exp(-u^2)}{\gamma_{l+1} + \int_0^u t^{2(l+1)}\exp(-t^2)\,\mathrm{d}t}\right). \tag{5.203}$$

For $\tilde{H}_l$ to be Hermitian with respect to $|El\rangle$ it is necessary that γ_l be real.

Let $|Nl\rangle$ denote an eigenket of H_l with eigenvalue E_{Nl} given by (5.103). Then with $D_l^\dagger = D_l'$ the kets

$$|\phi_{N+1,l}\rangle = \tfrac{1}{2}[(N+1)M\hbar\omega]^{-1/2}D_{l+1}^\dagger|N, l+1\rangle \qquad (N = 0, 1, 2, \ldots) \tag{5.204}$$

are orthonormal. It follows from the identity

$$\tilde{H}_l D_{l+1}^\dagger = D_{l+1}^\dagger(H_{l+1} + \hbar\omega) \tag{5.205}$$

that the kets (5.204) satisfy the eigenvalue equation

$$\tilde{H}_l|\phi_{N+1,l}\rangle = E_{N+1,l}|\phi_{N+1,l}\rangle \qquad (N = 0, 1, 2, \ldots). \tag{5.206}$$

If there exists a ket $|\phi_{0l}\rangle$ such that

$$D_{l+1}|\phi_{0l}\rangle = 0, \tag{5.207}$$

then

$$\langle\phi_{N+1,l}|\phi_{0l}\rangle = 0 \qquad (N = 0, 1, 2, \ldots) \tag{5.208}$$

and

$$\tilde{H}_l|\phi_{0l}\rangle = E_{0l}|\phi_{0l}\rangle. \tag{5.209}$$

Working in the coordinate representation and using

$$0 \le \int_0^u t^{2l}\exp(-t^2)\,\mathrm{d}t \le \frac{1}{2}\Gamma\left(l + \frac{1}{2}\right), \tag{5.210}$$

we see that for

$$\gamma_l > 0 \quad \text{or} \quad \gamma_l < -\tfrac{1}{2}\Gamma(l + \tfrac{1}{2}), \tag{5.211}$$

the potential $\tilde{V}$ has no singularities and $\tilde{V} \to \frac{1}{2}M\omega^2 r^2$ as $r \to \infty$. For these values of γ_l, the operators (5.165) are a one-parameter family of radial Hamiltonians. The parameter-dependent potentials (5.177) and (5.203), respectively, are also known as isospectral deformations of the Coulomb and oscillator potentials.

Equation (5.204) can be written

$$|\phi_{N+1,l}\rangle = |N+1, l\rangle + \tfrac{1}{2}[(N+1)M\hbar\omega]^{-1/2}g_{l+1}(r)|N, l+1\rangle. \tag{5.212}$$

Hence we obtain the radial coordinate-space wavefunctions

$$\phi_{N+1,l}(r) = R_{N+1,l}(r) + \tfrac{1}{2}[(N+1)M\hbar\omega]^{-1/2}g_{l+1}(r)R_{N,l+1}(r), \tag{5.213}$$

where R_{Nl} is the radial coordinate-space wavefunction of H_l (see (6.79)). If γ_l satisfies (5.211), the functions ϕ_{Nl} are normalized. Similarly, (5.207) yields

$$\phi_{0l}(r) = b_l r^l \exp\left(-\frac{r^2}{2a^2} - \frac{1}{\hbar} \int g_{l+1}(r)\, \mathrm{d}r \right), \tag{5.214}$$

where b_l is a constant.

The factorizations discussed above can be extended to an arbitrary number of dimensions [5], and they can also be obtained for other potentials [6].

5.9 The ground state

Suppose $\tilde{H}_l$ (eqn (5.165)) has a discrete spectrum and p_r is Hermitian. By taking the expectation value of both sides of (5.164) we have

$$2ME_l + \tilde{C}_l \geq 0, \tag{5.215}$$

where E_l is an eigenvalue of $\tilde{H}_l$. Hence $-\tilde{C}_l/2M$ is a lower bound for E_l. Consider s states, $|E_0\rangle$. If

$$D_1 |E_0\rangle = 0, \tag{5.216}$$

then (5.164) gives

$$D_1^\dagger D_1 = 2M(\tilde{H}_0 - E_0), \tag{5.217}$$

and $|E_0\rangle$ is the ground state. Substituting (5.160) in (5.217) we find that

$$-\hbar \frac{\mathrm{d}f_0}{\mathrm{d}r} + f_0^2 = 2M(\tilde{V} - E_0). \tag{5.218}$$

In the coordinate representation (5.216) is

$$\left(\frac{\mathrm{d}}{\mathrm{d}r} + \frac{1}{r} + \frac{f_0}{\hbar} \right) R_0(r) = 0, \tag{5.219}$$

which has the solution

$$R_0(r) = c_0 r^{-1} \exp\left(-\hbar^{-1} \int f_0\, \mathrm{d}r \right), \tag{5.220}$$

where c_0 is a constant. For the validity of (5.216) it is necessary that (5.220) be an acceptable radial wavefunction (see also Section 3.8).

For certain potentials (5.218) and (5.220) provide a simple method of determining the energy and radial coordinate-space wavefunction of the ground state. Consider, for example, the Hulthén potential [7]

$$\tilde{V}(r) = -U_0 \frac{1}{\exp(r/a) - 1}, \tag{5.221}$$

where U_0 and a are positive constants. For this potential (5.218) has the solution

$$f_0 = -\frac{\hbar}{a}\frac{1}{\exp(r/a) - 1} + (-2ME_0)^{1/2}, \tag{5.222}$$

where

$$E_0 = -\frac{\hbar^2}{8Ma^2}(\nu^2 - 1)^2, \tag{5.223}$$

and

$$\nu = (2Ma^2\hbar^{-2}U_0)^{1/2}. \tag{5.224}$$

The Hellmann–Feynman theorem shows that the condition for the existence of this state is (see Section 5.10)

$$\nu^2 > 1; \tag{5.225}$$

that is,

$$U_0 > \frac{\hbar^2}{2Ma^2}. \tag{5.226}$$

Substituting (5.222) in (5.220) and performing the integration we find

$$R_0(r) = c_0 u^{-1}[1 - \exp(-u)] \exp[-\tfrac{1}{2}(\nu^2 - 1)u], \tag{5.227}$$

where c_0 is a constant and $u = r/a$. For the Hulthén potential (5.223) is the ground state energy, (5.227) is the radial wavefunction for the ground state, and (5.225) is the condition for the existence of this state.* Other systems to which the above method can be applied have been discussed in the previous section.

For a given potential $\tilde{V}$ there is usually more than one solution to (5.218). For example, in the case of the oscillator ($\tilde{V} = \frac{1}{2}M\omega^2 r^2$), eqn (5.218) has the solutions

$$f_0 = M\omega r, \qquad E_0 = \tfrac{1}{2}\hbar\omega, \tag{5.228}$$

and

$$f_0 = M\omega r - \hbar r^{-1}, \qquad E_0 = \tfrac{3}{2}\hbar\omega. \tag{5.229}$$

The solution (5.228) must be rejected because the corresponding function (5.220) is not a wavefunction: it does not satisfy (F.5) and hence it is not a solution to Schrödinger's equation in Cartesian coordinates. The acceptable solution is (5.229). In the simple case of a constant potential $\tilde{V}$ we have the solutions

---

* Excited bound s states of the Hulthén potential are considered in Section 5.10.

$$f_0 = -\hbar k \cot kr, \qquad \hbar k \tan kr, \qquad f_0 = \pm i\hbar k, \tag{5.230}$$

where

$$k = [2M(E_0 - \tilde{V})\hbar^{-2}]^{1/2}. \tag{5.231}$$

Each of these four solutions is acceptable because for the corresponding operators D_1 there exist states that satisfy (5.216). (See Section 6.5. Of course, $|E_0\rangle$ is not a ground state in this case.)

5.10 Other spherically symmetric potentials

If we consider just s states ($l = 0$), the centrifugal term in (5.25) is zero and consequently there are additional potentials that possess analytic solutions for these states. These solutions can be obtained using the factorization method of Section 3.2 and we consider three examples.

5.10.1 The three-dimensional Morse potential

Here

$$V(r) = D\left[\exp\left(-2\frac{r - r_0}{a}\right) - 2\exp\left(-\frac{r - r_0}{a}\right)\right], \tag{5.232}$$

where D, r_0, and a are positive constants. In the limit*

$$r_0/a \to \infty, \tag{5.233}$$

the discussion given for the one-dimensional potential in Section 3.4 can be extended to the three-dimensional case by replacing x with $r - r_0$. Hence we let

$$X \to u = (r - r_0)/a - \ln(2k + 1) \tag{5.234}$$

(see (3.103)), and

$$p_X \to p_u = \frac{a}{\hbar} p_r. \tag{5.235}$$

The latter substitution ensures that $[p_u, u] = -i$.

Thus for bound s states the energy eigenvalues are given by (3.112) and the coordinate-space wavefunctions are†

---

* If r_0/a is finite, the analysis is more complicated [8].

† Note that in the coordinate representation

$$p_r \frac{1}{r} \to -i\hbar \frac{1}{r}\frac{d}{dr}$$

(see (5.35)). Hence $\chi_N(r)$ and the wavefunctions for the corresponding one-dimensional potentials satisfy the same differential equations.

$$\psi_N(r) = \frac{1}{r}\chi_N(r), \tag{5.236}$$

where $\chi_N(r)$ are obtained from (3.163) by making the substitution (5.234). The number of bound s states is the same as the number of bound states for the one-dimensional potential (see Section 3.4).

Implicit in the above is the assumption that p_r is Hermitian with respect to the bound states, and this requires

$$\chi_N(0) = 0; \tag{5.237}$$

this condition also ensures that (5.236) is a solution to the Schrödinger equation in Cartesian coordinates (see Appendix F). The wavefunctions calculated above satisfy (5.237) in the limit (5.233).

5.10.2 The three-dimensional symmetric Rosen–Morse potential

Here

$$V(r) = -U_0 \operatorname{sech}^2 \frac{r}{a}, \tag{5.238}$$

where U_0 and a are positive constants. We can extend the analysis for the corresponding one-dimensional potential (Section 3.5) to the three-dimensional potential (Section 3.5) to the three-dimensional case by making the substitutions

$$X \rightarrow u = r/a \tag{5.239}$$

and (5.235).

Thus in the coordinate-space wavefunctions (5.236) for bound s states, $\chi_N(r)$ are obtained from the even and odd wavefunctions (3.211) and (3.216) by making the substitution (5.239). There are two points to notice here. First, it is clear from (3.211) that the even wavefunctions do not satisfy (5.237) and hence they must be excluded. Second, because $0 \leq r \leq \infty$ whereas $-\infty \leq x \leq \infty$, we must include an extra factor of $2^{1/2}$ in the normalization constant in (3.216).

The energy eigenvalues for bound s states are (see (3.185) and (3.176))

$$E = -\frac{\hbar^2}{2Ma^2}(k - N)^2, \tag{5.240}$$

where k is given by the positive solution in (3.173) and $N = 1, 3, \ldots < k$ (see (3.177) with $U_1 = 0$). The condition for at least one bound state to exist is $k > 1$, that is

$$Ma^2\hbar^{-2}U_0 > 1. \tag{5.241}$$

5.10.3 *The Hulthén potential*

As a final example we consider bound s states of the potential (5.221) [9]. The energy eigenvalue equation is

$$\left(\frac{1}{2M}p_r^2 - U_0\frac{1}{\exp(r/a)-1}\right)|E\rangle = E|E\rangle, \tag{5.242}$$

where $E < 0$. It is convenient to introduce the dimensionless real quantity

$$\lambda = (-2Ma^2\hbar^{-2}E)^{1/2} \tag{5.243}$$

and to use the dimensionless operators (5.235) and (5.239). By multiplying both sides of (5.242) on the left by $(1 - e^{-u})^2$ and rearranging terms, we have

$$O_n|n\lambda\rangle = -\lambda^2|n\lambda\rangle \tag{5.244}$$

where $|n\lambda\rangle$ denotes $|E\rangle$ and

$$O_n = (1 - e^{-u})^2 p_u^2 - (n^2 + \lambda^2)e^{-u} + n^2 e^{-2u}, \tag{5.245}$$

$$n = (\nu^2 + \lambda^2)^{1/2}. \tag{5.246}$$

Inspection of (5.245) suggests that O_n can be factorized by

$$Q_n^{\pm} = \pm i(1 - e^{-u})p_u + ne^{-u} + K_n^{\pm}, \tag{5.247}$$

where $K_n^{\pm}$ are independent of u and p_u. A short calculation shows that

$$Q_{n+1}^{-}Q_n^{+} = O_n + K_{n+1}^{-}K_n^{+} \tag{5.248}$$

and

$$Q_{n-1}^{+}Q_n^{-} = O_n + K_{n-1}^{+}K_n^{-}, \tag{5.249}$$

where

$$K_n^{+} = K_{n+1}^{-} = -(2n+1)^{-1}(n^2 + n + \lambda^2). \tag{5.250}$$

It follows from (5.248), (5.249), and (5.244) that

$$Q_n^{\pm}|n\lambda\rangle = \beta_n^{\pm}|n \pm 1, \lambda\rangle. \tag{5.251}$$

According to (5.251) and (5.246) the shift operators $Q_n^{\pm}$ transform between eigenkets of the same λ belonging to different potentials (different ν). (Thus the kets $|n'\lambda\rangle$ and $|n\lambda\rangle$ are not necessarily orthogonal.)

Equations (5.223) and (5.225) show that for the ground state

$$\nu^2 = 2\lambda + 1, \tag{5.252}$$

and hence $n = \lambda + 1$. By applying $Q_{\lambda+1}^{+}, Q_{\lambda+2}^{+}Q_{\lambda+1}^{+}, \ldots$ to the ground state $|\lambda + 1, \lambda\rangle$ one generates $|\lambda + 2, \lambda\rangle, |\lambda + 3, \lambda\rangle, \ldots$. These kets are normalizable if the ground state is normalizable (see below). Thus for a given Hulthén potential (ν fixed) the values of n are

$$n = N + \lambda + 1 \qquad (N = 0, 1, 2, \ldots). \tag{5.253}$$

(There is an upper limit to the value of N for which the potential will bind the particle—see (5.256).) Equations (5.253), (5.246), and (5.243) yield the energy eigenvalues for bound s states

$$E_N = -\frac{\hbar^2}{2Ma^2}\left(\frac{v^2 - (N+1)^2}{2(N+1)}\right)^2. \tag{5.254}$$

The condition for the existence of these states can be obtained by treating U_0 as a continuous variable and applying the Hellmann–Feynman theorem to (5.242). Using (5.254) and (5.224) this yields

$$\langle (\mathrm{e}^u - 1)^{-1} \rangle = \frac{v^2 - (N+1)^2}{2(N+1)^2}. \tag{5.255}$$

For bound states the left-hand side of (5.255) is positive because the average value of the potential (5.221) is negative. Therefore $v^2 > (N+1)^2$. Thus for a given potential the number of bound s states is finite and determined by

$$U_0 > \frac{\hbar^2}{2Ma^2}(N+1)^2, \tag{5.256}$$

which is the extension of (5.226) to excited s states.

If the kets in (5.251) are normalized

$$|\beta_n^{\pm}|^2 = \langle n\lambda | (Q_n^{\pm})^{\dagger} Q_n^{\pm} | n\lambda \rangle. \tag{5.257}$$

Now

$$(Q_n^{+})^{\dagger} = Q_{n+1}^{-} - 2\mathrm{e}^{-u}, \qquad (Q_n^{-})^{\dagger} = Q_{n-1}^{+} + 2\mathrm{e}^{-u}. \tag{5.258}$$

Substituting (5.258) in (5.257) and using (5.248), (5.249), and (5.244) we find

$$|\beta_n^{\pm}|^2 = (K_n^{\pm})^2 - \lambda^2 \mp 2\langle \mathrm{e}^{-u} Q_n^{\pm} \rangle. \tag{5.259}$$

To evaluate the expectation values in (5.259) we require the expectation values of $\mathrm{e}^{-u}(1 - \mathrm{e}^{-u})p_u$, e^{-2u}, and e^{-u}. With the aid of the hypervirial theorem these can be expressed in terms of (5.255). Equation (6.4) with $f = \mathrm{e}^{-u}(1 - \mathrm{e}^{-u})$ yields

$$\langle \mathrm{i}\mathrm{e}^{-u}(1 - \mathrm{e}^{-u})p_u \rangle = \tfrac{1}{2}\langle \mathrm{e}^{-u} - 2\mathrm{e}^{-2u} \rangle. \tag{5.260}$$

Similarly, from (6.5) with $l = 0$ and $f = (1 - \mathrm{e}^{-u})^2$ and $f = 1 - \mathrm{e}^{-u}$, respectively, we obtain

$$\langle \mathrm{e}^{-2u} \rangle = \frac{n^2 + 3\lambda^2 - 1}{4(n^2 - 1)}\langle \mathrm{e}^{-u} \rangle \tag{5.261}$$

and

$$\langle \mathrm{e}^{-u} \rangle = \frac{2v^2}{4n^2 - 1}\langle (\mathrm{e}^u - 1)^{-1} \rangle. \tag{5.262}$$

In terms of n and λ, eqn. (5.255) is

$$\langle(\mathrm{e}^{u}-1)^{-1}\rangle=\frac{\lambda}{n-\lambda}. \tag{5.263}$$

Substituting (5.247), (5.250), and (5.260)–(5.263) in (5.259) we obtain after some algebra, and setting a phase factor equal to -1,

$$\beta_n^{\pm}=-\frac{n-\lambda\pm 1}{2n\pm 1}\left(\frac{n}{n\pm 1}(n+\lambda)(n+\lambda+1)\right)^{1/2}. \tag{5.264}$$

It is helpful to adopt a simpler notation by letting $|N\rangle$ denote $|n\lambda\rangle$. Then (5.251) becomes

$$Q_N^{\pm}|N\rangle=\beta_N^{\pm}|N\pm 1\rangle, \tag{5.265}$$

where

$$Q_N^{\pm}=\pm\mathrm{i}(1-\mathrm{e}^{-u})p_u+(N+\lambda+1)\mathrm{e}^{-u}+K_N^{\pm}, \tag{5.266}$$

and $\beta_N^{\pm}$ and $K_N^{\pm}$ are given by (5.264) and (5.250) with n replaced by (5.253). Because $|\beta_N^{+}|^2>0$ for $N=1,2,\ldots$, the kets $|1\rangle, |2\rangle, \ldots$ are normalizable if the ground state $|0\rangle$ is normalizable. Also $\beta_0^{-}=0$ and so

$$Q_0^{-}|0\rangle=0, \tag{5.267}$$

which is just (5.216) with D_1 given by (5.161) and (5.222). In Section 6.4 the shift operators are used to obtain coordinate-space wavefunctions.

5.11 Supersymmetry and shape invariance of spherically symmetric potentials

The basic theory of supersymmetry and shape invariance of one-dimensional systems given in Sections 3.8 and 3.9 can be extended to spherically symmetric systems [10] by replacing x with r and p with p_r. Thus, for example, (3.300) and (3.303) become

$$A_{\pm}=\pm\frac{\mathrm{i}}{(2M)^{1/2}}p_r+W(r) \tag{5.268}$$

and

$$V_{\pm}(r)=W^2\pm\frac{\hbar}{(2M)^{1/2}}\frac{\mathrm{d}W}{\mathrm{d}r}. \tag{5.269}$$

In a similar manner one can transcribe the other equations in Sections 3.8 and 3.9. Here we consider two applications of the solutions (3.337) and (3.343) for shape-invariant potentials.

(i) Corresponding to

$$W=\left(\frac{M}{2}\right)^{1/2}\omega r-\frac{\hbar(l+1)}{(2M)^{1/2}r} \tag{5.270}$$

we have

$$V_-(r, a_0) = \frac{1}{2}M\omega^2 r^2 + \frac{\hbar^2 l(l+1)}{2Mr^2} - \hbar\omega\left(l + \frac{3}{2}\right),$$
$$V_+(r, a_0) = \frac{1}{2}M\omega^2 r^2 + \frac{\hbar^2 (l+1)(l+2)}{2Mr^2} - \hbar\omega\left(l + \frac{1}{2}\right). \tag{5.271}$$

If we identify

$$a_s = l + s \qquad (s = 0, 1, \ldots), \tag{5.272}$$

then $V_\pm(r, a_s)$ are given by (5.271) with l replaced by a_s. These potentials satisfy (3.334) with remainder

$$R(a_s) = 2\hbar\omega \qquad (s = 1, 2, \ldots). \tag{5.273}$$

Hence (3.337) and (3.306) yield the eigenvalues

$$E_N^- = 2N\hbar\omega \qquad (N = 0, 1, \ldots). \tag{5.274}$$

The Hamiltonian H_- given by (3.302) and (5.271) is equal to the radial Hamiltonian (5.25) for the oscillator minus $\hbar\omega(l + \frac{3}{2})$ and E_N^- is equal to the eigenvalue (5.103) minus $\hbar\omega(l + \frac{3}{2})$.

(ii) With

$$W = \left(\frac{M}{2}\right)^{1/2} \frac{k}{\hbar(l+1)} - \frac{\hbar(l+1)}{(2M)^{1/2} r} \tag{5.275}$$

(k = positive constant), we have

$$V_-(r, a_0) = -\frac{k}{r} + \frac{\hbar^2 l(l+1)}{2Mr^2} + \frac{Mk^2}{2\hbar^2(l+1)^2},$$
$$V_+(r, a_0) = -\frac{k}{r} + \frac{\hbar^2 (l+1)(l+2)}{2Mr^2} + \frac{Mk^2}{2\hbar^2(l+1)^2}. \tag{5.276}$$

For a_s given by (5.272), $V_\pm(r, a_s)$ satisfy (3.334) with

$$R(a_s) = \frac{Mk^2}{2\hbar^2}\left(\frac{1}{(a_{s-1}+1)^2} - \frac{1}{(a_s+1)^2}\right) \qquad (s = 1, 2, \ldots). \tag{5.277}$$

Hence (3.337) and (3.306) give

$$E_N^- = -\frac{Mk^2}{2\hbar^2}\left(\frac{1}{(N+l+1)^2} - \frac{1}{(l+1)^2}\right) \qquad (N = 0, 1, \ldots). \tag{5.278}$$

The Hamiltonian H_- is equal to the radial Hamiltonian (5.25) for the Coulomb potential plus $Mk^2/[2\hbar^2(l+1)^2]$, and E_N^- are the eigenvalues (5.135) for bound states plus $Mk^2/[2\hbar^2(l+1)^2]$.

We leave it as an exercise for the reader to determine the radial coordinate-space wavefunctions from (3.308) and (3.343). Explicit expressions for these wavefunctions are given in Section 6.4. Other examples of spherically symmetric shape-invariant potentials are discussed in Ref. 11.

References

1. Newmarch, J.D. and Golding, R.M. (1978). Ladder operators for some spherically symmetric potentials. *American Journal of Physics*, **46**, 658–60.
2. See, for example, Landau, L.D. and Lifschitz, E.M. (1977). *Quantum mechanics*, 3rd edn, Section 21. Pergamon, Oxford.
3. Fernández, D.J. (1984). New hydrogen-like potentials. *Letters in Mathematical Physics*, **8**, 337–43.
4. Dongpei, Zhu (1987). A new potential with the spectrum of an isotonic oscillator. *Journal of Physics A: Math. Gen.*, **20**, 4331–6.
5. Alves, N.A. and Drigo Filho, E. (1988). The factorization method and supersymmetry. *Journal of Physics A: Math. Gen.*, **21**, 3215–25.
6. Drigo Filho, E. (1988). The Morse oscillator generalized from supersymmetry. *Journal of Physics A: Math. Gen.*, **21**, L1025–8.
7. The Hulthén potential and its applications are discussed by Hulthén, Lamek and Sugawara, Masao (1957). The two-nucleon problem. In *Encyclopedia of Physics*, Vol. 39, pp. 1–143, (ed. S. Flügge). Springer, Berlin. See also Matthys, P. and De Meyer, H. (1988). Dynamical-group approach to the Hulthén potential. *Physical Review*, A **38**, 1168–71 and references therein. In the latter paper solutions for the non-zero angular momentum states are discussed.
8. See Flügge, S. (1974). *Practical quantum mechanics*, pp. 182–86. Springer, New York.
9. de Lange, O.L. (1991). An operator solution for the Hulthén potential. *American Journal of Physics*, **59**, 151–54.
10. See Sukumar, C.V. (1985). Supersymmetric quantum mechanics and the inverse scattering method. *Journal of Physics A: Math. Gen.*, **18**, 2937–55.
11. Dutt, R., Khare, A., and Sukhatme, U.P. (1988). Sypersymmetry, shape invariance, and exactly solvable potentials. *American Journal of Physics*, **56**, 163–8.

6

APPLICATIONS

We consider some applications of results obtained in the previous chapter. In Sections 6.1–6.3 operator methods are used to calculate matrix elements for the three-dimensional isotropic harmonic oscillator and the Coulomb potential. The shift operators also provide a simple method of constructing normalized coordinate-space wavefunctions and some examples are given in Section 6.4. In Sections 6.5–6.8 we discuss applications to constant and sectionally constant potentials.

6.1 Use of the hypervirial theorem [1]

In the particular case of eigenkets $|El\rangle$ of a radial Hamiltonian H_l the hypervirial theorem (eqn (1.193)) becomes

$$\langle El|[H_l, W]|El\rangle = 0. \tag{6.1}$$

Here it is assumed that H_l is Hermitian with respect to $|El\rangle$ and $W|El\rangle$; otherwise, the operator W is arbitrary. By considering various forms for W certain expectation values can be obtained from (6.1). These calculations are similar to those in Section 3.3 for one-dimensional problems.

Using (5.25) and the commutation relation (5.22) we can show that

$$[g(r), H_l] = \frac{\mathrm{i}\hbar}{M}\frac{\mathrm{d}g}{\mathrm{d}r}p_r + \frac{\hbar^2}{2M}\frac{\mathrm{d}^2 g}{\mathrm{d}r^2} \tag{6.2}$$

and

$$[f(r)p_r, H_l] = \mathrm{i}\hbar\left(\frac{\hbar^2 l(l+1)}{Mr^3} - \frac{\mathrm{d}V}{\mathrm{d}r}\right)f + 2\mathrm{i}\hbar\frac{\mathrm{d}f}{\mathrm{d}r}\left(H_l - \frac{\hbar^2 l(l+1)}{2Mr^2} - V(r)\right) + \frac{\hbar^2}{2M}\frac{\mathrm{d}^2 f}{\mathrm{d}r^2}p_r. \tag{6.3}$$

Thus, with $W = g(r)$ and $\mathrm{d}g/\mathrm{d}r = f(r)$, (6.1) and (6.2) give

$$\langle nl|f(r)p_r|nl\rangle = \tfrac{1}{2}\mathrm{i}\hbar\langle nl|\mathrm{d}f/\mathrm{d}r|nl\rangle. \tag{6.4}$$

Similarly, with

$$W = f(r)p_r + \tfrac{1}{2}\mathrm{i}\hbar\,\mathrm{d}f/\mathrm{d}r,$$

(6.1) and (6.3) yield

$$\langle nl|\frac{\hbar^2}{4M}\frac{\mathrm{d}^3 f}{\mathrm{d}r^3} + 2\left(E - V(r) - \frac{\hbar^2 l(l+1)}{2Mr^2}\right)\frac{\mathrm{d}f}{\mathrm{d}r} - \frac{\mathrm{d}V}{\mathrm{d}r}f + \frac{\hbar^2 l(l+1)}{Mr^3}f|nl\rangle = 0. \tag{6.5}$$

Equations (6.4) and (6.5) are used below.

6.2 Matrix elements for the oscillator

The radial Hamiltonian for the oscillator,

$$H_l = \frac{1}{2M}p_r^2 + \frac{\hbar^2 l(l+1)}{2Mr^2} + \frac{1}{2}M\omega^2 r^2, \tag{6.6}$$

is a function of the parameters l and ω. We treat these parameters as continuous variables and apply the Hellmann–Feynman theorem (1.203) to the eigenvalue equation (5.26):

$$\left\langle \frac{\partial H_l}{\partial l} \right\rangle = \frac{\partial E}{\partial l} \tag{6.7}$$

and

$$\left\langle \frac{\partial H_l}{\partial \omega} \right\rangle = \frac{\partial E}{\partial \omega}, \tag{6.8}$$

where $\langle A \rangle$ denotes $\langle nl|A|nl\rangle$. Substituting (6.6) and (5.103) in (6.7) and (6.8) we find

$$\langle r^{-2}\rangle = (l + \tfrac{1}{2})^{-1}a^{-2} \tag{6.9}$$

and

$$\langle r^2\rangle = (n + \tfrac{3}{2})a^2, \tag{6.10}$$

where a is the constant (5.95). The result (6.10) can also be obtained from the virial theorem (1.192) and by using (5.103).

Next we apply (6.4) and (6.5). With $f = r^s$ (s a constant), (6.4) becomes

$$\langle r^s p_r\rangle = \tfrac{1}{2}\mathrm{i}\hbar s\langle r^{s-1}\rangle, \tag{6.11}$$

and with $f = r^{s+1}$, (6.5) becomes

$$(s+2)\langle r^{s+2}\rangle - (s+1)(2n+3)a^2\langle r^s\rangle - sa^4[\tfrac{1}{4}(s^2-1) - l(l+1)]\langle r^{s-2}\rangle = 0. \tag{6.12}$$

Setting $s = 0, 1, 2, \ldots$ in (6.11) gives

$$\langle p_r\rangle = 0, \tag{6.13}$$

$$\langle rp_r\rangle = \tfrac{1}{2}\mathrm{i}\hbar, \tag{6.14}$$

$$\langle r^2 p_r\rangle = \mathrm{i}\hbar\langle r\rangle, \tag{6.15}$$

etc. Note that (6.11), and hence (6.13)–(6.15), apply to both the oscillator and bound states of the Coulomb potential (and to any spherically symmetric problem for which the assumptions underlying (6.4) are valid, such as (5.157) and (5.158)).

Next we consider expectation values of powers of r (these expectation values have useful applications in, for example, molecular and nuclear physics

[2, 3]). From the recurrence relation (6.12), and using either (6.9) or (6.10), we can write down the expectation values of even-integer powers of r:

$$\langle r^4 \rangle = \tfrac{1}{8}[12(n+\tfrac{3}{2})^2 - 4l(l+1) + 3]a^4, \tag{6.16}$$

etc. Equation (6.12) can also be solved for the expectation value of any odd-integer power of r in terms of $\langle r \rangle$ and $\langle r^{-1} \rangle$:

$$\langle r^3 \rangle = \tfrac{2}{3}(2n+3)a^2\langle r \rangle + \tfrac{1}{3}l(l+1)a^4\langle r^{-1} \rangle, \tag{6.17}$$

etc.

To calculate $\langle r \rangle$ and $\langle r^{-1} \rangle$ we use shift operators obtained in the previous chapter. Consider first the case $n = l$. From (5.89)

$$\langle ll|\, R_l^+(-\omega)\,|ll\rangle = 0. \tag{6.18}$$

Substituting (5.85) and (6.13) this gives

$$\langle ll|\, r^{-1}\,|ll\rangle = [a^2(l+1)]^{-1}\langle ll|\, r\,|ll\rangle. \tag{6.19}$$

We use the operator $R_l^+(\omega)$ to obtain a recurrence relation for $\langle ll|\,r\,|ll\rangle$. From (5.83)

$$|\alpha^+(\omega)|^2\langle l+1, l+1|\, r\,|l+1, l+1\rangle = \langle ll|\,[R_l^+(\omega)]^\dagger r R_l^+(\omega)\,|ll\rangle. \tag{6.20}$$

Now

$$[R_l^+(\omega)]^\dagger r = r[R_l^+(\omega)]^\dagger - \hbar \tag{6.21}$$

(see (5.85) and (5.22)), and

$$[R_l^+(\omega)]^\dagger R_l^+(\omega)\,|ll\rangle = |\alpha^+(\omega)|^2\,|ll\rangle \tag{6.22}$$

(see (5.83), (5.85), and (5.86)). Substituting (6.21), (6.22), (5.85), (5.86), (6.19), and (6.13) in (6.20) yields the desired relation

$$\langle l+1, l+1|\, r\,|l+1, l+1\rangle = \frac{2l+4}{2l+3}\langle ll|\,r\,|ll\rangle. \tag{6.23}$$

Thus

$$\langle ll|\,r\,|ll\rangle = \frac{(2l+2)!!}{2(2l+1)!!}\langle 00|\,r\,|00\rangle, \tag{6.24}$$

where

$$\begin{aligned} N!! &= N(N-2)\ldots 2 \qquad (N \text{ even}) \\ &= N(N-2)\ldots 1 \qquad (N \text{ odd}). \end{aligned} \tag{6.25}$$

From (6.24) and (6.19)

$$\langle ll|\,r^{-1}\,|ll\rangle = \frac{(2l)!!}{a^2(2l+1)!!}\langle 00|\,r\,|00\rangle. \tag{6.26}$$

The average value of r in the ground state is calculated below (eqn (6.37)).

Next we calculate the expectation values of r and r^{-1} for kets with $n > l$. First we use the operator Q_n^+ to obtain recurrence relations. From (5.98)

$$|\beta^+|^2\langle n+2, l|r|n+2, l\rangle = \langle nl|(Q_n^+)^\dagger r Q_n^+|nl\rangle \tag{6.27}$$

and

$$|\beta^+|^2\langle n+2, l|r^{-1}|n+2, l\rangle = \langle nl|(Q_n^+)^\dagger r^{-1} Q_n^+|nl\rangle. \tag{6.28}$$

From (5.99) and (5.22) it follows that

$$(Q_n^+)^\dagger r = r(Q_n^+)^\dagger - \hbar r \tag{6.29}$$

and

$$(Q_n^+)^\dagger r^{-1} = r^{-1}(Q_n^+)^\dagger + \hbar r^{-1}. \tag{6.30}$$

Equations (6.27)–(6.30), (5.96), (5.101), (6.13), (6.15), and (6.17) yield the recurrence relations

$$r_{n+2} = \left(1 + \frac{n+3}{(n-l+2)(n+l+3)}\right) r_n - a^2 \frac{l(l+1)}{(n-l+2)(n+l+3)} (r^{-1})_n \tag{6.31}$$

and

$$(r^{-1})_{n+2} = \left(1 - \frac{n+2}{(n-l+2)(n+l+3)}\right)(r^{-1})_n + a^{-2}\frac{1}{(n-l+2)(n+l+3)} r_n, \tag{6.32}$$

where $r_n = \langle nl|r|nl\rangle$ and $(r^{-1})_n = \langle nl|r^{-1}|nl\rangle$.

Starting with the expectation values for $n = l$ (eqns (6.24) and (6.26)), the values of r_n and $(r^{-1})_n$ for $n = l+2, l+4, \ldots$ can be obtained from (6.31) and (6.32). The results can be expressed in the form

$$\langle nl|r|nl\rangle = \langle 00|r|00\rangle \frac{1}{2} \sum_{s=0}^{N} \frac{(2l+2+2s)!!}{(2l+1+2s)!!} a_s(N) \tag{6.33}$$

and

$$\langle nl|r^{-1}|nl\rangle = a^{-2}\langle 00|r|00\rangle \sum_{s=0}^{N} \frac{(2l+2s)!!}{(2l+1+2s)!!} b_s(N), \tag{6.34}$$

where $N = \frac{1}{2}(n-l) = 0, 1, 2, \ldots$, and $a_s(N)$ and $b_s(N)$ are independent of l. In Table 6.1(a) values of

$$\frac{(2s)!!}{(2s+1)!!} a_s(N)$$

are given for the first four values of N; by inspection, these entries are equal to

TABLE 6.1 Values of (a) $[(2s)!!/(2s+1)!!]a_s(N)$ and (b) $[(2s)!!/(2s-1)!!]b_s(N)$ where $a_s(N)$ and $b_s(N)$ are the quantities in (6.33) and (6.34). These values are obtained by using (6.24) and (6.26) in the recurrence relations (6.31) and (6.32)

	(a)				(b)			
N $s=$	0	1	2	3	0	1	2	3
0	1				1			
1	$-\frac{1}{2!!}$	1			$\frac{1}{2!!}$	1		
2	$-\frac{1}{4!!}$	$-\frac{1}{2!!}$	1		$\frac{3!!}{4!!}$	$\frac{1}{2!!}$	1	
3	$-\frac{3!!}{6!!}$	$-\frac{1}{4!!}$	$-\frac{1}{2!!}$	1	$\frac{5!!}{6!!}$	$\frac{3!!}{4!!}$	$\frac{1}{2!!}$	1

$$(-1)^{1+\delta_{sN}}\frac{(2N-2s-3)!!}{(2N-2s)!!},$$

with the proviso that $(-3)!!=1$ and $(-1)!!=1$. Thus

$$a_s(N)=(-1)^{1+\delta_{sN}}\frac{(2s+1)!!}{(2s)!!}\frac{(2N-2s-3)!!}{(2N-2s)!!}. \qquad (6.35)$$

The values of

$$\frac{(2s)!!}{(2s-1)!!}b_s(N)$$

in Table 6.1(b) are equal to

$$\frac{(2N-2s-1)!!}{(2N-2s)!!}.$$

Hence

$$b_s(N)=\frac{(2s-1)!!(2N-2s-1)!!}{(2s)!!(2N-2s)!!}. \qquad (6.36)$$

Finally, to calculate the factor $\langle 00|r|00\rangle$ in (6.33) and (6.34) we use the normalized ground state wavefunction (6.68):

$$\langle 00|r|00\rangle=\frac{4}{\pi^{1/2}a^3}\int_0^\infty r^3\exp(-r^2/a^2)\,dr=2\pi^{-1/2}a. \qquad (6.37)$$

This completes the calculation of expectation values of integral powers of r. The reader might find it instructive to obtain the results (6.33)–(6.36) by a wave-mechanical calculation, that is by evaluating

$$\int_0^\infty r^{k+2}(R_{nl}(r))^2\,dr, \qquad (k = \pm 1)$$

where R_{nl} is the radial wavefunction (6.79).

As a second example consider the matrix elements $\langle n'l'|r|nl\rangle$ where $l' = l \pm 1$. (These radial dipole matrix elements are important in the theory of rotation–vibration spectra of polyatomic molecules [2]. The selection rule $\Delta l = \pm 1$ is a consequence of (4.83) and (4.84); see also Section 8.3.) By taking linear combinations of the shift operators (5.85) we can express r in terms of these operators:

$$2M\omega r = R_l^+(-\omega) - R_l^+(\omega) \tag{6.38}$$

and

$$2M\omega r = R_l^-(-\omega) - R_l^-(\omega). \tag{6.39}$$

Hence using (5.83), (5.84), and (5.86) we obtain

$$r|nl\rangle = a[\tfrac{1}{2}(n-l)]^{1/2}|n-1, l+1\rangle - a[\tfrac{1}{2}(n+l+3)]^{1/2}|n+1, l+1\rangle \tag{6.40}$$

and

$$r|nl\rangle = a[\tfrac{1}{2}(n-l+2)]^{1/2}|n+1, l-1\rangle - a[\tfrac{1}{2}(n+l+1)]^{1/2}|n-1, l-1\rangle. \tag{6.41}$$

Multiplying (6.40) on the left with $\langle n', l+1|$ and (6.41) with $\langle n', l-1|$, and using the orthonormality condition $\langle n'l|nl\rangle = \delta_{n'n}$, we have

$$\langle n \pm 1, l \pm 1|r|nl\rangle = -a[\tfrac{1}{2}(n+l+2\pm 1)]^{1/2}, \tag{6.42}$$

$$\langle n \pm 1, l \mp 1|r|nl\rangle = a[\tfrac{1}{2}(n-l+1\pm 1)]^{1/2}, \tag{6.43}$$

and the selection rule $\Delta n = \pm 1$. If $n = l$, (6.40) reduces to the simple shift operation

$$r|ll\rangle = -a(l+\tfrac{3}{2})^{1/2}|l+1, l+1\rangle. \tag{6.44}$$

In a similar manner we can evaluate the radial factor $\langle n'l'|r^s|nl\rangle$ ($s = 2, 3, \ldots$) of multipole matrix elements. In Chapter 8 we generalize this approach to the total multipole matrix element for an oscillator.

6.3 Matrix elements for bound states of the Coulomb potential

The radial Hamiltonian with a Coulomb potential,

$$H_l = \frac{1}{2M}p_r^2 + \frac{\hbar^2 l(l+1)}{2Mr^2} - \frac{\hbar^2}{Mar}, \tag{6.45}$$

is a function of the parameters l and a. We treat these parameters as continuous variables and apply the Hellmann–Feynman theorem (1.203) to the eigenvalue equation (5.26):

$$\left\langle \frac{\partial H_l}{\partial l} \right\rangle = \frac{\partial E}{\partial l} \tag{6.46}$$

and

$$\left\langle \frac{\partial H_l}{\partial a} \right\rangle = \frac{\partial E}{\partial a}. \tag{6.47}$$

Substituting (6.45) and (5.135) in (6.46) and (6.47) we obtain

$$\langle r^{-2} \rangle = [a^2 n^3 (l + \tfrac{1}{2})]^{-1} \tag{6.48}$$

and

$$\langle r^{-1} \rangle = (an^2)^{-1}. \tag{6.49}$$

The last result also follows from the virial theorem (1.192) and (5.81).

Next we use the hypervirial theorem. From (6.4) with $f = r^s$ (s a constant) it follows that (6.13)–(6.15) are valid also for bound states of a Coulomb potential. Similarly, from (6.5) with $f = r^{s+1}$ we have the Kramers–Pasternak recursion relation

$$(s+1)\langle r^s \rangle - (2s+1)n^2 a \langle r^{s-1} \rangle - sn^2 a^2 [\tfrac{1}{4}(s^2 - 1) - l(l+1)] \langle r^{s-2} \rangle = 0. \tag{6.50}$$

With $s = 0$, (6.50) yields (6.49). Taking $s = 1, 2, \ldots$ we obtain

$$\langle r \rangle = \tfrac{1}{2} a [3n^2 - l(l+1)], \tag{6.51}$$

$$\langle r^2 \rangle = \tfrac{1}{2} n^2 a^2 [5n^2 - 3l(l+1) + 1], \tag{6.52}$$

etc. Substituting $s = -1, -2, \ldots$ in (6.50) and using (6.48) we find

$$\langle r^{-3} \rangle = [a^3 n^3 l(l + \tfrac{1}{2})(l+1)]^{-1}, \tag{6.53}$$

$$\langle r^{-4} \rangle = \frac{3n^2 - l(l+1)}{2an^2(l - \tfrac{1}{2})(l + \tfrac{3}{2})} \langle r^{-3} \rangle, \tag{6.54}$$

etc. The formulae (6.53), (6.54), etc., are restricted to their range of validity; thus, for example, $l \neq 0$ in these equations. These expectation values are important in, for example, the theory of Rydberg atoms [4].

There is a simple general relation between the expectation values of positive and negative powers of r; we leave it as an exercise for the reader to show that [5]

$$\langle r^{-(s+3)} \rangle = \left(\frac{2}{na} \right)^{2s+3} \frac{(2l - s - 1)!}{(2l + s + 2)!} \langle r^s \rangle \qquad (s = -1, 0, \ldots, 2l - 1).$$

The above applications of the Hellmann–Feynman and hypervirial theorems yield expectation values, with respect to the kets $|El\rangle$, of functions of r and p_r. In these calculations we used the radial Hamiltonian H_l. We can also use the Hamiltonian H and evaluate expectation values with respect to the kets $|Elm\rangle$. Because $H|Elm\rangle = H_l|Elm\rangle$ this yields the same expectation values, that is

$$\langle Elm|f(r)p_r^s|Elm\rangle = \langle El|f(r)p_r^s|El\rangle \tag{6.55}$$

(s a non-negative integer).

Next we consider the radial dipole matrix elements*

$$\langle n'l'|r|nl\rangle \qquad (n' \neq n, l' = l \pm 1) \tag{6.56}$$

for a Coulomb potential. We first deduce a recurrence relation for these matrix elements. From (5.78)

$$2lR_{l-1}^{+}r = (2l+1)R_l^{+}r + R_{l+1}^{-}r + 2\hbar(2l+1). \tag{6.57}$$

Taking matrix elements of both sides of (6.57) between $\langle n'l|$ and $|nl\rangle$ yields for $n' \neq n$

$$\begin{aligned}\left[1-\left(\frac{l}{n'}\right)^2\right]^{1/2}&\langle n', l-1|r|nl\rangle \\ &= \frac{2l+1}{2l+2}\left[1-\left(\frac{l+1}{n}\right)^2\right]^{1/2}\langle n'l|r|n, l+1\rangle \\ &\quad + \frac{1}{2l+2}\left[1-\left(\frac{l+1}{n'}\right)^2\right]^{1/2}\langle n', l+1|r|nl\rangle,\end{aligned} \tag{6.58}$$

where we have used (5.72), (5.22), (5.77), and (5.79). It is useful to note that (6.58) is also valid if n and n' are interchanged.

The recurrence relation (6.58) enables us to determine all matrix elements (6.56) in terms of a pair of such elements. A simple choice for this initial pair is $\langle n'n'|r|n, n'+1\rangle$ and $\langle n', n'-1|r|nn'\rangle$. Because $|n'n'\rangle = 0$ (see Section 5.5) the former matrix element is zero

$$\langle n'n'|r|n, n'+1\rangle = 0. \tag{6.59}$$

To calculate the latter matrix element we choose the coordinate representation

$$\langle n', n'-1|r|nn'\rangle = \int_0^\infty r^3 R_{n',n'-1}^{*}(r) R_{nn'}(r)\,dr, \tag{6.60}$$

where $R_{nl}(r) = \langle r|nl\rangle$. In the next section the radial functions $R_{nl}(r)$ are obtained by applying shift operators for angular momentum and energy to the

* These are required for a calculation of the electromagnetic transition probability in a Coulomb potential [6].

ground state $R_{10}(r)$; the result is (6.98). Substituting (6.98) in (6.60) and using (B.4), (B.5), and (3.152) to perform the integration we find

$$\langle n', n'-1|r|nn'\rangle = -a\frac{2^{2n'+1}(n')^{n'+2}n^{n'+1}}{(n+n')^{2n'+3}}\left(\frac{(n+n')!}{(2n'-1)!(n-n'-1)!}\right)^{1/2}S, \tag{6.61}$$

where

$$S = \sum_{j=0}^{n-n'-1}(2n'+2+j)\frac{(n-n'-1)!}{j!(n-n'-1-j)!}u^j \tag{6.62}$$

and $u = -2n'/(n+n')$. This sum is readily evaluated:

$$S = (2n'+2)S_{n-n'-1}(u) + (n-n'-1)uS_{n-n'-2}(u), \tag{6.63}$$

where

$$S_p(u) = \sum_{j=0}^{p}\frac{p!}{j!(p-j)!}u^j = (1+u)^p.$$

Hence

$$\langle n', n'-1|r|nn'\rangle = -a2^{2n'+2}(nn')^{n'+2}\left(\frac{(n+n')!}{(2n'-1)!(n-n'-1)!}\right)^{1/2} \times\frac{(n-n')^{n-n'-2}}{(n+n')^{n+n'+2}}. \tag{6.64}$$

From (6.58), (6.59), and (6.64) we can calculate the matrix elements (6.56). The results for $n' = 1$ and 2 are given in Table 6.2.

Alternatively, one can evaluate the matrix elements directly from

TABLE 6.2 Radial dipole matrix elements $\langle n', l \pm 1|r|nl\rangle$ $(n' \neq n)$ for bound states of the Coulomb potential with $n' = 1$ and 2, calculated as described in the text

$\langle 10\|r\|n1\rangle = -a2^4n^{7/2}\frac{(n-1)^{n-(5/2)}}{(n+1)^{n+(5/2)}}$
$\langle 21\|r\|n2\rangle = -a2^{19/2}3^{-1/2}(n^2-1)^{1/2}n^{9/2}\frac{(n-2)^{n-(7/2)}}{(n+2)^{n+(7/2)}}$
$\langle 21\|r\|n0\rangle = -a2^{15/2}3^{-1/2}n^{9/2}\frac{(n-2)^{n-3}}{(n+2)^{n+3}}$
$\langle 20\|r\|n1\rangle = -a2^{17/2}(n^2-1)^{1/2}n^{7/2}\frac{(n-2)^{n-3}}{(n+2)^{n+3}}$

$$\langle n'l'|r|nl\rangle = \int_0^\infty r^3 R^*_{n'l'}(r) R_{nl}(r)\,\mathrm{d}r. \tag{6.65}$$

Proceeding as above we find that the right-hand side of (6.65) reduces to a double sum which can be evaluated in terms of hypergeometric functions [7]. The same method can be applied to matrix elements of $r^s \exp(-\alpha r)$ where α is a positive constant [7].

The properties of radial matrix elements for the Coulomb potential have attracted a great deal of attention, and in addition to the above results there are several other relationships that have been derived for these matrix elements. For a study of these and their applications we refer the reader to the literature [5].

6.4 Radial coordinate-space wavefunctions

In the coordinate representation of wave mechanics the abstract operator equations derived in Chapter 5 reduce to differential equations (see Section 5.2). These differential forms can be obtained by making the substitutions (5.33)–(5.35) and $|Elm\rangle \to \psi_{Elm}(\mathbf{r})$ and $|El\rangle \to R_{El}(r)$ in the operator equations. Here $R_{El}(r)$ is the radial coordinate-space wavefunction; it is the normalized solution to (5.38). In the following we use the notation $R_{nl}(r) = \langle r|nl\rangle$ to denote this radial wavefunction.

In this section we use shift operations in differential form to calculate radial coordinate-space wavefunctions. These calculations involve the solution to a simple differential equation, and an elementary integration and differentiation; they are similar to those given in Chapters 3 and 4.

6.4.1 The oscillator

According to (5.98) and (5.101), $Q_0^-|00\rangle = 0$ and hence

$$\left(\frac{\mathrm{d}}{\mathrm{d}r} + \frac{r}{a^2}\right) R_{00}(r) = 0, \tag{6.66}$$

where R_{00} is the ground state wavefunction. The normalization condition is

$$\int_0^\infty r^2 |R_{00}(r)|^2\,\mathrm{d}r = 1. \tag{6.67}$$

The solution to (6.66) and (6.67) is (we set a phase factor equal to unity)

$$R_{00}(r) = \frac{2}{\pi^{1/4} a^{3/2}} \exp\left(-\frac{r^2}{2a^2}\right). \tag{6.68}$$

From the ground state we obtain all normalized radial wavefunctions $R_{nl}(r)$. Consider first the case $n = l$. From (6.44)

$$R_{l+1,l+1}(r) = -\left(\frac{2}{2l+3}\right)^{1/2}\frac{r}{a}R_{ll}(r). \tag{6.69}$$

Equations (6.68) and (6.69) give

$$R_{ll}(r) = (-1)^l \frac{2}{\pi^{1/4}a^{3/2}}\left(\frac{2^l}{(2l+1)!!}\right)^{1/2}\left(\frac{r}{a}\right)^l \exp\left(-\frac{r^2}{2a^2}\right). \tag{6.70}$$

If we write

$$R_{nl}(r) = R_{ll}(r)\phi_n(r), \tag{6.71}$$

then (5.98), (5.96), (5.101), and (6.70) show that

$$\phi_n(u) = 2[(n-l)(n+l+1)]^{-1/2}\left(u\frac{\mathrm{d}}{\mathrm{d}u} - u + \frac{1}{2}(n+l+1)\right)\phi_{n-2}(u) \tag{6.72}$$

where $u = r^2/a^2$. Now $\phi_l = 1$ and hence (6.72) yields

$$\phi_{2N+l}(u) = \gamma_N \dots \gamma_2\gamma_1 P_N(u) \qquad (N = 1, 2, \dots). \tag{6.73}$$

Here

$$\gamma_N = \left(\frac{2N+2l+1}{2N}\right)^{1/2} \tag{6.74}$$

and

$$P_N(u) = D_N \dots D_2 D_1 1, \tag{6.75}$$

where

$$D_N = 1 - \frac{u}{N+l+\frac{1}{2}} + \frac{u}{N+l+\frac{1}{2}}\frac{\mathrm{d}}{\mathrm{d}u}. \tag{6.76}$$

By inspection

$$\gamma_N \dots \gamma_2\gamma_1 = \left(\frac{(2N+2l+1)!!}{N!2^N(2l+1)!!}\right)^{1/2}, \tag{6.77}$$

where $N!!$ is defined by (6.25). Also, from (B.1) we have

$$P_N(u) = {}_1F_1(-N; l+\tfrac{3}{2}; u). \tag{6.78}$$

Substituting these results in (6.71) we obtain the normalized radial wavefunctions for the oscillator

$$R_{nl}(r) = (-1)^l\left(\frac{2^{l-N+2}(2N+2l+1)!!}{a^3\pi^{1/2}N![(2l+1)!!]^2}\right)^{1/2} {}_1F_1\left(-N; l+\frac{3}{2}; \frac{r^2}{a^2}\right)$$
$$\times \left(\frac{r^2}{a^2}\right)^{l/2}\exp\left(-\frac{r^2}{2a^2}\right). \tag{6.79}$$

Here

$$N = \tfrac{1}{2}(n - l) = 0, 1, 2, \ldots \tag{6.80}$$

is the radial quantum number; it is equal to the number of times that Q_n^+ must be applied to transform R_{ll} into R_{nl}. The radial wavefunction vanishes N times for finite values of r; thus the number of nodes of this function is equal to N.

The phase factor in (6.79) is determined by our choice of unity for the phase factors in (5.86) and (5.101); if we take these factors to be $\exp(\pm i\theta)$ and $\exp(\pm i\theta')$, respectively, an additional factor $\exp[-i(l\theta + N\theta')]$ would appear on the right-hand side of (6.79).

6.4.2 *The Coulomb potential*

Consider bound states. From (5.132), (5.133), and (5.121)

$$\left(\frac{d}{dr} + \frac{1}{a}\right) R_{10}(r) = 0, \tag{6.81}$$

where R_{10} is the ground state wavefunction. The normalization condition is

$$\int_0^\infty r^2 |R_{10}(r)|^2 \, dr = 1. \tag{6.82}$$

The solution to (6.81) and (6.82) is

$$R_{10}(r) = 2a^{-3/2} \exp\left(-\frac{r}{a}\right), \tag{6.83}$$

where we have set a phase factor equal to unity.

From the ground state we generate all normalized radial wavefunctions $R_{nl}(r)$. Consider first the case $n = l + 1$. From (5.137) and (5.114)

$$r\Delta_{l+1}^+ R_{l+1,l}(r) = -a[\tfrac{1}{2}(2l + 3)(l + 2)^3]^{1/2} R_{l+2,l+1}(r), \tag{6.84}$$

where (see (5.126) and (5.127))

$$\Delta_{l+1}^+ = \exp\left[\left(r\frac{d}{dr} + \frac{3}{2}\right) \ln \frac{l+1}{l+2}\right]. \tag{6.85}$$

To determine the effect of these differential scaling operators on wavefunctions we write

$$\Delta_n^\pm R_{n'l}(r) = \delta_n^\pm R_{n'l}\left(\frac{n}{n \pm 1} r\right) \tag{6.86}$$

where $\delta_n^\pm$ are independent of r. Now

$$\int_0^\infty R_{n'l}^*(r) R_{n'l}(r) r^2 \, dr = \int_0^\infty [\Delta_n^\pm R_{n'l}(r)]^* [\Delta_n^\pm R_{n'l}(r)] r^2 \, dr$$

(see (5.116)), and thus

$$\delta_n = \left(\frac{n}{n \pm 1}\right)^{3/2}, \tag{6.87}$$

where a phase factor has been set equal to unity. Using (6.86) and (6.87) in (6.84) we have

$$R_{l+2,l+1}(r) = -\left(\frac{2(l+1)^3}{(2l+3)(l+2)^4}\right)^{1/2} \frac{r}{a(l+2)} R_{l+1,l}\left(\frac{l+1}{l+2} r\right). \tag{6.88}$$

From (6.88) and (6.83)

$$R_{l+1,l}(r) = c_l \left(\frac{2r}{a(l+1)}\right)^l \exp\left(-\frac{r}{a(l+1)}\right), \tag{6.89}$$

where

$$c_l = \frac{(-1)^l}{[(2l+2)!]^{1/2}} \left(\frac{2}{a(l+1)}\right)^{3/2}. \tag{6.90}$$

Next we use the energy shift operation (5.128). In differential form this can be written (see (5.131), (5.114), (5.115), (6.86), and (6.87))

$$R_{nl}(u) = \left(\frac{n-1}{n}\right)^2 [(n-l-1)(n+l)]^{-1/2} \left(u \frac{\mathrm{d}}{\mathrm{d}u} - \frac{1}{2} u + n\right) R_{n-1,l}(u), \tag{6.91}$$

where $u = 2r/(na)$. Let

$$R_{nl}(u) = c_l u^l \exp(-\tfrac{1}{2}u) \phi_n(u). \tag{6.92}$$

Then $\phi_{l+1}(u) = 1$ and hence (6.91) and (6.92) yield

$$\phi_{N+l+1}(u) = \gamma_N \cdots \gamma_2 \gamma_1 P_N(u) \qquad (N = 1, 2, \ldots), \tag{6.93}$$

where

$$\gamma_N = \left(\frac{N+l}{N+l+1}\right)^2 \left(\frac{N+2l+1}{N}\right)^{1/2} \tag{6.94}$$

and P_N is given by (6.75) with

$$D_N = 1 - \frac{u}{N+2l+1} + \frac{u}{N+2l+1} \frac{\mathrm{d}}{\mathrm{d}u}. \tag{6.95}$$

By inspection

$$\gamma_N \cdots \gamma_2 \gamma_1 = \left(\frac{l+1}{N+l+1}\right)^2 \left[\frac{(N+2l+1)!}{N!(2l+1)!}\right]^{1/2}. \tag{6.96}$$

Also, from (B.1)

$$P_N(u) = {}_1F_1(-N; 2l+2; u). \tag{6.97}$$

Substituting these results in (6.92) we obtain the normalized radial coordinate-space wavefunctions for bound states of an attractive Coulomb potential

$$R_{nl}(r) = (-1)^l \left(\frac{2^2(n+l)!}{a^3 n^4 (n-l-1)![(2l+1)!]^2}\right)^{1/2} {}_1F_1\left(-n+l+1; 2l+2; \frac{2r}{na}\right)$$
$$\times \left(\frac{2r}{na}\right)^l \exp\left(-\frac{r}{na}\right), \tag{6.98}$$

where

$$n = N + l + 1. \tag{6.99}$$

The radial quantum number N $(= 0, 1, 2, \ldots)$ is equal to the number of times that Q_n^+ must be applied to transform $R_{l+1,l}$ into R_{nl}. The radial wavefunction has N nodes.

If we had chosen the phase factors in (5.79) and (5.131) to be $\exp(\pm i\theta)$ and $\exp(\pm i\theta')$, respectively, then an additional factor $\exp[-i(l\theta + n\theta')]$ would have appeared on the right-hand side of (6.98).

The radial wavefunctions (6.98) are often expressed in terms of the generalized Laguerre polynomials

$$L_n^{(\alpha)}(u) = \sum_{j=0}^{n} \binom{n+\alpha}{n-j} \frac{(-u)^j}{j!}$$
$$= \binom{n+\alpha}{n} {}_1F_1(-n; \alpha+1; u),$$

or the associated Laguerre polynomials

$$L_n^\alpha(u) = (-1)^\alpha n! L_{n-\alpha}^{(\alpha)}(u)$$

(α a non-negative integer).

In the operator method presented above normalized wavefunctions were calculated by using first-order differential operators (and, for the Coulomb potential, the scaling operation (6.86)). It is instructive to compare this method with the wave-mechanical approach in which one solves a second-order differential equation.

Consider bound states of an attractive Coulomb potential. The radial equation (5.38) can be written

$$\frac{1}{u^2}\frac{d}{du}\left(u^2 \frac{dR}{du}\right) + \left(\frac{\lambda}{u} - \frac{1}{4} - \frac{l(l+1)}{u^2}\right) R = 0, \tag{6.100}$$

where $u = 2r/(\lambda a)$ and

$$\lambda = (-2Ma^2\hbar^{-2}E)^{-1/2}. \tag{6.101}$$

Let

$$R(u) = u^s \exp(-u/2) F(u). \tag{6.102}$$

Then if s is a solution to

$$s(s+1) - l(l+1) = 0, \tag{6.103}$$

(6.100) becomes

$$u\frac{d^2F}{du^2} + (2s+2-u)\frac{dF}{du} - (s+1-\lambda)F = 0. \tag{6.104}$$

The task of finding a general solution to the confluent hypergeometric equation (6.104) is not trivial. The solution that is valid for all values of λ is [8]

$$F(u) = A\Psi(s+1-\lambda, 2s+2; u) + Be^u\Psi(s+1+\lambda, 2s+2; -u), \tag{6.105}$$

where A and B are arbitrary constants and

$$\begin{aligned}\Psi(a, 2s+2; u) = \frac{(-1)^{2s}}{(2s+1)!\Gamma(a-2s-1)}\Bigg[& {}_1F_1(a; 2s+2; u)\ln u \\ & + \sum_{r=0}^{\infty}\frac{\Gamma(a+r)\Gamma(2s+2)}{\Gamma(a)\Gamma(2s+2+r)} \\ & \times (\chi(a+r) - \chi(1+r) - \chi(2s+2+r))\frac{u^r}{r!}\Bigg] \\ & + \frac{(2s)!}{\Gamma(a)}\sum_{r=0}^{2s}\frac{\Gamma(a-2s-1+r)\Gamma(-2s)}{\Gamma(a-2s-1)\Gamma(-2s+r)}\frac{u^{r-2s-1}}{r!}.\end{aligned} \tag{6.106}$$

Here

$$\chi(z) = \frac{d}{dz}\ln\Gamma(z).$$

From this solution we can conclude:

(i) If $B \neq 0$, $R(u)$ diverges at $u = \infty$. Thus square-integrability of R requires $B = 0$.

(ii) Because [8]

$$\Psi(a, c; u) = u^{1-c}\Psi(a-c+1, 2-c; u),$$

the solution (6.102) is the same for both roots of (6.103), namely $s = l$ and $s = -l-1$. In the following we suppose $s = l$.

(iii) If $l = 0$, R is square-integrable for all λ. However, (6.102) with $l = 0$ is not a solution to the Schrödinger equation in Cartesian coordinates unless [9] $\lambda = N + 1$, where $N = 0, 1, 2, \ldots$.

(iv) If $l > 0$ and $\lambda \neq N + l + 1$ we see from the last term in (6.106) that

$$R \sim \frac{1}{\Gamma(l+1-\lambda)}u^{-(2l+1)} \qquad \text{as } u \to 0, \tag{6.107}$$

and consequently R is not square-integrable.

(v) If

$$\lambda = N + l + 1, \tag{6.108}$$

$\Gamma(l + 1 - \lambda)$ has a pole and the singularity (6.107) does not occur. Instead [9]

$$\begin{aligned}\Psi(l + 1 - \lambda, 2l + 2; u) &= \Psi(-N, 2l + 2; u) \\ &= (-1)^N \frac{\Gamma(N + 2l + 2)}{\Gamma(2l + 2)} {}_1F_1(-N; 2l + 2; u),\end{aligned} \tag{6.109}$$

and R is square-integrable.

(vi) If the radial wavefunction (6.102) with $F = A\Psi(-N, 2l + 2; u)$ is normalized, the result is identical (to within a phase factor) to (6.98). Also (6.101) and (6.108) yield the energy eigenvalue (5.135).

Next we consider the states of a repulsive Coulomb potential and the scattering states of an attractive Coulomb potential. For these the spectrum of energy eigenvalues is continuous and non-negative (Section 5.5). Thus (5.93) can be written

$$n = \pm \frac{\mathrm{i}}{k|a|}, \tag{6.110}$$

where the wavenumber

$$k = (2ME/\hbar^2)^{1/2} \tag{6.111}$$

is real. Radial coordinate-space wavefunctions can be obtained (apart from a normalization constant) by making the substitutions (6.110) in (6.98). Taking the lower sign in (6.110) we have

$$R_{kl}(r) = C_{kl}\,{}_1F_1\left(l + 1 + \frac{\mathrm{i}}{k|a|}; 2l + 2; 2\mathrm{i}\hat{a}kr\right)(2kr)^l \exp(-\mathrm{i}\hat{a}kr), \tag{6.112}$$

where C_{kl} is a constant and $\hat{a} = \operatorname{sign} a$ is $+1$ for an attractive potential and -1 for a repulsive potential (see Section 5.3). A convenient normalization is [10]

$$\int_0^\infty r^2 R_{k'l}(r) R_{kl}(r)\,\mathrm{d}r = 2\pi\delta(k' - k). \tag{6.113}$$

It can be shown that this requires [10]

$$C_{kl} = \frac{2k}{(2l + 1)!}\left|\Gamma\left(l + 1 - \frac{\mathrm{i}}{ka}\right)\right| \exp\left(\frac{\pi}{2ka}\right), \tag{6.114}$$

where a phase factor has been set equal to zero. Because these wavefunctions are real [11], our choice of sign in (6.110) is unimportant. The radial wavefunctions R_{El} that are normalized according to (5.28) are related to the above by [10]

$$R_{El}(r) = \left(\frac{M}{2\pi\hbar^2 k}\right)^{1/2} R_{kl}(r). \tag{6.115}$$

If the shift operations (5.77) are expressed in the coordinate representation (see Section 5.2) they yield recurrence relations for the wavefunctions (6.112); thus from one normalized wavefunction one can obtain all the normalized wavefunctions by application of these shift operations [11]. The solution (6.112) is the regular solution* to the radial equation (5.38) with a Coulomb potential and for $E \geq 0$. This radial equation also possesses a solution which is irregular at $r = 0$ and the differential shift operations obtained from (5.77) apply also to this irregular solution [11].

6.4.3 *The Hulthén potential*

As a final example we consider bound s states of the Hulthén potential (5.221). The ground state wavefunction is the normalized solution to the differential form of (5.267); that is (5.227) with

$$c_0 = 2a^{-3/2}[\lambda(\lambda + \tfrac{1}{2})(\lambda + 1)]^{1/2}. \tag{6.116}$$

To determine the coordinate-space wavefunction $\psi_N(u) = \langle u|N\rangle$ for excited s states we let (see (5.227) and (5.252))

$$\psi_N(u) = c_0 u^{-1}(1 - \mathrm{e}^{-u})\mathrm{e}^{-\lambda u}\phi_N(u), \tag{6.117}$$

and we use the raising operation (5.265) expressed in the coordinate representation. This yields

$$\phi_N(u) = \gamma_N \cdots \gamma_2\gamma_1 P_N(u) \qquad (N = 1, 2, \ldots), \tag{6.118}$$

where

$$\gamma_N = \left(\frac{(N + \lambda + 1)(N + 2\lambda)(N + 2\lambda + 1)}{(N + \lambda)(N + 1)^2}\right)^{1/2} \tag{6.119}$$

and $P_N(u)$ is given by (6.75) with

$$D_N = 1 - \frac{2N + 2\lambda + 1}{N + 2\lambda}x + \frac{2N + 2\lambda + 1}{(N + 2\lambda)(N + 2\lambda + 1)}x(1 - x)\frac{\mathrm{d}}{\mathrm{d}x} \tag{6.120}$$

and $x = \mathrm{e}^{-u}$.

It is elementary to show that

$$\gamma_N \cdots \gamma_2\gamma_1 = \frac{\Gamma(N + 2\lambda + 1)}{(N + 1)!\Gamma(2\lambda + 1)}\left(\frac{(N + \lambda + 1)(N + 2\lambda + 1)}{(\lambda + 1)(2\lambda + 1)}\right)^{1/2}, \tag{6.121}$$

* It is regular at $r = 0$.

where we have used (3.162). Also, with the aid of (B.3) we see that

$$P_N(u) = {}_2F_1(-N, N + 2\lambda + 2; 2\lambda + 1; x). \tag{6.122}$$

Hence we obtain the wavefunctions for bound s states of the Hulthén potential

$$\psi_N(u) = a^{-3/2}\frac{\Gamma(N + 2\lambda + 1)}{(N + 1)!\Gamma(2\lambda + 1)}[2\lambda(N + \lambda + 1)(N + 2\lambda + 1)]^{1/2}$$
$$\times {}_2F_1(-N, N + 2\lambda + 2; 2\lambda + 1; e^{-u})u^{-1}(1 - e^{-u})e^{-\lambda u}. \tag{6.123}$$

The condition for the existence of these states is (5.256).

It is a useful exercise to extend the calculations of Sections 6.1–6.4 to the potentials (5.157) and (5.158): in particular, one can use the Hellmann–Feynman and hypervirial theorems and shift operators for energy to obtain expectation values and wavefunctions for these potentials.

6.5 A sectionally constant potential

In the remainder of this chapter we consider a particle in a sectionally constant potential. We use an angular momentum basis and discuss applications of the shift operations (5.75) for angular momentum. For this purpose it is convenient to use the coordinate representation.

Consider a particle with energy E in a potential which is equal to a constant V in some interval. Suppose $E > V$ and define the wavenumber

$$k = [2M(E - V)/\hbar^2]^{1/2}. \tag{6.124}$$

In this interval the radial equation (5.38) is

$$\left(\frac{d^2}{dr^2} + \frac{2}{r}\frac{d}{dr} + k^2 - \frac{l(l + 1)}{r^2}\right)R_{kl}(r) = 0, \tag{6.125}$$

where $R_{kl}(r)$ is a radial coordinate-space wavefunction.

We obtain the general solution to (6.125) by using the differential form of the operator equations (5.75) and (5.216), namely [12],

$$\left(\frac{d}{dr} + \frac{1}{r} + \frac{1}{\hbar}f_0\right)R_{k0}(r) = 0, \tag{6.126}$$

$$\hbar\left(\frac{d}{dr} - \frac{l}{r}\right)R_{kl}(r) = \alpha^+ R_{k,l+1}(r), \tag{6.127}$$

$$\hbar\left(-\frac{d}{dr} - \frac{l + 1}{r}\right)R_{kl}(r) = \alpha^- R_{k,l-1}(r). \tag{6.128}$$

Here f_0 is given by (5.230) and $\alpha^\pm$ are, in general, functions of k and l.

For $f_0 = -\hbar k \cot kr$ and $f_0 = \hbar k \tan kr$ the solutions to (6.126) are

$$R_{k0} = A_k \frac{\sin kr}{kr} \tag{6.129}$$

and

$$R_{k0} = -B_k \frac{\cos kr}{kr}, \tag{6.130}$$

respectively, while for $f_0 = \mp i\hbar k$

$$R_{k0}^{\pm} = C_k^{\pm} \frac{\exp(\pm ikr)}{kr}. \tag{6.131}$$

Here A_k, B_k, and $C_k^{\pm}$ are independent of r and the minus sign in (6.130) is for convenience. Clearly $R_{k0}^{\pm}$ are linear combinations of (6.129) and (6.130): with the time dependence $\exp(-i\omega t)$ $(\omega > 0)$ the latter two equations describe a stationary spherical wave, whereas R_{k0}^{+} and R_{k0}^{-} have the form of outgoing and incoming spherical waves respectively.

We use (6.127) to generate radial wavefunctions with $l > 0$ from the s states (6.129)–(6.131). First note that because p_r is not Hermitian with respect to $R_{kl}(r)$ (see Appendix F), it is not necessary that the coefficients $\alpha^{\pm}$ be given by (5.73) and (5.74). However, at this point these two expressions are convenient choices for $\alpha^{\pm}$: thus we take (see (5.52), (5.53), (5.73), and (5.74))

$$\alpha^{\pm} = -[2M(E - V)]^{1/2} = -\hbar k, \tag{6.132}$$

where a phase factor has been set equal to -1. Let $R_{kl} = r^l F_{kl}$. Then (6.127) and (6.132) give

$$F_{k,l+1} = -\frac{1}{kr}\frac{\mathrm{d}}{\mathrm{d}r} F_{kl}, \tag{6.133}$$

and hence

$$F_{kl} = (-1)^l \frac{1}{k^l}\left(\frac{1}{r}\frac{\mathrm{d}}{\mathrm{d}r}\right)^l F_{k0}. \tag{6.134}$$

From (6.134) and (6.129)

$$R_{kl} = A_{kl} j_l(kr), \tag{6.135}$$

where

$$j_l(\rho) = (-1)^l \rho^l \left(\frac{1}{\rho}\frac{\mathrm{d}}{\mathrm{d}\rho}\right)^l \frac{\sin \rho}{\rho} \tag{6.136}$$

are spherical Bessel functions of the first kind. We have generalized (6.135) to cases where (6.132) does not apply by allowing A to depend on l; this would require a factor $A_{kl}/A_{k,l\pm1}$ in (6.132). An example is given in Section 6.7.

From (6.134) and (6.130)

$$R_{kl} = B_{kl} n_l(kr), \tag{6.137}$$

where

$$n_l(\rho) = (-1)^{l+1}\rho^l\left(\frac{1}{\rho}\frac{\mathrm{d}}{\mathrm{d}\rho}\right)^l\frac{\cos\rho}{\rho} \tag{6.138}$$

are spherical Bessel functions of the second kind (spherical Neumann functions). Equations (6.136) and (6.138) are Rayleigh's formulae for the spherical Bessel functions.

From (6.134) and (6.131)

$$R^+_{kl} = \mathrm{i}C^+_{kl}h^{(1)}_l(kr) \tag{6.139}$$

and

$$R^-_{kl} = -\mathrm{i}C^-_{kl}h^{(2)}_l(kr), \tag{6.140}$$

where

$$h^{(1)}_l(\rho) = \mathrm{i}(-1)^{l+1}\rho^l\left(\frac{1}{\rho}\frac{\mathrm{d}}{\mathrm{d}\rho}\right)^l\frac{\mathrm{e}^{\mathrm{i}\rho}}{\rho} \tag{6.141}$$

and

$$h^{(2)}_l(\rho) = \mathrm{i}(-1)^{l}\rho^l\left(\frac{1}{\rho}\frac{\mathrm{d}}{\mathrm{d}\rho}\right)^l\frac{\mathrm{e}^{-\mathrm{i}\rho}}{\rho} \tag{6.142}$$

are spherical Bessel functions of the third kind (spherical Hankel functions). The solutions j_l and n_l are linearly independent for every l, as are $h^{(1)}_l = j_l + \mathrm{i}n_l$ and $h^{(2)}_l = (h^{(1)}_l)^* = j_l - \mathrm{i}n_l$.

Equation (6.141) can be rewritten

$$h^{(1)}_l(\rho) = -\mathrm{i}(2l-1)!!\frac{\mathrm{e}^{\mathrm{i}\rho}}{\rho^{l+1}}P_l(\rho), \tag{6.143}$$

where $(2l-1)!!$ is defined by (6.25),

$$P_l(\rho) = D_l \dots D_2 D_1 1, \tag{6.144}$$

and

$$D_l = 1 - \frac{\mathrm{i}}{2l-1}\rho - \frac{\rho}{2l-1}\frac{\mathrm{d}}{\mathrm{d}\rho}. \tag{6.145}$$

Equation (B.2) with $N = l$ and $B = -2l$ shows that

$$P_l(\rho) = {}_1F_1(-l;\,-2l;\,-2\mathrm{i}\rho), \tag{6.146}$$

and hence

$$h^{(1)}_l(\rho) = -\mathrm{i}(2l-1)!!\frac{\mathrm{e}^{\mathrm{i}\rho}}{\rho^{l+1}}\,{}_1F_1(-l;\,-2l;\,-2\mathrm{i}\rho). \tag{6.147}$$

The spherical Bessel functions j_l and n_l are equal to the real and imaginary parts of (6.147) respectively. Thus for $l \geq 1$

$$j_l(\rho) = \frac{\sin \rho}{\rho}(\text{polynomial of order } l \text{ in } \rho^{-1})$$

$$+ \frac{\cos \rho}{\rho}(\text{polynomial of order } l-1 \text{ in } \rho^{-1}); \qquad (6.148)$$

$n_l(\rho)$ has a similar form but with $\sin \rho$ and $\cos \rho$ interchanged.

The limiting values as $\rho \to 0$ and $\rho \to \infty$ follow from (6.136) and (6.138). Near the origin

$$j_l(\rho) \approx \frac{\rho^l}{(2l+1)!!} \qquad (6.149)$$

and

$$n_l(\rho) \approx -(2l-1)!!\rho^{-l-1}, \qquad (6.150)$$

where $(-1)!! = 1$. Thus j_l is regular and n_l irregular at $\rho = 0$. As $\rho \to \infty$ the leading term in (6.136) is $\rho^{-1}(-\mathrm{d}/\mathrm{d}\rho)^l \sin \rho = \rho^{-1} \sin(\rho - \frac{1}{2}l\pi)$. A similar limit follows from (6.138). Hence the asymptotic forms are

$$j_l(\rho) \approx \rho^{-1} \sin(\rho - \tfrac{1}{2}l\pi), \qquad (6.151)$$

$$n_l(\rho) \approx -\rho^{-1} \cos(\rho - \tfrac{1}{2}l\pi), \qquad (6.152)$$

$$h_l^{(1)}(\rho) \approx (-\mathrm{i})^{l+1}\rho^{-1}\mathrm{e}^{\mathrm{i}\rho}. \qquad (6.153)$$

The general solution to (6.125) is a linear combination of any two of j_l, n_l, $h_l^{(1)}$, and $h_l^{(2)}$. For example,

$$R_{kl}(r) = A_{kl}j_l(kr) + B_{kl}n_l(kr). \qquad (6.154)$$

Other linear combinations are considered below.

If $E < V$ the solutions to (6.125) can be obtained by making the substitution

$$k \to \mathrm{i}\kappa = \mathrm{i}[2M(V-E)/\hbar^2]^{1/2} \qquad (6.155)$$

in (6.135)–(6.142). This yields

$$R_{\kappa l} = A_{\kappa l}j_l(\mathrm{i}\kappa r), \qquad (6.156)$$

$$R_{\kappa l} = B_{\kappa l}n_l(\mathrm{i}\kappa r), \qquad (6.157)$$

$$R_{\kappa l}^{+} = \mathrm{i}C_{\kappa l}^{+}h_l^{(1)}(\mathrm{i}\kappa r), \qquad (6.158)$$

and

$$R_{\kappa l}^{-} = -\mathrm{i}C_{\kappa l}^{-}h_l^{(2)}(\mathrm{i}\kappa r). \qquad (6.159)$$

Of these only (6.156) is finite at $r = 0$ and only (6.158) is finite at $r = \infty$. The general solution to (6.125) is a linear combination of any two of (6.156)–(6.159).

This completes our discussion of the general solutions to (6.125). In practical applications these solutions must satisfy certain boundary conditions, and we consider three examples.

6.6 The free particle

Suppose $V(r) = V$ for $0 \leq r \leq \infty$. Physical solutions to (6.125) must satisfy the boundary condition (see Appendix F)

$$rR_{kl} \to 0 \qquad \text{as } r \to 0, \tag{6.160}$$

and they must not diverge at $r = \infty$. This excludes the solutions (6.156)–(6.159) and consequently E must be greater than V. For the general solution (6.154) the boundary condition (6.160) requires $B_{kl} = 0$. Hence

$$R_{kl}(r) = A_{kl}\, j_l(kr). \tag{6.161}$$

Here the only restriction on the energy eigenvalue is $E > V$: the energy spectrum is continuous.

To calculate the constant A_{kl} consider first s states. Then

$$\begin{aligned}\int_0^\infty r^2 R_{k'0} R_{k0}\, \mathrm{d}r &= (k'k)^{-1} A_{k0}^2 \int_0^\infty \sin k'r \sin kr\, \mathrm{d}r \\ &= (2k'k)^{-1} A_{k0}^2 \pi[\delta(k' - k) - \delta(k' + k)],\end{aligned} \tag{6.162}$$

where

$$\delta(u) = \lim_{r\to\infty} \frac{\sin ur}{\pi u}$$

is a representation of a Dirac δ function (see Section 1.3). Now $k' + k \neq 0$ and thus if $A_{k0} = 2k$ we have the orthonormality condition for s states

$$\int_0^\infty r^2 R_{k'0} R_{k0}\, \mathrm{d}r = 2\pi\delta(k' - k). \tag{6.163}$$

In fact, with $A_{kl} = 2k$ all the wavefunctions (6.161) satisfy*

$$\int_0^\infty r^2 R_{k'l} R_{kl}\, \mathrm{d}r = 2\pi\delta(k' - k). \tag{6.164}$$

Wavefunctions that are normalized according to (5.28) are given in terms of R_{kl} by (6.115).

We can also use the radial wavefunctions (6.139) and (6.140). Consider the probability current density

$$\mathbf{J}^\pm = \frac{i\hbar}{2M}[\psi^\pm \nabla(\psi^\pm)^* - (\psi^\pm)^* \nabla\psi^\pm] \tag{6.165}$$

where $\psi^\pm = R_{kl}^\pm Y_{lm}$. As $r \to \infty$

---

* In (6.164) all but a negligible contribution to the integral comes from the region $r \to \infty$. Hence one can use the asymptotic form (6.151) in (6.164) and the calculation is the same as that leading to (6.163).

$$R_{kl}^{\pm} \to (\mp i)^l C_{kl}^{\pm} \frac{\exp(\pm ikr)}{kr}$$

and hence

$$\mathbf{J}^{\pm} \to \pm \frac{\hbar k}{M} |\psi^{\pm}|^2 \hat{\mathbf{r}}. \tag{6.166}$$

We can normalize to unit flux by requiring

$$\oint \mathbf{J}^{\pm} . \mathrm{d}\mathbf{S} = \pm 1, \tag{6.167}$$

where the integration is over any closed surface surrounding the origin 0. Let this surface be a sphere of radius r centred on 0. Then $\mathrm{d}\mathbf{S} = \hat{\mathbf{r}} r^2 \, \mathrm{d}\Omega$ and (6.167) yields

$$\frac{\hbar}{Mk} |C_{kl}^{\pm}|^2 \oint |Y_{lm}^{\pm}|^2 \, \mathrm{d}\Omega = \pm 1. \tag{6.168}$$

The integral in (6.168) is unity. So

$$C_{kl}^{\pm} = (Mk/\hbar)^{1/2}. \tag{6.169}$$

The solutions R_{kl}^{+} and R_{kl}^{-} apply respectively if there is a flux of particles away from or towards the origin.

We see that for a free particle the constants A_{kl} and $C_{kl}^{\pm}$ in the radial wavefunction are independent of l and thus (6.132) is an acceptable choice for $\alpha^{\pm}$; this is not true in the following example.

6.7 Particle confined in a sphere

Suppose the potential inside a sphere of radius a is equal to a constant V. For confinement of a particle to this sphere, its probability of being at the surface must be zero:

$$\langle a|El \rangle = 0. \tag{6.170}$$

Suppose $E > V$. (The solutions (6.156)–(6.159) for $E < V$ cannot satisfy both (6.160) and (6.170).) The solution that satisfies (6.160) is

$$R_{kl}(r) = A_{kl} j_l(kr). \tag{6.171}$$

Hence

$$j_l(ka) = 0. \tag{6.172}$$

Equation (6.172) determines the energy eigenvalues. For $l = 0, 1, 2$ the first four roots are given in Table 6.3. The energy eigenvalues are given in terms of these by

TABLE 6.3 The roots ρ_{nl} of $j_l(\rho_{nl}) = 0$ for $l = 0, 1, 2$ and $n = 1$ to 4. (The values of ρ_{nl} are given to two decimal figures)

n	$l =$ 0	1	2
1	3.14	4.49	5.76
2	6.28	7.72	9.10
3	9.42	10.90	12.32
4	12.57	14.07	15.51

$$E_{nl} = V + \frac{\hbar^2}{2Ma^2}\rho_{nl}^2. \tag{6.173}$$

If we let $r = ax$ and $ka = \rho_{nl}$, the normalization condition for the wavefunction (6.171) is

$$A_{kl}^2 a^3 \int_0^1 x^2[j_l(\rho_{nl}x)]^2 \, dx = 1. \tag{6.174}$$

The integral in (6.174) can be evaluated with the aid of the identity*

$$\frac{d}{d\rho}\left[\rho^2\left(\frac{d}{d\rho}\rho^{1/2}j_l\right)^2 - \left(l + \frac{1}{2}\right)^2 \rho j_l^2 + \rho^3 j_l^2\right] = 2\rho^2 j_l^2. \tag{6.175}$$

In (6.175) let $\rho = \rho_{nl}x$ and integrate between $\rho = 0$ and $\rho = \rho_{nl}$. Using (6.127), (6.172), and (6.149), we obtain

$$2\int_0^1 x^2[j_l(\rho_{nl}x)]^2 \, dx = [j_{l+1}(\rho_{nl})]^2, \tag{6.176}$$

and hence

$$A_{kl} = \sqrt{2}[a^{3/2}j_{l+1}(ka)]^{-1}. \tag{6.177}$$

6.8 Particle in a spherically symmetric rectangular well

Let

$$\begin{aligned} V(r) &= -V_0 \qquad (0 \le r < a) \\ &= 0 \qquad (r > a), \end{aligned} \tag{6.178}$$

where V_0 is a positive constant. The radial wavefunctions must satisfy

$$\lim_{\varepsilon \to 0} R_{kl}(a - \varepsilon) = \lim_{\varepsilon \to 0} R_{kl}(a + \varepsilon) \tag{6.179}$$

* Equation (6.175) can be verified using (6.125).

and

$$\lim_{\varepsilon\to 0}\frac{1}{R_{kl}}\frac{dR_{kl}}{dr}\bigg|_{a-\varepsilon} = \lim_{\varepsilon\to 0}\frac{1}{R_{\kappa l}}\frac{dR_{\kappa l}}{dr}\bigg|_{a+\varepsilon}. \tag{6.180}$$

Equation (6.179) ensures continuity of the probability density of the particle at $r = a$; (6.180) follows from (6.179) and the shift operations (6.127) and (6.128).

6.8.1 *Bound states* $(-V_0 < E < 0)$

For $r < a$, $E - V > 0$ and the solution satisfying (6.160) is

$$R_{kl} = A_{kl}\, j_l(kr); \tag{6.181}$$

for $r > a$, $E - V < 0$ and the solution that is zero at $r = \infty$ is

$$R_{\kappa l} = B_{\kappa l} h_l^{(1)}(\mathrm{i}\kappa r), \tag{6.182}$$

(see Section 6.5) where κ is given by (6.155) with $V = 0$. Thus (6.180) requires

$$\frac{1}{j_l(kr)}\frac{dj_l(kr)}{dr}\bigg|_{r=a} = \frac{1}{h_l^{(1)}(\mathrm{i}\kappa r)}\frac{dh_l^{(1)}(\mathrm{i}\kappa r)}{dr}\bigg|_{r=a}. \tag{6.183}$$

This can be written in a more convenient form by using the shift operations (6.127) and (6.128):

$$kX_l = \mathrm{i}\kappa Y_l, \tag{6.184}$$

where $X_l = j_{l+1}(ka)/j_l(ka)$ and $Y_l = h_{l+1}^{(1)}(\mathrm{i}\kappa a)/h_l^{(1)}(\mathrm{i}\kappa a)$ satisfy the recurrence relations

$$X_{l+1} = \frac{2l+3}{ka} - \frac{1}{X_l} \tag{6.185}$$

and

$$Y_{l+1} = \frac{2l+3}{\mathrm{i}\kappa a} - \frac{1}{Y_l}. \tag{6.186}$$

Using (6.136) and (6.141) we have

$$X_0 = \frac{\tan ka - ka}{ka \tan ka} \tag{6.187}$$

and

$$Y_0 = \frac{\kappa a + 1}{\mathrm{i}\kappa a}. \tag{6.188}$$

Define the dimensionless parameters

$$\beta = (2Ma^2V_0/\hbar^2)^{1/2} \tag{6.189}$$

and

$$x = \left(1 - \frac{|E|}{V_0}\right)^{1/2}. \tag{6.190}$$

Then $ka = \beta x$ and $\kappa a = \beta(1 - x^2)^{1/2}$ where $0 < x < 1$. From (6.185)–(6.190) it follows that (6.184) can be expressed as

$$\tan \beta x = F_l(\beta, x), \tag{6.191}$$

where

$$F_0(\beta, x) = -\frac{x}{(1 - x^2)^{1/2}}, \tag{6.192}$$

$$F_1(\beta, x) = \frac{\beta x(1 - x^2)}{1 + \beta x^2(1 - x^2)^{1/2}}, \tag{6.193}$$

$$F_2(\beta, x) = \beta x \frac{1 + \beta(1 - x^2)^{1/2} + \frac{1}{3}\beta^2 x^2(1 - x^2)}{1 + \beta(1 - x^2)^{1/2} - \frac{1}{3}\beta^3 x^2(1 - x^2)^{3/2}}, \tag{6.194}$$

etc.

The transcendental equation (6.191) can be solved graphically [13] and hence the energy eigenvalues obtained as functions of β. Results for $l = 0$, 1, and 2 are shown in Fig. 6.1 where $N = 1, 2, 3, \ldots$ labels the solutions to (6.191) (for a given β) in order of increasing x. It can be shown that as V_0 is increased the order in which the various levels Nl appear is [14] 10, 11, 12, 20, 13, 21, 14, 22, 30, 15, 23, ...

From (6.191) and (6.192) a simple criterion for the existence of bound s states can be given: there are no bound s states if $\beta < \frac{1}{2}\pi$, that is if

$$V_0 < \frac{\pi^2 \hbar^2}{8Ma^2}. \tag{6.195}$$

Also, if

$$(N - \tfrac{1}{2})\pi < \beta < (N + \tfrac{1}{2})\pi \qquad (N = 1, 2, \ldots) \tag{6.196}$$

there are N bound s states.

6.8.2 *Scattering states* ($E > 0$)

The radial wavefunctions are*

$$R_{kl} = A_l j_l(kr) - B_l n_l(kr) \tag{6.197}$$

for $r > a$, and

$$R_{kl} = C_l j_l(Kr) \tag{6.198}$$

* In the following we omit the index k on A, B, and C.

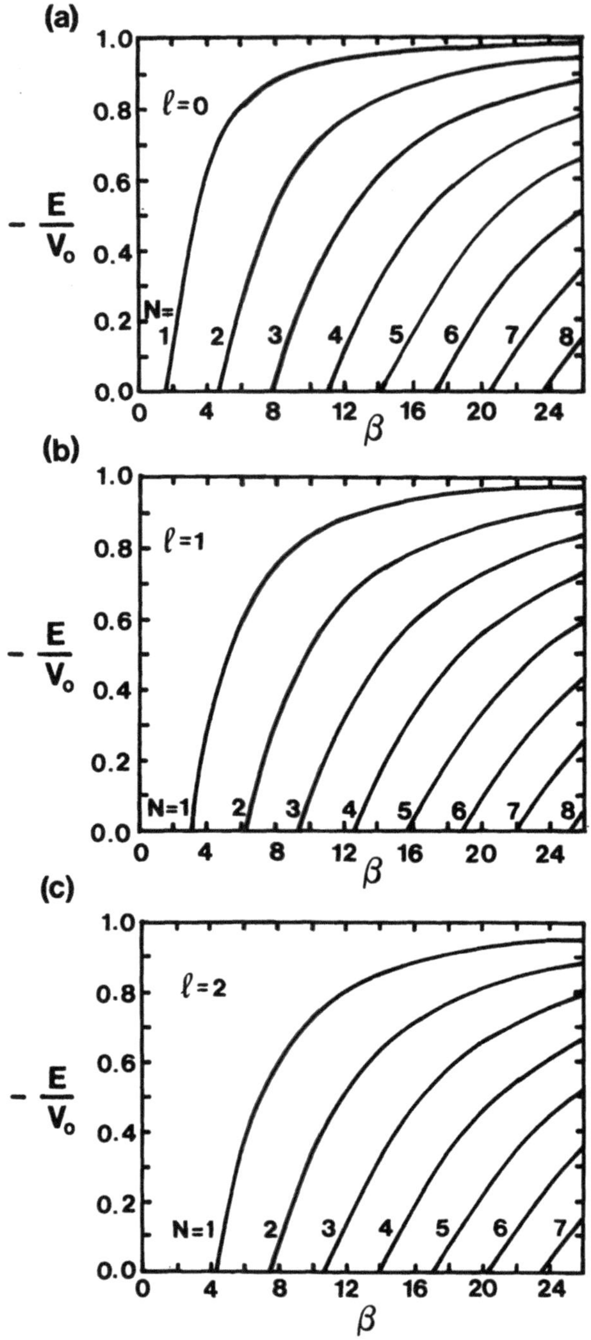

FIG. 6.1 Energy eigenvalues for bound states of the potential (6.178). Eigenvalues for $l = 0$, 1, and 2 are plotted as a function of the parameter β defined in (6.189).

for $r < a$. Here $k = (2ME/\hbar^2)^{1/2}$ and $K = [2M(E + V_0)/\hbar^2]^{1/2}$. The relative values of the constants A_l, B_l, and C_l are determined by (6.179) and (6.180). The ratio B_l/A_l determines the asymptotic behaviour of the wavefunction: as $r \to \infty$, (6.197) becomes (see (6.151) and (6.152))

$$R_{kl} = A_l(kr)^{-1} \sin(kr - \tfrac{1}{2}l\pi) + B_l(kr)^{-1} \cos(kr - \tfrac{1}{2}l\pi) \tag{6.199}$$

$$= (A_l^2 + B_l^2)^{1/2}(kr)^{-1} \sin(kr - \tfrac{1}{2}l\pi + \delta_l), \tag{6.200}$$

where

$$\delta_l = \tan^{-1}(B_l/A_l) \tag{6.201}$$

is the phase shift. It is clear that δ_l must be real. (Because the operator in (6.125) is real, there exists a regular solution that is real. Any other regular solution of (6.125) must equal this real solution times a complex constant. Thus B_l/A_l is real.)

To calculate B_l/A_l, and hence the phase shift, we substitute (6.197) and (6.198) in (6.180). This gives

$$\frac{B_l}{A_l} = \frac{\gamma_l j_l(ka) - j_l'(ka)}{\gamma_l n_l(ka) - n_l'(ka)} \tag{6.202}$$

where

$$\gamma_l = j_l'(Ka)/j_l(Ka) \tag{6.203}$$

and

$$j_l'(ka) = \left.\frac{\mathrm{d}j_l(kr)}{\mathrm{d}r}\right|_{r=a},$$

etc. The value of the phase shift computed from (6.201)–(6.203) is important because it determines the total cross-section for scattering by the potential [15].

Some of the above results are more generally applicable than to the simple example considered here. Thus for any potential $V(r)$ that is negligible for r greater than some distance a, the asymptotic form of the radial wavefunction is given by (6.200). Also, the phase shift can be obtained from (6.201) and (6.202) with

$$\gamma_l = \lim_{\varepsilon \to 0} \frac{1}{R_{kl}} \left.\frac{\mathrm{d}R_l(kr)}{\mathrm{d}r}\right|_{a-\varepsilon}. \tag{6.204}$$

Here R_{kl} is the regular solution of (6.125) for $r < a$: in general this solution has to be determined numerically.

Scattering by a conservative potential such as $V(r)$ is elastic. There are no sources and sinks of particles and hence conservation of probability is a necessary property of such scattering. One can readily verify that the solution obtained above is consistent with conservation of probability. From (6.165) and (6.199) we find

$$\mathbf{J} \to \frac{\hbar}{Mkr^2}(\mathrm{Im}\ A_l B_l^*)|Y_{lm}|^2\hat{\mathbf{r}} \tag{6.205}$$

as $r \to \infty$. Here Im denotes the imaginary part and therefore the right-hand side of (6.205) is zero because A_l/B_l, and hence $A_l B_l^*$, is real.

Equation (6.197) is not the only way of expressing the radial wavefunction outside the range of the potential. One can equally well choose (see Section 6.5)

$$R_{kl} = A'_l j_l(kr) + \mathrm{i}B'_l h_l^{(1)}(kr) \tag{6.206}$$

and

$$R_{kl} = A''_l h_l^{(1)}(kr) - B''_l h_l^{(2)}(kr). \tag{6.207}$$

If we use $h_l^{(1)} = j_l + \mathrm{i}n_l$ and $h_l^{(2)} = j_l - \mathrm{i}n_l$ in (6.206) and (6.207), we find by a similar calculation to the above that conservation of probability requires

$$B'_l = A'_l \exp(\mathrm{i}\delta_l) \sin \delta_l \quad \text{and} \quad B''_l = A''_l \exp(\mathrm{i}2\delta_l).$$

The solutions (6.197), (6.206), and (6.207) are, respectively, simple examples of the K matrix (reaction matrix), T matrix (transition matrix), and S matrix (scattering matrix) formalism of scattering theory [15].

References

1. Epstein, J.H. and Epstein, S.T. (1962). Some applications of hypervirial theorems to the calculation of average values. *American Journal of Physics*, **30**, 266–8.
2. See Shaffer, W.H. (1944). Degenerate modes of vibration and perturbation in polyatomic molecules. *Reviews of Modern Physics*, **16**, 245–59; Shaffer, W.H. (1976). Radial matrix elements of r^n for twofold and threefold isotropic simple harmonic oscillators. *Journal of Molecular Spectroscopy*, **63**, 323–47.
3. For example, Lawson, R.D. (1985). *Theory of the nuclear shell model*. Oxford University Press. Eisenberg, J.M. and Greiner, W. (1987). *Nuclear models*. North-Holland, Amsterdam.
4. Drachman, R.J. (1982). Rydberg states of helium: An optical–potential analysis. *Physical Review*, A **26**, 1228–38.
5. See Swainson, R.A. and Drake, G.W.F. (1990). An alternative proof of some relations between hydrogenic matrix elements. *Journal of Physics B: At. Mol. Opt. Phys.*, **23**, 1079–84. Drake, G.W.F. and Swainson, R.A. (1990). Expectation values of r^p for arbitrary hydrogenic states. *Physical Review*, A **42**, 1123–6 and references therein. Moreno, B., Lopez Pineiro, A., and Tipping, R.H. (1991). Algebraic solution for the Schrödinger equation: matrix elements for arbitrary powers of several r-dependent operators. Journal of Physics A: Math. Gen., **24**, 385–90.
6. See, for example, Condon, E.U. and Shortley, G.H. (1967). *The theory of atomic spectra*, Chs. 4 and 5. Cambridge University Press.
7. Menzel, D.H. (1964). Generalized radial integrals with hydrogenic functions. *Reviews of Modern Physics*, **36**, 613–17.
8. See, for example, Erdélyi, A. Magnus, W., Oberhettinger, F., and Tricomi, F.G. (1953). *Higher transcendental functions*, Vol. 1, Ch. 6. McGraw-Hill, New York.

9. See Armstrong, B.H. and Power, E.A. (1963). Acceptable solutions and boundary conditions for the Schrödinger and Dirac equations. *American Journal of Physics*, **31**, 262–8.
10. See, for example, Landau, L.D. and Lifschitz, E.M. (1977). *Quantum mechanics*, 3rd edn, Ch. 5. Pergamon, Oxford.
11. Abramowitz, M. and Stegun, I.A. (1972). *Handbook of mathematical functions*, Ch. 14. Dover, New York. Hull, M.H. Jr and Breit, G. (1959). Coulomb wave functions. In *Encyclopedia of Physics*, Vol. 41/1, pp. 408–65, (ed. S. Flügge). Springer, Berlin.
12. The counterpart of (6.126) for a particle in a constant one-dimensional potential is discussed in Schrödinger, E. (1941). Further studies on solving eigenvalue problems by factorization. *Proceedings of the Royal Irish Academy*, **46A**, 183–206. Similar remarks apply to (6.126).
13. See Flügge, S. (1974). *Practical quantum mechanics*, p. 161. Springer, New York.
14. See Ref. 10. p. 111.
15. See, for example, Rodberg, L.S. and Thaler, R.M. (1971). *Introduction to the quantum theory of scattering*. Academic Press, New York.

7

FACTORIZATIONS WITH APPLICATION TO THE MOMENTUM REPRESENTATION

In Chapter 5 we considered the factorization of a radial Hamiltonian for a particle in a spherically symmetric potential which is independent of angular momentum. This factorization provides shift operators for energy and angular momentum in the eigenkets of the radial Hamiltonian for the oscillator and Coulomb potentials. These operators are linear functions of the momentum operator* **p** and they are also non-linear functions of the position operator **r**: in the coordinate representation of wave mechanics they yield first-order differential operators.

In this chapter we show that for the three-dimensional isotropic harmonic oscillator and the Coulomb potential, one can also construct shift operators for energy and angular momentum that are linear functions* of **r** and non-linear functions of **p**. In the momentum representation of wave mechanics the shift operators yield first-order differential operators. Applications to the calculation of energy eigenvalues, matrix elements, and radial momentum-space wavefunctions are considered.

7.1 Introduction

The method used below to construct shift operators is similar to that given in Chapter 5. However, instead of using the radial operators r and p_r, we introduce the radial operators

$$p = (\mathbf{p}\cdot\mathbf{p})^{1/2} \tag{7.1}$$

and

$$\begin{aligned} r_p &= \tfrac{1}{2}(\hat{\mathbf{p}}\cdot\mathbf{r} + \mathbf{r}\cdot\hat{\mathbf{p}}) \\ &= \frac{1}{p}(\mathbf{p}\cdot\mathbf{r} + \mathrm{i}\hbar). \end{aligned} \tag{7.2}$$

It follows from (7.2) and (1.137) that

$$[r_p, f(p)] = \mathrm{i}\hbar(\mathrm{d}f/\mathrm{d}p). \tag{7.3}$$

Thus if r_p is Hermitian (see Appendix F), p and r_p are conjugate operators.

In Chapter 5 we used the identity (5.23) to express the Hamiltonian in a

* Except the energy shift operators for the Coulomb potential; these involve scaling operators which are non-linear functions of **p** and **r** (see Section 5.6).

form which is quadratic in p_r. For the problems considered below the corresponding identity that leads to operators which are quadratic in r_p is

$$\mathbf{L}^2 = \mathbf{p}^2(\mathbf{r}^2 - r_p^2) \tag{7.4}$$

(see Appendix G). By factorizing these quadratic forms we obtain operators which are linear in r_p.

7.2 The isotropic harmonic oscillator

Because the Hamiltonian

$$H = \frac{1}{2M}\mathbf{p}^2 + \frac{1}{2}M\omega^2\mathbf{r}^2 \tag{7.5}$$

is quadratic in $\mathbf{r}^2$, it is obvious how to obtain the desired factorizations: we use (7.4) to eliminate $\mathbf{r}^2$ in favour of r_p^2 in (7.5). Thus

$$H = \frac{1}{2}M\omega^2 r_p^2 + \frac{1}{2}M\omega^2\frac{1}{p^2}\mathbf{L}^2 + \frac{1}{2M}\mathbf{p}^2. \tag{7.6}$$

The effect of H on a ket $|Elm\rangle$ is the same as that of the radial Hamiltonian

$$H_l = \frac{1}{2}M\omega^2 r_p^2 + \frac{1}{2}M\omega^2\hbar^2 l(l+1)\frac{1}{p^2} + \frac{1}{2M}\mathbf{p}^2, \tag{7.7}$$

which is obtained from (7.6) by replacing $\mathbf{L}^2$ with its eigenvalues. Let $|El\rangle$ be an eigenket of H_l with eigenvalue E

$$H_l|El\rangle = E|El\rangle. \tag{7.8}$$

By factorizing H_l we obtain shift operators for the eigenvalues E and l. These factorizations can be derived in a similar manner to those performed in Chapter 5 for the radial Hamiltonian (5.25). However, it is not necessary to repeat the calculations because the results can be written down by making simple substitutions in the formulae derived in Chapter 5.

To see this we note that the radial Hamiltonian (7.7) and the radial Hamiltonian (6.6) transform into each other under the substitutions

$$r \leftrightarrow \frac{1}{M\omega}p \tag{7.9}$$

and

$$p_r \leftrightarrow -M\omega r_p. \tag{7.10}$$

(The minus sign in (7.10) is required by the commutation relations (5.22) and (7.3).)

These substitutions in (5.83)–(5.86) yield the shift operations

$$P_l^{\pm}(\omega)|El\rangle = \alpha^{\pm}(\omega)|E \pm \hbar\omega, l \pm 1\rangle \tag{7.11}$$

and

$$P_l^{\pm}(-\omega)|El\rangle = \alpha^{\pm}(-\omega)|E \mp \hbar\omega, l \pm 1\rangle, \tag{7.12}$$

where

$$P_l^{\pm}(\omega) = \mp \mathrm{i} r_p - \frac{p}{M\omega} - \hbar\left(l + \frac{1}{2} \pm \frac{1}{2}\right)\frac{1}{p}, \tag{7.13}$$

$$\alpha^{\pm}(\omega) = \pm\frac{\mathrm{i}}{\omega}\left(\frac{2}{M}\left[E + \hbar\omega\left(l + \frac{1}{2} \pm 1\right)\right]\right)^{1/2}, \tag{7.14}$$

and $|El\rangle$ is the ket in (7.8). The phase factors $\pm\mathrm{i}$ in (7.14) are required by the operator identities (8.7) and (8.32) and by our choice of unity for the phase factors associated with $\mathbf{U}^{\pm}$, $\mathbf{V}^{\pm}$ (see Chapter 4), and $R_l^{\pm}(\omega)$ (see Chapter 5).

Similarly, we obtain from (5.96) and (5.98) the shift operations

$$\tilde{Q}_n^{\pm}|nl\rangle = \tilde{\beta}^{\pm}|n \pm 2, l\rangle \tag{7.15}$$

where

$$\tilde{Q}_n^{\pm} = \mp\frac{\mathrm{i}}{\hbar} p r_p - \frac{p^2}{M\hbar\omega} + n + \frac{3}{2} \pm \frac{1}{2} \tag{7.16}$$

and n is defined by (5.92). For normalized kets in (7.15) the magnitudes of $\tilde{\beta}^{\pm}$ are given by (5.101). From (7.11), (7.12), and (7.15) it is clear that there must be relationships between $\tilde{Q}_n^{\pm}$ and products of P_l^{+} and P_l^{-}. These are provided by the identities

$$P_{l\pm1}^{\mp}(-\omega)P_l^{\pm}(\omega) = \frac{2}{M\omega}\left(\hbar\tilde{Q}_n^{\pm} + \frac{1}{2\omega}(H_l - E)\right). \tag{7.17}$$

Using (7.17) and (7.14) it follows that the phase factors of $\tilde{\beta}^{\pm}$ must be -1 (see also Section 7.5). Thus $\tilde{\beta}^{\pm} = -\beta^{\pm}$ where $\beta^{\pm}$ are given by (5.101).

7.3 The Coulomb potential [1]

The Hamiltonian is

$$H = \frac{1}{2M}\mathbf{p}^2 - \frac{\hbar^2}{Mar}, \tag{7.18}$$

where a is given by (5.55). We first rewrite the eigenvalue equation (5.11) in a form that is quadratic in $\mathbf{r}$. From (7.18)

$$\mathbf{r}^2(\mathbf{p}^2 - 2MH)^2 = 4\hbar^4 a^{-2}. \tag{7.19}$$

Operating with both sides of (7.19) on a ket $|Elm\rangle$, and using the commutators (1.151), (1.138), and (1.143) to expand the left-hand side of (7.19), we obtain the Hylleraas equation [2]

$$\Lambda_E|Elm\rangle = 4\hbar^2 a^{-2}|Elm\rangle, \tag{7.20}$$

where

$$\Lambda_E = \hbar^{-2}\mathbf{r}^2(\mathbf{p}^2 - 2ME)^2 + 2(2 - i\hbar^{-1}\mathbf{p}.\mathbf{r})(\mathbf{p}^2 - 2ME) \tag{7.21}$$

is quadratic in $\mathbf{r}$.

In (7.21) we use (7.2) and (7.4) to eliminate $\mathbf{r}$ in favour of r_p. Then, for the purposes of operating on $|Elm\rangle$, $\mathbf{L}^2$ is replaced* with $\hbar^2 l(l+1)$. This yields the radial operator that we wish to factorize†

$$\begin{aligned}\Lambda_{El} &= [\hbar^{-2}r_p^2 + l(l+1)p^{-2}](\mathbf{p}^2 - 2ME)^2 - 2i\hbar^{-1}r_p p(\mathbf{p}^2 - 2ME) \\ &= [\hbar^{-1}r_p(\mathbf{p}^2 - 2ME)]^2 + l(l+1)p^{-2}(\mathbf{p}^2 - 2ME)^2.\end{aligned} \tag{7.22}$$

The effect of Λ_{El} on $|Elm\rangle$ is the same as that of the operator Λ_E. If $|El\rangle$ denotes an eigenket of Λ_{El} then

$$\Lambda_{El}|El\rangle = 4\hbar^2 a^{-2}|El\rangle. \tag{7.23}$$

To obtain the factorizations

$$P'_{El}P_{El} = \Lambda_{El} + F_{El} \tag{7.24}$$

and

$$P_{El}P'_{El} = \Lambda_{E,l+1} + G_{El}, \tag{7.25}$$

where F_{El} and G_{El} are independent of p and r_p, inspection of (7.22) suggests the ansatz

$$P_{El} = -i\hbar^{-1}r_p(\mathbf{p}^2 - 2ME) + f_{El}(p) \tag{7.26}$$

and

$$P'_{El} = i\hbar^{-1}r_p(\mathbf{p}^2 - 2ME) + f_{El}(p). \tag{7.27}$$

From (7.26), (7.27), and (7.3)

$$-(\mathbf{p}^2 - 2ME)\frac{df_{El}}{dp} + f_{El}^2 = l(l+1)p^{-2}(\mathbf{p}^2 - 2ME)^2 + F_{El} \tag{7.28}$$

and

$$(\mathbf{p}^2 - 2ME)\frac{df_{El}}{dp} + f_{El}^2 = (l+1)(l+2)p^{-2}(\mathbf{p}^2 - 2ME)^2 + G_{El}. \tag{7.29}$$

It is straightforward to solve (7.28) and (7.29): the results are

$$F_{El} = G_{El} = 8ME(l+1)^2 \tag{7.30}$$

---

* $\mathbf{L}^2$ commutes with $\mathbf{p}^2$; thus $\mathbf{L}^2$ can be placed to the right of the factor $(\mathbf{p}^2 - 2ME)^2$ in (7.21).

† In (7.22) the radial operator r_p occurs to the left of other operators. This ordering is convenient because it simplifies subsequent calculations, particularly (7.28) and (7.29). Using (7.3) one can readily change the ordering in any of the operators presented in this chapter.

and

$$f_{El} = (l + 1)p^{-1}(\mathbf{p}^2 + 2ME). \tag{7.31}$$

From the factorizations (7.24) and (7.25) and the eigenvalue equation (7.23) we can deduce, by calculations that are similar to those in Section 5.4, that P_{El} and P'_{El} are shift operators for l in the kets $|El\rangle$. These shift operations can be written

$$P^{\pm}_{El}|El\rangle = \alpha^{\pm}(E, l)|E, l \pm 1\rangle. \tag{7.32}$$

Here $P^{+}_{El} = P_{El}$ and $P^{-}_{El} = P'_{E,l-1}$; that is

$$P^{\pm}_{El} = \mp i\hbar^{-1}r_p(\mathbf{p}^2 - 2ME) + (l + \tfrac{1}{2} \pm \tfrac{1}{2})p^{-1}(\mathbf{p}^2 + 2ME). \tag{7.33}$$

The shift operations (7.32) are valid for both an attractive and a repulsive Coulomb potential.

For bound states of an attractive Coulomb potential, r_p is Hermitian (see Appendix F) and therefore

$$(P^{+}_{El})^{\dagger} = P^{-}_{E,l+1} + 2p = P'_{El} + 2p. \tag{7.34}$$

Thus for normalized kets in (7.32)

$$\begin{aligned} |\alpha^{+}|^2 &= \langle (P'_{El} + 2p)P_{El}\rangle \\ &= 4\hbar^2 a^{-2} + 8ME(l + 1)^2 + 2\langle pP_{El}\rangle \end{aligned} \tag{7.35}$$

and

$$\begin{aligned} |\alpha^{-}|^2 &= \langle (P_{E,l-1} - 2p)P'_{E,l-1}\rangle \\ &= 4\hbar^2 a^{-2} + 8MEl^2 - 2\langle pP'_{E,l-1}\rangle. \end{aligned} \tag{7.36}$$

The expectation values in (7.35) and (7.36) are $\langle El|pP^{\pm}_{El}|El\rangle$, which are both equal to zero.* Hence

$$\alpha^{\pm} = \mp i2\hbar a^{-1}[1 + 2Ma^2\hbar^{-2}E(l + \tfrac{1}{2} \pm \tfrac{1}{2})^2]^{1/2}. \tag{7.37}$$

The phase factors $\mp i$ in (7.37) are required by the operator identities (9.8) and (9.26) and our choice of unity for the phase factors associated with $\mathbf{U}^{\pm}$, $\mathbf{V}^{\pm}$ (see Chapter 4), and $R^{\pm}_{l}$ (see Chapter 5).

Equations (7.32), (7.33), and (7.37) should be compared with (5.77)–(5.79). By similar reasoning to that given in Chapter 5, (7.37) yields the energy eigenvalues (5.81). If we let $|nl\rangle$ denote the ket $|El\rangle$, (7.32) and (7.37) give

$$P^{+}_{l+1,l}|l + 1, l\rangle = 0; \tag{7.38}$$

that is†

---

* This follows from (7.69) and (see Section 6.3)

$$\langle Elm|pr_p(\mathbf{p}^2 - 2ME)|Elm\rangle = 2\hbar^2 a^{-1}\langle pr_p r^{-1}\rangle = 2\hbar^2 a^{-1}\langle p_r\rangle = 0,$$

$$\langle Elm|\mathbf{p}^2 + 2ME|Elm\rangle = 0.$$

† It is clear from the above discussion that (7.38) and (7.39) apply also to the ket $|l + 1, l, l\rangle$.

$$\left(-\frac{\mathrm{i}}{\hbar}pr_p - \frac{(2l+4)p_0^2}{(l+1)^2\mathbf{p}^2 + p_0^2} + l + 3\right)|l+1, l\rangle = 0, \tag{7.39}$$

where

$$p_0 = \hbar/a. \tag{7.40}$$

We can also derive shift operators that change the energy eigenvalue. First we operate with $\mathbf{p}^2(\mathbf{p}^2 - 2ME)^{-2}$ on the left of (7.23) and use (7.3). For bound states this yields

$$\tilde{O}_n|nl\rangle = -l(l+1)|nl\rangle, \tag{7.41}$$

where n is defined by (5.93) and

$$\tilde{O}_n = \frac{1}{\hbar^2}\mathbf{p}^2 r_p^2 + \frac{6in^2}{\hbar}\frac{\mathbf{p}^2}{n^2\mathbf{p}^2 + p_0^2}(pr_p + \mathrm{i}\hbar) - 4n^2(n^2-1)p_0^2\frac{\mathbf{p}^2}{(n^2\mathbf{p}^2 + p_0^2)^2}. \tag{7.42}$$

The factorizations of $\tilde{O}_n$ are

$$\tilde{Q}_{n+1}^{-}\tilde{Q}_n^{+} = \tilde{O}_n + n(n+1) \tag{7.43}$$

and

$$\tilde{Q}_{n-1}^{+}\tilde{Q}_n^{-} = \tilde{O}_n + n(n-1) \qquad (n > 1). \tag{7.44}$$

Here

$$\tilde{Q}_n^{\pm} = \Delta_n^{\pm}\tilde{T}_n^{\pm}, \tag{7.45}$$

$$\tilde{T}_n^{\pm} = \mp\frac{\mathrm{i}}{\hbar}pr_p + 2(n \mp 1)\frac{p_0^2}{n^2\mathbf{p}^2 + p_0^2} - n \pm 2, \tag{7.46}$$

and $\Delta_n^{\pm}$ are scaling operators defined by (cf. (5.114)–(5.116))

$$\Delta_n^{\pm} np = (n \pm 1)p\Delta_n^{\pm}, \tag{7.47}$$

$$\Delta_n^{\pm}\frac{r_p}{n} = \frac{r_p}{n \pm 1}\Delta_n^{\pm}, \tag{7.48}$$

and

$$(\Delta_n^{\pm})^{\dagger}\Delta_n^{\pm} = 1. \tag{7.49}$$

($n \neq 1$ in Δ_n^{-}.)

Explicit expressions for $\Delta_n^{\pm}$ can be derived by calculations that are similar to those for the scaling operators in Section 5.6. The results are

$$\Delta_n^{\pm} = \exp\left[\left(\mathrm{i}\hbar^{-1}pr_p - \frac{1}{2}\right)\ln\frac{n}{n \pm 1}\right]. \tag{7.50}$$

Because

$$pr_p = rp_r - \mathrm{i}\hbar \tag{7.51}$$

(see (5.21), (7.2), and (1.138)), the scaling operators (7.50) are identical to those given by (5.126) and (5.127).

From (7.43), (7.44), and (7.41),

$$\tilde{Q}_n^{\pm}|nl\rangle = \tilde{\beta}^{\pm}(n, l)|n \pm 1, l\rangle. \tag{7.52}$$

The coefficients $\tilde{\beta}^{\pm}$ can be determined by calculations that are similar to, but lengthier than, those for the energy shift operators $Q_n^{\pm}$(see (5.129)–(5.131)). For normalized kets in (7.52)

$$\begin{aligned}|\tilde{\beta}^{+}(n, l)|^2 &= \langle nl|(\tilde{Q}_n^{+})^{\dagger}\tilde{Q}_n^{+}|nl\rangle \\ &= \langle nl|(\tilde{T}_n^{+})^{\dagger}\tilde{T}_n^{+}|nl\rangle,\end{aligned} \tag{7.53}$$

where we have used (7.49). Substituting (7.46) in (7.53) and multiplying out, we find

$$\begin{aligned}|\tilde{\beta}^{+}(n, l)|^2 = \langle nl|&\frac{1}{\hbar^2}\mathbf{p}^2 r_p^2 + \frac{2i}{\hbar} p r_p + (n-1)(n-2) \\ &- 2(n-1)(2n-5)\frac{p_0^2}{n^2\mathbf{p}^2 + p_0^2} \\ &+ 4(n-1)(n-2)\frac{p_0^4}{(n^2\mathbf{p}^2 + p_0^2)^2}|nl\rangle.\end{aligned} \tag{7.54}$$

The expectation values in (7.54) are given by (7.69)–(7.71), (7.74), and (7.77). From $|\tilde{\beta}^{+}|^2$ we can calculate $|\tilde{\beta}^{-}|^2$ because (7.41), (7.43), (7.44), and (7.52) require the relation

$$\tilde{\beta}^{-}(n+1, l)\tilde{\beta}^{+}(n, l) = n(n+1) - l(l+1).$$

The results are

$$\tilde{\beta}^{\pm}(n, l) = -\{n(n \pm 1)^{-1}[n(n \pm 1) - l(l+1)]\}^{1/2}. \tag{7.55}$$

In (7.55) a phase factor has been set equal to -1; the reason for this choice is given below (see Section 7.6).

In the above $\tilde{Q}_1^{-}$ is not defined. Because $\tilde{Q}_n^{-}$ annihilates $|n, n-1\rangle$ if $n > 1$ (see (7.52) and (7.55)), it is desirable to define $\tilde{Q}_1^{-}$ such that

$$\tilde{Q}_1^{-}|10\rangle = 0. \tag{7.56}$$

From (7.39) and (7.46) we see that (7.56) is satisfied if

$$\tilde{Q}_1^{-} = \tilde{T}_1^{-}. \tag{7.57}$$

Using $P_{nl}^{\pm}$ and $\tilde{Q}_n^{\pm}$ we can construct operators that change both n and l. When $n = l + 1$ these operators have a simple form because we can use (7.39) to eliminate terms that are linear in r_p. For example, the shift operation

$$P_{n+1,l}^{+}\tilde{Q}_n^{+}|nl\rangle = \alpha^{+}(n+1, l)\beta^{+}(n, l)|n+1, l+1\rangle \tag{7.58}$$

simplifies to

$$\Delta_{l+1}^{+}\frac{(l+1)p}{(l+1)^2\mathbf{p}^2+p_0^2}|l+1,l\rangle = -\frac{\mathrm{i}}{2p_0}\left(\frac{2l+3}{2l+4}\right)^{1/2}|l+2,l+1\rangle. \quad (7.59)$$

An alternative derivation of (7.59) can be given by showing that

$$\Lambda_{l+2,l+1}\left\{\Delta_{l+1}^{+}\frac{p}{(l+1)^2\mathbf{p}^2+p_0^2}|l+1,l\rangle\right\}$$
$$= 4p_0^2\left\{\Delta_{l+1}^{+}\frac{p}{(l+1)^2\mathbf{p}^2+p_0^2}|l+1,l\rangle\right\}. \quad (7.60)$$

Comparison of (7.23) and (7.60) shows that the ket in braces is proportional to $|l+2,l+1\rangle$. The proportionally constant can be evaluated by using (7.49) and the expectation values (7.74) and (7.77) to calculate the norm of the ket in braces.

7.4 Expectation values and matrix elements

7.4.1 *The oscillator*

For the oscillator the algebraic methods used in Chapter 6 (the Hellmann–Feynman theorem, hypervirial theorem, and shift operations) can also be applied to the operators and kets discussed above. Alternatively, the results can be written down simply by making the substitutions (7.9) and (7.10) in the formulae derived in Chapter 6. Thus from (6.9)–(6.12) we have

$$\langle p^{-2}\rangle = (l+\tfrac{1}{2})^{-1}p_0^{-2}, \quad (7.61)$$

$$\langle \mathbf{p}^2\rangle = (n+\tfrac{3}{2})p_0^2, \quad (7.62)$$

and the recurrence relations

$$\langle p^s r_p\rangle = -\tfrac{1}{2}\mathrm{i}\hbar s\langle p^{s-1}\rangle \quad (7.63)$$

and

$$(s+2)\langle p^{s+2}\rangle - (s+1)(2n+3)p_0^2\langle p^s\rangle - sp_0^4[\tfrac{1}{4}(s^2-1)-l(l+1)]\langle p^{s-2}\rangle$$
$$= 0, \quad (7.64)$$

where $p_0 = \hbar/a$ and a is defined by (5.95). Equations (7.62) and (7.64) give the expectation values of even-integer powers of p. For odd-integer powers of p we need also $\langle p\rangle$ and $\langle p^{-1}\rangle$. These are found by making the substitution (7.9) in (6.33) and (6.34).

By taking linear combinations of $P_l^{\pm}(\omega)$ and $P_l^{\pm}(-\omega)$ we can evaluate the matrix elements $\langle n'l'|p|nl\rangle$ where $l' = l \pm 1$. These calculations are similar to those leading to (6.40) and (6.41) and the results are

$$p|nl\rangle = ip_0[\tfrac{1}{2}(n-l)]^{1/2}|n-1, l+1\rangle - ip_0[\tfrac{1}{2}(n+l+3)]^{1/2}|n+1, l+1\rangle \tag{7.65}$$

and

$$\begin{aligned} p|nl\rangle = &+ip_0[\tfrac{1}{2}(n-l+2)]^{1/2}|n+1, l-1\rangle \\ &+ ip_0[\tfrac{1}{2}(n+l+1)]^{1/2}|n-1, l-1\rangle. \end{aligned} \tag{7.66}$$

Hence we have

$$\langle n \pm 1, l \pm 1|p|nl\rangle = \mp ip_0[\tfrac{1}{2}(n+l+2 \pm 1)]^{1/2}, \tag{7.67}$$

$$\langle n \pm 1, l \mp 1|p|nl\rangle = \pm ip_0[\tfrac{1}{2}(n-l+1 \pm 1)]^{1/2}, \tag{7.68}$$

and the selection rule $\Delta n = \pm 1$ for these matrix elements of p.

7.4.2 *The Coulomb potential*

For the Coulomb potential the operator Λ_{El} in (7.23) is not Hermitian.* Consequently, we cannot apply the Hellmann–Feynman theorem or the hypervirial theorem to the radial equation (7.23). However, the expectation values of certain functions of p and r_p can be expressed in terms of expectation values of functions of r and p_r: for bound states the latter can be evaluated as in Chapter 6. Some examples are given below.

In these calculations the expectation values are with respect to the eigenkets $|Elm\rangle$ rather than $|El\rangle$; this simplifies the calculations but does not affect the result because

$$\langle Elm|f(p, r_p)|Elm\rangle = \langle El|f(p, r_p)|El\rangle, \tag{7.69}$$

where $f(p, r_p)$ denotes a power series in p and r_p (see Appendix H).

Consider a simple example. From the identity (7.51) and the expectation values (6.14) and (6.55)

$$\langle pr_p\rangle = -\tfrac{1}{2}i\hbar. \tag{7.70}$$

Similarly, using the identity (7.4)

$$\langle \mathbf{p}^2 r_p^2\rangle = \langle \mathbf{p}^2\mathbf{r}^2 - \mathbf{L}^2\rangle = \langle 2ME\mathbf{r}^2 + 2\hbar^2 a^{-1} r - \mathbf{L}^2\rangle,$$

where in the last step we have used (7.18) and the Hermitian property of H. For bound states the expectation values of r and $\mathbf{r}^2$ are given by (6.51), (6.52), and (6.55). Hence

$$\langle \mathbf{p}^2 r_p^2\rangle = \tfrac{1}{2}\hbar^2[n^2 - l(l+1) - 1]. \tag{7.71}$$

Expectation values can also be calculated with the aid of the Hylleraas

* Taking the adjoint of (7.22) one can show that

$$\Lambda_{El}^\dagger = \Lambda_{El} - 4i\hbar^{-1}(\mathbf{p}^2 - 2ME)pr_p + 6(\mathbf{p}^2 - 2ME) + 8ME.$$

equation (7.20). For example, operating on the left of both sides of (7.20) with $(\mathbf{p}^2 - 2ME)^{-1}\Omega$ (where Ω is some operator) gives

$$(\mathbf{p}^2 - 2ME)^{-1}\Omega|Elm\rangle = \tfrac{1}{4}p_0^{-2}(\mathbf{p}^2 - 2ME)^{-1}\Omega[\hbar^{-2}(\mathbf{p}^2 - 2ME)^2\mathbf{r}^2 + 6(\mathbf{p}^2 - 2ME)(i\hbar^{-1}pr_p - 1) - 8ME]|Elm\rangle. \tag{7.72}$$

In writing down (7.72) we have used (1.140) and (1.147) to commute $\mathbf{r}^2$ and $\mathbf{p}.\mathbf{r}$ to the right of other operators in (7.21). By making various choices for Ω in (7.72) we can obtain useful expectation values. In these calculations it is convenient to use the identity (7.51) and the relation

$$\langle(\mathbf{p}^2 - 2ME)A\rangle = \langle(\mathbf{p}^2 - 2MH)A\rangle = 2\hbar p_0\langle r^{-1}A\rangle, \tag{7.73}$$

where A is some operator.

Multiplying (7.72) on the left by $\langle Elm|$, letting $\Omega = 1$, and using (5.81), we find

$$\langle(n^2\mathbf{p}^2 + p_0^2)^{-1}\rangle = [4p_0^2(n^2 - 1)]^{-1}\langle 2a^{-1}r - 6 + 6i\hbar^{-1}pr_p\rangle = [4p_0^2(n^2 - 1)]^{-1}[3n^2 - l(l+1) - 3], \tag{7.74}$$

where in the last step we have substituted (6.51) and (7.70).

Similarly, with $\Omega = (\mathbf{p}^2 - 2ME)^{-1}$ we have

$$\langle(n^2\mathbf{p}^2 + p_0^2)^{-2}\rangle = [4p_0^2(n^2 - 1)]^{-1}\langle \hbar^{-2}\mathbf{r}^2 + 6n^2(n^2\mathbf{p}^2 + p_0^2)^{-1}(i\hbar^{-1}pr_p - 1)\rangle. \tag{7.75}$$

The expectation value of $(n^2\mathbf{p}^2 + p_0^2)^{-1}pr_p$ can be evaluated by letting $\Omega = pr_p$ in (7.72): after some manipulation we obtain

$$\langle(n^2\mathbf{p}^2 + p_0^2)^{-1}pr_p\rangle = -i\hbar[4p_0^2(n^2 - 1)]^{-1}[n^2 - l(l+1) - 1]. \tag{7.76}$$

Substituting (7.74), (7.76), and (6.52) in (7.75), we find

$$\langle(n^2\mathbf{p}^2 + p_0^2)^{-2}\rangle = n^2[8p_0^4(n^2 - 1)]^{-1}[5n^2 - 3l(l+1) - 5]. \tag{7.77}$$

Note that (7.74)–(7.77) are indeterminate if $n = 1$. These expectation values must be calculated separately; for example, using (7.56), (7.57), (7.46), and (7.70) we find

$$\langle 100|(\mathbf{p}^2 + p_0^2)^{-1}|100\rangle = \tfrac{5}{8}p_0^{-2}. \tag{7.78}$$

7.5 Discussion

The operators $P_l^{\pm}$, $P_{El}^{\pm}$, and $\tilde{Q}_n^{\pm}$ derived in this chapter are shift operators for the eigenkets $|nl\rangle$ (or $|El\rangle$) of the radial equations (7.8) and (7.23). Of these, $P_l^{\pm}$ and $P_{El}^{\pm}$ are not, by themselves, shift operators for the eigenkets $|nlm\rangle$ of the Hamiltonian H (see Chapters 8 and 9). Consider the effect of $\tilde{Q}_n^{\pm}$ (eqns (7.16), (7.45), and (7.57)) on $|nlm\rangle$.

For the oscillator the calculations are similar to those in Section 5.7, and we simply state the results:

$$\tilde{Q}_n^{\pm}|nlm\rangle = -\beta^{\pm}(n, l)|n \pm 2, l, m\rangle, \tag{7.79}$$

where $\tilde{Q}_n^{\pm}$ and $\beta^{\pm}$ are given by (7.16) and (5.101), and n is defined by (5.92). From $\tilde{Q}_n^{\pm}$ we construct operators*

$$\tilde{Q}^{\pm} = \mp \frac{i}{\hbar} pr_p - \frac{p^2}{M\hbar\omega} + \frac{H}{\hbar\omega} \pm \frac{1}{2} \tag{7.80}$$

that satisfy

$$\tilde{Q}^{\pm}|nlm\rangle = -\beta^{\pm}(n, l)|n \pm 2, l, m\rangle. \tag{7.81}$$

For bound states of a Coulomb potential first express the Hylleraas equation (7.20) in terms of n (eqn (5.93)):

$$\Lambda_n|nlm\rangle = 4\hbar^2 a^{-2}|nlm\rangle, \tag{7.82}$$

where

$$\Lambda_n = \hbar^{-2}\mathbf{r}^2(\mathbf{p}^2 + n^{-2}p_0^2)^2 + 2(2 - i\hbar^{-1}\mathbf{p}.\mathbf{r})(\mathbf{p}^2 + n^{-2}p_0^2). \tag{7.83}$$

Then one can show that

$$\Lambda_{n\pm 1}\{\tilde{Q}_n^{\pm}|nlm\rangle\} = 4\hbar^2 a^{-2}\{\tilde{Q}_n^{\pm}|nlm\rangle\}, \tag{7.84}$$

where $\tilde{Q}_n^{\pm}$ are given by (7.45) and (7.57). From (7.84), and because $\tilde{Q}_n^{\pm}$ are scalars,

$$\tilde{Q}_n^{\pm}|nlm\rangle = \tilde{\beta}^{\pm}(n, l)|n \pm 1, l, m\rangle. \tag{7.85}$$

From the calculations leading to (7.55), and noting the relation (7.69), we see that the coefficients in (7.85) are given by (7.55).

Finally, we mention that it is possible to eliminate n from $\tilde{Q}_n^{\pm}$, although the resulting shift operators $\tilde{Q}^{\pm}$ are rather complicated. To obtain $\tilde{Q}^{\pm}$ we cannot simply replace n in (7.45) with the operator $\mathcal{N}$ (eqn (5.146)) because $\mathcal{N}$ does not commute with $\mathbf{r}$ and $\mathbf{p}$. Instead we use (7.72) with $\Omega = 1$ to eliminate the term involving $(n^2\mathbf{p}^2 + p_0^2)^{-1}|nlm\rangle$ in (7.85) (see (7.46)). Then we rearrange operators using (7.47) and (7.48), and replace n with $\mathcal{N}$. The result is

$$\tilde{Q}^{\pm} = \frac{1}{2\hbar^2}\mathbf{p}^2\mathbf{r}^2\Delta^{\pm}(\mathcal{N} \pm 1)^{-1} + \frac{1}{2a^2}\mathbf{r}^2\Delta^{\pm}(\mathcal{N} \pm 1)^{-3}$$

$$\mp \frac{i}{\hbar}\mathbf{r}.\mathbf{p}\Delta^{\pm}[1 \mp 3(\mathcal{N} \pm 1)^{-1}] + \Delta^{\pm}[3(\mathcal{N} \pm 1)^{-1} - \mathcal{N}], \tag{7.86}$$

* Note that (7.80) and (5.141) are related by $\tilde{Q}^{\pm} = -Q^{\pm}$; this accounts for the minus sign in (7.79).

provided $\tilde{Q}^-$ does not act on $|100\rangle$, in which case

$$\tilde{Q}^- = \frac{1}{6\hbar^2}\mathbf{p}^2\mathbf{r}^2 + \frac{1}{6a^2}\mathbf{r}^2 + \frac{\mathrm{i}}{\hbar}\mathbf{r}.\mathbf{p} + \frac{3}{2} \tag{7.87}$$

and $\tilde{Q}^-|100\rangle = 0$. The effect of $\tilde{Q}^\pm$ on $|nlm\rangle$ is the same as that of $\tilde{Q}_n^\pm$, namely

$$\tilde{Q}^\pm|nlm\rangle = \tilde{\beta}^\pm(n, l)|n \pm 1, l, m\rangle. \tag{7.88}$$

7.6 Radial momentum-space wavefunctions

It is straightforward to express the abstract operator equations derived in this chapter in terms of a particular representation. In the momentum representation the resulting equations are differential equations.

For example, the eigenvalue equations for the oscillator become

$$H\bar{\psi}_{Elm}(\mathbf{p}') = E\bar{\psi}_{Elm}(\mathbf{p}'), \tag{7.89}$$

$$\mathbf{L}^2\bar{\psi}_{Elm}(\mathbf{p}') = \hbar^2 l(l+1)\bar{\psi}_{Elm}(\mathbf{p}'), \tag{7.90}$$

$$L_z\bar{\psi}_{Elm}(\mathbf{p}') = \hbar m\bar{\psi}_{Elm}(\mathbf{p}'), \tag{7.91}$$

while for the Coulomb potential (7.89) is replaced by

$$\Lambda_E\bar{\psi}_{Elm}(\mathbf{p}') = 4p_0^2\bar{\psi}_{Elm}(\mathbf{p}'). \tag{7.92}$$

Here the momentum-space wavefunction is given by

$$\bar{\psi}_{Elm}(\mathbf{p}') = \langle\mathbf{p}'|Elm\rangle, \tag{7.93}$$

where $|\mathbf{p}'\rangle$ is an eigenket of $\mathbf{p}$ with eigenvalue $\mathbf{p}'$, and H, Λ_E, $\mathbf{L}^2$, and L_z are the differential forms obtained by making the substitutions

$$\mathbf{p} \to \mathbf{p}', \qquad \mathbf{r} \to \mathrm{i}\hbar\nabla_{\mathbf{p}'} \tag{7.94}$$

in the corresponding abstract operators (see Sections 1.4 and 1.7). From (7.2) and (7.94),

$$r_p \to \mathrm{i}\hbar\left(\frac{\mathrm{d}}{\mathrm{d}p'} + \frac{1}{p'}\right). \tag{7.95}$$

In the following applications to the momentum representation we omit the prime on $\mathbf{p}$.

The normalizable solutions to the partial differential equations (7.89)–(7.92) are given by the separation of variables

$$\bar{\psi}_{Elm}(\mathbf{p}) = \bar{R}_{El}(p)\,Y_{lm}(\hat{\mathbf{p}}) \tag{7.96}$$

(see Chapters 8 and 9). Here

$$Y_{lm}(\hat{\mathbf{p}}) = \langle\hat{\mathbf{p}}|lm\rangle \tag{7.97}$$

is a spherical harmonic (see Section 4.10) and

$$\bar{R}_{El}(p) = \langle p|El\rangle \tag{7.98}$$

is the radial momentum-space wavefunction. This radial wavefunction is the normalized solution to the differential form of (7.8) or (7.23); for example, the differential equation corresponding to (7.23) is

$$\left[-\left(\frac{\mathrm{d}}{\mathrm{d}p}+\frac{1}{p}\right)^2+\frac{l(l+1)}{p^2}\right](\mathbf{p}^2-2ME)^2\bar{R}_{El}(p)+2\left(\frac{\mathrm{d}}{\mathrm{d}p}+\frac{1}{p}\right)p(\mathbf{p}^2-2ME)\bar{R}_{El}(p)$$
$$= 4p_0^2\bar{R}_{El}(p). \tag{7.99}$$

We can summarize the above discussion: to write an operator equation in the momentum representation we make the substitutions (7.94), (7.95), and $|Elm\rangle \to \bar{\psi}_{Elm}(\mathbf{p})$ and $|El\rangle \to \bar{R}_{El}(p)$. We now use the shift operators in differential form to calculate the radial momentum-space wavefunctions for the oscillator and bound states of the Coulomb potential. The calculations are similar to those in Chapter 6 for the radial coordinate-space wavefunctions.

7.6.1 *The oscillator*

In Chapter 6 we used shift operations involving r and Q_n^+ to calculate the radial coordinate-space wavefunctions. Similarly, to determine the radial momentum-space wavefunctions we can use shift operations involving p and $\tilde{Q}_n^+$ (see (7.65) and (7.15)). The two calculations are closely related: the latter can be obtained from the former by (i) replacing r with p and a with $p_0 = \hbar/a$ in (6.66)–(6.72), and (ii) by taking account of the phase factors in (7.65) and (7.15). The phase factors $-\mathrm{i}$ in the second term in (7.65) and -1 in (7.15) produce a phase factor $(-\mathrm{i})^l(-1)^N$ in $\bar{R}_{nl}(p)$; this factor replaces the phase factor $(-1)^l$ in (6.79). In this manner we obtain

$$\bar{R}_{nl}(p) = \mathrm{i}^l(-1)^{N+l}\left(\frac{2^{l-N+2}(2N+2l+1)!!}{p_0^3\pi^{1/2}N![(2l+1)!!]^2}\right)^{1/2} {}_1F_1\left(-N; l+\frac{3}{2}; \frac{p^2}{p_0^2}\right)\left(\frac{p^2}{p_0^2}\right)^{l/2}$$
$$\times \exp\left(-\frac{p^2}{2p_0^2}\right). \tag{7.100}$$

Here $N = \frac{1}{2}(n-l) = 0, 1, \ldots$ is the number of times shift operators $\tilde{Q}_n^+$ must be applied to transform $\bar{R}_{ll}(p)$ into $\bar{R}_{nl}(p)$. The wavefunction (7.100) is the normalized solution to the differential form of (7.8).

7.6.2 *The Coulomb potential*

Consider bound states. From (7.56) we obtain the differential form (we use the notation $p^2 = \mathbf{p}^2$)

$$\left(\frac{\mathrm{d}}{\mathrm{d}p}+\frac{4p}{p^2+p_0^2}\right)\bar{R}_{10}(p) = 0, \tag{7.101}$$

where $\bar{R}_{10}$ is the radial wavefunction for the ground state. The solution to (7.101) that is normalized according to

$$\int_0^\infty p^2 |\bar{R}_{10}(p)|^2 \, dp = 1 \tag{7.102}$$

is

$$\bar{R}_{10}(p) = -\left(\frac{32}{\pi p_0^3}\right)^{1/2} \frac{p_0^4}{(p^2 + p_0^2)^2}. \tag{7.103}$$

In (7.103) a phase factor has been set equal to -1; the reason for this choice is given below.

From $\bar{R}_{10}$ we generate all normalized wavefunctions $\bar{R}_{nl}$. Consider first the wavefunctions with $n = l + 1$: from (7.59) and (7.47)

$$\frac{(l+2)p}{(l+2)^2 p^2 + p_0^2} \Delta_{l+1}^+ \bar{R}_{l+1,l}(p) = -\frac{\mathrm{i}}{2p_0}\left(\frac{2l+3}{2l+4}\right)^{1/2} \bar{R}_{l+2,l+1}(p), \tag{7.104}$$

where Δ_{l+1}^+ is given by (7.50) and (7.95). The effect of these differential scaling operators on radial momentum-space wavefunctions is given by

$$\Delta_n^\pm \bar{R}_{n'l}(p) = \delta_n^\pm \bar{R}_{n'l}\left(\frac{n \pm 1}{n} p\right), \tag{7.105}$$

where $\delta_n^\pm$ are independent of p. According to (7.49)

$$\int_0^\infty \bar{R}_{n'l}^*(p) \bar{R}_{n'l}(p) p^2 \, dp = \int_0^\infty [\Delta_n^\pm \bar{R}_{n'l}(p)]^* [\Delta_n^\pm \bar{R}_{n'l}(p)] p^2 \, dp.$$

It follows that (we set a phase factor equal to unity)

$$\delta_n^\pm = \left(\frac{n \pm 1}{n}\right)^{3/2}. \tag{7.106}$$

Substituting (7.105) and (7.106) in (7.104) we obtain the recurrence relation

$$\bar{R}_{l+2,l+1}(p) = \mathrm{i}2p_0 \left(\frac{2l+4}{2l+3}\right)^{1/2} \left(\frac{l+2}{l+1}\right)^{3/2} \frac{(l+2)p}{(l+2)^2 p^2 + p_0^2} \bar{R}_{l+1,l}\left(\frac{l+2}{l+1} p\right). \tag{7.107}$$

Starting with (7.103) this yields

$$\bar{R}_{l+1,l}(p) = c_l \frac{[(l+1)p/p_0]^l}{[(l+1)^2 (p/p_0)^2 + 1]^{l+2}}, \tag{7.108}$$

where

$$c_l = -\mathrm{i}^l \frac{2^{l+3}(l+1)!(l+1)^{3/2}}{[p_0^3 \pi (2l+2)!]^{1/2}}. \tag{7.109}$$

Next we express (7.52) in differential form: with the aid of (7.47), (7.48), (7.55),

(7.105), and (7.106) we readily find

$$\bar{R}_{nl}(v) = -\left(\frac{n}{n-1}\right)^2 [(n-l-1)(n+l)]^{-1/2}$$
$$\times \left(-n+4+\frac{2(n-2)}{v^2+1}+v\frac{\mathrm{d}}{\mathrm{d}v}\right)\bar{R}_{n-1,l}(v), \tag{7.110}$$

where

$$v = np/p_0. \tag{7.111}$$

Let

$$\bar{R}_{nl}(v) = c_l \frac{v^l}{(v^2+1)^{l+2}} \bar{\phi}_n(v) \tag{7.112}$$

and

$$z = (v^2+1)^{-1}. \tag{7.113}$$

Substituting (7.112) and (7.113) in (7.110) we obtain

$$\bar{\phi}_{N+l+1}(z) = \gamma_N D_N \bar{\phi}_{N+l}(z) \qquad (N = 1, 2, \ldots), \tag{7.114}$$

where

$$\gamma_N = \left(\frac{N+l+1}{N+l}\right)^2 \left(\frac{N+2l+1}{N}\right)^{1/2} \tag{7.115}$$

and

$$D_N = 1 - 2z + \frac{2}{N+2l+1} z(1-z)\frac{\mathrm{d}}{\mathrm{d}z}. \tag{7.116}$$

Now $\bar{\phi}_{l+1} = 1$ and hence (7.114) yields

$$\bar{\phi}_{N+l+1}(z) = \gamma_N \ldots \gamma_2 \gamma_1 P_N(z), \tag{7.117}$$

where

$$P_N(z) = D_N \ldots D_2 D_1 1. \tag{7.118}$$

By inspection

$$\gamma_N \ldots \gamma_2 \gamma_1 = \left(\frac{N+l+1}{l+1}\right)^2 \left(\frac{(N+2l+1)!}{(2l+1)!N!}\right)^{1/2}. \tag{7.119}$$

The polynomial $P_N(z)$ can be identified by using (B.3) with N replaced by $N-1$. This shows that

$$P_N(z) = {}_2F_1(-N, N+2l+2; l+\tfrac{3}{2}; z). \tag{7.120}$$

Collecting the above results we obtain the normalized radial momentum-space wavefunctions for bound states of the Coulomb potential

$$\bar{R}_{nl}(p) = -\mathrm{i}^l 2^{2l+3}\left(\frac{n^4(n+l)!(l!)^2}{p_0^3 2\pi(n-l-1)![(2l+1)!]^2}\right)^{1/2}$$
$$\times\, {}_2F_1\left(-n+l+1, n+l+1; l+\frac{3}{2}; \frac{1}{(np/p_0)^2+1}\right)$$
$$\times \frac{(np/p_0)^l}{[(np/p_0)^2+1]^{l+2}}, \tag{7.121}$$

where

$$n = N + l + 1. \tag{7.122}$$

The radial quantum number N $(= 0, 1, 2, \ldots)$ is the number of times the shift operators $\tilde{Q}_n^+$ must be applied to transform $\bar{R}_{l+1,l}(p)$ into $\bar{R}_{nl}(p)$.

The wavefunction (7.121) is the normalized solution to the radial equation (7.99). This solution is often expressed in terms of the Gegenbauer polynomial T_n^β. The latter is related to the hypergeometric function by*

$${}_2F_1(-n, n+2\beta+1; 1+\beta; u) = \frac{2^\beta n!\Gamma(\beta+1)}{\Gamma(n+2\beta+1)} T_n^\beta(1-2u). \tag{7.123}$$

The phase factors in the wavefunctions are determined by the choice of phase factors in the shift operations used to generate these wavefunctions. For the momentum-space wavefunction $\bar{\psi}_{nlm}(\mathbf{p}) = \bar{R}_{nl}(p)Y_{lm}(\hat{\mathbf{p}})$ of the oscillator and Coulomb potentials, our phase conventions are stated in Chapter 4 and the present chapter. For the coordinate-space wavefunction $\psi_{nlm}(\mathbf{r}) = R_{nl}(r)Y_{lm}(\hat{\mathbf{r}})$ see Chapters 4 and 5. Our phase choices are consistent with the property that $\psi_{nlm}(\mathbf{r})$ and $\bar{\psi}_{nlm}(\mathbf{p})$ are Fourier transforms† of each other (see Section 1.4).

References

1. de Lange, O.L. and Raab, R.E. (1987). An operator solution for the hydrogen atom with application to the momentum representation. *American Journal of Physics*, **55**, 913–17.
2. Hylleraas, E.A. (1932). Die Wellegleichung des Keplerproblems in Impulsraum. *Zeitschrift für Physik*, **74**, 216–24.
3. Morse, P.M. and Feshbach, H. (1953). *Methods of theoretical physics*, Part I, p. 782. McGraw-Hill, New York.
4. Morse, P.M. and Feshbach, H. (1953). *Methods of theoretical physics*, Part II, Ch. 12. McGraw-Hill, New York.

---

* The Gegenbauer polynomial in (7.123) is that defined in Ref. 3.

† These Fourier transforms are evaluated in Ref. 4.

8

THE ISOTROPIC HARMONIC OSCILLATOR IN AN ANGULAR MOMENTUM BASIS

In this chapter we use an algebraic method to solve the simultaneous eigenvalue problem for H, $\mathbf{L}^2$, and L_z (eqns (5.11)–(5.13)), where

$$H = \frac{1}{2M}\mathbf{p}^2 + \frac{1}{2}M\omega^2\mathbf{r}^2 \tag{8.1}$$

is the non-relativistic Hamiltonian for the three-dimensional isotropic harmonic oscillator. This method is based on shift operators for energy and angular momentum.

In Chapter 4 we considered a 'two-dimensional' problem in which $\mathbf{L}^2$ and L_z constitute a c.s.c.o. and we derived shift operators for the angular momentum quantum numbers l and m in the eigenkets $|lm\rangle$ of these observables. In Chapter 5 we discussed a 'one-dimensional' problem, namely the eigenvalue equation for the radial Hamiltonian (eqn (5.25)). Here we use results from Chapters 4 and 5 to construct shift operators for the eigenkets of the three-dimensional problem (5.11)–(5.13).

The shift operators are used to obtain eigenvalues and matrix elements, and to construct a spectrum-generating algebra for the oscillator in an angular momentum basis. The calculations are independent of any particular representation; they illustrate that the eigenvalues and matrix elements of an operator are determined by the algebra of operators and not by the choice of a representation. We also consider the coordinate and momentum representations: by repeated application of shift operators (in differential form) to a ground state wavefunction, we generate the normalized coordinate-space and momentum-space wavefunctions of the oscillator.

The oscillator is an example of a system that possesses 'accidental degeneracy' associated with additional constants of the motion. The constants of the motion are discussed and they are used to derive an invariance algebra for the oscillator.

8.1 Shift operations

From the shift operators $R_l^{\pm}(\omega)$ (eqn (5.85)) we construct the operators

$$R^{\pm}(\omega) = \pm i p_r - \hbar r^{-1}(S \pm \tfrac{1}{2}) - M\omega r, \tag{8.2}$$

where S is given by (4.48). Because

$$S|Elm\rangle = (l + \tfrac{1}{2})|Elm\rangle, \tag{8.3}$$

we see that

$$R^{\pm}(\omega)|Elm\rangle = R_l^{\pm}(\omega)|Elm\rangle. \tag{8.4}$$

It is useful to note that because S is a function of $\mathbf{L}^2$, it commutes with scalars such as r, p_r, etc. So, for example, in (8.2) the order of the operators r^{-1} and S is unimportant. The adjoints of (8.2) are

$$[R^{\pm}(\omega)]^{\dagger} = \mp \mathrm{i}p_r - \hbar r^{-1}(S \pm \tfrac{1}{2}) - M\omega r, \tag{8.5}$$

and one can readily show that*

$$[R^{\pm}(\omega)]^{\dagger} R^{\pm}(\omega) = 2MH + 2M\hbar\omega(S \pm 1). \tag{8.6}$$

We see that $R^{\pm}(\omega)$ factorize the operators on the right-hand side of (8.6).

Consider the operators

$$\mathbf{D}^{\pm}(\omega) = \mathbf{U}^{\pm} R^{\pm}(\omega), \tag{8.7}$$

where

$$\mathbf{U}^{\pm} = \pm \mathrm{i}r^{-1}\mathbf{r} \times \mathbf{L} + \hbar r^{-1}\mathbf{r}(S \pm \tfrac{1}{2}) \tag{8.8}$$

are the shift operators (4.66). Note that the order of $\mathbf{r}$ and S in (8.8) is important because S does not commute with the vector operator $\mathbf{r}$. If the products in (8.7) are multiplied out and simplified, they yield (see Appendix I)

$$\mathbf{D}^{\pm}(\omega) = \mathbf{F}^{\pm}(\omega) \pm \mathrm{i}\mathbf{G}^{\pm}(\omega), \tag{8.9}$$

where

$$\mathbf{F}^{\pm}(\omega) = -M\hbar\omega\mathbf{r}(S \pm \tfrac{1}{2}) - \mathbf{p} \times \mathbf{L} \tag{8.10}$$

and

$$\mathbf{G}^{\pm}(\omega) = -M\omega\mathbf{r} \times \mathbf{L} + \hbar\mathbf{p}(S \pm \tfrac{1}{2}). \tag{8.11}$$

In terms of the boson annihilation and creation operators $\mathbf{a}$ and $\mathbf{a}^{\dagger}$ defined by (2.253) and (2.254)

$$\mathbf{D}^{+}(\omega) = -(2M\hbar\omega)^{1/2}[\mathrm{i}\mathbf{a}^{\dagger} \times \mathbf{L} + \hbar\mathbf{a}^{\dagger}(S + \tfrac{1}{2})], \tag{8.12}$$

$$\mathbf{D}^{-}(\omega) = (2M\hbar\omega)^{1/2}[\mathrm{i}\mathbf{a} \times \mathbf{L} - \hbar\mathbf{a}(S - \tfrac{1}{2})]. \tag{8.13}$$

Also

$$\mathbf{D}^{+}(-\omega) = (2M\hbar\omega)^{1/2}[\mathrm{i}\mathbf{a} \times \mathbf{L} + \hbar\mathbf{a}(S + \tfrac{1}{2})], \tag{8.14}$$

$$\mathbf{D}^{-}(-\omega) = -(2M\hbar\omega)^{1/2}[\mathrm{i}\mathbf{a}^{\dagger} \times \mathbf{L} - \hbar\mathbf{a}^{\dagger}(S - \tfrac{1}{2})]. \tag{8.15}$$

The four operators (8.12)–(8.15) are particular cases of the vector operators

$$\pm \mathrm{i}\mathbf{v} \times \mathbf{L} + \hbar\mathbf{v}(S \pm \tfrac{1}{2}) \tag{8.16}$$

* To prove (8.6) we can either multiply out (8.2) and (8.5), or allow both sides of (8.6) to operate on $|Elm\rangle$ and then use (8.4) and the factorizations (5.66) and (5.67).

derived in Section 4.4; they correspond to the choices $\mathbf{a}$ and $\mathbf{a}^\dagger$ (apart from a constant factor) for the vector operator $\mathbf{v}$. Therefore

$$[L_i, D_j^\pm(\omega)] = i\hbar\varepsilon_{ijk}D_k^\pm(\omega) \tag{8.17}$$

and

$$[\mathbf{L}^2, \mathbf{D}^\pm(\omega)] = \pm 2\hbar^2\mathbf{D}^\pm(\omega)(S \pm \tfrac{1}{2}). \tag{8.18}$$

(See (4.45) and (4.51).) Also, because

$$[H, \mathbf{a}^\dagger] = \hbar\omega\mathbf{a}^\dagger, \tag{8.19}$$

$$[H, \mathbf{a}] = -\hbar\omega\mathbf{a}, \tag{8.20}$$

(Section 2.10), and H commutes with $\mathbf{L}$, we have

$$[H, \mathbf{D}^\pm(\omega)] = \pm\hbar\omega\mathbf{D}^\pm(\omega). \tag{8.21}$$

It follows from (8.17), (8.18), and (8.21) that

$$D_\mu^\pm(\omega)|Elm\rangle = \gamma_\mu^\pm(\omega)|E \pm \hbar\omega, l \pm 1, m + \mu\rangle \tag{8.22}$$

and

$$D_\mu^\pm(-\omega)|Elm\rangle = \gamma_\mu^\pm(-\omega)|E \mp \hbar\omega, l \pm 1, m + \mu\rangle, \tag{8.23}$$

where $\mu = \pm 1$ or 0, and

$$D_{\pm 1}^+ = D_x^+ \pm iD_y^+, \qquad D_{\pm 1}^- = D_x^- \pm iD_y^-, \qquad D_0^\pm = D_z^\pm. \tag{8.24}$$

In the shift operations (8.22) and (8.23) the superscript and subscript on D indicate the shift in l and m, respectively.

The coefficients $\gamma_\mu^\pm(\omega)$ can be obtained from previous results: for normalized kets in (8.22)

$$\begin{aligned}|\gamma_\mu^\pm(\omega)|^2 &= \langle Elm|[R_l^\pm(\omega)]^\dagger[U_\mu^\pm]^\dagger U_\mu^\pm R_l^\pm(\omega)|Elm\rangle \\ &= \langle Elm|[U_\mu^\pm]^\dagger U_\mu^\pm[R_l^\pm(\omega)]^\dagger R_l^\pm(\omega)|Elm\rangle,\end{aligned} \tag{8.25}$$

because $\mathbf{U}^\pm$ commute with $R_l^\pm$.* Now use (8.6). Then

$$|\gamma_\mu^\pm(\omega)|^2 = |\alpha^\pm(\omega)|^2\langle Elm|[U_\mu^\pm]^\dagger U_\mu^\pm|Elm\rangle, \tag{8.26}$$

where $\alpha^\pm(\omega)$ are given by (5.86). In Chapter 4 we evaluated the average of $[U_\mu^\pm]^\dagger U_\mu^\pm$ with respect to the kets $|lm\rangle$; these results apply also if the average is with respect to the kets $|Elm\rangle$ (see Appendix E). Hence

$$\gamma_\mu^\pm(\omega) = \alpha^\pm(\omega)\beta_\mu^\pm, \tag{8.27}$$

where $\beta_\mu^\pm$ are given by (4.70)–(4.72). In (8.27) the phase factors of $\gamma_\mu^\pm$ correspond to the choices of phase made in Chapters 4 and 5.

---

* (i) L_i commute with the scalars $R_l^\pm$ and (ii) $\hat{x}$, $\hat{y}$, and $\hat{z}$ commute with $R_l^\pm$ because $[\mathbf{r}\cdot\mathbf{p}, \hat{r}_i] = r_j\{[p_j, r_i]r^{-1} + r_i[p_j, r^{-1}]\} = 0$.

The construction in (8.7) for the operators $\mathbf{D}^{\pm}$ is not unique. The shift operators $P_l^{\pm}(\omega)$ (eqn (7.13)) lead us to define

$$P^{\pm}(\omega) = \mp \mathrm{i}r_p - \frac{p}{M\omega} - \frac{\hbar}{p}\left(S \pm \frac{1}{2}\right). \tag{8.28}$$

Then

$$[P^{\pm}(\omega)]^{\dagger} = \pm \mathrm{i}r_p - \frac{p}{M\omega} - \frac{\hbar}{p}\left(S \pm \frac{1}{2}\right), \tag{8.29}$$

and we have the factorizations

$$[P^{\pm}(\omega)]^{\dagger}P^{\pm}(\omega) = (M\omega)^{-2}[2MH + 2M\hbar\omega(S \pm 1)]. \tag{8.30}$$

Consider the operators $\mathbf{V}^{\pm}P^{\pm}(\omega)$, where

$$\mathbf{V}^{\pm} = \pm \mathrm{i}p^{-1}\mathbf{p} \times \mathbf{L} + \hbar p^{-1}\mathbf{p}(S \pm \tfrac{1}{2}) \tag{8.31}$$

are the shift operators (4.74). By calculations similar to those in Appendix I we find

$$\mathbf{D}^{\pm}(\omega) = \mp \mathrm{i}M\omega\mathbf{V}^{\pm}P^{\pm}(\omega). \tag{8.32}$$

Equations (8.7) and (8.32) show that the operators $\mathbf{D}^{\pm}$ are bifactorizable. There are other systems with this property, namely, the Coulomb problem (Section 9.1) and the Dirac oscillator (Section 12.9).

From (8.7) and (8.32) the effect of $\mathbf{D}^{\pm}(\omega)$ on $|Elm\rangle$ is the same as that of the operators $\mathbf{U}_l^{\pm}R_l^{\pm}(\omega)$ and $\mp \mathrm{i}M\omega\mathbf{V}_l^{\pm}P_l^{\pm}(\omega)$, where $\mathbf{U}_l^{\pm}$ and $\mathbf{V}_l^{\pm}$ are given by (4.73) and (4.76), and $R_l^{\pm}$ and $P_l^{\pm}$ by (5.85) and (7.13). We note that $\mathbf{U}_l^{\pm}$ and $R_l^{\pm}(\omega)$ are linear in $\mathbf{p}$; thus in the coordinate representation they yield first-order differential operators. Similarly, $\mathbf{V}_l^{\pm}$ and $P_l^{\pm}(\omega)$ are linear in $\mathbf{r}$ and in the momentum representation they also yield first-order differential operators.

In the limit $\omega \to 0$ the particle is free and (8.7)–(8.11) and (8.31) yield

$$\mathbf{D}^{\pm} = \pm \mathrm{i}p\mathbf{V}^{\pm} = \mathbf{U}^{\pm}R^{\pm}, \tag{8.33}$$

where

$$R^{\pm} = \pm \mathrm{i}p_r - \hbar r^{-1}(S \pm \tfrac{1}{2}).$$

The shift operators (8.33) transform the free-particle eigenkets $|Elm\rangle$ according to (see (8.22))

$$D_{\mu}^{\pm}|Elm\rangle = \gamma_{\mu}^{\pm}|E, l \pm 1, m + \mu\rangle, \tag{8.34}$$

where the coefficients $\gamma_{\mu}^{\pm}$ are given by (8.27), (4.70)–(4.72), and (6.132).

It may be helpful, in comparing the above results with those derived in Chapters 4, 5, and 7, to imagine a 'factorization' or 'direct product' of kets $|Elm\rangle = |El\rangle \otimes |lm\rangle$. If $|El\rangle$ is the ket in (5.26), we suppose that the radial operators r and p_r act on $|El\rangle$ only, while the angular momentum operator $\mathbf{L}$

and the vector operator $\hat{\mathbf{r}}$ act on $|lm\rangle$. Then (8.22) and (8.23) follow directly from (4.58), (4.70)–(4.72), and (5.83)–(5.86). Similarly, if $|El\rangle$ is the ket in (7.8) and if the radial operators p and r_p act on $|El\rangle$ only, while $\mathbf{L}$ and $\hat{\mathbf{p}}$ act on $|lm\rangle$, then (8.22) and (8.23) follow from (4.75), (4.70)–(4.72), and (7.11)–(7.14). However, no such assumptions are used in our derivation of (8.22) and (8.23).

8.2 Eigenvalues

The kets $|Elm\rangle$ satisfy the shift operations

$$L_\pm|Elm\rangle = \hbar[l(l+1) - m(m\pm 1)]^{1/2}|E, l, m\pm 1\rangle, \tag{8.35}$$

and therefore $m = -l, -l+1, \ldots, l$ and $l = 0, \frac{1}{2}, 1, \frac{3}{2}, \ldots$ (Section 4.3). We first show that the half-integral values of l must be excluded. This proof is an extension of that in Section 4.6 for kets $|lm\rangle$ of the c.s.c.o. $\mathbf{L}^2$ and L_z.

Consider a ket $|E'll\rangle$ where $E' > 0$ and l is a half integer greater than $\frac{1}{2}$. Suppose $\mathbf{L}$ is Hermitian with respect to kets with half-integral l. Then repeated application of $D^-_{-1}(-\omega)$ to $|E'll\rangle$ will yield the kets $|E\frac{1}{2}\frac{1}{2}\rangle$, where

$$E > E' > 0, \tag{8.36}$$

(see (8.23)). Also

$$|E + \hbar\omega, -\tfrac{1}{2}, -\tfrac{1}{2}\rangle \sim D^-_{-1}(-\omega)|E\tfrac{1}{2}\tfrac{1}{2}\rangle \neq 0. \tag{8.37}$$

(The result (8.37) can be proved by contradiction: suppose $D^-_{-1}(-\omega)|E\frac{1}{2}\frac{1}{2}\rangle = 0$. Then

$$R^-_{1/2}(-\omega)[\hbar(\hat{x} - i\hat{y}) - \hat{z}L_-]|E\tfrac{1}{2}\tfrac{1}{2}\rangle = 0, \tag{8.38}$$

where we have used the factorization (8.7) and the fact that $R^-_{1/2}$ commutes with the operator in brackets. Multiply (8.38) on the left by $\hat{x} + i\hat{y}$ and commute the latter operator with $R^-_{1/2}$. Then using the calculation below (4.80) we find that (8.38) becomes

$$R^-_{1/2}(-\omega)|E\tfrac{1}{2}\tfrac{1}{2}\rangle = 0. \tag{8.39}$$

Taking the norm of both sides of (8.39) and using (8.4) and (8.6) we obtain $2ME = 0$. This contradicts (8.36).)

But

$$\mathbf{L}^2|E + \hbar\omega, -\tfrac{1}{2}, -\tfrac{1}{2}\rangle = -\tfrac{1}{4}\hbar^2|E + \hbar\omega, -\tfrac{1}{2}, -\tfrac{1}{2}\rangle, \tag{8.40}$$

which contradicts the requirement that the eigenvalues of $\mathbf{L}^2$ be non-negative if $\mathbf{L}$ is Hermitian (Section 4.1). We conclude that $\mathbf{L}$ is not Hermitian with respect to the space spanned by $|Elm\rangle$ if l is half-integral. Hence

$$l = 0, 1, 2, \ldots. \tag{8.41}$$

Next we determine the energy eigenvalues. If the kets generated by the shift operations (8.22) and (8.23) are to have non-negative norm, then $|\gamma^\pm_\mu|^2$ must be non-negative. Consider the shift operation

$$D_0^+(-\omega)|Elm\rangle = \alpha^+(-\omega)\beta_0^+|E - \hbar\omega, l+1, m\rangle. \tag{8.42}$$

Because $|\beta_0^+|^2 > 0$ (see (4.70)), it is necessary that $|\alpha^+(-\omega)|^2 \geq 0$. As discussed in Section 5.5, this requires that l be bounded above. Then

$$E = (n + \tfrac{3}{2})\hbar\omega, \tag{8.43}$$

where

$$n = l_{\max} = 0, 1, 2, \ldots. \tag{8.44}$$

In the following we label the eigenkets $|nlm\rangle$.

8.3 Matrix elements

We can use (8.9)–(8.11) to express $\mathbf{r}$ and $\mathbf{p}$ in terms of S and the four operators $\mathbf{D}^\pm(\omega)$ and $\mathbf{D}^\pm(-\omega)$. These expressions together with (8.22), (8.23), and (8.3) yield the useful results

$$\begin{aligned} r_\mu|nlm\rangle = [2M\hbar\omega(2l+1)]^{-1}[&\gamma_\mu^+(-\omega)|n-1, l+1, m+\mu\rangle \\ &+ \gamma_\mu^-(-\omega)|n+1, l-1, m+\mu\rangle - \gamma_\mu^+(\omega)|n+1, l+1, m+\mu\rangle \\ &- \gamma_\mu^-(\omega)|n-1, l-1, m+\mu\rangle] \end{aligned} \tag{8.45}$$

and

$$\begin{aligned} p_\mu|nlm\rangle = [2i\hbar(2l+1)]^{-1}[&\gamma_\mu^+(-\omega)|n-1, l+1, m+\mu\rangle \\ &- \gamma_\mu^-(-\omega)|n+1, l-1, m+\mu\rangle + \gamma_\mu^+(\omega)|n+1, l+1, m+\mu\rangle \\ &- \gamma_\mu^-(\omega)|n-1, l-1, m+\mu\rangle]. \end{aligned} \tag{8.46}$$

Here $\gamma_\mu^\pm$ are given by (8.27) with

$$\alpha^\pm(\omega) = (2M\hbar\omega)^{1/2}(n + l + 2 \pm 1)^{1/2} \tag{8.47}$$

and

$$\alpha^\pm(-\omega) = (2M\hbar\omega)^{1/2}(n - l + 1 \mp 1)^{1/2}. \tag{8.48}$$

(Equation (8.48) is obtained from (5.86) by replacing ω with $-\omega$ and then substituting the energy eigenvalue (5.88).)

Equations (8.45) yield the matrix elements

$$\begin{aligned} \langle n'l'm'|z|nlm\rangle = [2M\hbar\omega(2l+1)]^{-1}[&\gamma_0^+(-\omega)\delta_{n',n-1}\delta_{l',l+1} \\ &+ \gamma_0^-(-\omega)\delta_{n',n+1}\delta_{l',l-1} - \gamma_0^+(\omega)\delta_{n',n+1}\delta_{l',l+1} \\ &- \gamma_0^-(\omega)\delta_{n',n-1}\delta_{l',l-1}]\delta_{m'm} \end{aligned} \tag{8.49}$$

and

$$\begin{aligned} \langle n'l'm'|x \pm iy|nlm\rangle = [2M\hbar\omega(2l+1)]^{-1}[&\gamma_{\pm1}^+(-\omega)\delta_{n',n-1}\delta_{l',l+1} \\ &+ \gamma_{\pm1}^-(-\omega)\delta_{n',n+1}\delta_{l',l-1} - \gamma_{\pm1}^+(\omega)\delta_{n',n+1}\delta_{l',l+1} \\ &- \gamma_{\pm1}^-(\omega)\delta_{n',n-1}\delta_{l',l-1}]\delta_{m',m\pm1}. \end{aligned} \tag{8.50}$$

By repeated use of (8.45) we can evaluate any matrix element of the form $\langle n'l'm'|x^{n_1}y^{n_2}z^{n_3}|nlm\rangle$ where n_i are non-negative integers. Similarly, by using (8.46) we can determine the matrix elements of $p_x^{n_1}p_y^{n_2}p_z^{n_3}$.

8.4 Eigenkets generated from the ground state $|000\rangle$

For the eigenkets $|nlm\rangle$ we have derived

(i) shift operators $Q^{\pm}$ and $\tilde{Q}^{\pm}$ that change only the energy (see (5.143) and (7.81));

(ii) shift operators $D_\mu^{\pm}$ that change both the energy and the angular momentum (see (8.22) and (8.23));

(iii) shift operators $L_{\pm}$ that change the z component of angular momentum (see (8.35)).

It is clear that starting from a given eigenket we can build up all the eigenkets of H, $\mathbf{L}^2$, and L_z by applying the appropriate shift operators. For example, all the eigenkets $|nlm\rangle$ are generated by applying Q^+, $D_{+1}^{+}(\omega)$, and L_- to the ground state or 'vacuum' $|000\rangle$.

This procedure can be simplified by selecting a particular sequence of transformations. Consider (8.45) with $n = l = m$ and $\mu = 1$. Of the four coefficients $\gamma_1^{\pm}(\omega)$ and $\gamma_1^{\pm}(-\omega)$ that appear on the right-hand side of (8.45), all are zero except $\gamma_1^{+}(\omega) = -2\hbar[M\hbar\omega(l+1)]^{1/2}(2l+1)$. (See (8.27), (8.47), (8.48), and (4.71).) Hence

$$r\hat{r}_+|lll\rangle = a(l+1)^{1/2}|l+1, l+1, l+1\rangle, \tag{8.51}$$

where $\hat{r}_+ = \hat{x} + i\hat{y}$. In a similar manner (8.46) yields

$$p\hat{p}_+|lll\rangle = ip_0(l+1)^{1/2}|l+1, l+1, l+1\rangle, \tag{8.52}$$

where $\hat{p}_+ = \hat{p}_x + i\hat{p}_y$.

Now perform the following operations

1. Apply $(r\hat{r}_+)^l$ to $|000\rangle$. This generates (see (8.51))

$$|lll\rangle = a^{-l}(l!)^{-1/2}(r\hat{r}_+)^l|000\rangle. \tag{8.53}$$

2. Apply $(Q^+)^N$, where $N = 1, 2, \ldots$, to $|lll\rangle$. This generates (see (5.143) and (5.101))

$$|nll\rangle = \left(\frac{(2l+1)!!}{2^N N!(2l+2N+1)!!}\right)^{1/2}(Q^+)^N|lll\rangle, \tag{8.54}$$

where $n = l + 2N$.

3. Apply L_-^{l-m} to $|nll\rangle$. This generates (see (8.35))

$$|nlm\rangle = \left(\frac{(l+m)!}{\hbar^{2(l-m)}(2l)!(l-m)!}\right)^{1/2} L_-^{l-m}|nll\rangle. \tag{8.55}$$

In this manner all the eigenkets of H, $\mathbf{L}^2$, and L_z are obtainable from $|000\rangle$.

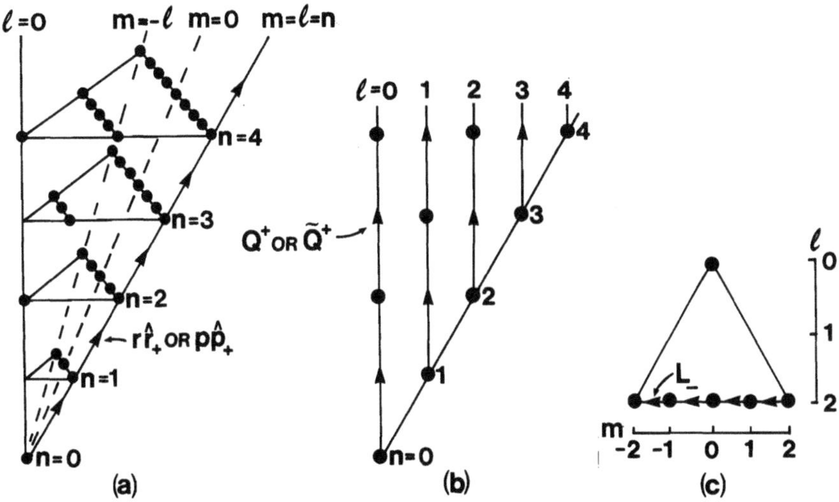

FIG. 8.1 (a) The states $|nlm\rangle$ of the isotropic harmonic oscillator. The transformations (8.53) and (8.53′) are depicted. (b) In the plane $m = l$ (the plane containing the lines $m = l$ and $l = 0$ in (a)). The transformations (8.54) and (8.54′) are depicted. (c) In the plane $n = 2$. The transformations effected by L_- are depicted.

An alternative and equally simple procedure would be:

1′. $|lll\rangle = (\mathrm{i}p_0)^{-l}(l!)^{-1/2}(p\hat{p}_+)^l|000\rangle.$ (8.53′)

(See (8.52).)

2′. $$|nll\rangle = \left(\frac{(2l+1)!!}{2^N N!(2l+2N+1)!!}\right)^{1/2}(\tilde{Q}^+)^N|lll\rangle, \qquad (8.54')$$

where $n = l + 2N$ and $N = 1, 2, \ldots$. (See (7.81) and (5.101).)

3′. Use (8.55).

The transformations (8.53)–(8.55), (8.53′), and (8.54′) are depicted in Fig. 8.1. The analogous diagram for the bound states of a Coulomb potential is shown in Fig. 9.1.

8.5 Wavefunctions

We apply the theory discussed above to a calculation of the coordinate-space wavefunctions

$$\psi_{nlm}(\mathbf{r}) = \langle\mathbf{r}|nlm\rangle. \qquad (8.56)$$

From (5.141) and (5.143) expressed in the coordinate representation

$$\left(\frac{\partial}{\partial r} + \frac{r}{a^2}\right)\psi_{000}(\mathbf{r}) = 0. \qquad (8.57)$$

Similarly, from (5.13) and (8.35) $L_i|000\rangle = 0$ and hence (see (4.121))

$$\frac{\partial}{\partial x^2}\psi_{000} = \frac{\partial}{\partial y^2}\psi_{000} = \frac{\partial}{\partial z^2}\psi_{000}. \tag{8.58}$$

According to (8.58), $\psi_{000}(\mathbf{r})$ must be a function of r only, so that (8.57) gives

$$\psi_{000}(\mathbf{r}) = C\exp\left(-\frac{r^2}{2a^2}\right), \tag{8.59}$$

where C is a constant. The normalization

$$\iint |\psi_{000}|^2 r^2\, \mathrm{d}r\, \mathrm{d}\Omega = 4\pi\int_0^\infty |\psi_{000}|^2 r^2\, \mathrm{d}r = 1 \tag{8.60}$$

requires

$$C = \frac{1}{(4\pi)^{1/2}}\frac{2}{\pi^{1/4}a^{3/2}}. \tag{8.61}$$

Thus in terms of the radial wavefunction $R_{00}(r)$ (eqn (6.68)) and the spherical harmonic $Y_{00}(\hat{r})$ (eqn (4.122))

$$\psi_{000}(\mathbf{r}) = R_{00}(r)Y_{00}(\hat{\mathbf{r}}). \tag{8.62}$$

From $\psi_{000}(\mathbf{r})$ we obtain the wavefunctions ψ_{nlm} by using the operator equations (8.53)–(8.55) expressed in the coordinate representation. In these calculations the shift operators act either on $R_{00}(r)$ or on $Y_{00}(\hat{\mathbf{r}})$; because the effect of these operators on R_{00} and Y_{00} has been calculated in Sections 6.4 and 4.10, we can now simply write down the results.

From (8.53) and (8.62)

$$\psi_{lll}(\mathbf{r}) = R_{ll}(r)Y_{ll}(\hat{\mathbf{r}}), \tag{8.63}$$

where Y_{ll} and R_{ll} are given by (4.123) and (6.70).

In applying (8.54) we use the result

$$\langle\mathbf{r}|(Q^+)^N|lll\rangle = Q^+_{l+2(N-1)}\cdots Q^+_{l+2}Q^+_l\psi_{lll}(\mathbf{r}), \tag{8.64}$$

where

$$Q^+_k = r\frac{\partial}{\partial r} - \frac{r^2}{a^2} + k + 3 \tag{8.65}$$

is the differential operator obtained from (5.141) and (5.142). Substituting (8.63) in (8.64) and noting that Q^+_k commutes with Y_{ll}, we obtain from (8.54)

$$\psi_{nll}(\mathbf{r}) = R_{nl}(r)Y_{ll}(\hat{\mathbf{r}}), \tag{8.66}$$

where $R_{nl}(r)$ is the radial coordinate-space wavefunction (6.79). In applying (8.55) we note that

$$\langle\mathbf{r}|L_-^{l-m}|nll\rangle = L_-^{l-m}\psi_{nll}(\hat{\mathbf{r}}) = R_{nl}(r)L_-^{l-m}Y_{ll}(\hat{\mathbf{r}}), \tag{8.67}$$

where the differential operator L_- is given by (4.117). In the last step we have used (8.66) and the fact that L_- commutes with functions of r (see Appendix C). Hence (8.55) and the calculations in Section 4.10 yield

$$\psi_{nlm}(\mathbf{r}) = R_{nl}(r)\, Y_{lm}(\hat{\mathbf{r}}), \tag{8.68}$$

where $Y_{lm}(\hat{\mathbf{r}})$ is the spherical harmonic (4.129).

In a similar manner we can calculate the momentum-space wavefunctions

$$\bar{\psi}_{nlm}(\mathbf{p}) = \langle \mathbf{p}|nlm\rangle. \tag{8.69}$$

In these calculations we use (8.53′), (8.54′), and (8.55) expressed in the momentum representation. As in the above, the result is given by combining the shift operations considered in Sections 4.10 and 7.6. We find

$$\bar{\psi}_{nlm}(\mathbf{p}) = \bar{R}_{nl}(p)\, Y_{lm}(\hat{\mathbf{p}}), \tag{8.70}$$

where $\bar{R}_{nl}(p)$ is the radial momentum-space wavefunction (7.100).

The 'separation of variables' in (8.68) and (8.70) is an important property of these solutions. In the above calculations it results from:

(i) The separation of variables that occurs in the ground state; in the coordinate representation, for example, this property is an obvious consequence of the first-order equations (8.57) and (8.58) that, together with the normalization (8.60), completely specify $\psi_{000}(\mathbf{r})$ (see (8.62)).

(ii) The fact that separation of variables is preserved by the differential forms of the raising operations (8.53)–(8.55) (or (8.53′), (8.54′), and (8.55)).

In the basis H_1, H_2, H_3 the coordinate-space wavefunctions for the oscillator, $\tilde{\psi}_{n_1n_2n_3}(\mathbf{r})$, are given by (2.269) and (2.157) (we use a tilde on $\psi_{n_1n_2n_3}$ to distinguish it from (8.68)). It can be shown that there are linear relations between $\psi_{nlm}(\mathbf{r})$ and $\tilde{\psi}_{n_1n_2n_3}(\mathbf{r})$ [1]:

$$\psi_{nlm}(\mathbf{r}) = \sum_{n_1,n_2,n_3=0} a^{nlm}_{n_1n_2n_3}\tilde{\psi}_{n_1n_2n_3}(\mathbf{r}), \tag{8.71}$$

where the sums over n_1, n_2, and n_3 are subject to $n_1 + n_2 + n_3 = n$. For example, when $n = 1$

$$\begin{aligned}
\psi_{111}(\mathbf{r}) &= \frac{1}{\sqrt{2}}(\tilde{\psi}_{100} + \mathrm{i}\tilde{\psi}_{010}),\\
\psi_{110}(\mathbf{r}) &= -\tilde{\psi}_{001}(\mathbf{r}),\\
\psi_{1,1,-1}(\mathbf{r}) &= -\frac{1}{\sqrt{2}}(\tilde{\psi}_{100} - \mathrm{i}\tilde{\psi}_{010}).
\end{aligned} \tag{8.72}$$

8.6 An invariance algebra

Every non-relativistic quantum-mechanical Hamiltonian with a spherically symmetric potential commutes with the three components L_i of the orbital angular momentum operator. The commutation relations of the L_i define the Lie algebra so(3) (see Section 4.9), which is an invariance algebra for the system. The Hermitian operators L_i are generators of the Lie group SO(3), which is a symmetry group of the system. Because $L_\pm$ transform between states with different quantum numbers m (see Section 4.4), the SO(3) symmetry explains the degeneracy of the energy eigenvalues with respect to m. This degeneracy is sometimes called a 'geometric degeneracy', signifying that it is a consequence of the 'geometric symmetry' of the system (in this case symmetry with respect to rotation in three-space).

For certain spherically symmetric potentials there may be additional operators that commute with the Hamiltonian, and consequently the invariance algebra is larger than so(3). If the additional constants of the motion transform between different states of the system, the energy eigenvalues will possess a greater amount of degeneracy than that due to spherical symmetry alone. This additional degeneracy is often referred to as 'accidental degeneracy' and the associated symmetry is called a 'dynamical symmetry' [2]. Two important examples of a system that possesses such dynamical symmetry are the non-relativistic two-body problem with an attractive potential that is proportional to r^2 (the oscillator) or r^{-1} (the Coulomb problem). In this section we discuss the additional constants of the motion and the invariance algebra for the oscillator. For the Coulomb potential this topic is treated in Section 9.6.

For the oscillator we have already seen that the nine operators $a_i^\dagger a_j$ $(i, j = 1, 2, 3)$, where $a_i^\dagger$ and a_j are the boson creation and annihilation operators (2.253) and (2.254), are constants of the motion (see Section 2.10). An invariance algebra for the oscillator can be obtained by taking suitable linear combinations of these constants of the motion. For this purpose it is convenient to define the nine operators

$$T_{ij} = \tfrac{1}{2}(a_i^\dagger a_j + a_j a_i^\dagger), \tag{8.73}$$

which are also constants of the motion. First we express $\mathbf{L}$ in terms of T_{ij}. Using (2.253) and (2.254),

$$\mathbf{L} = -i\hbar \mathbf{a}^\dagger \times \mathbf{a}. \tag{8.74}$$

Thus

$$\begin{aligned} L_i &= -i\hbar\varepsilon_{ijk} a_j^\dagger a_k \\ &= -i\hbar\varepsilon_{ijk}\tfrac{1}{2}(a_j^\dagger a_k + a_k a_j^\dagger) && (8.75) \\ &= -i\hbar\varepsilon_{ijk} T_{jk}; && (8.76) \end{aligned}$$

we see that L_i is proportional to the difference $T_{jk} - T_{kj}$, where i, j, k are cyclic. The remaining six constants of the motion are elements of the symmetric second-rank tensor

$$A_{ij} = \tfrac{1}{2}\hbar\omega(T_{ij} + T_{ji}). \tag{8.77}$$

In terms of $\mathbf{r}$ and $\mathbf{p}$, (8.77) is

$$A_{ij} = \frac{1}{2M}(p_i p_j + M^2\omega^2 r_i r_j). \tag{8.78}$$

We use L_i and A_{ij} to construct an invariance algebra. To evaluate the commutation relations of these operators we first use (2.255) and (8.73) to show that

$$[T_{ij}, T_{kl}] = \delta_{jk} T_{il} - \delta_{il} T_{kj}. \tag{8.79}$$

From (8.76), (8.77), and (8.79) we find

$$[L_i, A_{jk}] = \mathrm{i}\hbar\varepsilon_{ilm}(\delta_{jl} A_{km} + \delta_{kl} A_{jm}) \tag{8.80}$$

and

$$[A_{ij}, A_{kl}] = \tfrac{1}{4}\mathrm{i}\hbar\omega^2(\delta_{ik}\varepsilon_{jlm} + \delta_{il}\varepsilon_{jkm} + \delta_{jk}\varepsilon_{ilm} + \delta_{jl}\varepsilon_{ikm})L_m. \tag{8.81}$$

Equations (8.80), (8.81), and (4.4) show that the operators L_i and A_{ij} close under commutation. Their Lie algebra can be determined by taking suitable linear combinations of these operators.

Because

$$H = A_{11} + A_{22} + A_{33}, \tag{8.82}$$

(see (8.78) and (8.1)) only five of the six operators A_{ij} are additional constants of the motion, namely the five linearly independent elements of a symmetric, traceless second-rank tensor: A_{12}, A_{13}, A_{23}, and any *two* of $3A_{11} - H$, $3A_{22} - H$, and $3A_{33} - H$. An indication of how to proceed is provided by the commutators of $3A_{33} - H = 2A_{33} - A_{11} - A_{22}$ and $L_\pm$. Use of (8.80) shows that

$$[L_\pm, 3A_{33} - H] = 6\hbar[\mp(A_{13} \pm \mathrm{i}A_{23})]. \tag{8.83}$$

Also

$$[L_\pm, \mp(A_{13} \pm \mathrm{i}A_{23})] = \hbar(A_{11} - A_{22} \pm 2\mathrm{i}A_{12}) \tag{8.84}$$

and

$$[L_\pm, A_{11} - A_{22} \pm 2\mathrm{i}A_{12}] = 0. \tag{8.85}$$

These results suggest that suitable linear combinations for the five additional constants of the motion are*

$$Q_0 = \omega^{-1}(2A_{33} - A_{11} - A_{22}), \tag{8.86}$$

$$Q_{\pm 1} = \mp\omega^{-1}(A_{13} \pm \mathrm{i}A_{23}), \tag{8.87}$$

$$Q_{\pm 2} = \omega^{-1}(A_{11} - A_{22} \pm 2\mathrm{i}A_{12}). \tag{8.88}$$

---

* The label Q signifies that these operators are elements of a quadrupole tensor. The factors ω^{-1} in (8.86)–(8.88) give Q_λ the dimensions of angular momentum.

Indeed, using (8.80), (8.81), and (4.4) we obtain the 28 commutation relations

$$\begin{aligned}
&[L_3, Q_0] = [Q_0, Q_{\pm 2}] = [Q_{\pm 1}, Q_{\pm 2}] = [L_\pm, Q_{\pm 2}] = 0,\\
&[L_\pm, L_\mp] = -4[Q_{\pm 1}, Q_{\mp 1}] = \tfrac{1}{2}[Q_{\pm 2}, Q_{\mp 2}] = \pm 2\hbar L_3,\\
&[L_\pm, Q_{\mp 1}] = \hbar Q_0,\\
&\pm[L_3, L_\pm] = -\tfrac{2}{3}[Q_0, Q_{\pm 1}] = [Q_{\mp 1}, Q_{\pm 2}] = \hbar L_\pm,\\
&\pm[L_3, Q_{\pm 1}] = \tfrac{1}{6}[L_\pm, Q_0] = \tfrac{1}{4}[L_\mp, Q_{\pm 2}] = \hbar Q_{\pm 1},\\
&\pm[L_3, Q_{\pm 2}] = 2[L_\pm, Q_{\pm 1}] = 2\hbar Q_{\pm 2}.
\end{aligned} \tag{8.89}$$

The identification of this Lie algebra is made in Section 8.9.

To calculate the effect of the operators L_i and Q_λ ($\lambda = 0, \pm 1, \pm 2$) of this invariance algebra on the eigenkets of H, we first express these operators in terms of the boson operators $\mathbf{a}$ and $\mathbf{a}^\dagger$. For the angular momentum operator this has been done in (8.75). For the Q_λ we substitute (8.77) and (8.73) in (8.86)–(8.88). This gives

$$Q_0 = \hbar(2a_3^\dagger a_3 - a_1^\dagger a_1 - a_2^\dagger a_2), \tag{8.90}$$

$$Q_{\pm 1} = \tfrac{1}{2}(\mp a_3^\dagger a_1 \mp a_1^\dagger a_3 - \mathrm{i}a_2^\dagger a_3 - \mathrm{i}a_3^\dagger a_2), \tag{8.91}$$

$$Q_{\pm 2} = \hbar(a_1^\dagger a_1 - a_2^\dagger a_2 \pm \mathrm{i}a_1^\dagger a_2 \pm \mathrm{i}a_2^\dagger a_1). \tag{8.92}$$

The effect of these operators on the eigenkets $|n_1 n_2 n_3\rangle$ of the c.s.c.o. H_1, H_2, H_3 can be written down using (2.263) and (2.264); it is clear that L_i and Q_λ transform within the subspaces of degenerate eigenkets (subspaces with $n_1 + n_2 + n_3 = \text{constant}$).

For the eigenkets $|nlm\rangle$ of an angular momentum basis, the effect of L_i is given by (8.35) and (5.13). To calculate the effect of Q_λ on $|nlm\rangle$ we express Q_λ in terms of the operators

$$\begin{aligned}
a_{\pm 1} &= a_1 \pm \mathrm{i}a_2, & (a^\dagger)_{\pm 1} &= a_1^\dagger \pm \mathrm{i}a_2^\dagger,\\
a_0 &= a_3, & a_0^\dagger &= a_3^\dagger.
\end{aligned} \tag{8.93}$$

Then

$$Q_0 = \hbar\{2a_0^\dagger a_0 - \tfrac{1}{2}[(a^\dagger)_{-1}a_{+1} + (a^\dagger)_{+1}a_{-1}]\}, \tag{8.90$'$}$$

$$Q_{\pm 1} = \mp\tfrac{1}{2}\hbar[a_0^\dagger a_{\pm 1} + (a^\dagger)_{\mp 1}a_0], \tag{8.91$'$}$$

$$Q_{\pm 2} = \hbar(a^\dagger)_{\pm 1}a_{\pm 1}. \tag{8.92$'$}$$

From (8.45)–(8.48) and (8.27)

$$\begin{aligned}
a_\mu|nlm\rangle = \hbar^{-1}(2l+1)^{-1}[&(n-l)^{1/2}\beta_\mu^+\,|n-1, l+1, m+\mu\rangle\\
&-(n+l+1)^{1/2}\beta_\mu^-\,|n-1, l-1, m+\mu\rangle]
\end{aligned} \tag{8.94}$$

and

$$(a^\dagger)_\mu|nlm\rangle = \hbar^{-1}(2l+1)^{-1}[(n-l+2)^{1/2}\beta_\mu^-|n+1, l-1, m+\mu\rangle$$
$$-(n+l+3)^{1/2}\beta_\mu^+|n+1, l+1, m+\mu\rangle], \tag{8.95}$$

where $\mu = \pm 1$ or 0, and $\beta_\mu^\pm$ are given by (4.70)–(4.72). Using (8.90′)–(8.95) we can calculate $Q_\lambda|nlm\rangle$. It is clear that Q_λ transforms $|nlm\rangle$ into linear combinations of degenerate eigenkets $|nl'm'\rangle$ where $l' = l, l \pm 2$ and $m' = m + \lambda$.

8.7 A spectrum-generating algebra

We refer to a Lie algebra that covers the states of a system in one (or more) irreducible representations as a spectrum-generating algebra for the system. For example, for the oscillator the operators $a_i a_j + a_i^\dagger a_j^\dagger$, $a_i^\dagger a_j + a_i a_j^\dagger$, $i(a_i a_j - a_i^\dagger a_j^\dagger)$, and $i(a_i^\dagger a_j - a_j^\dagger a_i)$ are a Hermitian basis for the Lie algebra sp(6, R). The eigenkets $|n_1 n_2 n_3\rangle$ of the c.s.c.o. H_1, H_2, H_3 span two representations of this algebra, one containing all the states with $n_1 + n_2 + n_3$ odd, the other with $n_1 + n_2 + n_3$ even [3]. Thus sp(6, R) is a spectrum-generating algebra for the oscillator in the basis H_1, H_2, H_3.

Similarly, one can construct a spectrum-generating algebra for the eigenkets $|nlm\rangle$ of an angular momentum basis [4]. In this case it is preferable to work with shift operators for the kets $|nlm\rangle$ rather than with the boson annihilation and creation operators. In this section we use the shift operators derived in Section 8.1 to construct a spectrum-generating algebra for the oscillator in an angular momentum basis.

First we make the following changes:

(i) The eigenkets $|nlm\rangle$ are labelled $|Nlm\rangle$ where $N = \frac{1}{2}(n - l)$ is the radial quantum number (see Section 5.6).

(ii) Define*

$$\begin{aligned}
\mathbf{D} &= (2M\hbar^2)^{-1/2} S^{1/2} \mathbf{D}^-(\omega)\{S[H + \hbar\omega(S-1)]\}^{-1/2},\\
\mathbf{D}^\dagger &= (2M\hbar^2)^{-1/2} S^{1/2} \mathbf{D}^+(\omega)\{S[H + \hbar\omega(S+1)]\}^{-1/2},\\
Q &= Q^-[2(H/\hbar\omega + S - 1)]^{-1/2},\\
Q^\dagger &= Q^+[2(H/\hbar\omega + S + 1)]^{-1/2},
\end{aligned} \tag{8.96}$$

where $Q^\pm$ are given by (5.141) and $\mathbf{D}^\pm$ by (8.9)–(8.11). In terms of these the shift operations (8.22) and (5.143) are

$$D_z|Nlm\rangle = [(l+m)(l-m)]^{1/2}|N, l-1, m\rangle, \tag{8.97}$$

$$D_{\pm 1}|Nlm\rangle = \pm[(l \mp m)(l \mp m - 1)]^{1/2}|N, l-1, m \pm 1\rangle, \tag{8.98}$$

$$D_z^\dagger|Nlm\rangle = [(l+m+1)(l-m+1)]^{1/2}|N, l+1, m\rangle, \tag{8.99}$$

---

* We arrive at these definitions by requiring that the resulting shift operations have the simple coefficients given in (8.97)–(8.102).

$$(D^\dagger)_{\pm 1}|Nlm\rangle = \mp[(l \pm m + 1)(l \pm m + 2)]^{1/2}|N, l + 1, m \pm 1\rangle, \quad (8.100)$$

$$Q|Nlm\rangle = N^{1/2}|N - 1, l, m\rangle, \quad (8.101)$$

$$Q^\dagger|Nlm\rangle = (N + 1)^{1/2}|N + 1, l, m\rangle. \quad (8.102)$$

Here $D_{\pm 1} = D_x \pm iD_y$ and $(D^\dagger)_{\pm 1} = D_x^\dagger \pm iD_y^\dagger$. (Using (8.97)–(8.102) one can verify that $\mathbf{D}^\dagger$ and $Q^\dagger$, respectively, are the adjoints of $\mathbf{D}$ and Q. The reasoning is similar to that in Section 4.9.) We note that $\mathbf{D}$ and $\mathbf{D}^\dagger$ change only the angular momentum quantum numbers while Q and $Q^\dagger$ change only the radial quantum number.

Define the Hermitian operators

$$\mathbf{X} = \tfrac{1}{2}(\mathbf{D} + \mathbf{D}^\dagger) \quad (8.103)$$

and

$$\mathbf{Y} = \tfrac{1}{2}i(\mathbf{D} - \mathbf{D}^\dagger). \quad (8.104)$$

By comparing eqns (8.97)–(8.100) and (4.102)–(4.105) we can readily identify a subalgebra: as in Section 4.9, we find that the 10 operators L_i, X_i, Y_i ($i = 1, 2, 3$), and S are a Hermitian basis for the Lie algebra so(3, 2).

To identify the other subalgebra define

$$Z_1 = \tfrac{1}{4}(QQ + Q^\dagger Q^\dagger), \quad Z_2 = \tfrac{1}{4}i(QQ - Q^\dagger Q^\dagger), \quad Z_3 = \tfrac{1}{2}(Q^\dagger Q + \tfrac{1}{2}). \quad (8.105)$$

Then using (8.101) and (8.102) we find

$$[Z_1, Z_2] = -iZ_3, \qquad [Z_2, Z_3] = iZ_1, \qquad [Z_3, Z_1] = iZ_2. \quad (8.106)$$

Thus the three operators Z_i are a Hermitian basis for the Lie algebra so(2, 1). Also, the Z_i commute with $\mathbf{L}$, $\mathbf{X}$, $\mathbf{Y}$, and S. Hence

$$\mathrm{so}(2, 1) \oplus \mathrm{so}(3, 2)$$

is a spectrum-generating algebra for the three-dimensional isotropic harmonic oscillator in an angular momentum basis. The eigenkets $|Nlm\rangle$ span two irreducible representations of this algebra [4].

The four operators D_i and Q commute (see (8.97), (8.98), and (8.101)) and therefore they possess common eigenkets

$$D_i|z, \boldsymbol{\zeta}\rangle = \zeta_i|z, \boldsymbol{\zeta}\rangle, \qquad Q|z, \boldsymbol{\zeta}\rangle = z|z, \boldsymbol{\zeta}\rangle. \quad (8.107)$$

Similarly, because the three operators a_i commute they also possess common eigenkets

$$a_i|\boldsymbol{\eta}\rangle = \eta_i|\boldsymbol{\eta}\rangle. \quad (8.108)$$

The properties of the 'coherent states' $|\boldsymbol{\eta}\rangle$, the 'coherent angular momentum states' $|z, \boldsymbol{\zeta}\rangle$, and the relationship between them are an interesting topic that has been studied in detail by Bracken and Leemon [5]. They have shown that $|\boldsymbol{\eta}\rangle$ and $|z, \boldsymbol{\zeta}\rangle$ provide two different quasi-classical descriptions of the oscillator; only in the special case where the classical motion is circular, rather than elliptical, do these two descriptions coincide. In particular the uncertainty in angular momentum, $(\langle \mathbf{L}^2 \rangle - \langle \mathbf{L} \rangle^2)^{1/2}$, is smaller for the states $|z, \boldsymbol{\zeta}\rangle$ than for $|\boldsymbol{\eta}\rangle$ [5].

8.8 The classical limit

In the classical limit the equation of motion is

$$\frac{\mathrm{d}\mathbf{p}}{\mathrm{d}t} = -M\omega^2\mathbf{r}. \tag{8.109}$$

(In this section $\mathbf{r}$ and $\mathbf{p}$ denote the classical position and momentum vectors.) The energy

$$H \equiv E = \frac{1}{2M}(\mathbf{p}^2 + M^2\omega^2\mathbf{r}^2) \tag{8.110}$$

and the angular momentum

$$\mathbf{L} = \mathbf{r} \times \mathbf{p} \tag{8.111}$$

are constants of the motion.

Consider the classical limits of the operators derived in Sections 8.1 and 8.6. We start with the symmetric tensor (8.78). In the limit $\hbar \to 0$ this becomes the classical tensor

$$A_{ij} = \frac{1}{2M}(p_i p_j + M^2\omega^2 r_i r_j). \tag{8.112}$$

Using (8.109) it follows that

$$\frac{\mathrm{d}A_{ij}}{\mathrm{d}t} = 0, \tag{8.113}$$

and hence A_{ij} are also constants of the motion.

To understand the significance of this tensor it is helpful to consider the eigenvalue equations [6]

$$\sum_{j=1}^{3} A_{ij}u_j^{(\alpha)} = A^{(\alpha)}u_i^{(\alpha)}, \tag{8.114}$$

where $\alpha = 1, 2, 3$. From the secular determinant one obtains the eigenvalues

$$A^{(1)} = \tfrac{1}{2}[E + (E^2 - \omega^2L^2)^{1/2}], \quad A^{(2)} = \tfrac{1}{2}[E - (E^2 - \omega^2L^2)^{1/2}], \quad A^{(3)} = 0. \tag{8.115}$$

Now

$$L_i A_{ij} = \frac{1}{2M}\varepsilon_{ikl}r_k p_l(p_i p_j + M^2\omega^2 r_i r_j) = 0, \tag{8.116}$$

because $\varepsilon_{ikl}p_l p_i = 0$ and $\varepsilon_{ikl}r_k r_i = 0$. Therefore the scalar product of $\mathbf{L}$ with both sides of (8.114) yields $\mathbf{L}.\mathbf{u}^{(1)} = 0$ and $\mathbf{L}.\mathbf{u}^{(2)} = 0$: the eigenvectors $\mathbf{u}^{(1)}$ and $\mathbf{u}^{(2)}$ lie in the plane of the orbit. Choose coordinate axes such that $\mathbf{L} = (0, 0, L)$. Then with an appropriate choice of initial conditions the solution to (8.109) is

$$\mathbf{r} = (a \cos \omega t, b \sin \omega t, 0), \tag{8.117}$$

where a and b are constants. Hence

$$\frac{x^2}{a^2} + \frac{y^2}{b^2} = 1, \tag{8.118}$$

and the orbit is an ellipse with centre at the origin. In this coordinate system A_{ij} is diagonal

$$A_{ij} = \frac{1}{2}M\omega^2 \begin{bmatrix} a^2 & 0 & 0 \\ 0 & b^2 & 0 \\ 0 & 0 & 0 \end{bmatrix}. \tag{8.119}$$

The eigenfunctions are $\mathbf{u}^{(1)} = \hat{\mathbf{x}}$ and $\mathbf{u}^{(2)} = \hat{\mathbf{y}}$, with eigenvalues $A^{(1)} = \frac{1}{2}M\omega^2 a^2$ and $A^{(2)} = \frac{1}{2}M\omega^2 b^2$. Thus the major and minor axes of the elliptical orbit are along the eigenvectors $\mathbf{u}^{(1)}$ and $\mathbf{u}^{(2)}$ of A_{ij}, while the lengths of these axes are equal to $(8A^{(1)}M^{-1}\omega^{-2})^{1/2}$ and $(8A^{(2)}M^{-1}\omega^{-2})^{1/2}$. In this sense the tensor A_{ij} specifies the orbit of the oscillator.

Consider next the classical limits of the shift operators $\mathbf{D}^{\pm}(\omega)$ (eqns (8.9)–(8.11)). First note that $\hbar S = (\mathbf{L}^2 + \frac{1}{4}\hbar^2)^{1/2}$ (see (4.48)) and therefore the limit as $\hbar \to 0$ is

$$\hbar S = (\mathbf{L}^2)^{1/2} = L. \tag{8.120}$$

Hence we obtain the classical vectors

$$\mathbf{D}_c^{\pm}(\omega) = \mathbf{F}_c(\omega) \pm \mathrm{i}\mathbf{G}_c(\omega), \tag{8.121}$$

where

$$\mathbf{F}_c(\omega) = -M\omega L\mathbf{r} - \mathbf{p} \times \mathbf{L} \tag{8.122}$$

and

$$\mathbf{G}_c(\omega) = -M\omega\mathbf{r} \times \mathbf{L} + \mathbf{p}L \tag{8.123}$$

$$= \mathbf{F}_c(\omega) \times \hat{\mathbf{L}}. \tag{8.124}$$

The vectors $\mathbf{F}_c$ and $\mathbf{G}_c$ are perpendicular to $\mathbf{L}$ and the magnitudes

$$F_c = G_c = L[2M(H + \omega L)]^{1/2} \tag{8.125}$$

are constant. Using (8.109) we can show that

$$\frac{\mathrm{d}\mathbf{F}_c(\omega)}{\mathrm{d}t} = \boldsymbol{\omega} \times \mathbf{F}_c(\omega), \tag{8.126}$$

where $\boldsymbol{\omega} = \omega\hat{\mathbf{L}}$ is the angular velocity of the oscillator. Thus $\mathbf{F}_c$ and $\mathbf{G}_c$ perform circular motion with angular velocity $\boldsymbol{\omega}$.

For the Coulomb problem in the classical limit there are similar constants of the motion and rotating vectors (see Section 9.8).

8.9 The Lie algebra su(3)

In this section we define a matrix representation of the Lie algebra su(3) and we show that the commutation relations (8.89) of an invariance algebra for the oscillator can be put in the same form as those for su(3). We also give a brief discussion of the applications of su(3) in particle physics.

Let λ_i denote a Hermitian traceless 3×3 matrix. A simple method of constructing a set of linearly independent such matrices is:*

(i) Write down the off-diagonal Hermitian 3×3 matrices obtained by putting 1 (or i) in any one off-diagonal position and 1 (or $-$i) in the transposed position, and zero everywhere else. This yields six linearly independent matrices.

(ii) Write down the two diagonal Hermitian traceless 3×3 matrices with diagonal elements $1, -1, 0$ and $1, 1, -2$.

Hence we obtain the Gell-Mann matrices

$$\lambda_1 = \begin{bmatrix} 0 & 1 & 0 \\ 1 & 0 & 0 \\ 0 & 0 & 0 \end{bmatrix}, \quad \lambda_2 = \begin{bmatrix} 0 & -\mathrm{i} & 0 \\ \mathrm{i} & 0 & 0 \\ 0 & 0 & 0 \end{bmatrix}, \quad \lambda_3 = \begin{bmatrix} 1 & 0 & 0 \\ 0 & -1 & 0 \\ 0 & 0 & 0 \end{bmatrix},$$

$$\lambda_4 = \begin{bmatrix} 0 & 0 & 1 \\ 0 & 0 & 0 \\ 1 & 0 & 0 \end{bmatrix}, \quad \lambda_5 = \begin{bmatrix} 0 & 0 & -\mathrm{i} \\ 0 & 0 & 0 \\ \mathrm{i} & 0 & 0 \end{bmatrix}, \quad \lambda_6 = \begin{bmatrix} 0 & 0 & 0 \\ 0 & 0 & 1 \\ 0 & 1 & 0 \end{bmatrix}, \qquad (8.127)$$

$$\lambda_7 = \begin{bmatrix} 0 & 0 & 0 \\ 0 & 0 & -\mathrm{i} \\ 0 & \mathrm{i} & 0 \end{bmatrix}, \quad \lambda_8 = \frac{1}{\sqrt{3}} \begin{bmatrix} 1 & 0 & 0 \\ 0 & 1 & 0 \\ 0 & 0 & -2 \end{bmatrix}.$$

Here the normalization

$$\text{Trace } \lambda_a \lambda_b = 2\delta_{ab} \qquad (8.128)$$

has been adopted.

Using (8.127) we can readily show that

---

* This method can be used to construct a matrix representation for su(N), $N = 2, 3, \ldots$. When $N = 2$ it yields the three Pauli matrices (4.190). In general it yields $N^2 - 1$ Hermitian $N \times N$ matrices that are a fundamental representation of the generators of the group SU(N). $N - 1$ of these matrices are diagonal.

$$\left[\frac{1}{2}\lambda_a, \frac{1}{2}\lambda_b\right] = \mathrm{i}\sum_{c=1}^{8} f_{abc}\frac{1}{2}\lambda_c, \tag{8.129}$$

where f_{abc} is totally antisymmetric with non-vanishing elements

$$\begin{aligned}
&f_{123} = 1 \qquad && f_{246} = \frac{1}{2} \qquad && f_{367} = -\frac{1}{2} \\
&f_{147} = \frac{1}{2} && f_{257} = \frac{1}{2} && f_{458} = \frac{\sqrt{3}}{2} \\
&f_{156} = -\frac{1}{2} && f_{345} = \frac{1}{2} && f_{678} = \frac{\sqrt{3}}{2}.
\end{aligned} \tag{8.130}$$

The commutation relations (8.129) are those of the Lie algebra su(3) expressed in the basis of the Gell-Mann matrices. (One can make a different choice of basis, for example, by taking linear combinations of the matrices λ_i. The values of f_{abc} in (8.129) depend on the choice of basis, but they are independent of the representation one uses. The standard form (Cartan–Weyl basis) is the usual means of classifying the classical Lie algebras. [7])

The unitary 3×3 matrices

$$U = \exp\left(\mathrm{i}\sum_{a=1}^{8} \varepsilon_a \frac{1}{2}\lambda_a\right), \tag{8.131}$$

where ε_a are real parameters that commute with λ_a, are a matrix representation of the Lie group SU(3). The Hermitian matrices $\frac{1}{2}\lambda_a$ are a 3×3 matrix representation of the group generators. There are higher-order matrix representations of these generators; the 3×3 representation is called a fundamental representation. The quantities f_{abc} are the structure constants of SU(3). Because the λ_a are traceless, the determinant of U must be unity (see (4.199)). The nomenclature SU(3) signifies the group whose matrix representations are special (determinant = 1), unitary matrices, the lowest order of these matrices being 3×3.

We can now identify the invariance algebra of the three-dimensional isotropic harmonic oscillator (the Lie algebra of eqns (8.89)). Define operators γ_i, $i = 1, 2, \ldots, 8$, by

$$\begin{aligned}
&L_x = \hbar\gamma_7, \qquad L_y = -\hbar\gamma_5, \qquad L_z = \hbar\gamma_2, \\
&Q_0 = -\sqrt{3}\hbar\gamma_8, \quad Q_{\pm 1} = \mp\tfrac{1}{2}\hbar(\gamma_4 \pm \mathrm{i}\gamma_6), \quad Q_{\pm 2} = \hbar(\gamma_3 \pm \mathrm{i}\gamma_1).
\end{aligned} \tag{8.132}$$

Then using (8.89) we can show that

$$\left[\frac{1}{2}\gamma_a, \frac{1}{2}\gamma_b\right] = \mathrm{i}\sum_{c=1}^{8} f_{abc}\frac{1}{2}\gamma_c, \tag{8.133}$$

where f_{abc} are given by (8.130). Thus the invariance algebra of the oscillator is (isomorphic to) su(3); its symmetry group is SU(3).

The identification of the symmetry group of the three-dimensional isotropic harmonic oscillator was first made by Jauch in 1939 [8]. In addition to its intrinsic interest this result has some useful applications, notably to the theory of nuclear structure [9]. However, the most famous use of SU(3) has occurred in the theory of elementary particles [10]. While this topic is beyond the scope of this book, we conclude this section with a discussion that illustrates the use in particle physics of some of the material presented above.

For SU(3) there are eight group generators and these satisfy the commutation relations

$$[G_a, G_b] = \mathrm{i} \sum_{c=1}^{8} f_{abc} G_c. \tag{8.134}$$

Consider their matrix representations

$$G_a = \tfrac{1}{2}\lambda_a, \tag{8.135}$$

where λ_a are given by (8.127). The two diagonal generators G_3 and G_8 together with $G^2 = G_1^2 + G_2^2 + G_3^2$ are a set of commuting matrices and they possess the simultaneous eigenvectors

$$v_1 = \begin{pmatrix} 1 \\ 0 \\ 0 \end{pmatrix}, \qquad v_2 = \begin{pmatrix} 0 \\ 1 \\ 0 \end{pmatrix}, \qquad v_3 = \begin{pmatrix} 0 \\ 0 \\ 1 \end{pmatrix}. \tag{8.136}$$

The eigenvalues can be calculated by operating with G_3, G_8, and G^2 on v_i. For example,

$$G_3 v_1 = \frac{1}{2} \begin{pmatrix} 1 & 0 & 0 \\ 0 & -1 & 0 \\ 0 & 0 & 0 \end{pmatrix} \begin{pmatrix} 1 \\ 0 \\ 0 \end{pmatrix} = \frac{1}{2} \begin{pmatrix} 1 \\ 0 \\ 0 \end{pmatrix} = \frac{1}{2} v_1.$$

Thus v_1 has eigenvalue $G_3 = \frac{1}{2}$. (For eigenvalues we use the same symbol as the operator.) In this way we find the eigenvalues given in Table 8.1. In a plot

TABLE 8.1 The eigenvalues of G_3, G_8, and $G^2 = G_1^2 + G_2^2 + G_3^2$, where G_a are defined by (8.135) and the eigenvectors v_i are given in (8.136)

	G_3	G_8	$G^2 = g(g+1)$	g
v_1	$\frac{1}{2}$	$\frac{1}{2\sqrt{3}}$	$\frac{1}{2}\left(\frac{1}{2}+1\right)$	$\frac{1}{2}$
v_2	$-\frac{1}{2}$	$\frac{1}{2\sqrt{3}}$	$\frac{1}{2}\left(\frac{1}{2}+1\right)$	$\frac{1}{2}$
v_3	0	$-\frac{1}{\sqrt{3}}$	0	0

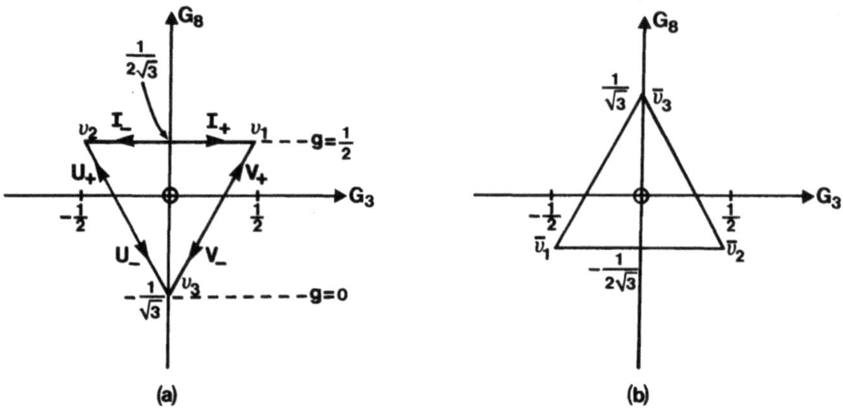

FIG. 8.2 (a) The weight diagram for the eigenvectors v_i (the fundamental representation **3** of SU(3)). Also shown are the transformations effected by the shift operators (8.137). (b) The weight diagram for the eigenvectors $\bar{v}_i$ (the fundamental representation $\bar{\mathbf{3}}$ of SU(3)).

of G_8 versus G_3 the v_i are represented by three points that form an equilateral triangle. This so-called weight diagram is shown in Fig. 8.2(a).

Apart from G_3 and G_8 there are six off-diagonal generators of SU(3). From these we can obtain shift operators by a construction similar to that for the shift operators $J_\pm$ of angular momentum. (See Section 4.2. The reason for this similarity is that SU(3) possesses three SU(2) subgroups.) The linear combinations

$$I_\pm = G_1 \pm \mathrm{i}G_2, \qquad U_\pm = G_6 \pm \mathrm{i}G_7, \qquad V_\pm = G_4 \pm \mathrm{i}G_5, \tag{8.137}$$

have the representations

$$\begin{aligned} I_+ &= \begin{bmatrix} 0 & 1 & 0 \\ 0 & 0 & 0 \\ 0 & 0 & 0 \end{bmatrix}, \qquad & I_- &= \begin{bmatrix} 0 & 0 & 0 \\ 1 & 0 & 0 \\ 0 & 0 & 0 \end{bmatrix}, \\ U_+ &= \begin{bmatrix} 0 & 0 & 0 \\ 0 & 0 & 1 \\ 0 & 0 & 0 \end{bmatrix}, \qquad & U_- &= \begin{bmatrix} 0 & 0 & 0 \\ 0 & 0 & 0 \\ 0 & 1 & 0 \end{bmatrix}, \\ V_+ &= \begin{bmatrix} 0 & 0 & 1 \\ 0 & 0 & 0 \\ 0 & 0 & 0 \end{bmatrix}, \qquad & V_- &= \begin{bmatrix} 0 & 0 & 0 \\ 0 & 0 & 0 \\ 1 & 0 & 0 \end{bmatrix}. \end{aligned} \tag{8.138}$$

The effect of these six operators on v_i is readily calculated. For example,

$$I_+ v_1 = \begin{bmatrix} 0 & 1 & 0 \\ 0 & 0 & 0 \\ 0 & 0 & 0 \end{bmatrix} \begin{bmatrix} 1 \\ 0 \\ 0 \end{bmatrix} = 0$$

and

$$I_+ v_2 = \begin{bmatrix} 0 & 1 & 0 \\ 0 & 0 & 0 \\ 0 & 0 & 0 \end{bmatrix} \begin{bmatrix} 0 \\ 1 \\ 0 \end{bmatrix} = \begin{bmatrix} 1 \\ 0 \\ 0 \end{bmatrix} = v_1.$$

In this way we find the transformations associated with these operators:

	ΔG_3	ΔG_8	Δg
$I_\pm$	± 1	0	0
$U_\pm$	$\mp\frac{1}{2}$	$\pm\frac{3}{2\sqrt{3}}$	± 1
$V_\pm$	$\pm\frac{1}{2}$	$\pm\frac{3}{2\sqrt{3}}$	± 1

The shift operations are depicted in Fig. 8.2(a). Note that the shift operators do not transform to states with eigenvalues outside the triangle because I_+ and V_+ annihilate v_1, I_- and U_+ annihilate v_2, and U_- and V_- annihilate v_3.

The eigenvectors v_i are referred to as a fundamental representation, or the fundamental triplet **3**, of the eigenkets of SU(3). There exists a second fundamental representation, the $\mathbf{\bar{3}}$. This is clear if we write the complex conjugate of (8.134) as

$$[-G_a^*, -G_b^*] = \mathrm{i} \sum_{c=1}^{8} f_{abc}(-G_c^*); \tag{8.139}$$

therefore if G_a are a representation of a Lie group, so are $-G_a^*$. Hence for the eigenvectors $\bar{v}_i$ ($i = 1, 2, 3$) of this representation of SU(3) the eigenvalues corresponding to G_3 and G_8 are the negatives of those for v_i, while the values of g are unchanged. This gives the weight diagram* shown in Fig. 8.2(b). In certain applications of SU(3) the **3** represents particle states (say) and the $\mathbf{\bar{3}}$ represents antiparticle states.

In the above calculations we have used a particular representation of the eigenvectors and operators. The same results can be derived using the commutation relations (8.134). For example, for the operators (8.137) it follows from (8.134) and (8.130) that

$$[G_3, I_\pm] = \pm I_\pm, \qquad [G_3, U_\pm] = \mp\tfrac{1}{2}U_\pm, \qquad [G_3, V_\pm] = \pm\tfrac{1}{2}V_\pm \tag{8.140}$$

and

$$[G_8, I_\pm] = 0, \qquad [G_8, U_\pm] = \pm\frac{3}{2\sqrt{3}}U_\pm, \qquad [G_8, V_\pm] = \pm\frac{3}{2\sqrt{3}}V_\pm. \tag{8.141}$$

Hence $I_\pm$, $U_\pm$, and $V_\pm$ are shift operators that change the eigenvalues G_3 and G_8 in the manner given above. These shift operations are generally valid;

* Note that for SU(2) the weight diagrams of the fundamental representations 2 and $\bar{2}$ are the same; the 2 and $\bar{2}$ are equivalent representations of SU(2).

thus they apply also to more complicated eigenkets of SU(3) such as any direct products $v_i v_j$, $v_i \bar{v}_j$, etc. (The shift in the value of g is more difficult to calculate in general. For example, the operators $U_{\pm}$ do not always produce eigenkets of G^2; to obtain such eigenkets it may be necessary to form linear combinations of the kets produced by shift operations. [11])

The representations **3** and $\bar{\mathbf{3}}$ have important applications in particle physics, particularly to the fundamental particles known as quarks:

(i) In the original quark model it was proposed that all hadrons (strongly interacting particles) are built out of three fundamental spin $\frac{1}{2}$ particles, the u, d, and s quarks, and their antiparticles, the $\bar{\text{u}}$, $\bar{\text{d}}$, and $\bar{\text{s}}$ quarks. The quarks are represented by the **3** of SU(3); that is, their state vector contains a factor* v_1 for the u quark, v_2 for the d quark, and v_3 for the s quark. Similarly, the antiquarks are represented by the $\bar{\mathbf{3}}$ of SU(3).

Here the eigenvalues of G_3, G_8, and G^2 correspond to the observables $I_3 = G_3$, $Y = (2/\sqrt{3})G_8$, and $I^2 = G^2$, where I and Y are known as isospin and hypercharge. The baryons (strongly interacting fermions) are bound states of three quarks or of three antiquarks: for example, the proton is $v_1 v_1 v_1$. The mesons (strongly interacting bosons) are bound states of a quark and an antiquark: for example, the π^+ is $v_1 \bar{v}_1$.

(ii) The u, d, and s quarks are referred to as different flavours of quark, and the associated transformation group is the flavour group $\text{SU}_\text{f}(3)$. The state vector of each quark contains as a factor a second eigenvector of an SU(3) group, the colour group $\text{SU}_\text{c}(3)$. For this eigenvector the eigenvalues corresponding to the diagonal generators G_3 and G_8 of $\text{SU}_\text{c}(3)$ represent an additional property known as colour charge. Although the flavour group $\text{SU}_\text{f}(3)$ and the colour group $\text{SU}_\text{c}(3)$ have the same mathematical properties, their respective eigenvectors represent physically different properties of quarks. For example,

$$\begin{pmatrix} 1 \\ 0 \\ 0 \end{pmatrix}_\text{f} \begin{pmatrix} 0 \\ 1 \\ 0 \end{pmatrix}_\text{c}$$

represents a u quark ($I_3 = \frac{1}{2}$, $Y = \frac{1}{3}$, $I = \frac{1}{2}$) with colour charge quantum numbers $G_3 = -\frac{1}{2}$ and $G_8 = (2\sqrt{3})^{-1}$.

The shift operators of $\text{SU}_\text{f}(3)$ change the flavour of a quark (u → d, etc.) without changing its colour charge, whereas the shift operators of $\text{SU}_\text{c}(3)$ change the colour charge of a quark without changing its flavour. The colour charge of a quark is thought to be an attribute analogous to electric charge, in that colour charge is responsible for the gauge field (the gluon field) that causes the strong interaction between quarks. The reader who wishes to learn about this fascinating subject is referred to the literature cited below [10].

---

* Other factors of the state vector are the spin eigenket, the space part of the wavefunction, etc.

References

1. See, for example, Cohen-Tannoudji, C., Diu, B., and Laloë, F. (1977). *Quantum mechanics*, Vol. 1, pp. 814–23. Wiley, New York. Elliott, J.P. and Dawber, P.G. (1979). *Symmetry in physics*, Vol. 2, Ch. 19. Macmillan, London.
2. For a recent discussion of accidental degeneracy see Moshinsky, M., Quesne, C. and Loyola, G. (1990). Art or science, the determination of the symmetry Lie algebra for a Hamiltonian with accidental degeneracy. *Annals of Physics*, **198**, 103–31.
3. Hwa, R.C. and Nuyts, J. (1966). Group embedding for the harmonic oscillator. *Physical Review*, **145**, 1188–95.
4. Bracken, A.J. and Leemon, H.I. (1980). The isotropic harmonic oscillator in an angular momentum basis: An algebraic formulation. *Journal of Mathematical Physics*, **21**, 2170–81.
5. Bracken, A.J. and Leemon, H.I. (1981). A new set of coherent states for the isotropic harmonic oscillator: Coherent angular momentum states. *Journal of Mathematical Physics*, **22**, 719–32. Bracken, A.J., McAnally, D.S., and Odundun, O.A. (1987). Coherent angular momentum states for the two-dimensional oscillator. *Journal of Mathematical Physics*, **28**, 397–403.
6. Fradkin, D.M. (1965). Three-dimensional isotropic harmonic oscillator and SU_3. *American Journal of Physics*, **33**, 207–11.
7. See, for example, Wybourne, B.G. (1974). *Classical groups for physicists*. Wiley, New York, and Lichtenberg, D.B. (1978). *Unitary symmetry and elementary particles*. Academic Press, New York.
8. Jauch, J.M. (1939). Groups of quantum mechanical contact transformations and the degeneracy of energy levels. *Physical Review*, **55**, 1132 (A). Jauch, J.M. and Hill, E.L. (1940). On the problem of degeneracy in quantum mechanics. *Physical Review*, **57**, 641–5.
9. See Elliott, J.P. (1963). *Selected topics in nuclear theory*, (ed. F. Janouch). International Atomic Energy Agency, Vienna; and Harvey, M. (1968). *Advances in Nuclear Physics*, 1. Plenum, New York.
10. See, for example, Cheng, T.P. and Li, L.F. (1984). *Gauge theory of elementary particle physics*. Clarendon, Oxford. Coleman, S. (1985). *Aspects of symmetry*. Cambridge University Press, London. Georgi, H. (1982). *Lie algebras in particle physics*. Benjamin, Reading, MA. Gibson, W.M. and Pollard, B.R. (1980). *Symmetry principles in elementary particle physics*. Cambridge University Press, London. Gottfried, K. and Weisskopf, V.F. (1986). *Concepts of particle physics*, Vols I and II. Clarendon, Oxford. Halzen, F. and Martin, A.D. (1984). *Quarks and leptons: An introductory course in modern particle physics*. Wiley, New York. Quigg, C. (1983). *Gauge theories of the strong, weak, and electromagnetic interactions*. Benjamin, Reading, MA.
11. See Lichtenberg, D.B. Ref. 7, p. 146.

9

THE COULOMB PROBLEM IN AN ANGULAR MOMENTUM BASIS

In this chapter an algebraic method is used to solve the simultaneous eigenvalue problem for H, $\mathbf{L}^2$, and L_z (eqns (5.11)–(5.13)), where

$$H = \frac{1}{2M}\mathbf{p}^2 - \frac{\hbar^2}{Mar} \tag{9.1}$$

is the non-relativistic Hamiltonian for an attractive Coulomb potential. This method is similar to that presented in Chapter 8 for the oscillator: thus we use results from Chapters 4 and 5 to construct shift operators for energy and angular momentum in the eigenkets of (5.11)–(5.13).

The constants of the motion possessed by the non-relativistic Coulomb problem are discussed and invariance algebras for this problem are derived. The shift operators are used to obtain a spectrum-generating algebra for bound states of the Coulomb problem in an angular momentum basis.

Because only abstract operators are used, the analysis is independent of any representation. We also use the shift operators, expressed in differential forms corresponding to the coordinate and momentum representations, to calculate normalized coordinate-space and momentum-space wavefunctions for bound states of the Coulomb problem.

9.1 Shift operations

The effect of the operators $R_l^\pm$ (eqn (5.78)) on $|Elm\rangle$ is the same as that of

$$\pm \mathrm{i}p_r - \hbar r^{-1}(S \pm \tfrac{1}{2}) + \hbar a^{-1}(S \pm \tfrac{1}{2})^{-1}. \tag{9.2}$$

Here S is given by (4.48), so that

$$S|Elm\rangle = (l + \tfrac{1}{2})|Elm\rangle. \tag{9.3}$$

Because $S - \frac{1}{2}$ annihilates s states, the operator corresponding to the lower sign in (9.2) has the defect that its action on $|E00\rangle$ is undefined. We can avoid this inconvenience by multiplying (9.2) by $(S \pm \frac{1}{2})$. Thus we define

$$R^\pm = \pm \mathrm{i}p_r(S \pm \tfrac{1}{2}) - \hbar r^{-1}(S \pm \tfrac{1}{2})^2 + \hbar a^{-1}. \tag{9.4}$$

Then

$$R^\pm |Elm\rangle = (l + \tfrac{1}{2} \pm \tfrac{1}{2})R_l^\pm |Elm\rangle. \tag{9.5}$$

Because p_r, r^{-1}, and S are Hermitian with respect to $|Elm\rangle$, and because S commutes with p_r and r^{-1}, the adjoints of (9.4) are

$$(R^\pm)^\dagger = \mp \mathrm{i}p_r(S \pm \tfrac{1}{2}) - \hbar r^{-1}(S \pm \tfrac{1}{2})^2 + \hbar a^{-1}. \tag{9.6}$$

Using (9.4) and (9.6) we can show that $R^{\pm}$ provide the factorizations*

$$(R^{\pm})^{\dagger}R^{\pm} = 2MH(S \pm \tfrac{1}{2})^2 + \hbar^2 a^{-2}. \tag{9.7}$$

Consider the operators

$$\mathbf{C}^{\pm} = a\hbar^{-1}\mathbf{U}^{\pm}R^{\pm}, \tag{9.8}$$

where $\mathbf{U}^{\pm}$ are given by (8.8). Multiplying out the products in (9.8) we find (see Appendix J)

$$\mathbf{C}^{\pm} = \hbar\mathbf{A}(S \pm \tfrac{1}{2}) \pm \mathrm{i}\mathbf{F}, \tag{9.9}$$

where

$$\mathbf{A} = \tfrac{1}{2}a\hbar^{-2}(\mathbf{L} \times \mathbf{p} - \mathbf{p} \times \mathbf{L}) + r^{-1}\mathbf{r} \tag{9.10}$$

and

$$\mathbf{F} = r^{-1}\mathbf{r} \times \mathbf{L} + a\hbar^{-2}\mathbf{p}\mathbf{L}^2. \tag{9.11}$$

The vector operator† $\mathbf{A}$ is the Pauli–Lenz vector. The vector operator† $\mathbf{F}$ is related to $\mathbf{A}$: use of the operator identity

$$\mathbf{L} \times \mathbf{p} = -\mathbf{p} \times \mathbf{L} + 2\mathrm{i}\hbar\mathbf{p} \tag{9.12}$$

and (4.42), with $\mathbf{v} = \mathbf{p}$, shows that

$$\mathbf{F} = \mathbf{A} \times \mathbf{L}. \tag{9.13}$$

Note that (9.9) corresponds to (4.50) with $\mathbf{v} = \mathbf{A}$. Also, $\mathbf{A}$ is Hermitian because $(\mathbf{p} \times \mathbf{L})^{\dagger} = -\mathbf{L}^{\dagger} \times \mathbf{p}^{\dagger} = -\mathbf{L} \times \mathbf{p}$. However $\mathbf{F}$ is not Hermitian because $\mathbf{A}$ does not commute with $\mathbf{L}$.

Next we evaluate the commutator of H with $\mathbf{A}$. Now H commutes with $\mathbf{L}$. Thus using (9.12), the commutators (1.139), (1.150), (1.152), and the identity (I4) we obtain

$$[H, \mathbf{A}] = 0. \tag{9.14}$$

From (9.13) and (9.14)

$$[H, \mathbf{F}] = 0, \tag{9.15}$$

and hence

$$[H, \mathbf{C}^{\pm}] = 0. \tag{9.16}$$

Because $\mathbf{C}^{\pm}$ are vector operators

$$[L_i, C_j^{\pm}] = \mathrm{i}\hbar\varepsilon_{ijk}C_k^{\pm}. \tag{9.17}$$

Also because $R^{\pm}$ are scalar operators

---

* Equation (9.7) can also be verified by allowing both sides to operate on $|Elm\rangle$ and then using (9.5) and the factorizations (5.59) and (5.60).

† Each term in (9.10) and (9.11) is a vector operator (see Section 4.5).

$$[\mathbf{L}^2, \mathbf{C}^{\pm}] = a\hbar^{-1}[\mathbf{L}^2, \mathbf{U}^{\pm}]R^{\pm} = \pm 2a\hbar\mathbf{C}^{\pm}(S \pm \tfrac{1}{2}). \tag{9.18}$$

(See (4.51) and (4.49).)

It follows from (9.16)–(9.18) and (9.3) that

$$C_{\mu}^{\pm}|Elm\rangle = \gamma_{\mu}^{\pm}|E, l \pm 1, m + \mu\rangle, \tag{9.19}$$

where $\mu = \pm 1$ or 0 and

$$C_{\pm 1}^{+} = C_x^{+} \pm \mathrm{i}C_y^{+}, \qquad C_{\pm 1}^{-} = C_x^{-} \pm \mathrm{i}C_y^{-}, \qquad C_0^{\pm} = C_z^{\pm}. \tag{9.20}$$

The shift operations (9.19) are valid for the states of a repulsive Coulomb potential and for the bound states ($E < 0$) and the scattering states ($E \geq 0$) of an attractive Coulomb potential.

The coefficients $\gamma_{\mu}^{\pm}$ can be deduced from previous results: by calculations similar to those leading to (8.27) we find*

$$\gamma_{\mu}^{\pm} = a\hbar^{-1}(l + \tfrac{1}{2} \pm \tfrac{1}{2})\alpha^{\pm}\beta_{\mu}^{\pm}, \tag{9.21}$$

where $\alpha^{\pm}$ and $\beta_{\mu}^{\pm}$ are given by (5.79) and (4.70)–(4.72). In (9.21) the phase factors of $\gamma_{\mu}^{\pm}$ correspond to the choices of phase made in Chapters 4 and 5.

The product (9.8) is not the only way of constructing the operators $\mathbf{C}^{\pm}$. From the shift operators $P_{El}^{\pm}$ (eqn (7.33)) we construct operators

$$P^{\pm} = \mp \mathrm{i}\hbar^{-1}r_p(\mathbf{p}^2 - 2MH) + p^{-1}(\mathbf{p}^2 + 2MH)(S \pm \tfrac{1}{2}) \tag{9.22}$$

that satisfy

$$P^{\pm}|Elm\rangle = P_{El}^{\pm}|Elm\rangle. \tag{9.23}$$

The adjoints of (9.22) are

$$(P^{\pm})^{\dagger} = \pm \mathrm{i}\hbar^{-1}(\mathbf{p}^2 - 2MH)r_p + (\mathbf{p}^2 + 2MH)p^{-1}(S \pm \tfrac{1}{2}), \tag{9.24}$$

where we have used the fact that r_p, p^2, p^{-1}, and S are Hermitian with respect to $|Elm\rangle$ and S commutes with H and functions of p. The operators $P^{\pm}$ provide the factorizations

$$(P^{\pm})^{\dagger}P^{\pm} = 4[2MH(S \pm \tfrac{1}{2})^2 + \hbar^2 a^{-2}]. \tag{9.25}$$

(See Appendix J.) By evaluating the products $\mathbf{V}^{\pm}P^{\pm}$, where $\mathbf{V}^{\pm}$ are the vector operators (8.31), we find that

$$\mathbf{C}^{\pm} = \pm \mathrm{i}\tfrac{1}{2}a\hbar^{-1}\mathbf{V}^{\pm}P^{\pm}, \tag{9.26}$$

where $\mathbf{C}^{\pm}$ are given by (9.9)–(9.11). (See Appendix J.) Equations (9.8) and (9.26) show that the operators $\mathbf{C}^{\pm}$ are bifactorizable.

The effect of $\mathbf{C}^{\pm}$ on $|Elm\rangle$ is the same as that of the operators†

$$a\hbar^{-1}(l + \tfrac{1}{2} \pm \tfrac{1}{2})\mathbf{U}_l^{\pm}R_l^{\pm} \quad \text{or} \quad \pm \mathrm{i}\tfrac{1}{2}a\hbar^{-1}\mathbf{V}_l^{\pm}P_{El}^{\pm}, \tag{9.27}$$

---

* The factor $l + \frac{1}{2} \pm \frac{1}{2}$ in (9.21) is due to the same factor in (9.5).

† This follows from (9.8), (9.5), and the fact that $\mathbf{U}^{\pm}$ commute with $R_l^{\pm}$ because $\mathbf{L}$ and $\hat{\mathbf{r}}$ commute with $R_l^{\pm}$ (see Section 8.1); similarly for the factorization (9.26).

where $\mathbf{U}_l^\pm$ and $\mathbf{V}_l^\pm$ are given by (4.73) and (4.76). Because $\mathbf{U}_l^\pm$ and $\mathbf{R}_l^\pm$ are linear in $\mathbf{p}$, in the coordinate representation they yield simple operators (first-order partial differential operators). Similarly, because $\mathbf{V}_l^\pm$ and $P_{El}^\pm$ are linear in $\mathbf{r}$, they yield first-order partial differential operators in the momentum representation.

The shift operators $\mathbf{C}^\pm$ change the angular momentum quantum numbers l and m of the kets $|Elm\rangle$. To change the energy eigenvalue E we must use the shift operators $Q^\pm$ or $\tilde{Q}^\pm$ (see (5.152), (5.153), and (7.86), (7.88)).

9.2 Eigenvalues

The shift operations

$$L_\pm|Elm\rangle = \hbar[l(l+1) - m(m \pm 1)]^{1/2}|E, l, m \pm 1\rangle, \tag{9.28}$$

that were derived in Chapter 4, show that $m = -l, -l+1, \ldots, l$ and $l = 0, \frac{1}{2}, 1, \frac{3}{2}, \ldots$. We first prove that the half-integral values of l must be excluded. This proof is similar to that given in Section 8.2 for the eigenkets of the oscillator.

Denote the eigenkets by $|nlm\rangle$, where the dimensionless quantity n is given by (5.93) for bound states and $n = (2Ma^2E/\hbar^2)^{-1/2}$ for positive energy states. For positive energy states n can have any real positive value. For bound states the requirement that the kets generated by Q^- have non-negative norm restricts the values of n to (see (5.153) and (5.131))

$$n = l+1, l+2, \ldots. \tag{9.29}$$

Assume that $\mathbf{L}$ is Hermitian with respect to a Hilbert space spanned by $|nlm\rangle$ where l is half-integral. Then by application of $C_\mu^\pm$ we can obtain kets $|n\frac{1}{2}\frac{1}{2}\rangle$. Also

$$|n, -\tfrac{1}{2}, -\tfrac{1}{2}\rangle \sim C_{-1}^-|n\tfrac{1}{2}\tfrac{1}{2}\rangle \neq 0. \tag{9.30}$$

(If $C_{-1}^-|n\frac{1}{2}\frac{1}{2}\rangle = 0$, the factorization (9.8) shows that

$$R_{1/2}^-[\hbar(\hat{x} - i\hat{y}) - \hat{z}L_-]|n\tfrac{1}{2}\tfrac{1}{2}\rangle = 0. \tag{9.31}$$

Operate with $\hat{x} + i\hat{y}$ on the left of (9.31) and commute this operator with $R_{1/2}^-$. Then using the calculation below (4.80) we find that (9.31) becomes $R_{1/2}^-|n\frac{1}{2}\frac{1}{2}\rangle = 0$; that is

$$R^-|n\tfrac{1}{2}\tfrac{1}{2}\rangle = 0. \tag{9.32}$$

Taking the norm of both sides of (9.32) and using (9.7) we obtain $n^2 = \frac{1}{4}$, which contradicts (9.29). Thus C_{-1}^- cannot annihilate any of the bound states $|n\frac{1}{2}\frac{1}{2}\rangle$ or all the positive energy states $|n\frac{1}{2}\frac{1}{2}\rangle$.)

The existence of a ket $|n, -\frac{1}{2}, -\frac{1}{2}\rangle$ contradicts the assumption that $\mathbf{L}$ is Hermitian. (See also Sections 8.2 and 4.6.) We conclude that $\mathbf{L}$ is not Hermitian with respect to the space spanned by $|nlm\rangle$ if l is half-integral. Hence for bound states and positive energy states of a Coulomb potential

$$l = 0, 1, 2, \ldots. \tag{9.33}$$

An alternative proof of (9.33) is discussed in Section 9.5.

Finally, for bound states the energy eigenvalues are given by (5.93), (9.29), and (9.33) as

$$E = -\frac{\hbar^2}{2Ma^2n^2} \qquad (n = 1, 2, \ldots). \tag{9.34}$$

9.3 Eigenkets generated from the ground state $|100\rangle$

For the eigenkets $|nlm\rangle$ we have derived

(i) shift operators $Q^{\pm}$ and $\tilde{Q}^{\pm}$ that change the energy (see (5.153) and (7.88));

(ii) shift operators $C_\mu^{\pm}$ that change the angular momentum (see (9.19));

(iii) shift operators $L_{\pm}$ that change the z component of angular momentum (see (9.28)).

In the case of bound states we can generate all the eigenkets of H, $\mathbf{L}^2$, and L_z from a given eigenket by applying the appropriate shift operators. For example, the operators Q^+, C_0^+, and $L_{\pm}$ generate all eigenkets from the ground state or 'vacuum' $|100\rangle$.

We can simplify some of these shift operations by selecting a particular sequence of transformations. Consider (5.153) with $n = l + 1$. Substituting (5.120), (5.121), and (5.131) in (5.153) we have

$$\Delta_{l+1}^{+}\left(\frac{\mathrm{i}}{\hbar}rp_r - \frac{r}{(l+1)a} + l + 1\right)|l+1, l, m\rangle = (2l+4)^{1/2}|l+2, l, m\rangle. \tag{9.35}$$

Now

$$\left(\mathrm{i}p_r - \hbar(l+1)\frac{1}{r} + \frac{\hbar}{(l+1)a}\right)|l+1, l, m\rangle = 0 \tag{9.36}$$

because the norm of the left-hand side of (9.36) is zero. From (9.35) and (9.36)

$$2\Delta_{l+1}^{+}\left(-\frac{r}{(l+1)a} + l + 1\right)|l+1, l, m\rangle = (2l+4)^{1/2}|l+2, l, m\rangle. \tag{9.37}$$

Operate on both sides of (9.37) with $C_{+1}^{+} = a\hbar^{-1}U_{+1}^{+}R^{+}$, use (5.114) and (5.115) to commute R^+ with Δ_{l+1}^{+}, and then use (9.36) to eliminate p_r. This yields

$$(2l+4)^{1/2}C_{+1}^{+}|l+2, l, m\rangle = -\frac{2}{(l+2)}\Delta_{l+1}^{+}\frac{r}{a(l+1)}U_{+1}^{+}|l+1, l, m\rangle, \tag{9.38}$$

where we have used the fact that* U_{+1}^{+} commutes with r and Δ_{l+1}^{+}. Because

* $\mathrm{U}^{\pm}$ commute with r because $\mathbf{L}$ and S commute with r. If we expand (5.126) as a power series in $\mathbf{r}\cdot\mathbf{p}$ and note that $\mathbf{L}$, S, and $\hat{\mathbf{r}}$ commute with $\mathbf{r}\cdot\mathbf{p}$, it follows that $[\mathrm{U}^{\pm}, \Delta_{l+1}^{+}] = 0$.

$L_+|l+1, l, l\rangle = 0$ (see (9.28)), eqn (4.59) yields

$$U_{+1}^{-}|l+1, l, l\rangle = 0.$$

This result and (4.66) show that

$$U_{+1}^{+}|l+1, l, l\rangle = \hbar(2l+1)\hat{r}_+|l+1, l, l\rangle, \tag{9.39}$$

where $\hat{r}_+ = \hat{x} + \mathrm{i}\hat{y}$. Also from (9.19) and (9.21)

$$C_{+1}^{+}|l+1, l, l\rangle = -\hbar\frac{(2l+1)(2l+2)^{1/2}}{l+2}|l+2, l+1, l+1\rangle. \tag{9.40}$$

Equations (9.38)–(9.40) provide the simple shift operations

$$\Delta_{l+1}^{+}\frac{r}{l+1}\hat{r}_+|l+1, l, l\rangle = a[(l+1)(l+2)]^{1/2}|l+2, l+1, l+1\rangle. \tag{9.41}$$

In a similar manner, using the shift operators $\tilde{Q}_{l+1}^{+}$ and $C_{+1}^{+} = \mathrm{i}\frac{1}{2}a\hbar^{-1}V_{+1}^{+}P^{+}$ and (7.39) applied to $|l+1, l, l\rangle$, one can derive the analogous result

$$\Delta_{l+1}^{+}\frac{(l+1)p}{(l+1)^2p^2+p_0^2}\hat{p}_+|l+1, l, l\rangle = \frac{\mathrm{i}}{2p_0}\left(\frac{2l+3}{2l+4}\right)^{1/2}|l+2, l+1, l+1\rangle, \tag{9.42}$$

where $\hat{p}_+ = \hat{p}_x + \mathrm{i}\hat{p}_y$. We note that the operators and the coefficients in (9.41) and (9.42) are products of shift operators and coefficients derived in Chapters 4, 5, and 7 (see (4.61), (4.63), (4.64), (5.137), and (7.59)).

Now perform the following operations:

1. Apply

$$\left(\Delta_l^{+}\frac{r}{l}\right)\cdots\left(\Delta_2^{+}\frac{r}{2}\right)\left(\Delta_1^{+}\frac{r}{1}\right)\hat{r}_+^{l}$$

to $|100\rangle$. This generates (see (9.41))

$$|l+1, l, l\rangle = a^{-l}(l!)^{-1}(l+1)^{-1/2}\left(\Delta_l^{+}\frac{r}{l}\right)\cdots\left(\Delta_2^{+}\frac{r}{2}\right)\left(\Delta_1^{+}\frac{r}{1}\right)\hat{r}_+^{l}|100\rangle. \tag{9.43}$$

2. Apply $(Q^+)^N$, where $N = 1, 2, \ldots$, to $|l+1, l, l\rangle$. This generates (see (5.153) and (5.131))

$$|nll\rangle = \left(\frac{N+l+1}{l+1}N!(2l+2)(2l+3)\ldots(2l+N+1)\right)^{-1/2}$$
$$\times (Q^+)^N|l+1, l, l\rangle, \tag{9.44}$$

where $n = l+1+N$.

3. Apply L_-^{l-m} to $|nll\rangle$. This generates (see (9.28))

$$|nlm\rangle = \left(\frac{(l+m)!}{\hbar^{2(l-m)}(2l)!(l-m)!}\right)^{1/2} L_-^{l-m}|nll\rangle. \tag{9.45}$$

These shift operations enable us to obtain all the eigenkets of H, $\mathbf{L}^2$, and L_z from $|100\rangle$.

An alternative procedure would be:

1′.
$$|l+1, l, l\rangle = (-\mathrm{i}2p_0)^l \left(\frac{(2l+2)!!}{2(2l+1)!!}\right)^{1/2} \left(\Delta_l^+ \frac{lp}{l^2p^2+p_0^2}\right)\cdots$$
$$\times \left(\Delta_2^+ \frac{2p}{2^2p^2+p_0^2}\right)\left(\Delta_1^+ \frac{p}{p^2+p_0^2}\right)\hat{p}_+^l\,|100\rangle. \tag{9.43′}$$

(See (9.42).)

2′.
$$|nll\rangle = (-1)^N \left(\frac{l+1}{N+l+1}(2l+2)(2l+3)\dots(2l+N+1)\right)^{-1/2}$$
$$\times (\tilde{Q}^+)^N |l+1, l, l\rangle, \tag{9.44′}$$

where $n = l + 1 + N$ and $N = 1, 2, \dots$. (See (7.88).)

3′. Use (9.45).

The transformations (9.43)–(9.45), (9.43′), and (9.44′) are depicted in Fig. 9.1.

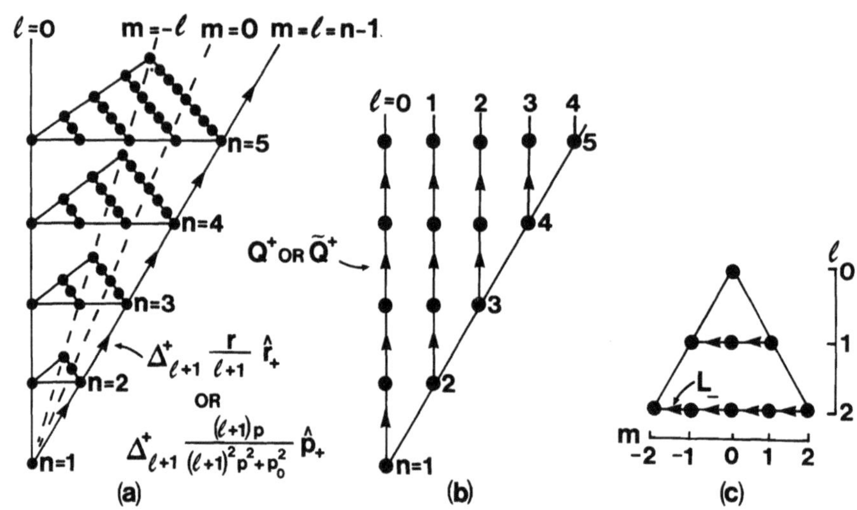

FIG. 9.1 (a) The bound states $|nlm\rangle$ for an attractive Coulomb potential. The transformations (9.43) and (9.43′) are depicted. (b) In the plane $m = l$ (the plane containing the lines $m = l$ and $l = 0$ in (a)). The transformations (9.44) and (9.44′) are depicted. (c) In the plane $n = 3$. The transformations effected by L_- are depicted.

9.4 Wavefunctions

We apply the results derived in the previous section to a calculation of the coordinate-space wavefunctions for bound states

$$\psi_{nlm}(\mathbf{r}) = \langle \mathbf{r}|nlm\rangle. \tag{9.46}$$

These calculations are similar to those in Section 8.5 for the oscillator.

In the coordinate representation the operator equations $Q_1^-|100\rangle = 0$ (see (5.145)) and $L_i|100\rangle = 0$ (see (5.13) and (9.28)) yield the first-order partial differential equations satisfied by the ground state wavefunction:

$$\left(\frac{\partial}{\partial r} + \frac{1}{a}\right)\psi_{100}(\mathbf{r}) = 0 \tag{9.47}$$

and

$$\mathbf{r} \times \nabla_{\mathbf{r}}\psi_{100}(\mathbf{r}) = 0. \tag{9.48}$$

According to (9.48), ψ_{100} is a function of r (cf. Section 8.5); the normalized solution to (9.47) is

$$\psi_{100}(\mathbf{r}) = \frac{1}{(4\pi)^{1/2}}\frac{2}{a^{3/2}}\exp\left(-\frac{r}{a}\right). \tag{9.49}$$

This can be expressed in terms of the radial wavefunction $R_{10}(r)$ (eqn (6.83)) and the spherical harmonic $Y_{00}(\hat{\mathbf{r}})$ (eqn (4.122)) as

$$\psi_{100}(\mathbf{r}) = R_{10}(r)\,Y_{00}(\hat{\mathbf{r}}). \tag{9.50}$$

From ψ_{100} we obtain ψ_{nlm} by using the operator equations (9.43)–(9.45) expressed in the coordinate representation. In these calculations the shift operators act either on $R_{10}(r)$ or on $Y_{00}(\hat{\mathbf{r}})$. Consequently ψ_{nlm} can be written down using results obtained in Sections 4.10 and 6.4.

From (9.43) and (9.50)

$$\psi_{l+1,l,l}(\mathbf{r}) = R_{l+1,l}(r)\,Y_{ll}(\hat{\mathbf{r}}), \tag{9.51}$$

where Y_{ll} and $R_{l+1,l}$ are given by (4.123) and (6.89). In applying (9.44) we have

$$\begin{aligned}\langle \mathbf{r}|(Q^+)^N|l+1, l, l\rangle &= \langle \mathbf{r}|Q^+_{l+N}\cdots Q^+_{l+2}Q^+_{l+1}|l+1, l, l\rangle \\ &= [Q^+_{l+N}\cdots Q^+_{l+2}Q^+_{l+1}R_{l+1,l}(r)]\,Y_{ll}(\hat{\mathbf{r}}),\end{aligned} \tag{9.52}$$

where

$$Q_k^+ = \Delta_k^+\left(r\frac{\mathrm{d}}{\mathrm{d}r} - \frac{r}{ka} + k + 1\right), \tag{9.53}$$

(see (5.151)) and we have used the fact that Q_k^+ commutes with Y_{ll}. The factor in brackets in (9.52) has been evaluated in Section 6.4. Hence we obtain

$$\psi_{nll}(\mathbf{r}) = R_{nl}(r)\,Y_{ll}(\hat{\mathbf{r}}), \tag{9.54}$$

where $R_{nl}(r)$ is the radial coordinate-space wavefunction (6.98).

In applying (9.45) we use (cf. (8.67))

$$\langle \mathbf{r}|L_-^{l-m}|nll\rangle = R_{nl}(r)L_-^{l-m}Y_{ll}(\hat{\mathbf{r}}). \tag{9.55}$$

Thus (9.45) and the results in Section 4.10 yield

$$\psi_{nlm}(\mathbf{r}) = R_{nl}(r)Y_{lm}(\hat{\mathbf{r}}), \tag{9.56}$$

where $Y_{lm}(\hat{\mathbf{r}})$ is the spherical harmonic (4.129). In the case of a repulsive Coulomb potential and scattering states of an attractive Coulomb potential, we must replace R_{nl} in (9.56) with the radial wavefunctions R_{kl} given by (6.112).

To construct the momentum-space wavefunctions for bound states

$$\bar{\psi}_{nlm}(\mathbf{p}) = \langle \mathbf{p}|nlm\rangle \tag{9.57}$$

we can use the property that (9.43′) and (9.44′) have a simple form when expressed in the momentum representation. These calculations are similar to those above and $\bar{\psi}_{nlm}$ can be written down using results obtained in Sections 4.10 and 7.6. The result is

$$\bar{\psi}_{nlm}(\mathbf{p}) = \bar{R}_{nl}(p)Y_{lm}(\hat{\mathbf{p}}), \tag{9.58}$$

where $\bar{R}_{nl}(p)$ is the radial momentum-space wavefunction (7.121) and $Y_{lm}(\hat{\mathbf{p}})$ is the spherical harmonic (4.145). The separation of variables in (9.56) and (9.58) occurs for the same reasons that are given in Section 8.5.

9.5 Discussion

1. In Section 4.4 we derived the operators

$$\pm i\mathbf{v} \times \mathbf{L} + \hbar\mathbf{v}(S \pm \tfrac{1}{2}), \tag{9.59}$$

where $\mathbf{v}$ is a vector operator with respect to $\mathbf{L}$. The simplest choices for $\mathbf{v}$ are

$$\text{(I) } \hat{\mathbf{r}} = r^{-1}\mathbf{r}, \quad \text{(II) } \hat{\mathbf{p}} = p^{-1}\mathbf{p}, \quad \text{(III) } \mathbf{r}, \quad \text{(IV) } \mathbf{p}.$$

The operators (9.59) corresponding to (I)–(IV) are

$$\text{(I′) } \mathbf{U}^{\pm}, \quad \text{(II′) } \mathbf{V}^{\pm}, \quad \text{(III′) } r\mathbf{U}^{\pm}, \quad \text{(IV′) } p\mathbf{V}^{\pm},$$

where $\mathbf{U}^{\pm}$ and $\mathbf{V}^{\pm}$ are given by (8.8) and (8.31).

We have seen that the operators (I′)–(IV′) have the following applications:

(i) $\mathbf{U}^{\pm}$ and $\mathbf{V}^{\pm}$ are shift operators for l and m in the eigenkets $|lm\rangle$ of the c.s.c.o. $\mathbf{L}^2$ and L_z. In the coordinate and momentum representations respectively we can obtain first-order partial differential operators from $\mathbf{U}^{\pm}$ and $\mathbf{V}^{\pm}$ (see Chapter 4).

(ii) $\pm ip\mathbf{V}^{\pm}$ are the operators (8.33); they are shift operators for l and m in the eigenkets $|Elm\rangle$ of a free particle (or a particle in a constant potential). In the coordinate representation we can obtain first-order partial differential operators from $p\mathbf{V}^{\pm}$ (see (8.33) and Sections 4.5 and 6.4). The operators

$r\mathbf{U}^{\pm}$ have similar properties, but with respect to the eigenkets of a 'Hamiltonian' which is proportional to $\mathbf{r}^2$ plus a constant.

(iii) The linear combinations

$$\mathbf{C}^{\pm} = -\frac{r}{a}\mathbf{U}^{\pm} \pm \mathrm{i}\frac{p}{p_0}\mathbf{V}^{\pm}, \tag{9.60}$$

where a is the constant (5.95) and $p_0 = \hbar a^{-1}$, are equal to $(M\hbar\omega)^{-1/2}$ times the operators (8.7). They are shift operators for energy and angular momentum in the eigenkets $|Elm\rangle$ of the c.s.c.o. H, $\mathbf{L}^2$, and L_z, where H is the Hamiltonian of the three-dimensional isotropic harmonic oscillator (eqn (8.1)).

(iv) The combinations

$$\mathbf{C}^{\pm} = \mathbf{U}^{\pm} \pm \mathrm{i}\frac{p}{p_0}\mathbf{V}^{\pm}(S \pm \tfrac{1}{2}), \tag{9.61}$$

where $p_0 = \hbar a^{-1}$ and a is the constant (5.55), are the operators (9.9). They are shift operators for angular momentum in the eigenkets $|Elm\rangle$ of the c.s.c.o. $H, \mathbf{L}^2$, and L_z, where H is the Hamiltonian for a Coulomb potential (eqn (9.1)). In the coordinate and momentum representations we can obtain first-order partial differential operators from (9.60) (see Section 8.1) and from (9.61) (see Section 9.1).

2. Suppose Γ and L_z constitute a c.s.c.o., where Γ denotes some set of operators. Also assume that this problem contains parameters from which we can construct a real constant p_0 with the dimension of momentum.

It is useful to define the dimensionless operator

$$\mathbf{a} = \frac{1}{\sqrt{2p_0}}\left(\mathrm{i}\mathbf{p} + \hbar^{-1}p_0^2\mathbf{r}\right). \tag{9.62}$$

Then

$$\mathbf{a}^{\dagger} = \frac{1}{\sqrt{2p_0}}\left(-\mathrm{i}\mathbf{p} + \hbar^{-1}p_0^2\mathbf{r}\right). \tag{9.63}$$

(In the case of the oscillator, for example, $p_0 = (M\hbar\omega)^{1/2}$ and (9.62) and (9.63) are the familiar boson annihilation and creation operators (2.253) and (2.254).) From (9.62) and (9.63), and the canonical commutation relations for $\mathbf{r}$ and $\mathbf{p}$ (see (1.65)–(1.67)) it follows that

$$[a_i, a_j^{\dagger}] = \delta_{ij}, \tag{9.64}$$

$$[a_i, a_j] = 0, \tag{9.65}$$

$$[a_i^{\dagger}, a_j^{\dagger}] = 0. \tag{9.66}$$

As in the case of the oscillator, we can express the orbital angular momentum operator in terms of $\mathbf{a}$ and $\mathbf{a}^\dagger$ (cf. (8.74))

$$\mathbf{L} = -i\hbar\mathbf{a}^\dagger \times \mathbf{a}. \tag{9.67}$$

Thus

$$L_z = i\hbar(a_1 a_2^\dagger - a_1^\dagger a_2). \tag{9.68}$$

We can use (9.64)–(9.66) and (9.68) to prove that the eigenvalues of L_z must be integers. First make the transformations*

$$b_1 = \frac{1}{\sqrt{2}}(a_1 + ia_2), \qquad b_2 = \frac{1}{\sqrt{2}}(a_1 - ia_2), \qquad b_3 = a_3. \tag{9.69}$$

Then

$$[b_i, b_j^\dagger] = \delta_{ij}, \tag{9.70}$$

$$[b_i, b_j] = 0, \tag{9.71}$$

$$[b_i^\dagger, b_j^\dagger] = 0. \tag{9.72}$$

From (9.68) and (9.69)

$$L_z = \hbar(-N_1 + N_2) \tag{9.73}$$

where $N_i = b_i^\dagger b_i$. As in the theory of the oscillator, it follows from (9.70)–(9.72) that the eigenvalues of N_1 and N_2 are non-negative integers (see Section 2.2). Also, because N_1 and N_2 commute they have simultaneous eigenvalues.

Let $|\gamma m\rangle$ denote an eigenket of the c.s.c.o. Γ and L_z.† Then (9.73) and the remarks above show that m must have integral values. This, together with (4.30), shows that for the eigenkets $|\gamma l m\rangle$ of the c.s.c.o. Γ, $\mathbf{L}^2$, and L_z the values of l are restricted to $l = 0, 1, 2, \ldots$.

9.6 Invariance algebras

We have seen that the components of the Pauli–Lenz operator $\mathbf{A}$ are three additional Hermitian constants of the motion (Section 9.1). Because A_i, together with the constants of the motion L_i, provide shift operators that transform between eigenkets with different l (see (9.19), (9.9), and (9.13)), these additional constants of the motion account for the 'accidental degeneracy' that the energy eigenvalues (9.34) possess with respect to l. We now discuss the invariance algebras of the Coulomb problem.

Consider the nine commutation relations between the six operators L_i and A_i:

---

* Equations (9.69) correspond to the unitary transformation $\mathbf{b} = U\mathbf{a}U^\dagger$, where [1]

$$U = \exp(\tfrac{1}{2}i\pi a_2^\dagger a_2)\exp(\tfrac{1}{4}\pi(a_1^\dagger a_2 - a_2^\dagger a_1)).$$

† An example is discussed in Chapter 10.

$$[L_i, L_j] = i\hbar\varepsilon_{ijk}L_k, \tag{9.74}$$

$$[L_i, A_j] = i\hbar\varepsilon_{ijk}A_k, \tag{9.75}$$

and

$$[A_i, A_j] = -2i\hbar^{-3}Ma^2H\varepsilon_{ijk}L_k. \tag{9.76}$$

(Equation (9.75) follows directly from the fact that **A** is a vector operator; (9.76) is proved in Appendix J.) The appearance of the Hamiltonian in (9.76) complicates the interpretation of (9.74)–(9.76) in terms of a Lie algebra. Because H commutes with **A** we can remove H from (9.76) by redefining **A**. However, because the eigenvalues of H can be positive, negative, or zero, we cannot achieve this by a single definition. Instead we must consider separately the energy eigenspaces on which (9.74)–(9.76) are assumed to act.

Define

$$\mathscr{A} = (-2\hbar^{-2}Ma^2H)^{-1/2}\mathbf{A} \qquad \text{if } E < 0, \tag{9.77}$$

$$= \mathbf{A} \qquad \text{if } E = 0, \tag{9.78}$$

$$= (2\hbar^{-2}Ma^2H)^{-1/2}\mathbf{A} \qquad \text{if } E > 0. \tag{9.79}$$

Then (9.75) and (9.76) become

$$[L_i, \mathscr{A}_j] = i\hbar\varepsilon_{ijk}\mathscr{A}_k \tag{9.75$'$}$$

and

$$[\mathscr{A}_i, \mathscr{A}_j] = \sigma i\hbar^{-1}\varepsilon_{ijk}L_k, \tag{9.76$'$}$$

where

$$\begin{aligned} \sigma &= -\operatorname{sign} E \qquad (E \neq 0) \\ &= 0. \qquad\qquad (E = 0) \end{aligned} \tag{9.80}$$

Thus the six operators L_i and $\mathscr{A}_i$ close under commutation.

(i) For bound states ($E < 0$), $\sigma = 1$ and (9.74), (9.75′), and (9.76′) correspond to the Lie algebra so(4). In terms of the antisymmetric operators $L_{ab} = -L_{ba}$ ($a, b = 1, \ldots, 4$) where

$$(L_{ab}) = \begin{bmatrix} 0 & L_3 & -L_2 & \hbar\mathscr{A}_1 \\ & 0 & L_1 & \hbar\mathscr{A}_2 \\ & & 0 & \hbar\mathscr{A}_3 \\ & & & 0 \end{bmatrix}, \tag{9.81}$$

the commutation relations are

$$[L_{ab}, L_{bc}] = -i\hbar L_{ac} \tag{9.82}$$

with $a \neq b \neq c$. When $a, b = 1, 2,$ or 3,

$$L_{ab} = r_a p_b - r_b p_a,$$

and (9.82) is (9.74). If we introduce a fourth 'coordinate' r_4 and its conjugate momentum p_4 such that

$$L_{a4} = r_a p_4 - r_4 p_a = \hbar \mathscr{A}_a,$$

then (9.82) can be thought of as an extension of (9.74) to four dimensions. It can be shown that the six operators L_{ab} are generators of the group SO(4) of proper rotations in four dimensions [2]. Thus SO(4) is the symmetry group for bound states of an attractive Coulomb potential. (The coordinate system in which the problem exhibits four-dimensional rotational invariance is obtained by projecting the momentum coordinates in three-dimensional momentum space on to a four-dimensional hypersphere. In this coordinate system the momentum-space wavefunctions are hyperspherical harmonics. The operators L_i and $\mathscr{A}_i$ are generators of the infinitesimal rotations of this hypersphere [3].)

(ii) For the continuum states ($E > 0$), $\sigma = -1$ and (9.74), (9.75′), and (9.76′) correspond to the Lie algebra so(3, 1), which is the algebra of the Lorentz group SO(3, 1).

(iii) If $E = 0$, $\sigma = 0$ and (9.74), (9.75′), and (9.76′) are the Lie algebra of the generators of the three-dimensional Euclidean group E(3).

There are other operators than $\mathscr{A}$ that can be used to obtain an invariance algebra. To see this let us rewrite the shift operators (9.9) in forms that are mutually adjoint. From (9.9) and (9.13) we have

$$\mathbf{C}^{\pm}(\hbar S)^{-1} = \mathbf{A} \pm \mathrm{i}\mathbf{B}, \tag{9.83}$$

where

$$\mathbf{B} = (\mathbf{A} \times \mathbf{L} - \mathrm{i}\tfrac{1}{2}\hbar\mathbf{A})(\hbar S)^{-1}. \tag{9.84}$$

Consider bound states and define (cf. (9.77))

$$\mathscr{B} = (-2\hbar^{-2} M a^2 H)^{-1/2}\mathbf{B}. \tag{9.85}$$

The effect of the operators $\mathscr{A}$ and $\mathscr{B}$ on the kets $|nlm\rangle$ is given by (9.19)–(9.21), and (9.83), (9.85) as

$$\begin{aligned}\mathscr{A}_z|nlm\rangle &= a_l[(l+m)(l-m)]^{1/2}|n, l-1, m\rangle \\ &\quad + a_{l+1}[(l+m+1)(l-m+1)]^{1/2}|n, l+1, m\rangle,\end{aligned} \tag{9.86}$$

$$\begin{aligned}(\mathscr{A}_x \pm \mathrm{i}\mathscr{A}_y)|nlm\rangle &= \pm a_l[(l \mp m)(l \mp m - 1)]^{1/2}|n, l-1, m \pm 1\rangle \\ &\quad \mp a_{l+1}[(l \pm m + 1)(l \pm m + 2)]^{1/2}|n, l+1, m \pm 1\rangle,\end{aligned} \tag{9.87}$$

and

$$\begin{aligned}\mathscr{B}_z|nlm\rangle &= \mathrm{i}a_l[(l+m)(l-m)]^{1/2}|n, l-1, m\rangle \\ &\quad - \mathrm{i}a_{l+1}[(l+m+1)(l-m+1)]^{1/2}|n, l+1, m\rangle,\end{aligned} \tag{9.88}$$

$$(\mathscr{B}_x \pm i\mathscr{B}_y)|nlm\rangle = \pm ia_l[(l \mp m)(l \mp m - 1)]^{1/2}|n, l-1, m \pm 1\rangle$$
$$\pm ia_{l+1}[(l \pm m + 1)(l \pm m + 2)]^{1/2}|n, l+1, m \pm 1\rangle, \tag{9.89}$$

where

$$a_l = [(n^2 - l^2)/(4l^2 - 1)]^{1/2}.$$

Using (9.88) and (9.89) it is straightforward to show that $\mathscr{B}_i$ are Hermitian with respect to the basis kets $|nlm\rangle$. Hence $\mathscr{A} + i\mathscr{B}$ and $\mathscr{A} - i\mathscr{B}$ are mutually adjoint shift operators. Also, using (5.13), (9.28), (9.88), and (9.89), we can show that

$$[L_i, \mathscr{B}_j] = i\hbar\varepsilon_{ijk}\mathscr{B}_k \tag{9.90}$$

and

$$[\mathscr{B}_i, \mathscr{B}_j] = i\hbar^{-1}\varepsilon_{ijk}L_k. \tag{9.91}$$

Hence the six operators L_i and $\mathscr{B}_i$ are a Hermitian basis for the algebra so(4).

9.7 A spectrum-generating algebra

We can construct a spectrum-generating algebra for the bound states of a Coulomb potential in an angular momentum basis by using the shift operators for energy and angular momentum (the operators $Q^{\pm}$, $\mathbf{C}^{\pm}$, and $L_{\pm}$). We do this by analogy with the method used for the oscillator in Section 8.7.

Label the eigenkets $|Nlm\rangle$, where $N = n - l - 1$ is the radial quantum number, and define

$$\mathbf{D}^{+} = \mathbf{C}^{+}Q^{+} \tag{9.92}$$

and

$$\mathbf{D}^{-} = Q^{-}\mathbf{C}^{-}. \tag{9.93}$$

Then $D_{\mu}^{\pm}$ transform $|Nlm\rangle$ into $|N, l \pm 1, m + \mu\rangle$. Let

$$\mathbf{D} = F(H, \mathbf{L}^2)\mathbf{D}^{-} \tag{9.94}$$

and

$$\mathbf{D}^{\dagger} = \mathbf{D}^{+}G(H, \mathbf{L}^2), \tag{9.95}$$

where the functions of constants of the motion, F and G, are to be determined. Using the expressions (5.131) and (9.21) for the coefficients in (5.153) and (9.19), we can show that the effect of $\mathbf{D}$ and $\mathbf{D}^{\dagger}$ on $|Nlm\rangle$ will be the same as (8.97)–(8.100) provided

$$F|nlm\rangle = \hbar^{-1}\left(\frac{2l+1}{2l+3}\frac{(n+1)^3}{n(n-l)^2(n+l+1)(n+l+2)}\right)^{1/2}|nlm\rangle$$

and

$$G|nlm\rangle = \frac{n(2l+3)}{(n+1)(2l+1)} F|nlm\rangle.$$

Thus F and G are given by the formulae

$$F = \hbar^{-1}\left(\frac{S(\mathcal{N}+1)^3}{(S+1)\mathcal{N}(\mathcal{N}-S+\frac{1}{2})^2(\mathcal{N}+S+\frac{1}{2})(\mathcal{N}+S+\frac{3}{2})}\right)^{1/2}$$

and

$$G = \frac{\mathcal{N}(S+1)}{(\mathcal{N}+1)S} F,$$

where $\mathcal{N}$ is defined by (5.146) and S by (4.48). Similarly, we can show that the effect of

$$Q = \left(\frac{\mathcal{N}+1}{\mathcal{N}(\mathcal{N}+S+\frac{1}{2})}\right)^{1/2} Q^- \tag{9.96}$$

and

$$Q^\dagger = Q^+\left(\frac{\mathcal{N}}{(\mathcal{N}+1)(\mathcal{N}+S+\frac{1}{2})}\right)^{1/2} \tag{9.97}$$

on $|Nlm\rangle$ is given by (8.101) and (8.102).

Thus as in Section 8.7 we obtain a spectrum-generating algebra

$$\mathrm{so}(2, 1) \oplus \mathrm{so}(3, 2)$$

for the bound states of an attractive Coulomb potential. As in the case of the oscillator, the eigenkets $|Nlm\rangle$ span two irreducible representations of this algebra, one containing all the kets with N even, the other with N odd. The bound states of the Coulomb potential are described within a single unitary irreducible representation of the Lie algebras so(4, 1) or so(4, 2). The corresponding Lie groups SO(4, 1) and SO(4, 2) are respectively the quantum number group and the dynamical group of the Coulomb problem [4].

9.8 The classical limit

In this section we discuss the classical limits of the shift operators derived in Sections 9.1 and 9.7. Consider first the operators **A** and **F**: using (9.10)–(9.12) and (5.55) we obtain for the limits as $\hbar \to 0$

$$\mathbf{A}_c = (M\kappa)^{-1}\mathbf{L} \times \mathbf{p} + \hat{\mathbf{r}} \tag{9.98}$$

and

$$\mathbf{F}_c = \hat{\mathbf{r}} \times \mathbf{L} + (M\kappa)^{-1}\mathbf{L}^2\mathbf{p} \tag{9.99}$$

$$= \mathbf{A}_c \times \mathbf{L}. \tag{9.100}$$

Here, and in the remainder of this section, $\mathbf{r}$, $\mathbf{p}$, and $\mathbf{L} = \mathbf{r} \times \mathbf{p}$ denote the classical position, momentum, and angular momentum vectors. Also, M is the reduced mass (see (5.8)).

In the classical limit the shift operators $\mathbf{C}^{\pm}$ become (see (9.9), (8.120), and (9.98)–(9.100))

$$\mathbf{C}_c^{\pm} = L(\mathbf{A}_c \pm \mathrm{i}\mathbf{B}_c) \tag{9.101}$$

where

$$\mathbf{B}_c = \hat{\mathbf{r}} \times \hat{\mathbf{L}} + (M\ell)^{-1} L\mathbf{p} \tag{9.102}$$

$$= \mathbf{A}_c \times \hat{\mathbf{L}}. \tag{9.103}$$

The classical vectors $\mathbf{A}_c$ and $\mathbf{B}_c$ possess some interesting properties and we consider these below.

(i) The classical equation of motion is

$$\frac{\mathrm{d}\mathbf{p}}{\mathrm{d}t} = -\ell r^{-2}\hat{\mathbf{r}}, \tag{9.104}$$

and $\mathbf{L}$ is a conserved vector because

$$\frac{\mathrm{d}\mathbf{L}}{\mathrm{d}t} = \mathbf{r} \times \frac{\mathrm{d}\mathbf{p}}{\mathrm{d}t} = 0. \tag{9.105}$$

Similarly we can show that $\mathbf{A}_c$ and $\mathbf{B}_c$ are also conserved vectors. Using (9.104) and (9.105)

$$\frac{\mathrm{d}}{\mathrm{d}t}\mathbf{L} \times \mathbf{p} = -\ell r^{-2}\mathbf{L} \times \hat{\mathbf{r}}. \tag{9.106}$$

The unit vector $\hat{\mathbf{r}}$ rotates in a circle with angular velocity $\boldsymbol{\omega}$ ($= \dot{\theta}\hat{\mathbf{L}}$, where θ is the angular position of the particle in a plane perpendicular to $\mathbf{L}$). Therefore

$$\frac{\mathrm{d}\hat{\mathbf{r}}}{\mathrm{d}t} = \boldsymbol{\omega} \times \hat{\mathbf{r}} \tag{9.107}$$

and

$$\mathbf{L} = Mr^2\boldsymbol{\omega}. \tag{9.108}$$

So

$$\frac{\mathrm{d}\hat{\mathbf{r}}}{\mathrm{d}t} = (Mr^2)^{-1}\mathbf{L} \times \hat{\mathbf{r}}. \tag{9.109}$$

From (9.98), (9.106), and (9.109)

$$\frac{\mathrm{d}\mathbf{A}_c}{\mathrm{d}t} = 0. \tag{9.110}$$

Because $\mathbf{B}_c$ is a product of two conserved vectors (see (9.103))

$$\frac{d\mathbf{B}_c}{dt} = 0. \tag{9.111}$$

(ii) The magnitude of $\mathbf{A}_c$ can be calculated by squaring (9.98) and using the identities $(\mathbf{L} \times \mathbf{p})^2 = \mathbf{L}^2\mathbf{p}^2$ and

$$\mathbf{r}.(\mathbf{L} \times \mathbf{p}) = -\mathbf{L}.(\mathbf{r} \times \mathbf{p}) = -\mathbf{L}^2.$$

Thus*

$$\mathbf{A}_c^2 = 1 + (2\mathbf{L}^2E/M\ell^2), \tag{9.112}$$

where

$$E = \frac{1}{2M}\mathbf{p}^2 - \frac{\ell}{r} \tag{9.113}$$

is the energy.

(iii) The vectors $\hat{\mathbf{r}}$ and $\mathbf{L} \times \mathbf{p}$ are perpendicular to $\mathbf{L}$. Hence

$$\mathbf{A}_c.\mathbf{L} = 0, \tag{9.114}$$

and $\mathbf{A}_c$ lies in the plane of the motion. The scalar product of $\mathbf{r}$ with (9.98) gives

$$\mathbf{r}.\mathbf{A}_c = -(M\ell)^{-1}\mathbf{L}^2 + r. \tag{9.115}$$

But

$$\mathbf{r}.\mathbf{A}_c = rA_c \cos\theta, \tag{9.116}$$

where θ is the angle between $\mathbf{r}$ and $\mathbf{A}_c$. Solving (9.115) and (9.116) for r we obtain for the trajectory the polar equation of a conic section

$$r(\theta) = \frac{r_0}{1 - e\cos\theta}, \tag{9.117}$$

where

$$r_0 = (M\ell)^{-1}\mathbf{L}^2 \tag{9.118}$$

is the semilatus rectum and

$$e = A_c \tag{9.119}$$

is the eccentricity. The vector $\mathbf{A}_c$ points along a symmetry axis of the conic section (the line $\theta = 0$); it is directed from a focus away from the point of closest approach. This is illustrated for an elliptical orbit ($E < 0$) in Fig. 9.2(a): in the case of planetary motion $\mathbf{A}_c$ is directed towards the aphelion.

* The quantum-mechanical analogue of (9.112) is (10.6).

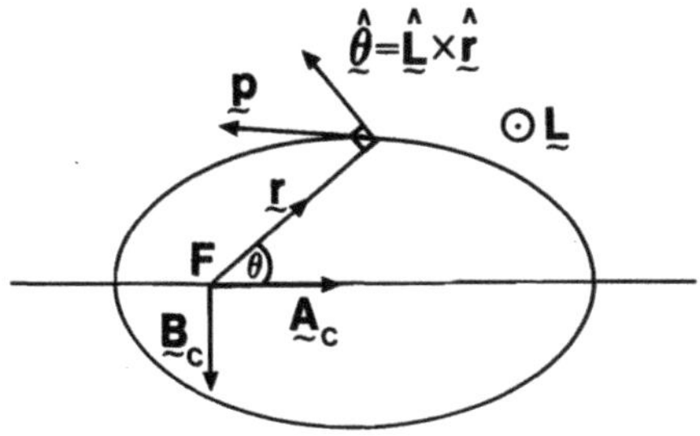

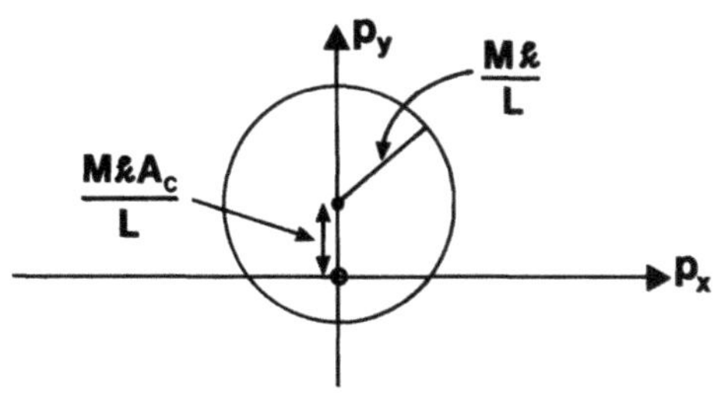

FIG. 9.2 (a) The trajectory in position space (for $E < 0$) and the conserved vectors $\mathbf{A}_c$ and $\mathbf{B}_c$. F is a focus of the elliptical orbit. (b) The trajectory in momentum space (for $E < 0$). The x axis is parallel to $\mathbf{A}_c$, the y axis is parallel to $-\mathbf{B}_c$.

(iv) Equation (9.102) can be solved for the momentum in terms of $\mathbf{B}_c$

$$\mathbf{p} = M\ell L^{-1}(\mathbf{B}_c + \hat{\boldsymbol{\theta}}), \tag{9.120}$$

where $\hat{\boldsymbol{\theta}}$ is the unit vector $\hat{\mathbf{L}} \times \hat{\mathbf{r}}$. If we choose Cartesian axes such that $\mathbf{A}_c = (A_c, 0, 0)$ and $\mathbf{B}_c = (0, -B_c, 0)$, where $B_c = M\ell L^{-1} A_c$, then (9.120) yields

$$p_x^2 + (p_y - M\ell L^{-1} A_c)^2 = (M\ell L^{-1})^2. \tag{9.121}$$

Thus the trajectory in momentum space (the so-called hodograph) is always a circle of radius $M\ell L^{-1}$ with centre at $(0, M\ell L^{-1} A_c)$ (see Fig. 9.2(b)).

(v) For an elliptical orbit the length of the semimajor axis is

$$a = \frac{1}{2}[r(0) + r(\pi)] = \frac{1}{2}\left(\frac{r_0}{1-e} + \frac{r_0}{1+e}\right).$$

Hence

$$1 - e^2 = (M\ell a)^{-1}\mathbf{L}^2. \tag{9.122}$$

Equations (9.112), (9.119), and (9.122) show that

$$E = -\frac{k}{2a}. \tag{9.123}$$

Thus orbits with the same semimajor axis and different eccentricity (different angular momentum) have the same energy. This degeneracy is associated with the symmetry provided by the constants of the motion $\mathbf{L}$ and $\mathbf{A}_c$.

(vi) The conserved vectors $\mathbf{L}$ and $\mathbf{A}_c$ yield generators of a symmetry group. One can show that

$$\{L_i, L_j\} = \varepsilon_{ijk} L_k, \quad \{L_i, \mathscr{A}_j\} = \varepsilon_{ijk} \mathscr{A}_k, \quad \{\mathscr{A}_i, \mathscr{A}_j\} = \sigma \varepsilon_{ijk} L_k, \tag{9.124}$$

where

$$\begin{aligned} \mathscr{A} &= \left(-\frac{2E}{Mk^2}\right)^{-1/2} \mathbf{A}_c, && (E < 0) \\ &= \mathbf{A}_c, && (E = 0) \\ &= \left(\frac{2E}{Mk^2}\right)^{-1/2} \mathbf{A}_c, && (E > 0) \end{aligned} \tag{9.125}$$

$$\begin{aligned} \sigma &= -\operatorname{sign} E, && (E \neq 0) \\ &= 0, && (E = 0) \end{aligned} \tag{9.126}$$

and { , } denotes a Poisson bracket [5]. Equations (9.124) correspond to the quantum-mechanical relations (9.74), (9.75′), and (9.76′) with [,] replaced by $i\hbar\{ \ , \ \}$. The algebras (9.124) and the associated symmetry groups are the same as those for the quantum-mechanical case (Section 9.6). The conserved vectors $\mathbf{L}$ and $\mathbf{B}_c$ also provide generators of a symmetry group: If $\mathscr{B}$ denotes the vector (9.125) with $\mathbf{A}_c$ replaced by $\mathbf{B}_c$, then eqns (9.124) remain valid when $\mathscr{A}$ is replaced by $\mathscr{B}$.

Next we consider the classical limits of the vector operators $\mathbf{D}^{\pm}$ that appear in the spectrum-generating algebra discussed in the previous section. From (9.92) and (9.93) we obtain for the limits as $\hbar \to 0$

$$\mathbf{D}_c = Q_c^+ \mathbf{C}_c^+ \tag{9.127}$$

and

$$\mathbf{D}_c^* = Q_c^- \mathbf{C}_c^-. \tag{9.128}$$

Here $\mathbf{C}_c^{\pm}$ are given by (9.101) and $Q_c^{\pm}$ are the classical limits of (5.152)*

* Note that the operator $\mathcal{N}$ in (5.152) tends to $Mk\hbar^{-1}(-2ME)^{-1/2}$ as $\hbar \to 0$ (see (5.146) and (5.55)).

$$Q_c^{\pm} = [\pm i\mathbf{r}.\mathbf{p} - (-2ME)^{1/2}(r - \tfrac{1}{2}k|E|^{-1})]e^{\pm i\delta}, \tag{9.129}$$

where

$$\delta = -(-2ME)^{1/2}(Mk)^{-1}\mathbf{r}.\mathbf{p} \tag{9.130}$$

and $E < 0$. Equation (9.127) can be rewritten

$$\mathbf{D}_c = \mathbf{F}_c + i\mathbf{G}_c. \tag{9.131}$$

Here $\mathbf{F}_c$ and $\mathbf{G}_c$ are the complex time-dependent vectors

$$\mathbf{F}_c = (f\mathbf{A}_c + g\mathbf{B}_c)e^{i\delta} \tag{9.132}$$

and

$$\mathbf{G}_c = (-g\mathbf{A}_c + f\mathbf{B}_c)e^{i\delta} \tag{9.133}$$

$$= \mathbf{F}_c \times \hat{\mathbf{L}}, \tag{9.134}$$

where

$$f = -L(-2ME)^{1/2}(r - \tfrac{1}{2}k|E|^{-1}) \tag{9.135}$$

and

$$g = -L\mathbf{r}.\mathbf{p}. \tag{9.136}$$

The vectors $\mathbf{F}_c$ and $\mathbf{G}_c$ are perpendicular to $\mathbf{L}$ and

$$\mathbf{F}_c.\mathbf{F}_c^* = \mathbf{G}_c.\mathbf{G}_c^* = L^4 A_c^4(1 - A_c^2)^{-1}, \tag{9.137}$$

where A_c is the eccentricity (see (9.119)). Let $\mathbf{X}$ denote the modulus

$$\mathbf{X} = \mathbf{F}_c e^{-i\delta} \quad \text{or} \quad \mathbf{G}_c e^{-i\delta}.$$

Since

$$\frac{dr}{dt} = (Mr^{-1})\mathbf{r}.\mathbf{p} \quad \text{and} \quad \frac{d}{dt}(\mathbf{r}.\mathbf{p}) = 2E + kr^{-1},$$

we have

$$\frac{d\mathbf{X}}{dt} = \boldsymbol{\omega}' \times \mathbf{X} \tag{9.138}$$

where

$$\boldsymbol{\omega}' = \frac{1}{r}\left(-\frac{2E}{M}\right)^{1/2}\hat{\mathbf{L}}. \tag{9.139}$$

Thus the moduli of $\mathbf{F}_c$ and $\mathbf{G}_c$ perform circular motion with angular velocity $\boldsymbol{\omega}'$ that is different from the angular velocity $\boldsymbol{\omega} = (Mr^2)^{-1}\mathbf{L}$ of the particle.*

---

* The classical vector $\boldsymbol{\omega}' \times \mathbf{r}$ has magnitude $(-2E/M)^{1/2}$; this velocity is the 'angular frequency' of a four-dimensional oscillator obtained from the Coulomb problem by the Kustaanheimo–Stiefel transformation (Section 9.11).

Its average value can be calculated with the aid of the classical virial theorem:

$$\langle \omega' \rangle = \frac{2E}{\ell}\left(-\frac{2E}{M}\right)^{1/2} = 2\pi T^{-1}, \tag{9.140}$$

where T is the period of the motion.

It is interesting to compare the classical limits of the operators derived in Sections 8.8 and 9.8. In this limit, the generators of an invariance algebra for the Coulomb potential are provided by two vector constants of the motion ($\mathbf{L}$ and $\mathbf{A}_c$ or $\mathbf{L}$ and $\mathbf{B}_c$) that specify the orientation and symmetry axis of the orbit. For the oscillator the corresponding generators are $\mathbf{L}$ and a conserved, symmetric second-rank tensor A_{ij}, whose eigenvectors specify the orientation and the symmetry axes of the orbit.

The classical limits of the vector operators $\mathbf{D}^{\pm}$ that appear in a spectrum-generating algebra are given by (8.121) for the oscillator and by (9.131) and (9.128) for the Coulomb potential. For the oscillator the time-dependent vectors $\mathbf{F}_c$ and $\mathbf{G}_c$ in (8.121) are real, perpendicular to $\mathbf{L}$, and they perform circular motion with the angular velocity of the oscillator. For the Coulomb potential the corresponding time-dependent vectors in (9.131) are complex. Otherwise they are similar to those for the oscillator: they are perpendicular to $\mathbf{L}$ and their moduli perform circular motion with the angular velocity (9.139).

9.9 Relativistic effects in the classical limit

The relativistic classical Coulomb problem can be solved exactly [6]. In this section we present an approximate analysis which is a useful introduction to the relativistic quantum-mechanical Coulomb problems discussed in Chapter 12. The relativistic Hamiltonian in an inertial frame Σ is

$$H \equiv E = (\mathbf{p}^2c^2 + m_0^2c^4)^{1/2} - \ell r^{-1}, \tag{9.141}$$

where m_0 is the rest mass. We rewrite (9.141) as

$$\mathscr{H} = \frac{1}{2m_0}\mathbf{p}^2 - k_1 r^{-1} - k_2 r^{-2} \tag{9.142}$$

where

$$\mathscr{H} = (E^2 - m_0^2c^4)/2m_0c^2, \tag{9.143}$$

$$k_1 = \ell E/m_0c^2, \tag{9.144}$$

$$k_2 = \ell^2/2m_0c^2. \tag{9.145}$$

Suppose $v \ll c$. Then $\mathbf{p} \approx m_0\mathbf{v}$ and $\mathscr{H}$ has the form of the non-relativistic Hamiltonian modified by an inverse-cube attractive force. To calculate the effects of this departure from the Coulomb potential it is helpful to introduce a rotating frame of reference.

In a frame Σ' rotating with angular velocity $\mathbf{\Omega}$ with respect to Σ, the position and velocity vectors are

$$\mathbf{r}' = \mathbf{r} \tag{9.146}$$

and

$$\frac{\mathrm{d}\mathbf{r}'}{\mathrm{d}t} = \frac{\mathrm{d}\mathbf{r}}{\mathrm{d}t} - \mathbf{\Omega} \times \mathbf{r}. \tag{9.147}$$

It is straightforward to show that if

$$\mathbf{\Omega} = (1 \pm \gamma)\boldsymbol{\omega}, \tag{9.148}$$

where $\boldsymbol{\omega} = \dot{\theta}\hat{\mathbf{L}}$ and

$$\gamma = \left[1 - \left(\frac{k}{cL}\right)^2\right]^{1/2}, \tag{9.149}$$

then (9.142) becomes

$$\mathscr{H} = \frac{1}{2m_0}(\mathbf{p}')^2 - k_1(r')^{-1}. \tag{9.150}$$

Here $\mathbf{p}' = m_0\, \mathrm{d}\mathbf{r}'/\mathrm{d}t$ and we are assuming that

$$cL > k. \tag{9.151}$$

(If $cL < k$ the particle falls in to the centre of force. This instability is a relativistic effect [6].) In the limit $c \to \infty$ we must have $\Omega = 0$; therefore the lower sign in (9.148) applies and

$$\mathbf{\Omega} = (1 - \gamma)\boldsymbol{\omega}. \tag{9.152}$$

Equation (9.150) shows that in a frame Σ' rotating with angular velocity $\mathbf{\Omega}$ the inverse-cube force is absent. By analogy with (9.98) we define

$$\mathbf{A}'_\mathrm{c} = (m_0 k_1)^{-1}\mathbf{L}' \times \mathbf{p}' + \hat{\mathbf{r}}' \tag{9.153}$$

where $\mathbf{L}' = \mathbf{r}' \times \mathbf{p}'$. One can easily show that $\mathbf{A}'_\mathrm{c}$ is conserved in Σ', and consequently if $\mathscr{H} < 0$ (i.e. $E < m_0c^2$) the trajectory in Σ' is an ellipse. Thus the trajectory has the non-relativistic form in a frame rotating in such a way that the perihelion appears fixed.* In the inertial frame Σ, $\mathbf{A}'_\mathrm{c}$ rotates with angular velocity $\mathbf{\Omega}$ and the trajectory is a precessing ellipse (see Fig. 9.3).† Substituting (9.146) and (9.147) in (9.153) we find that

$$\mathbf{A}'_\mathrm{c} = \gamma\frac{1}{m_0k_1}\mathbf{L} \times \mathbf{p} + \left(1 + \frac{\gamma(1-\gamma)\mathbf{L}^2}{m_0k_1}\frac{1}{r}\right)\hat{\mathbf{r}}. \tag{9.154}$$

* This is a general property of the relativistic solution [6].

† For the gravitational interaction a similar calculation can be performed in general relativity [7].

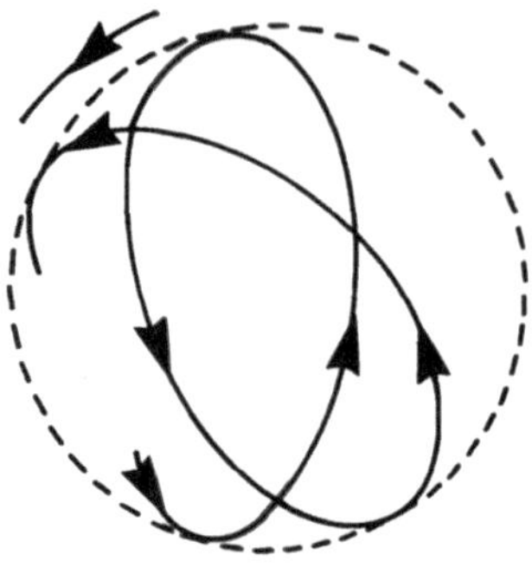

FIG. 9.3 Precession of an elliptical orbit due to the relativistic variation of mass with velocity. The radius of the circumscribed circle is equal to A'_c.

From this time-dependent vector we cannot obtain Poisson brackets such as (9.124): therefore the symmetry is reduced to that generated by **L**, namely SO(3). The inverse-cube force plays the role of a symmetry-breaking interaction and, as a result, the degeneracy is partially removed.

It can be shown that [5]

$$H = m_0 c^2 \left(1 + \frac{4\pi^2 k^2 c^{-2}}{\{J_r + [(J_\theta + J_\phi)^2 - 4\pi^2 k^2 c^{-2}]^{1/2}\}^2}\right)^{-1/2}. \qquad (9.155)$$

Here the J are action variables in spherical polar coordinates

$$J_r = \oint p_r \,\mathrm{d}r, \quad J_\theta = \oint p_\theta \,\mathrm{d}\theta, \quad J_\phi = \oint p_\phi \,\mathrm{d}\phi, \qquad (9.156)$$

where $p_r = m\dot{r}$, $p_\theta = mr\dot{\theta}$, and $p_\phi = mr \sin\theta\, \dot{\phi}$. In the non-relativistic limit $(c \to \infty)$

$$H \approx m_0 c^2 - \frac{2\pi^2 k^2 m_0}{(J_r + J_\theta + J_\phi)^2}, \qquad (9.157)$$

and the motion is completely degenerate in the sense that the frequencies $\nu_r = \partial H/\partial J_r$, $\nu_\theta = \partial H/\partial J_\theta$, and $\nu_\phi = \partial H/\partial J_\phi$ are all equal [5]. For the Hamiltonian (9.155) the degeneracy is incomplete because $\nu_\theta = \nu_\phi \neq \nu_r$. In the old quantum theory the formulae (9.155) and (9.156) yielded the correct result for the energy levels of the relativistic hydrogen-like atom (see Chapter 12).

9.10 Historical notes

The long and interesting history of the vectors $\mathbf{A}_c$ and $\mathbf{B}_c$ has been traced by Volk [8] and Goldstein [9]. The following notes are based on their findings. The magnitude A_c occurred in the first solution of the so-called inverse problem in 1710 by J. Hermann. Starting with an inverse-square law for the gravitational force, Hermann integrated the equation of motion in Cartesian form to obtain the orbit. This he expressed as a conic section in the xy plane with symmetry axis along $0x$. The eccentricity (i.e. A_c) is given in terms of integration constants from the x and y components of the equation of motion. Hermann's calculations were generalized in 1710 by J.I. Bernoulli who considered arbitrary orientation of the orbit in its plane.

In Volume I of *Traité de mécanique céleste* by P.S. de Laplace, which was published in 1798, seven constants of the motion are identified for an inverse-square force. These correspond essentially to the components of **L** and $\mathbf{A}_c$, and a function of the energy. Laplace pointed out that there could be only five independent constants of the motion and he derived two relations between the seven constants (cf. (9.112) and (9.114)). He also used constants corresponding to **L** and $\mathbf{A}_c$ to deduce the equation of the orbit (cf. (9.114)–(9.117)). Unlike the discoveries by Hermann and Bernoulli, which were largely forgotten, Laplace's contributions became well known, particularly in the field of celestial mechanics, and $\mathbf{B}_c$ and $\mathbf{A}_c$ are sometimes referred to as the first and second Laplace vectors.

The usefulness of the Laplace vectors was recognized. For example, J. Lagrange studied the effects of gravitational interaction between planets on their Laplace vectors. Also, in the nineteenth century the first Laplace vector was used by W. Hamilton and others in a proof that the trajectory in momentum space is circular (cf. (9.120), (9.121)). It appears that the earliest treatment of the second Laplace vector $\mathbf{A}_c$ in modern vector notation was given by J.W. Gibbs and E.B. Wilson in a text on vector analysis published in 1901. The same topic was discussed in a text by C. Runge published in 1919. The second Laplace vector is most frequently referred to as the Runge–Lenz vector although, as Goldstein remarks, a more appropriate name would be the Hermann–Bernoulli–Laplace vector [9].

The first applications in quantum mechanics were made by W. Lenz in 1924 and W. Pauli in 1926. Lenz used the vector $\mathbf{A}_c$ and the old quantum theory to calculate the effect of crossed electric and magnetic fields on atomic hydrogen. Soon after Heisenberg's discovery of matrix mechanics, Pauli used this mechanics to derive the Bohr formula for the energy levels of the non-relativistic hydrogen atom. This landmark publication was influential in the rapid acceptance of the new mechanics [10]. Although Pauli's treatment is presented in the language of matrix mechanics, it is essentially an abstract operator analysis based on the Pauli–Lenz operator (9.10). This vector operator was constructed as a quantum-mechanical analogue of the second Laplace vector [10]; it differs from the classical vector by a term $i\hbar(M\mathscr{k})^{-1}\mathbf{p}$ that is included to make **A** a Hermitian operator (Section 9.1).

It is remarkable that the magnitude of the vector $\mathbf{A}_c$ appeared in the first direct solution to the Kepler problem by the newly discovered classical dynamics, while two centuries later the corresponding Hermitian operator **A** was an essential ingredient in the first solution to the Coulomb problem by the new quantum mechanics. Subsequently there have been numerous publications dealing with the properties of the Pauli–Lenz operator and particularly with group theoretical properties of the Coulomb problem [11].

9.11 Comparison of the oscillator and Coulomb problems

This topic has attracted considerable interest, partly because these simple systems illustrate concepts such as accidental degeneracy, dynamical (hidden) symmetry, etc. [7, 12], and also because the relationship between them is of practical value, for example in the computation of matrix elements [13], the derivation of the Coulomb path integral [14], and the construction of coherent states [15]. Here we give a brief outline of this subject.

In the classical limit the orbits for the oscillator and the negative energy orbits for the Coulomb problem are both elliptical (see (8.118) and (9.117)). In the Coulomb case the origin is at a focus and there is one symmetry axis, whereas for the oscillator the origin is at the centre of the ellipse and there are two symmetry axes. It is nevertheless possible to devise a substitution that transforms the functional form of one orbit into that of the other [16]. Each problem possesses the maximum number of independent algebraic constants of the motion (five) for a system with three degrees of freedom. Also, they are the only spherically symmetric potentials that yield closed orbits for all bounded motion (Bertrand's theorem [17]).

In celestial mechanics it has been known for a long time that the Kepler problem (bounded motion in an inverse-square gravitational field) can be transformed into a four-dimensional harmonic oscillator with a constraint [18]. This procedure is known as regularization because the resulting oscillator equations do not possess explicit singularities at the origin. The details are as follows [18–20]. For the Kepler problem

$$M\ddot{\mathbf{r}} = -\ell r^{-2}\hat{\mathbf{r}} \tag{9.158}$$

and

$$E = \tfrac{1}{2}M\dot{\mathbf{r}}^2 - \ell r^{-1} < 0, \tag{9.159}$$

where $M = m_1 m_2/(m_1 + m_2)$ and $\ell = Gm_1 m_2$. The transformation

$$x_1 = u_1^2 - u_2^2 - u_3^2 + u_4^2, \quad x_2 = 2(u_1u_2 - u_3u_4), \quad x_3 = 2(u_1u_3 + u_2u_4), \tag{9.160}$$

subject to the constraint

$$u_4\,\mathrm{d}u_1 - u_3\,\mathrm{d}u_2 + u_2\,\mathrm{d}u_3 - u_1\,\mathrm{d}u_4 = 0, \tag{9.161}$$

changes the equation of motion (9.158) into

$$\frac{\mathrm{d}^2 u_i}{\mathrm{d}\tau^2} + \left(-\frac{E}{2M}\right)u_i = 0 \qquad (i = 1, \ldots, 4). \tag{9.162}$$

Here τ is the 'fictional time' defined by

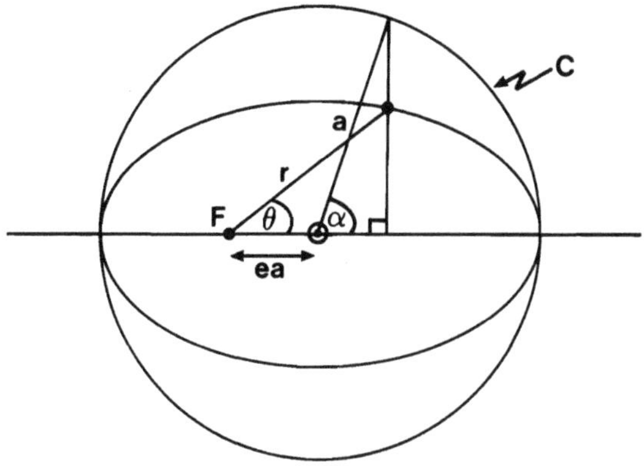

FIG. 9.4 The eccentric anomaly α for an elliptical orbit. The circle C circumscribes the ellipse.

$$\frac{\mathrm{d}}{\mathrm{d}\tau} = r\frac{\mathrm{d}}{\mathrm{d}t}. \tag{9.163}$$

Thus τ has the dimension of $(\text{velocity})^{-1}$. In terms of the eccentric anomaly α defined in Fig. 9.4 it can be shown that

$$\tau = (Ma/\ell)^{1/2}\alpha. \tag{9.164}$$

Equations (9.160) are known as the Kustaanheimo–Stiefel (KS) transformation. The transformed equation of motion (9.162) is formally that of a four-dimensional isotropic harmonic oscillator, where the velocity

$$v_{\mathrm{KS}} = \left(-\frac{E}{2M}\right)^{1/2} \tag{9.165}$$

plays the role of an angular frequency.

In the quantum-mechanical case the oscillator and Coulomb problems are also closely related. For example, in the coordinate representation of wave mechanics one can transform the radial equation for bound states of a Coulomb potential into the radial equation for a two-dimensional isotropic harmonic oscillator [21, 22]; alternatively, one can make a non-unitary transformation to a three-dimensional isotropic harmonic oscillator [1]. However, it is the relationship between the Coulomb problem and a constrained four-dimensional oscillator that has attracted most attention [23]. One method of establishing this connection is based on the KS transformation (9.160). Under this transformation the Schrödinger equation

$$\left(-\frac{\hbar^2}{2M}\nabla^2 - \frac{\ell}{r}\right)\psi = E\psi$$

becomes [23]

$$\left[-\frac{\hbar^2}{8M}\sum_{i=1}^{4}\frac{\partial^2}{\partial u_i^2}+\frac{\hbar^2}{8Mu^2}\left(u_4\frac{\partial}{\partial u_1}-u_3\frac{\partial}{\partial u_2}+u_2\frac{\partial}{\partial u_3}-u_1\frac{\partial}{\partial u_4}\right)^2-\mathscr{k}\right]\psi=u^2E\psi, \tag{9.166}$$

where $u^2 = \sum_{i=1}^{4} u_i^2$. If the constraint condition*

$$\left(u_4\frac{\partial}{\partial u_1}-u_3\frac{\partial}{\partial u_2}+u_2\frac{\partial}{\partial u_3}-u_1\frac{\partial}{\partial u_4}\right)\psi=0 \tag{9.167}$$

is imposed then (9.166) gives

$$\left[-\frac{\hbar^2}{2m}\sum_{i=1}^{4}\frac{\partial^2}{\partial u_i^2}+\frac{1}{2}m\left(-\frac{E}{2M}\right)u^2\right]\psi=\mathscr{k}\psi, \tag{9.168}$$

where $m = 4M$.

If $E < 0$, (9.168) is the equation of a four-dimensional isotropic oscillator with mass m, energy $\mathscr{k}$, and 'angular frequency'

$$\omega=\left(-\frac{E}{2M}\right)^{1/2}=v_{\mathrm{KS}}. \tag{9.169}$$

(See (9.165).) For this oscillator we can define the annihilation and creation operators (cf. Section 2.1)

$$a_j=\frac{1}{(2m\hbar\omega)^{1/2}}(\mathrm{i}p_j+m\omega u_j) \tag{9.170}$$

and

$$a_j^\dagger=\frac{1}{(2m\hbar\omega)^{1/2}}(-\mathrm{i}p_j+m\omega u_j), \tag{9.171}$$

where $p_j = -\mathrm{i}\hbar\partial/\partial u_j$ and $j = 1, 2, 3, 4$. These satisfy the commutation relations (2.255). Equation (9.168) can be written

$$H_0\psi=\mathscr{k}\psi \tag{9.172}$$

where

$$H_0=\hbar\omega\left(\sum_{i=1}^{4}a_i^\dagger a_i+2\right). \tag{9.173}$$

The eigenvalues of $a_i^\dagger a_i$ are $n_i = 0, 1, \ldots,$ for $i = 1$ to 4 (cf. Section 2.2). Hence

$$\mathscr{k}=(n_1+n_2+n_3+n_4+2)\hbar\omega. \tag{9.174}$$

Substituting (9.169) and (5.55) in (9.174) and solving for E we find

$$E=-\frac{\hbar^2}{2Ma^2n^2}, \tag{9.175}$$

* This condition is equivalent to the property (10.4) [24].

where

$$n = \tfrac{1}{2}(n_1 + n_2 + n_3 + n_4 + 2) = 1, \tfrac{3}{2}, 2, \tfrac{5}{2}, \ldots. \tag{9.176}$$

It can be shown that the constraint (9.167) requires [15] $n_1 + n_2 = n_3 + n_4$. This eliminates the half-integral values in (9.176) and hence (9.175) is the familiar Bohr formula for the energy eigenvalues.

The KS transformation has been applied to the path integral treatment of the hydrogen atom [14]. The same transformation has been used to derive radial coherent states for the Coulomb problem [15].

Finally we compare the Lie algebras of the oscillator and the negative energy Coulomb problem. In the classical limit these systems possess both su(3) and so(4) algebras [25]. For example, one can construct a conserved (Laplace) vector for the non-relativistic oscillator*

$$\begin{aligned}\mathbf{A}_c = \pm\left(\frac{(E^2 - \omega^2 L^2)^{1/2}}{2M\omega^2}\right)^{1/2} &\{L^2 - M[E - (E^2 - \omega^2 L^2)^{1/2}]r^2\} \\ &\times \{\mathbf{L} \times \mathbf{p} - [E - (E^2 - \omega^2 L^2)^{1/2}]\mathbf{r}\},\end{aligned} \tag{9.177}$$

that together with $\mathbf{L}$ provides generators of so(4). Similarly, for bounded orbits of the Coulomb potential there exists a conserved symmetric second-rank tensor A_{ij} that together with $\mathbf{L}$ yields the algebra su(3) [25].

The existence of the algebras su(3) and so(4) has been proved for all classical spherically symmetric potentials, both relativistic and non-relativistic [25]. However, it is only for the oscillator and Coulomb potentials in the non-relativistic limit that the vector and tensor constants of the motion associated with these algebras are single valued. In general $\mathbf{A}$ and A_{ij} are complicated infinite-valued functions of the particle motion that cannot be expressed as quantum-mechanical operators [25]. Only for the Laplace vectors $\mathbf{A}_c$ and $\mathbf{B}_c$ of the non-relativistic Coulomb problem and the tensor A_{ij} of the non-relativistic oscillator have quantum-mechanical counterparts been constructed.

An interesting variation of the Coulomb problem is the bisected hydrogen atom discussed by Ravenhall *et al.* [27]. This consists of a hydrogen atom with an impenetrable planar barrier passing through the nucleus. For such an atom all states of even parity with respect to reflections in the plane are excluded; the remaining states transform according to SU(3). This example illustrates the property that the irreducible representations of SO(4) that occur in the hydrogen atom can be made to span two irreducible representations of SU(3). Conversely, for the bisected oscillator the states transform according to SO(4) [27].

* If the $\pm$ sign is omitted in (9.177) the resulting vector is piecewise conserved [26].

References

1. See, for example, Plebanski, J. (1964). *Notes from lectures on elementary quantum mechanical operators.* Instituto Politecnico Nacional, Mèxico.
2. See, for example, Elliott, J.P. and Dawber, P.G. (1979). *Symmetry in physics*, Vol. 2. Macmillan, London.
3. See, for example, Englefield, M.J. (1972). *Group theory and the Coulomb problem.* Wiley, New York.
4. See, for example, Wybourne, B.G. (1974). *Classical groups for physicists.* Wiley, New York; and Adams, B.G., Cizek, J., and Paldus, J., (1982). Representation theory of so(4, 2) for the perturbation treatment of hydrogenic-type Hamiltonians by algebraic methods. *International Journal of Quantum Chemistry*, **31**, 153–71. The usefulness of the larger algebra so(4, 2) is discussed in the latter reference.
5. See, for example, Goldstein, H. (1980). *Classical mechanics*, 2nd edn. Addison-Wesley, Reading, MA. For recent work on the classical limit of Lie algebras see Bulgac, A. and Kusnezov, D. (1990). Classical limit for Lie algebras. *Annals of Physics*, **199**, 187–224.
6. See Landau, L.D. and Lifschitz, E.M. (1962). *The classical theory of fields*, 2nd edn. Pergamon, Oxford, and Biedenharn, L.C. (1983). The 'Sommerfeld puzzle' revisited and resolved. *Foundations of Physics*, **13**, 133–4.
7. See Greenberg, D.F. (1966). Accidental degeneracy. *American Journal of Physics*, **34**, 1101–9. For a discussion of Greenberg's method of computing the secular precessional frequency, see McDonald, K.T., Farina, C., and Tort, A. (1990). Right and wrong use of the Lenz vector for non-Newtonian potentials. *American Journal of Physics*, **58**, 540–2.
8. Volk, O. (1976). Miscellanea from the history of celestial mechanics. *Celestial Mechanics*, **14**, 365–82.
9. Goldstein, H. (1976). More on the prehistory of the Laplace or Runge–Lenz vector. *American Journal of Physics*, **44**, 1123–4.
10. See, for example, Mehra, J. and Rechenberg, H. (1982). *The historical development of quantum theory*, Vol. 3. Springer, New York, and van der Waerden, B.L. (ed.) (1968). *Sources in quantum mechanics.* Dover, New York.
11. See, for example, Refs 3 and 4.
12. McIntosh, H.V. (1971). Symmetry and degeneracy. In *Group theory and its applications*, Vol. II, (ed. E.M. Loebl). Academic Press, New York. This excellent article includes a comprehensive list of references. See also Ref. 2 of Chapter 8.
13. See, for example, Moshinsky, M., Seligmann, T.H., and Wolf, K.B. (1972). Canonical transformations and the radial oscillator and Coulomb problems. *Journal of Mathematical Physics*, **13**, 901–7, and Chacón, E., Levi, D., and Moshinsky, M. (1976). Lie algebras in the Schrödinger picture and radial matrix elements. *Journal of Mathematical Physics*, **17**, 1919–29.
14. Duru, I.H., and Kleinert, H. (1979). Solution of the path integral for the H-atom. *Physics Letters*, B **84**, 185–8. Ho, R. and Inomata, A. (1982). Exact-path-integral treatment of the hydrogen atom. *Physical Review Letters*, **48**, 231–4.
15. Gerry, C.G. (1986). Coherent states and the Kepler–Coulomb problem. *Physical Review*, A **33**, 6–11. Gerry, C.G., and Kiefer, J. (1988). Radial coherent states for the Coulomb problem. *Physical Review* A **37**, 665–71. For recent work on these states see McAnally, D.S. and Bracken, A.J. (1990). Quasiclassical states of the Coulomb system and so(4, 2). *Journal of Physics A: Math. Gen.*, **23**, 2027–47.

16. Cisneros, A. and McIntosh, H.V. (1970). Search for a universal symmetry group in two dimensions. *Journal of Mathematical Physics*, **11**, 870–95.
17. For references to and discussion of Bertrand's theorem see Refs 5, 12, and Tickochinsky, Y. (1988). A simplified proof of Bertrand's theorem. *American Journal of Physics*, **56**, 1073–5.
18. See Ref. 8 and Volk, O. (1973). Concerning the derivation of the KS-transformation. *Celestial Mechanics*, **8**, 297–305.
19. Kustaanheimo, P. and Stiefel, E.L. (1965). Perturbation theory of Kepler motion based on spinor regularization. *Journal für die reine und angewandte Mathematik*, **218**, 204–19.
20. Stiefel, E.L. and Scheifle, G. (1971). *Linear and regular celestial mechanics.* Springer, Berlin.
21. Bergman, D. and Frishnan, Y. (1965). A relation between the hydrogen atom and multidimensional harmonic oscillators. *Journal of Mathematical Physics*, **6**, 1855–6.
22. Moshinsky, M. (1971). Canonical transformations and quantum mechanics. Notes of the Latin American School of Physics, University of México, México.
23. For references see Kibler, M. and Négadi, T. (1984). Connection between the hydrogen atom and the harmonic oscillator: The zero-energy case. *Physical Review*, A **29**, 2891–4.
24. Chen, A.C. and Kibler, M. (1985). Connection between the hydrogen atom and the four-dimensional oscillator. *Physical Review*, A **31**, 3960–3.
25. Fradkin, D.M. (1967). Existence of the dynamic symmetries O_4 and SU_3 for all classical central potential problems. *Progress of Theoretical Physics*, **37**, 798–812. Bacry, H., Ruegg, H., and Soriau, J.M. (1966). Dynamical groups and spherical potentials in classical mechanics. *Communications in Mathematical Physics*, **3**, 323–33. For a recent discussion of Fradkin's work see Holas, A. and March, N.H. (1990). A generalization of the Runge–Lenz constant of classical motion in a central potential. *Journal of Physics A: Math. Gen.*, **23**, 735–49.
26. Buch, L.H. and Denman, H.H. (1975). Conserved and piecewise-conserved Runge vectors for the isotropic harmonic oscillator. *American Journal of Physics*, **43**, 1046–8.
27. Ravenhall, D.G., Sharp, R.T., and Pardee, W.J. (1967). A connection between SU(3) and O(4). *Physical Review*, **164**, 1950–6.

10

THE COULOMB PROBLEM IN THE BASIS $|H\mathscr{A}_z L_z\rangle$

For the non-relativistic Coulomb problem the operators H, $\mathscr{A}_z$, and L_z are also a set of commuting operators (see (9.14) and (9.75′)). In this chapter we consider the simultaneous eigenvalue problem for bound states

$$H|n\lambda\mu\rangle = E|n\lambda\mu\rangle, \tag{10.1}$$

$$\mathscr{A}_z|n\lambda\mu\rangle = \lambda|n\lambda\mu\rangle, \tag{10.2}$$

and

$$L_z|n\lambda\mu\rangle = \hbar\mu|n\lambda\mu\rangle. \tag{10.3}$$

Here $\mathscr{A}_z$ is given in terms of the Pauli–Lenz vector by (9.77) and n is defined by (5.93). The basis $|H\mathscr{A}_z L_z\rangle$ is useful in problems such as the Stark effect, where there is a perturbation that specifies a particular direction in space [1].

The discussion presented below illustrates the application of operator methods in a coordinate system that is not Cartesian. The analysis is based on shift operators for n, λ, and μ. The shift operators for λ and μ are easily constructed (Section 10.1) and their properties determine in a simple manner the possible values of n, λ, and μ (Section 10.2). In Section 10.3 operators are expressed in terms of parabolic coordinates, and shift operators for n are derived in Section 10.4. The hypervirial and Hellmann–Feynman theorems are used to obtain expectation values in Section 10.5. The shift operators generate all excited bound states from the ground state (Section 10.6) and this property is used to construct normalized coordinate-space wavefunctions in parabolic coordinates (Section 10.7).

10.1 Shift operators for λ and μ

The vector operator $\mathscr{A}$ has the useful properties*

$$\mathscr{A}\cdot\mathbf{L} = \mathbf{L}\cdot\mathscr{A} = 0, \tag{10.4}$$

$$\mathbf{L}\times\mathscr{A} + \mathscr{A}\times\mathbf{L} = 2i\hbar\mathscr{A}, \tag{10.5}$$

and

$$\mathscr{A}^2 = -\hbar^2(2Ma^2H)^{-1} - \hbar^{-2}\mathbf{L}^2 - 1. \tag{10.6}$$

The discussion in this section is based on these properties and the commutation relations (9.75′) and (9.76′).

---

* Equation (10.4) is obtained by taking the scalar product of (J.11) with $\mathbf{L}$ and using (C.12) and (C.13). Equation (10.5) follows from (9.75′); eqn (10.6) can be proved by squaring (J.11) and using the identity (G.2).

Define

$$W^{\pm} = \hbar^{-1}L_{\pm} + \mathscr{A}_{\pm}. \tag{10.7}$$

Because $\mathbf{L}$ and $\mathscr{A}$ are constants of the motion

$$[H, W^{\pm}] = 0. \tag{10.8}$$

From (9.75′) and (9.76′)

$$[\mathscr{A}_z, W^{\pm}] = \pm W^{\pm} \tag{10.9}$$

and

$$[L_z, W^{\pm}] = \pm\hbar W^{\pm}. \tag{10.10}$$

Equations (10.8)–(10.10) show that

$$W^{\pm}|n\lambda\mu\rangle = \alpha^{\pm}_{\lambda\mu}|n, \lambda \pm 1, \mu \pm 1\rangle. \tag{10.11}$$

Now $(W^{\pm})^{\dagger} = W^{\mp}$, and hence for normalized kets in (10.11)

$$|\alpha^{\pm}_{\lambda\mu}|^2 = \langle \hbar^{-2}L_{\mp}L_{\pm} + \hbar^{-1}(L_{\mp}\mathscr{A}_{\pm} + \mathscr{A}_{\mp}L_{\pm}) + \mathscr{A}_{\mp}\mathscr{A}_{\pm}\rangle. \tag{10.12}$$

From (9.75′)

$$\mathscr{A}_{\mp}\mathscr{A}_{\pm} = \mathscr{A}^2 - \mathscr{A}_z^2 \mp \hbar^{-1}L_z. \tag{10.13}$$

From (10.4)

$$\mathscr{A}_{\mp}L_{\pm} = -\mathscr{A}_z L_z \pm \mathrm{i}(\mathscr{A} \times \mathbf{L})_z \tag{10.14}$$

and

$$L_{\mp}\mathscr{A}_{\pm} = -L_z\mathscr{A}_z \pm \mathrm{i}(\mathbf{L} \times \mathscr{A})_z. \tag{10.15}$$

Substituting (10.13)–(10.15) and (4.17), with $\mathbf{J} = \mathbf{L}$, in (10.12) and using (10.5) and (10.6) we obtain

$$|\alpha^{\pm}_{\lambda\mu}|^2 = \langle -\hbar^{-2}(2Ma^2H)^{-1} - 1 - \mathscr{A}_z^2 - \hbar^{-2}L_z^2 \mp 2\hbar^{-1}(\hbar\mathscr{A}_z \pm \mathscr{A}_z L_z + L_z)\rangle. \tag{10.16}$$

Hence

$$\begin{aligned}\alpha^{\pm}_{\lambda\mu} &= \{n^2 - [1 \pm (\lambda + \mu)]^2\}^{1/2} \\ &= [n - 1 \mp (\lambda + \mu)]^{1/2}[n + 1 \pm (\lambda + \mu)]^{1/2},\end{aligned} \tag{10.17}$$

where we have chosen a phase factor to be unity.

For the operators

$$X^{\pm} = \hbar^{-1}L_{\pm} - \mathscr{A}_{\pm} \tag{10.18}$$

a similar analysis shows that*

$$X^{\pm}|n\lambda\mu\rangle = \tilde{\alpha}^{\pm}_{\lambda\mu}|n, \lambda \mp 1, \mu \pm 1\rangle, \tag{10.19}$$

* The shift operations (10.11) and (10.19) can also be derived using the theory of the coupling of two angular momenta [1].

where

$$\tilde{\alpha}^{\pm}_{\lambda\mu} = [n - 1 \mp (\mu - \lambda)]^{1/2}[n + 1 \pm (\mu - \lambda)]^{1/2}. \tag{10.20}$$

We can also write down shift operators that change only λ or μ. For example

$$D^{\pm}|n\lambda\mu\rangle = \tilde{\alpha}^{\mp}_{\lambda\pm1,\mu\pm1}\alpha^{\pm}_{\lambda\mu}|n, \lambda \pm 2, \mu\rangle \tag{10.21}$$

where

$$D^{\pm} = X^{\mp}W^{\pm} = W^{\pm}X^{\mp}. \tag{10.22}$$

For the purposes of operating on $|n\lambda\mu\rangle$, $D^{\pm}$ can be replaced by simpler operators (see (10.79)–(10.82)).

10.2 Eigenvalues

As an application of these shift operators we determine the eigenvalues in (10.1)–(10.3). Non-negativity of the norm in (10.11) and (10.19) requires $|\alpha^{\pm}_{\lambda\mu}|^2 \geq 0$ and $|\tilde{\alpha}^{\pm}_{\lambda\mu}|^2 \geq 0$. Hence

$$-(n - 1) \leq \mu + \lambda \leq n - 1 \tag{10.23}$$

and

$$-(n - 1) \leq \mu - \lambda \leq n - 1. \tag{10.24}$$

Therefore $n \geq 1$ and

$$\begin{aligned} -(n - 1) + |\lambda| \leq \mu \leq n - 1 - |\lambda|, \\ -(n - 1) + |\mu| \leq \lambda \leq n - 1 - |\mu|. \end{aligned} \tag{10.25}$$

The maximum and minimum values of λ are $n - 1$ and $-(n - 1)$. To transform between kets with these extreme eigenvalues requires N applications of D^+ or D^-, where $N = 0, 1, 2, \ldots$. Each application changes λ by two. Hence $2(n - 1) = 2N$ and the possible values of n are

$$n = 1, 2, \ldots. \tag{10.26}$$

Identical results are obtained when λ is kept constant and μ is shifted.

The transformations effected by the shift operators are illustrated for $n = 3$ in Fig. 10.1. For $n \geq 2$ this figure is a square with sides $\lambda = n - 1 - |\mu|$ and $\mu = n - 1 - |\lambda|$ (see (10.25)). The possible values of λ and μ are represented by points on and inside this square, namely

$$\lambda = n - 1 - \mu - 2N, \tag{10.27}$$

where

$$\mu = -N, -N + 1, \ldots, n - 1 - N \tag{10.28}$$

and

$$N = 0, 1, \ldots, n - 1. \tag{10.29}$$

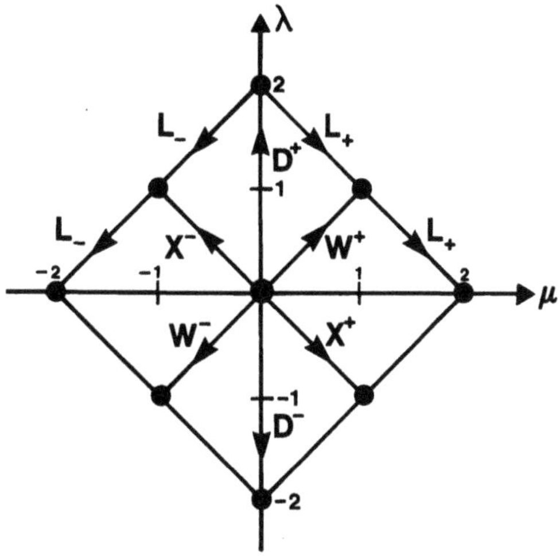

FIG. 10.1 Schematic illustration of the shift operations (10.11), (10.19), (10.21), (10.32), and (10.34) for $n = 3$.

On the boundary of the square certain shift operations have a simple form. To see this we use

$$L_\pm = \tfrac{1}{2}\hbar(W^\pm + X^\pm). \tag{10.30}$$

If

$$\lambda = n - 1 - \mu \qquad (\mu > 0), \tag{10.31}$$

then $W^+|n\lambda\mu\rangle = 0$ and (10.19), (10.20), and (10.30) yield the simple result

$$L_+|n\lambda\mu\rangle = \hbar[(n-1-\mu)(\mu+1)]^{1/2}|n, \lambda-1, \mu+1\rangle. \tag{10.32}$$

Similarly, if

$$\lambda = n - 1 + \mu \qquad (\mu < 0), \tag{10.33}$$

then

$$L_-|n\lambda\mu\rangle = \hbar[(n-1+\mu)(1-\mu)]^{1/2}|n, \lambda-1, \mu-1\rangle. \tag{10.34}$$

The transformations (10.32) and (10.34) are illustrated in Fig. 10.1.

10.3 Parabolic coordinates

It is useful at this point to introduce the parabolic coordinates [2]

$$\xi = r + z, \qquad \eta = r - z, \qquad \phi = \tan^{-1}(y/x), \tag{10.35}$$

where $r = (x^2 + y^2 + z^2)^{1/2}$. The conjugate momentum operator of ϕ is (see Section 4.11)

$$p_\phi = L_z. \tag{10.36}$$

To construct analogous momentum operators of ξ and η we proceed as follows.

We require that p_ξ have the properties

$$[p_\xi, \xi] = -i\hbar \tag{10.37}$$

and*

$$(\xi p_\xi)^\dagger = p_\xi \xi. \tag{10.38}$$

From (10.35) and using the commutators (1.65) and (1.67) we have

$$[\mathbf{r}.\mathbf{p}, \xi] = -i\hbar\xi, \qquad [rp_z, \xi] = -i\hbar\xi. \tag{10.39}$$

Hence

$$\left[\frac{1}{2(r+z)}(\mathbf{r}.\mathbf{p} + rp_z), \xi\right] = -i\hbar. \tag{10.40}$$

Although (10.40) has the form (10.37), nevertheless (10.38) is not satisfied because (see (1.22), (1.23), (1.136), and (1.138))

$$\tfrac{1}{2}(\mathbf{r}.\mathbf{p} + rp_z)^\dagger = \tfrac{1}{2}(\mathbf{r}.\mathbf{p} + rp_z) - i\hbar - \tfrac{1}{2}i\hbar\xi r^{-1}. \tag{10.41}$$

Instead (10.40) and (10.41) show that

$$p_\xi = \frac{1}{2(r+z)}(\mathbf{r}.\mathbf{p} + rp_z) - \frac{i\hbar}{4r} \tag{10.42}$$

satisfies both (10.37) and (10.38).

Similarly, starting with the commutators

$$[\mathbf{r}.\mathbf{p}, \eta] = -i\hbar\eta, \qquad [-rp_z, \eta] = -i\hbar\eta, \tag{10.43}$$

we can show that

$$p_\eta = \frac{1}{2(r-z)}(\mathbf{r}.\mathbf{p} - rp_z) - \frac{i\hbar}{4r} \tag{10.44}$$

has the properties

$$[p_\eta, \eta] = -i\hbar \tag{10.45}$$

and

$$(\eta p_\eta)^\dagger = p_\eta \eta. \tag{10.46}$$

For the functions $f(\xi, \eta)$ considered below one can use (1.136) to generalize (10.37) and (10.45) to

$$[p_\xi, f(\xi, \eta)] = -i\hbar\frac{\partial f}{\partial \xi} \tag{10.47}$$

* The more restrictive condition $p_\xi^\dagger = p_\xi$ is not satisfied (see Section 10.7).

and

$$[p_\eta, f(\xi, \eta)] = -i\hbar \frac{\partial f}{\partial \eta}. \tag{10.48}$$

(See also (10.105) and (10.106).)

By inverting (10.35), (10.36), (10.42), and (10.44) we obtain the Cartesian components of **r** and **p** in terms of the parabolic components

$$x \pm iy = (\xi\eta)^{1/2} \exp(\pm i\phi), \tag{10.49}$$

$$z = \tfrac{1}{2}(\xi - \eta), \tag{10.50}$$

$$p_x \pm ip_y = \frac{2(\xi\eta)^{1/2}}{\xi + \eta}\left(p_\xi + p_\eta + \frac{i\hbar}{\xi + \eta}\right)\exp(\pm i\phi)$$
$$\pm \frac{i}{(\xi\eta)^{1/2}} \exp(\pm i\phi) p_\phi, \tag{10.51}$$

$$p_z = \frac{2}{\xi + \eta}(\xi p_\xi - \eta p_\eta) + i\hbar \frac{\xi - \eta}{(\xi + \eta)^2}. \tag{10.52}$$

Using (10.47)–(10.52) we can express H in terms of the operators in parabolic coordinates

$$H = \frac{1}{2M}\left[\frac{4}{\xi + \eta}\left(\xi p_\xi^2 + \eta p_\eta^2 - \frac{i\hbar}{\xi + \eta}(\eta p_\xi + \xi p_\eta)\right)\right.$$
$$\left. + \frac{\hbar^2}{(\xi + \eta)^2} + \frac{1}{\xi\eta} L_z^2 - \frac{4\hbar^2}{a}\frac{1}{\xi + \eta}\right]. \tag{10.53}$$

Similarly, the z component of the Pauli–Lenz vector (9.10) can be written

$$A_z = \frac{2a}{\hbar^2}\left(\frac{\xi\eta}{\xi + \eta}(p_\xi^2 - p_\eta^2) + \frac{i\hbar}{(\xi + \eta)^2}(\xi^2 p_\eta - \eta^2 p_\xi)\right.$$
$$\left. + \frac{\hbar^2}{2a}\frac{\xi - \eta}{\xi + \eta} - \frac{1}{2}\hbar^2 \frac{\xi - \eta}{(\xi + \eta)^2} - \frac{\xi - \eta}{4\xi\eta} L_z^2\right). \tag{10.54}$$

Equation (10.54) can be used to eliminate p_η from the Hamiltonian. Thus

$$H = \frac{2}{M}\left(p_\xi^2 - i\hbar \frac{\eta}{\xi(\xi + \eta)} p_\xi + \frac{1}{4\xi^2} L_z^2 + \frac{1}{4}\hbar^2 \frac{2\eta - \xi}{\xi(\xi + \eta)^2} - \frac{\hbar^2}{2a\xi}(1 + A_z)\right). \tag{10.55}$$

Substituting (10.55), (9.77), (10.2), and (10.3) in (10.1) we find

$$H_{\lambda\mu}(\xi)|n\lambda\mu\rangle = E|n\lambda\mu\rangle \tag{10.56}$$

where

$$H_{\lambda\mu}(\xi) = \frac{2}{M}\left(p_\xi^2 - i\hbar \frac{\eta}{\xi(\xi + \eta)} p_\xi + \frac{\hbar^2\mu^2}{4\xi^2} + \frac{1}{4}\hbar^2 \frac{2\eta - \xi}{\xi(\xi + \eta)^2} - \frac{\hbar^2}{2an\xi}(n + \lambda)\right). \tag{10.57}$$

In a similar manner we can eliminate p_ξ from the Hamiltonian. This yields

$$H_{\lambda\mu}(\eta)|n\lambda\mu\rangle = E|n\lambda\mu\rangle, \tag{10.58}$$

where $H_{\lambda\mu}(\eta)$ is given by (10.57) with

$$\xi \leftrightarrow \eta, \qquad p_\xi \to p_\eta, \qquad \lambda \to -\lambda. \tag{10.59}$$

From (10.56) and (10.58) we obtain some useful shift operators.

10.4 Shift operators for energy

To construct shift operators for energy we follow the method used in Chapters 3 and 5. By multiplying both sides of (10.56) on the left with ξ^2 and rearranging terms we have

$$O_{n\lambda}(\xi)|n\lambda\mu\rangle = -\tfrac{1}{4}\mu^2|n\lambda\mu\rangle, \tag{10.60}$$

where

$$O_{n\lambda}(\xi) = \hbar^{-2}\xi^2 p_\xi^2 - i\hbar^{-1}\frac{\xi\eta}{\xi+\eta}p_\xi + \frac{\xi(2\eta-\xi)}{4(\xi+\eta)^2} + \frac{1}{4a^2n^2}\xi^2$$
$$- \frac{1}{2an}(n+\lambda)\xi. \tag{10.61}$$

This can be factorized:

$$T_{\lambda\pm 2}^{\mp}(\xi)T_\lambda^{\pm}(\xi) = O_{n\lambda}(\xi) + \tfrac{1}{4}(n+\lambda\pm 1)^2, \tag{10.62}$$

where

$$T_\lambda^{\pm}(\xi) = \pm i\hbar^{-1}\xi p_\xi \mp \frac{1}{2}\frac{\xi}{\xi+\eta} - \frac{1}{2an}\xi + \frac{1}{2}(n+\lambda\pm 1). \tag{10.63}$$

Consider the operators (cf. Section 5.6)

$$Q_{n\lambda}^{\pm}(\xi) = \Delta_n^{\pm}T_\lambda^{\pm}(\xi). \tag{10.64}$$

Here $\Delta_n^{\pm}$ are the scaling operators in Section 5.6; thus

$$\Delta_n^{\pm}(\xi \text{ or } \eta) = \frac{n}{n\pm 1}(\xi \text{ or } \eta)\Delta_n^{\pm}, \tag{10.65}$$

$$\Delta_n^{\pm}(p_\xi \text{ or } p_\eta) = \frac{n\pm 1}{n}(p_\xi \text{ or } p_\eta)\Delta_n^{\pm}, \tag{10.66}$$

and $\Delta_n^{\pm}$ commute with ϕ and p_ϕ. Using (10.63)–(10.66) and the property $\Delta_{n\pm 1}^{\mp}\Delta_n^{\pm} = 1$ we find

$$Q_{n\pm 1,\lambda\pm 1}^{\mp}(\xi)Q_{n\lambda}^{\pm}(\xi) = O_{n\lambda}(\xi) + \tfrac{1}{4}(n+\lambda\pm 1)^2, \tag{10.67}$$

which suggests* that $Q_{n\lambda}^{\pm}(\xi)$ change n and λ by ± 1. This can be verified by

* The space spanned by the eigenkets of $O_{n\lambda}(\xi)$ is a subspace of that spanned by $|n\lambda\mu\rangle$, and hence (10.72) does not follow from (10.67) alone.

evaluating the commutators of H, A_z, and L_z with $Q^{\pm}_{n\lambda}(\xi)$. It can be shown that

$$HQ^{\pm}_{n\lambda} = \left(\frac{n}{n \pm 1}\right)^2 Q^{\pm}_{n\lambda} H \pm 2\left(\frac{n}{n \pm 1}\right)^2 \Delta^{\pm}_n \Omega, \tag{10.68}$$

$$A_z Q^{\pm}_{n\lambda} = Q^{\pm}_{n\lambda}\left(\frac{n}{n \pm 1}(A_z - 1) + 1\right) \pm 2Ma\hbar^{-2}\frac{n}{n \pm 1}\Delta^{\pm}_n \eta\Omega, \tag{10.69}$$

and

$$[L_z, Q^{\pm}_{n\lambda}] = 0, \tag{10.70}$$

where

$$\Omega = \frac{\xi}{\xi + 1}\left(H - E + \frac{\hbar^2}{2Ma\xi}(A_z - \lambda n^{-1})\right). \tag{10.71}$$

According to (10.1), (10.2), and (9.77), Ω annihilates $|n\lambda\mu\rangle$. Therefore (10.68)–(10.70) and (9.77) show that

$$Q^{\pm}_{n\lambda}(\xi)|n\lambda\mu\rangle = \beta^{\pm}_{n\lambda}|n \pm 1, \lambda \pm 1, \mu\rangle. \tag{10.72}$$

For normalized kets in (10.72)

$$|\beta^{\pm}_{n\lambda}|^2 = \langle (T^{\pm}_\lambda)^\dagger (\Delta^{\pm}_n)^\dagger \Delta^{\pm}_n T^{\pm}_\lambda \rangle = \langle (T^{\pm}_\lambda)^\dagger T^{\pm}_\lambda \rangle. \tag{10.73}$$

From (10.63) and (10.38)

$$[T^{\pm}_\lambda(\xi)]^\dagger = T^{\mp}_{\lambda \pm 2}(\xi) \mp \left(1 + \frac{\xi}{\xi + \eta}\right). \tag{10.74}$$

Hence (10.62), (10.60), and (10.73) yield

$$|\beta^{\pm}_{n\lambda}|^2 = \frac{1}{4}[(n + \lambda \pm 1)^2 - \mu^2] \mp \left\langle \left(1 + \frac{\xi}{\xi + \eta}\right) T^{\pm}_\lambda(\xi) \right\rangle. \tag{10.75}$$

The expectation values in (10.75) are calculated below: using (10.87), (10.88), (10.94), (10.95), and (10.98), and setting a phase factor equal to unity, we find

$$\beta^{\pm}_{n\lambda} = \frac{1}{2}\left(\frac{n \pm 1}{n}(n + \lambda \pm 1 + \mu)(n + \lambda \pm 1 - \mu)\right)^{1/2}. \tag{10.76}$$

We could also have performed the above analysis starting with (10.58). This yields the shift operations

$$Q^{\pm}_{n\lambda}(\eta)|n\lambda\mu\rangle = \beta^{\pm}_{n,-\lambda}|n \pm 1, \lambda \mp 1, \mu\rangle. \tag{10.77}$$

Here $Q^{\pm}_{n\lambda}(\eta) = \Delta^{\pm}_n T^{\pm}_\lambda(\eta)$ and $T^{\pm}_\lambda(\eta)$ are obtained from (10.63) by making the substitutions (10.59), that is

$$T^{\pm}_\lambda(\eta) = \pm i\hbar^{-1}\eta p_\eta \mp \frac{1}{2}\frac{\eta}{\xi + \eta} - \frac{1}{2an}\eta + \frac{1}{2}(n - \lambda \pm 1). \tag{10.78}$$

The operators $T^{\pm}_\lambda(\xi)$ and $T^{\pm}_\lambda(\eta)$ have other uses. For the products

$$D^{\pm}_\lambda = T^{\mp}_\lambda(\eta) T^{\pm}_\lambda(\xi) \tag{10.79}$$

it is straightforward though lengthy to prove that

$$D_\lambda^\pm|n\lambda\mu\rangle = -\tfrac{1}{4}D^\pm|n\lambda\mu\rangle, \tag{10.80}$$

where $D^\pm$ are the shift operators (10.22). From (10.80), (10.21), (10.20), and (10.17)

$$D_\lambda^\pm|n\lambda\mu\rangle = \gamma_\lambda^\pm|n, \lambda \pm 2, \mu\rangle, \tag{10.81}$$

where

$$\gamma_\lambda^\pm = -\tfrac{1}{4}[(n+\lambda+\mu\pm 1)(n-\lambda-\mu\mp 1)(n+\lambda-\mu\pm 1)(n-\lambda+\mu\mp 1)]^{1/2}. \tag{10.82}$$

The operators $D_\lambda^\pm$ are second order in momentum and consequently they are simpler than $D^\pm$ which are fourth order in momentum. Also, the factored form (10.79) is advantageous in certain applications (see below).

10.5 Expectation values

By treating a as a continuous variable in (10.53) and applying the Hellmann–Feynman theorem we obtain

$$\left\langle \frac{1}{\xi+\eta} \right\rangle = \frac{1}{2an^2}, \tag{10.83}$$

which is just (6.49). Next we use the hypervirial theorem to show that certain expectation values can be expressed in terms of (10.83).

From (10.53), (10.47), and (10.48) it can be shown that

$$\frac{1}{2}M[H, f(\xi)] = -2i\hbar\frac{\xi}{\xi+\eta}\frac{df}{d\xi}p_\xi - \hbar^2\frac{\xi}{\xi+\eta}\frac{d^2 f}{d\xi^2} - \hbar^2\frac{\eta}{(\xi+\eta)^2}\frac{df}{d\xi}. \tag{10.84}$$

From (10.53) and (10.54) it follows by inspection that

$$\frac{\hbar^2}{2a}[A_z, f(\xi)] = \frac{1}{2}M\eta[H, f(\xi)]. \tag{10.85}$$

Applying the hypervirial theorem to (10.84) with $f = \xi^{N+1}$, and to (10.85) with $f = \xi^N$, and using the identity

$$\frac{\eta}{\xi(\xi+\eta)} = \frac{1}{\xi} - \frac{1}{\xi+\eta}, \tag{10.86}$$

one readily obtains

$$\langle \xi^N p_\xi \rangle = \tfrac{1}{2}i\hbar N\langle \xi^{N-1}\rangle. \tag{10.87}$$

Here N is a non-negative integer and the expectation values are with respect to $|n\lambda\mu\rangle$. Similarly, with $f = \xi^N$ in (10.84) and using (10.86), we find

$$\left\langle \frac{\xi^N}{\xi+\eta}p_\xi \right\rangle = \frac{1}{2}i\hbar\left\langle N\frac{\xi^{N-1}}{\xi+\eta} - \frac{\xi^N}{(\xi+\eta)^2}\right\rangle. \tag{10.88}$$

For the operator

$$R = \mathrm{i}p_\xi - \frac{1}{2}\frac{\hbar}{\xi + \eta} \tag{10.89}$$

it can be shown that

$$\frac{1}{2}M[H, R] = -\frac{\hbar^3}{2\xi(\xi+\eta)}\left(\frac{1}{a}A_z + \frac{2\mathrm{i}}{\hbar}p_\xi - \frac{1}{\hbar^2\xi}L_z^2 + \frac{1}{a} - \frac{1}{\xi+\eta}\right). \tag{10.90}$$

The commutator

$$[H, f(\xi)R] = [H, f(\xi)]R + f(\xi)[H, R] \tag{10.91}$$

can be written down using (10.84) and (10.90). Applying the hypervirial theorem to (10.91) with $f = \xi^N$ and using (10.88) to eliminate terms in p_ξ we find

$$\left\langle \frac{\xi^N}{\xi+\eta} G \right\rangle = 0, \tag{10.92}$$

where

$$G = \tfrac{1}{2}NME + \tfrac{1}{4}\hbar^2(N-1)[(N-1)^2 - \mu^2]\xi^{-2}$$
$$+ \hbar^2(4an)^{-1}(n+\lambda)(2N-1)\xi^{-1}. \tag{10.93}$$

For $N = 1, 2, \ldots$ the recurrence relation (10.92) provides expectation values in terms of (10.83); thus, for example, with $N = 1$ and 2 we obtain

$$\left\langle \frac{\xi}{\xi+\eta} \right\rangle = \frac{n+\lambda}{2n}, \tag{10.94}$$

$$\left\langle \frac{\xi^2}{\xi+\eta} \right\rangle = \frac{1}{4}a[3(n+\lambda)^2 + 1 - \mu^2]. \tag{10.95}$$

There are other useful recurrence relations. From (10.54), (10.53), and (10.84) it is straightforward to show that (10.85) is also valid if $f(\xi)$ is replaced by $f(\xi)R$. Because of the factor η in (10.85) this yields instead of (10.92) the recurrence relation

$$\left\langle \eta \frac{\xi^N}{\xi+\eta} G \right\rangle = 0. \tag{10.96}$$

Substituting (10.86) in (10.96) we obtain

$$\langle \xi^N G \rangle = \left\langle \frac{\xi^{N+1}}{\xi+\eta} G \right\rangle. \tag{10.97}$$

For $N = 1, 2, \ldots$ eqns (10.97) and (10.92) yield expectation values of ξ^N in terms of (10.83). For example, if $N = 1$

$$\langle \xi \rangle = \tfrac{1}{4}a[(n+\lambda)^2 + 4n(n+\lambda) + 1 - \mu^2]. \tag{10.98}$$

Analogous calculations show that the results (10.87), (10.88), and (10.92)–(10.98) are also valid if we make the substitutions (10.59). Thus, for example, (10.87) and (10.98) become

$$\langle \eta^N p_\eta \rangle = \tfrac{1}{2} i\hbar N \langle \eta^{N-1} \rangle \tag{10.99}$$

and

$$\langle \eta \rangle = \tfrac{1}{4} a[(n-\lambda)^2 + 4n(n-\lambda) + 1 - \mu^2]. \tag{10.100}$$

For a perturbing potential

$$H^{(1)} = kz = \tfrac{1}{2} k(\xi - \eta), \tag{10.101}$$

where k is a constant, (10.98) and (10.100) yield a first-order shift in the energy eigenvalues

$$\langle H^{(1)} \rangle = \tfrac{3}{2} ka\lambda n. \tag{10.102}$$

This result has application to the theory of the linear Stark effect [3].

10.6 Eigenkets generated from the ground state $|100\rangle$

All the bound state eigenkets of H, $\mathscr{A}_z$, and L_z can be generated from a given eigenket by applying the shift operators derived in Section 10.1 and 10.4. These operators can be used in various ways. For example, from (10.72) and (10.76)

$$|n+1, n, 0\rangle = [n(n+1)]^{-1/2} Q^+_{n,n-1}(\xi)|n, n-1, 0\rangle. \tag{10.103}$$

Thus operating with $Q^+_{10}(\xi)$, $Q^+_{21}(\xi)Q^+_{10}(\xi)$, ... on the ground state or 'vacuum' $|100\rangle$ generates $|n, n-1, 0\rangle$ for $n = 2, 3, \ldots$ (see Fig. 10.2(a)). According to

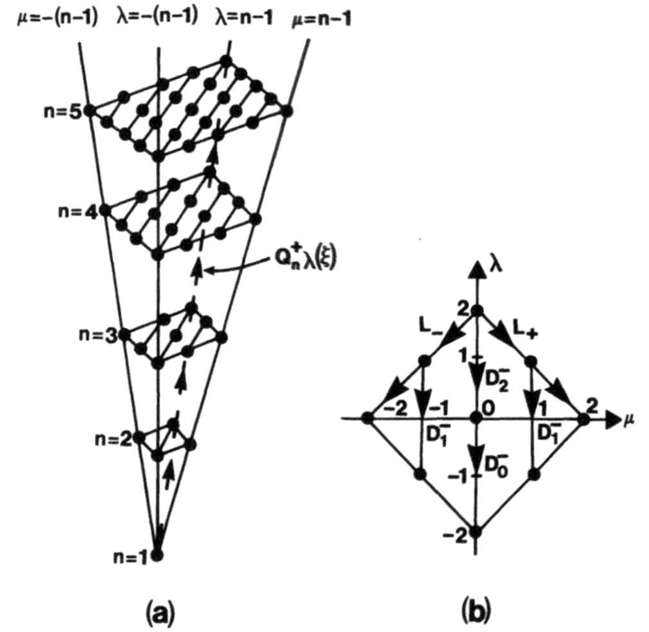

FIG. 10.2 (a) Use of the shift operators $Q^+_{n\lambda}(\xi)$ to generate $|n, n-1, 0\rangle$ from $|100\rangle$. (b) Use of the shift operators $L_\pm$ and D^-_λ to generate $|n\lambda\mu\rangle$ from $|n, n-1, 0\rangle$. The plane $n = 3$ is shown.

(10.32) and (10.34), repeated application of $L_\pm$ to these kets produces $|n\lambda\mu\rangle$ with $\lambda = n - 1 - |\mu|$ (see Fig. 10.2(b)). The remaining kets $|n\lambda\mu\rangle$ are generated by applying D_λ^- (see (10.81) and Fig. 10.2(b)). It is instructive to compare Fig. 10.2 with the corresponding diagram for the eigenkets $|nlm\rangle$ of the angular momentum basis (see Fig. 9.1).

If the ground state is normalizable then so are the eigenkets produced in the above manner: the simple sequence of transformations shown in Fig. 10.2 generates all excited states from the ground state. In the next section we use these transformations to construct the coordinate-space wavefunctions for bound states

$$\psi_{n\lambda\mu}(\mathbf{r}) = \langle \mathbf{r}|n\lambda\mu\rangle. \tag{10.104}$$

10.7 Wavefunctions

First we express the shift operators in differential form. Because

$$\begin{aligned}\frac{\partial}{\partial\xi} &= \frac{\partial x}{\partial\xi}\frac{\partial}{\partial x} + \frac{\partial y}{\partial\xi}\frac{\partial}{\partial y} + \frac{\partial z}{\partial\xi}\frac{\partial}{\partial z}\\ &= \frac{x}{2\xi}\frac{\partial}{\partial x} + \frac{y}{2\xi}\frac{\partial}{\partial y} + \frac{1}{2}\frac{\partial}{\partial z}\end{aligned}$$

(see (10.49) and (10.50)), in the coordinate representation (10.42) becomes

$$p_\xi = -i\hbar\left(\frac{\partial}{\partial\xi} + \frac{1}{2(\xi+\eta)}\right). \tag{10.105}$$

Similarly, from (10.44)

$$p_\eta = -i\hbar\left(\frac{\partial}{\partial\eta} + \frac{1}{2(\xi+\eta)}\right). \tag{10.106}$$

Thus the differential forms of (10.63) and (10.78) are

$$T_\lambda^\pm(\xi) = \pm\xi\frac{\partial}{\partial\xi} - \frac{\xi}{2an} + \frac{1}{2}(n + \lambda \pm 1) \tag{10.107}$$

and

$$T_\lambda^\pm(\eta) = \pm\eta\frac{\partial}{\partial\eta} - \frac{\eta}{2an} + \frac{1}{2}(n - \lambda \pm 1). \tag{10.108}$$

We can also express the scaling operators $\Delta_n^\pm$ in differential form. However, for our purposes we require only the effect of these operators on wavefunctions (see Section 6.4)

$$\Delta_n^\pm\psi(\mathbf{r}) = \left(\frac{n}{n\pm 1}\right)^{3/2}\psi\left(\frac{n}{n\pm 1}\mathbf{r}\right). \tag{10.109}$$

According to (10.72), (10.76), (10.77), and (10.3), the ground state is annihilated by $Q_{10}^-(\xi)$, $Q_{10}^-(\eta)$, and L_z. Hence

$$\left(\xi\frac{\partial}{\partial\xi}+\frac{\xi}{2a}\right)\psi_{100}=0,\quad \left(\eta\frac{\partial}{\partial\eta}+\frac{\eta}{2a}\right)\psi_{100}=0,\quad \frac{\partial}{\partial\phi}\psi_{100}=0. \quad (10.110)$$

The normalized solution is*

$$\psi_{100}=\frac{1}{(\pi a^3)^{1/2}}\exp\left(-\frac{\xi+\eta}{2a}\right). \quad (10.111)$$

From (10.103), (10.107), (10.109), (10.65), and (10.111) we have

$$\psi_{n,n-1,0}(u)=\frac{1}{(\pi a^3)^{1/2}}\exp\left(-\frac{\xi+\eta}{2an}\right)\chi_n(u), \quad (10.112)$$

where

$$u=\xi/an, \quad (10.113)$$

$$\chi_n(u)=\left(\frac{n-1}{n}\right)^2\left(1-\frac{u}{n-1}+\frac{u}{n-1}\frac{\mathrm{d}}{\mathrm{d}u}\right)\chi_{n-1}(u)\quad (n=2,3,\ldots), \quad (10.114)$$

and $\chi_1=1$. Thus

$$\chi_n(u)=\frac{1}{n^2}\prod_{j=0}^{n-2}\left(1-\frac{u}{j+1}+\frac{u}{j+1}\frac{\mathrm{d}}{\mathrm{d}u}\right)1$$
$$=\frac{1}{n^2}\,{}_1F_1(-(n-1);1;u) \quad (10.115)$$

(see (B.1)) and

$$\psi_{n,n-1,0}=\frac{1}{(\pi a^3)^{1/2}n^2}\exp\left(-\frac{\xi+\eta}{2an}\right){}_1F_1\left(-(n-1);1;\frac{\xi}{an}\right). \quad (10.116)$$

Now apply L_+ to (10.116). The exponential factor in (10.116) is a function of r and therefore it commutes with L_+ (Chapter 4). The polynomial in (10.116) is a function of $r+z$. Because

$$\frac{\partial}{\partial x_i}f(r+z)=\left(\frac{x_i}{r}+\delta_{i3}\right)\frac{\mathrm{d}f(\xi)}{\mathrm{d}\xi},$$

use of (4.117) shows that

$$L_+f(r+z)=-\hbar(x+\mathrm{i}y)\frac{\mathrm{d}f(\xi)}{\mathrm{d}\xi}. \quad (10.117)$$

Using the obvious identity

* Because ψ_{100} is a function of r only, the normalization can be performed using spherical polar coordinates.

$$\frac{\mathrm{d}}{\mathrm{d}u}\,{}_1F_1(-N;B;u) = -\frac{N}{B}\,{}_1F_1(-(N-1);B+1;u) \tag{10.118}$$

and the fact that L_+ commutes with $x + \mathrm{i}y$, eqns (10.116), (10.117), (10.32), and (10.49) yield after μ applications of L_+

$$\psi_{n,n-1-\mu,\mu} = \frac{1}{(\pi a^3)^{1/2}n^2}\left(\frac{(n-1)!}{(n-1-\mu)!(\mu!)^3}\right)^{1/2}\exp\left(-\frac{\xi+\eta}{2an}\right)\left(\frac{\xi\eta}{a^2n^2}\right)^{\mu/2}\exp(\mathrm{i}\mu\phi)$$
$$\times\,{}_1F_1\left(-(n-1-\mu);\mu+1;\frac{\xi}{an}\right), \tag{10.119}$$

where $\mu \geq 0$.

In (10.119) λ has its maximum value. To decrease λ we apply the lowering operator D_λ^- defined by (10.79), (10.107), and (10.108). This produces eigenfunctions $\psi_{n\lambda\mu}$ with λ given by (10.27). For these values of λ

$$D_\lambda^- \exp\left(-\frac{\xi+\eta}{2an}\right)(\xi\eta)^{\mu/2}f(\eta)g(\xi)$$
$$= \exp\left(-\frac{\xi+\eta}{2an}\right)(\xi\eta)^{\mu/2}\left(\eta\frac{\mathrm{d}}{\mathrm{d}\eta} - \frac{\eta}{an} + \mu + 1 + N\right)f(\eta)$$
$$\times\left(-\xi\frac{\mathrm{d}}{\mathrm{d}\xi} + n - 1 - \mu - N\right)g(\xi), \tag{10.120}$$

and the coefficient (10.82) is

$$\gamma_\lambda^- = -[(n-1-N)(n-1-\mu-N)(\mu+1+N)(1+N)]^{1/2}. \tag{10.121}$$

From (10.81) and (10.119)–(10.121) we find after N applications of the lowering operators

$$\psi_{n\lambda\mu} = \frac{(-1)^N}{(\pi a^3)^{1/2}n^2\mu!}\exp\left(-\frac{\xi+\eta}{2an}\right)\left(\frac{\xi\eta}{a^2n^2}\right)^{\mu/2}\exp(\mathrm{i}\mu\phi)$$
$$\times\prod_{j=0}^{N-1}\left(\frac{(n-1)!}{n-1-j}\,\frac{n-1-\mu-j}{(n-1-\mu)!}\,\frac{\mu+1+j}{\mu!}\,\frac{1}{1+j}\right)^{1/2}$$
$$\times\prod_{j=0}^{N-1}\left(1 - \frac{\eta}{(\mu+1+j)an} + \frac{\eta}{\mu+1+j}\frac{\mathrm{d}}{\mathrm{d}\eta}\right)1$$
$$\times\prod_{j=0}^{N-1}\left(1 - \frac{\xi}{n-1-\mu-j}\frac{\mathrm{d}}{\mathrm{d}\xi}\right)$$
$$\times\,{}_1F_1\left(-(n-1-\mu);\mu+1;\frac{\xi}{an}\right), \tag{10.122}$$

where $\mu \geq 0$ and λ is given by (10.27).

By inspection, the first product in (10.122) is equal to

$$\left(\frac{(n-1-N)!}{(n-1-\mu-N)!}\frac{(\mu+N)!}{(\mu!)^2}\frac{1}{N!}\right)^{1/2}.$$

Equation (B.1) shows that the second product in (10.122) is equal to ${}_1F_1(-N;\mu+1;\eta/an)$. The identity*

$$\left(1-\frac{u}{n}\frac{\mathrm{d}}{\mathrm{d}u}\right){}_1F_1(-n;b;u)={}_1F_1(-(n-1);b;u)$$

shows that the third product in (10.122) equals ${}_1F_1(-(n-1-\mu-N);\mu+1;\xi/an)$. Using these results and (10.27) in (10.122) we obtain the coordinate-space wavefunctions for $\mu\geq 0$

$$\begin{aligned}\psi_{n\lambda\mu}&=\frac{(-1)^{(1/2)(n-1-\mu-\lambda)}}{(\pi a^3)^{1/2}n^2(\mu!)^2}\left(\frac{[(n-1+\mu+\lambda)/2]![(n-1+\mu-\lambda)/2]!}{[(n-1-\mu+\lambda)/2]![(n-1-\mu-\lambda)/2]!}\right)^{1/2}\\&\quad\times\exp\left(-\frac{\xi+\eta}{2an}\right)\left(\frac{\xi\eta}{a^2n^2}\right)^{\mu/2}\exp(\mathrm{i}\mu\phi)\\&\quad\times{}_1F_1\left(-\frac{1}{2}(n-1-\mu-\lambda);\mu+1;\frac{\eta}{an}\right)\\&\quad\times{}_1F_1\left(-\frac{1}{2}(n-1-\mu+\lambda);\mu+1;\frac{\xi}{an}\right).\end{aligned}\tag{10.123}$$

If $\mu<0$ we must use L_- instead of L_+ in the above calculation. Then (10.117) is replaced by

$$L_-f(r+z)=\hbar(x-\mathrm{i}y)\frac{\mathrm{d}f(\xi)}{\mathrm{d}\xi},$$

and we find

$$\psi_{n,\lambda,-|\mu|}=(-1)^{\mu}\psi^*_{n\lambda|\mu|}.\tag{10.124}$$

The parabolic wavefunctions (10.123) and the spherical wavefunctions (6.98) are linear functions of each other [4]

$$\psi_{nlm}(\mathbf{r})=\sum_{\lambda=-(n-1)+|\mu|}^{n-1-|\mu|}c_\lambda(nlm)\psi_{n\lambda\mu}(\mathbf{r}),\tag{10.125}$$

where $m=\mu$. A discussion of the relative phase factors of the two sets of wavefunctions is given in Ref. 5.

In parabolic coordinates the volume element is [2]

$$\mathrm{d}^3\mathbf{r}=\tfrac{1}{4}(\xi+\eta)\,\mathrm{d}\xi\,\mathrm{d}\eta\,\mathrm{d}\phi.\tag{10.126}$$

---

* This result can be proved in a similar manner to (B.1).

Here $0 \le \xi \le \infty, 0 \le \eta \le \infty$, and $0 \le \phi \le 2\pi$. Thus the condition for (10.105) to be Hermitian with respect to $\psi_{n\lambda\mu}$ is

$$\psi_{n\lambda\mu} \to 0 \quad \text{as} \quad \xi \to 0. \tag{10.127}$$

If $\mu = 0$ the wavefunctions (10.123) do not satisfy (10.127) and hence p_ξ is not Hermitian with respect to $\psi_{n\lambda\mu}$; similarly p_η is not Hermitian with respect to these functions. Operators p_ξ and p_η that are Hermitian with respect to $\psi_{n\lambda\mu}$ can be obtained by modifying (10.42) and (10.44) [6].

References

1. Biedenharn, L.C. and Louck, J.D. (1981). *Angular momentum in quantum physics*, Section 7.4. Addison-Wesley, Reading, MA.
2. See, for example, Pauling, L. and Wilson E. (1935). *Introduction to quantum mechanics*, Appendix IV. McGraw-Hill, New York.
3. See, for example, Bethe, H.A. and Salpeter, E.E. (1957). *Quantum mechanics of one- and two-electron atoms*. In *Encyclopedia of Physics*, Vol. 35, (ed. Flügge, S.). Springer, Berlin.
4. Parks, D. (1960). Relation between the parabolic and spherical eigenfunctions of hydrogen. *Zeitschrift für Physik*, **159**, 155–7.
5. Englefield, M.J. (1972). *Group theory and the Coulomb problem*, Ch. 5. Wiley, New York.
6. Robinson, P.D. and Hirschfelder, J.O. (1963). Generalized momentum operators in quantum mechanics. *Journal of Mathematical Physics*, **4**, 338–47.

11

RELATIVISTIC QUANTUM MECHANICS

In this chapter we give a brief introduction to the relativistic equations for spin 0 and spin $\frac{1}{2}$ particles (the Klein–Gordon and Dirac equations), and we solve these equations for two simple problems. Throughout we consider only single-particle theory: thus we do not discuss 'second quantization' of the relativistic equations (i.e. the use of quantized fields that create and annihilate particles and antiparticles).

11.1 Relativistic equations

According to the special theory of relativity the classical Hamiltonian of a free particle is [1]

$$H = (c^2\mathbf{p}^2 + m_0^2c^4)^{1/2}, \tag{11.1}$$

where m_0 is the rest mass and

$$\mathbf{p} = (1 - v^2/c^2)^{-1/2}m_0\mathbf{v} \tag{11.2}$$

is the momentum of the particle. Because of the square root in (11.1) the transition to a quantum-mechanical Hamiltonian is not unique. We consider three possibilities. It is convenient to restrict the discussion initially to the coordinate representation.

If we use (11.1) in (1.122) we have

$$(-\hbar^2c^2\nabla^2 + m_0^2c^4)^{1/2}\Psi(\mathbf{r}, t) = i\hbar\frac{\partial}{\partial t}\Psi(\mathbf{r}, t). \tag{11.3}$$

Although this equation has some useful applications [2], it also has limitations: for example, it is unsymmetrical in the space and time coordinates, and it is difficult to use in the case of magnetic interactions. We do not consider (11.3) further here.

A different procedure that leads to a more symmetrical equation starts by squaring both sides of (11.1). One then allows the operators corresponding to this squared equation to operate on $\Psi(\mathbf{r}, t)$. Replacing H with $i\hbar\partial/\partial t$, this yields

$$\left(-\hbar^2c^2\nabla^2 + \hbar^2\frac{\partial^2}{\partial t^2} + m_0^2c^4\right)\Psi(\mathbf{r}, t) = 0. \tag{11.4}$$

If Ψ is a scalar (the same in all inertial frames), (11.4) is the Klein–Gordon equation for a free particle [3]. The fact that (11.4) is second order in time requires some care in the interpretation of the probability density of the particle [4].

With the notation

$$p_\mu = -i\hbar\partial/\partial x_\mu \qquad (\mu = 1, 2, 3, 4) \tag{11.5}$$

and

$$x_4 = ict, \tag{11.6}$$

(11.4) can be written

$$(p_\mu^2 + m_0^2c^2)\Psi(x_\nu) = 0. \tag{11.7}$$

(Here, and in what follows, a repeated Greek index implies summation from one to four.)

For a particle of charge q in an electromagnetic field with vector and scalar potentials $\mathbf{A}(\mathbf{r}, t)$ and $\Phi(\mathbf{r}, t)$, one must replace p_μ in (11.7) with the gauge-covariant derivative [5]

$$D_\mu = p_\mu - qA_\mu, \tag{11.8}$$

where A_μ is the four-vector potential

$$A_\mu = \left(\mathbf{A}, \frac{i}{c}\Phi\right). \tag{11.9}$$

Thus the Klein–Gordon equation for a charged particle in an electromagnetic field is

$$(D_\mu^2 + m_0^2c^2)\Psi(x_\nu) = 0. \tag{11.10}$$

An alternative relativistic equation was obtained by Dirac [6] who used the ingenious device of expressing the sum of four squares as a perfect square:

$$c^2\mathbf{p}^2 + m_0^2c^4 = (c\boldsymbol{\alpha}.\mathbf{p} + \beta m_0c^2)^2. \tag{11.11}$$

Here $\mathbf{p}$ is the momentum operator and $\boldsymbol{\alpha}$ and β are operators which are independent of $\mathbf{p}$ and m_0, and which commute with $\mathbf{p}$ and $\mathbf{r}$. Thus the right-hand side of (11.11) can be written

$$\tfrac{1}{2}c^2(\alpha_j\alpha_k + \alpha_k\alpha_j)p_jp_k + m_0c^3(\alpha_j\beta + \beta\alpha_j)p_j + m_0^2c^4\beta^2.$$

This is equal to the left-hand side of (11.11) if

$$\alpha_j\alpha_k + \alpha_k\alpha_j = 2\delta_{jk}, \tag{11.12}$$

$$\alpha_j\beta + \beta\alpha_j = 0, \tag{11.13}$$

$$\beta^2 = 1. \tag{11.14}$$

Hence the operators α_j and β must anticommute and have unit squares.

From (11.1) and the identity (11.11) we have the quantum-mechanical Hamiltonian

$$H = c\boldsymbol{\alpha}.\mathbf{p} + \beta m_0c^2. \tag{11.15}$$

If this Hamiltonian, with $\mathbf{p} = -i\hbar\nabla$, is used in (1.122) we obtain the Dirac equation for a free particle

$$(-i\hbar c\boldsymbol{\alpha}.\nabla + \beta m_0 c^2)\Psi(\mathbf{r}, t) = i\hbar\frac{\partial}{\partial t}\Psi(\mathbf{r}, t). \tag{11.16}$$

This equation can be written in a more symmetrical form by multiplying both sides on the left by β. Then

$$(i\gamma_\mu p_\mu + m_0 c)\Psi(x_\nu) = 0, \tag{11.17}$$

where p_μ is given by (11.5) and

$$\boldsymbol{\gamma} = -i\beta\boldsymbol{\alpha}, \qquad \gamma_4 = \beta. \tag{11.18}$$

One can readily show that the operators γ_μ must also anticommute and have unit squares

$$\gamma_\mu\gamma_\nu + \gamma_\nu\gamma_\mu = 2\delta_{\mu\nu}. \tag{11.19}$$

For a charged particle in an electromagnetic field the Dirac Hamiltonian (11.15) is

$$H = c\boldsymbol{\alpha}.\mathbf{D} + \beta m_0 c^2, \tag{11.20}$$

and the Dirac equation (11.17) is

$$(i\gamma_\mu D_\mu + m_0 c)\Psi(x_\nu) = 0, \tag{11.21}$$

where D_μ is given by (11.8).

The operators α_i, β, and γ_μ can be represented by matrices whose elements are real or complex numbers. Because H is Hermitian it is clear that these matrices must be Hermitian (see (11.15)) and therefore square, of order N say. Taking the determinant of both sides of (11.12) and (11.13) we have

$$|\alpha_j||\alpha_k| = (-1)^N|\alpha_k||\alpha_j| \qquad (j \neq k)$$

and

$$|\alpha_j||\beta| = (-1)^N|\beta||\alpha_j|.$$

Because $|\alpha_j| \neq 0$ and $|\beta| \neq 0$ (see (11.12) and (11.14)) it follows that N must be even.

Now if α_j and β satisfy (11.12)–(11.14), so do $U^\dagger\alpha_j U$ and $U^\dagger\beta U$, where U is any unitary $N \times N$ matrix. Hence one of the matrices α_j and β can be diagonalized: this is usually taken to be β. From (11.14) the eigenvalues of β are ± 1. Thus we take

$$\beta = \begin{pmatrix} 1 & 0 \\ 0 & -1 \end{pmatrix}. \tag{11.22}$$

Let

$$\alpha_j = \begin{pmatrix} A_j & C_j \\ D_j & B_j \end{pmatrix}. \tag{11.23}$$

The elements in (11.22) and (11.23) are matrices of order $\frac{1}{2}N$. Substituting (11.22) and (11.23) in (11.12) and (11.13) we find $A_j = B_j = 0$ and

$$C_j D_k + C_k D_j = D_j C_k + D_k C_j = 2\delta_{jk}. \tag{11.24}$$

For $j = k$ it follows from (11.24) that $D_j = C_j^{-1}$. However, (11.24) cannot be satisfied for $j \neq k$ if C_j and D_j are numbers, real or complex; that is, if $N = 2$. If $N = 4$, (11.24) is satisfied by

$$C_j = a\sigma_j \tag{11.25}$$

and

$$D_j = b\sigma_j, \tag{11.26}$$

where σ_j are the Pauli matrices and the numbers a and b satisfy

$$ab = 1. \tag{11.27}$$

(See (4.190) and (4.205).) Thus $N = 4$ is the smallest value for which a matrix representation is possible. (We are assuming $m_0 \neq 0$.)

A convenient choice is $a = 1$. Then

$$\boldsymbol{\alpha} = \begin{pmatrix} 0 & \boldsymbol{\sigma} \\ \boldsymbol{\sigma} & 0 \end{pmatrix}, \tag{11.28}$$

and (11.18) yields

$$\boldsymbol{\gamma} = \begin{pmatrix} 0 & -\mathrm{i}\boldsymbol{\sigma} \\ \mathrm{i}\boldsymbol{\sigma} & 0 \end{pmatrix}. \tag{11.29}$$

The matrix representations (11.22), (11.28), and (11.29) are often referred to as the Pauli representation. In the following we adopt this representation [7]. Note that eqns (11.22), (11.28), and (11.29) are written in block notation: the entries are 2×2 matrices. We will frequently employ this convenient notation. Thus, for example, (11.15) can be expressed as

$$H = c\begin{pmatrix} m_0 c & \boldsymbol{\sigma}.\mathbf{p} \\ \boldsymbol{\sigma}.\mathbf{p} & -m_0 c \end{pmatrix}. \tag{11.30}$$

The Dirac wavefunction is the column vector (bi-spinor)

$$\Psi = \begin{bmatrix} \Psi_1 \\ \Psi_2 \\ \Psi_3 \\ \Psi_4 \end{bmatrix} \tag{11.31}$$

and the Dirac equation consists of four coupled partial differential equations (see (11.40)).

In a second inertial frame the Dirac equation (11.16) is [8]

$$(-i\hbar c\boldsymbol{\alpha}.\boldsymbol{\nabla}' + \beta m_0 c^2)\Psi'(\mathbf{r}', t') = i\hbar\frac{\partial}{\partial t'}\Psi'(\mathbf{r}', t') \tag{11.32}$$

where

$$\Psi'(\mathbf{r}', t') = \exp(-\tfrac{1}{2}\boldsymbol{\alpha}.\hat{\mathbf{v}}\tanh^{-1}v/c)\Psi(\mathbf{r}, t). \tag{11.33}$$

Here $\mathbf{v}$ is the velocity of the second frame relative to the first and $\mathbf{r}'$, t' and $\mathbf{r}$, t are related by a Lorentz transformation.

The use of the coordinate representation in the above has the advantage that it exhibits the symmetry between space and time coordinates (see (11.7) and (11.17)). One can also write the Klein–Gordon and Dirac equations in abstract form by treating time as a parameter, and the space and momentum coordinates as operators that together with $\boldsymbol{\alpha}$ and β are the fundamental dynamical variables of the system [8]. In this approach the Klein–Gordon equation (11.10) is

$$(D_\mu^2 + m_0^2c^2)|\psi\rangle = 0, \tag{11.34}$$

and the Dirac equation (11.21) is

$$(i\gamma_\mu D_\mu + m_0 c)|\psi\rangle = 0, \tag{11.35}$$

where D_μ is given by (11.8) with

$$p_\mu = (\mathbf{p}, -i\hbar\partial/\partial x_4). \tag{11.36}$$

In the applications that follow we will use the Klein–Gordon and Dirac equations in these abstract forms.

The Dirac ket and bra are, respectively, the column and row vectors

$$|\psi\rangle = \begin{bmatrix} |\psi_1\rangle \\ |\psi_2\rangle \\ |\psi_3\rangle \\ |\psi_4\rangle \end{bmatrix}, \qquad \langle\psi| = (\langle\psi_1|, \langle\psi_2|, \langle\psi_3|, \langle\psi_4|). \tag{11.37}$$

The normalization

$$\langle\psi|\psi\rangle = 1 \tag{11.38}$$

requires

$$\sum_{\lambda=1}^{4} \langle\psi_\lambda|\psi_\lambda\rangle = 1. \tag{11.39}$$

Equation (11.35) consists of the four coupled equations

$$\begin{aligned}
m_0c^2|\psi_1\rangle + cD_3|\psi_3\rangle + c(D_1 - \mathrm{i}D_2)|\psi_4\rangle &= \left(\mathrm{i}\hbar\frac{\partial}{\partial t} - q\Phi\right)|\psi_1\rangle, \\
m_0c^2|\psi_2\rangle + c(D_1 + \mathrm{i}D_2)|\psi_3\rangle - cD_3|\psi_4\rangle &= \left(\mathrm{i}\hbar\frac{\partial}{\partial t} - q\Phi\right)|\psi_2\rangle, \\
-m_0c^2|\psi_3\rangle + cD_3|\psi_1\rangle + c(D_1 - \mathrm{i}D_2)|\psi_2\rangle &= \left(\mathrm{i}\hbar\frac{\partial}{\partial t} - q\Phi\right)|\psi_3\rangle, \\
-m_0c^2|\psi_4\rangle + c(D_1 + \mathrm{i}D_2)|\psi_1\rangle - cD_3|\psi_2\rangle &= \left(\mathrm{i}\hbar\frac{\partial}{\partial t} - q\Phi\right)|\psi_4\rangle.
\end{aligned} \tag{11.40}$$

If A_μ is time independent and if $|\psi\rangle$ is an energy eigenket (Section 1.5), (11.34) and (11.35) can be written

$$c^2(\mathbf{D}^2 + m_0^2c^2)|\psi\rangle = (E - q\Phi)^2|\psi\rangle \tag{11.41}$$

and

$$c(\boldsymbol{\alpha}.\mathbf{D} + \beta m_0 c)|\psi\rangle = (E - q\Phi)|\psi\rangle, \tag{11.42}$$

where E is the energy eigenvalue.

One can readily express any of the abstract equations in the coordinate representation: for example, (11.34) and (11.35) become (11.10) and (11.21) with $\Psi(\mathbf{r}, t) = \langle\mathbf{r}|\psi\rangle$. The elements of Ψ are $\Psi_\lambda(\mathbf{r}, t) = \langle\mathbf{r}|\psi_\lambda\rangle$ and the normalization condition (11.39) is

$$\sum_{\lambda=1}^{4}\int \Psi_\lambda^*\Psi_\lambda\, \mathrm{d}^3\mathbf{r} = 1. \tag{11.43}$$

Similarly, (11.40) reduce to a set of four coupled partial differential equations for Ψ_λ.

One can also use the momentum representation. Consider, for example, a Dirac particle in a spherically symmetric electrostatic field. Then (11.42) is

$$(c\boldsymbol{\alpha}.\mathbf{p} + \beta m_0c^2)|\psi\rangle = [E - q\Phi(r)]|\psi\rangle. \tag{11.44}$$

If we multiply (11.44) on the left by $\langle\mathbf{p}'|$ and use (1.71), (1.93), and(1.96) we find

$$(c\boldsymbol{\alpha}.\mathbf{p} + \beta m_0c^2 - E)\bar{\psi}(\mathbf{p}) = -q(2\pi\hbar)^{-3/2}\int \mathrm{d}^3\mathbf{p}'U(\mathbf{p}' - \mathbf{p})\bar{\psi}(\mathbf{p}'), \tag{11.45}$$

where

$$\bar{\psi}(\mathbf{p}') = \langle\mathbf{p}'|\psi\rangle = (2\pi\hbar)^{-3/2}\int \mathrm{d}^3\mathbf{r}\, \exp(-\mathrm{i}\mathbf{p}'.\mathbf{r}/\hbar)\psi(\mathbf{r}) \tag{11.46}$$

and

$$U(\mathbf{p}' - \mathbf{p}) = (2\pi\hbar)^{-3/2}\int \mathrm{d}^3\mathbf{r}\, \exp[\mathrm{i}(\mathbf{p}' - \mathbf{p}).\mathbf{r}/\hbar]\Phi(r). \tag{11.47}$$

Equation (11.45) is an integral equation for the momentum-space wavefunction $\bar{\psi}(\mathbf{p})$ [9].

11.2 Spin angular momentum

The Dirac equation has the remarkable property that it describes a particle with intrinsic, or spin, angular momentum. To see this one can evaluate the commutator of the Hamiltonian (11.15) and the orbital angular momentum operator

$$[H, L_j] = [c\alpha_k p_k, L_j] = c\alpha_k [p_k, L_j] = -i\hbar c\alpha_k \varepsilon_{jkl} p_l,$$

where we have used (C.11). Thus

$$[H, \mathbf{L}] = -i\hbar c\boldsymbol{\alpha} \times \mathbf{p}, \tag{11.48}$$

and $\mathbf{L}$ does not commute with the Hamiltonian of a free Dirac particle. To find an angular momentum operator that does commute with H, consider the matrices

$$\boldsymbol{\Sigma} = \begin{pmatrix} \boldsymbol{\sigma} & 0 \\ 0 & \boldsymbol{\sigma} \end{pmatrix}. \tag{11.49}$$

Using the block form (11.30) we have

$$[H, \boldsymbol{\Sigma}] = c\begin{pmatrix} 0 & [\boldsymbol{\sigma}.\mathbf{p}, \boldsymbol{\sigma}] \\ [\boldsymbol{\sigma}.\mathbf{p}, \boldsymbol{\sigma}] & 0 \end{pmatrix}.$$

Now

$$[\boldsymbol{\sigma}.\mathbf{p}, \sigma_j] = [\sigma_k p_k, \sigma_j] = [\sigma_k, \sigma_j] p_k = -2i\varepsilon_{jkl} p_k \sigma_l,$$

where we have used (4.204). Hence

$$[H, \boldsymbol{\Sigma}] = 2ic\boldsymbol{\alpha} \times \mathbf{p}. \tag{11.50}$$

Therefore if we define

$$\mathbf{S} = \tfrac{1}{2}\hbar\boldsymbol{\Sigma} \tag{11.51}$$

and

$$\mathbf{J} = \mathbf{L} + \mathbf{S}, \tag{11.52}$$

it follows from (11.48) and (11.50) that

$$[H, \mathbf{J}] = 0. \tag{11.53}$$

One can readily show that $\mathbf{S}$ and $\mathbf{J}$ satisfy the commutation relations for angular momentum operators

$$[S_i, S_j] = i\hbar\varepsilon_{ijk} S_k, \qquad [J_i, J_j] = i\hbar\varepsilon_{ijk} J_k. \tag{11.54}$$

The operators **S** and **J**, respectively, are referred to as the spin and total angular momentum operators [10]. Now

$$\mathbf{S}^2 = \frac{1}{4}\hbar^2\begin{pmatrix} \boldsymbol{\sigma}^2 & 0 \\ 0 & \boldsymbol{\sigma}^2 \end{pmatrix} = \frac{1}{2}\left(\frac{1}{2}+1\right)\hbar^2\begin{pmatrix} 1 & 0 \\ 0 & 1 \end{pmatrix}. \tag{11.55}$$

(See (4.205).) Hence the quantum number s in

$$\mathbf{S}^2|\psi\rangle = \hbar^2 s(s+1)|\psi\rangle \tag{11.56}$$

is equal to a half. Also

$$S_z^2 = \tfrac{1}{4}\hbar^2\mathbf{1}, \tag{11.57}$$

where **1** is the unit 4×4 matrix. Therefore the possible eigenvalues of S_z are $\pm\frac{1}{2}\hbar$. The Dirac equation describes particles with spin angular momentum equal to $\frac{1}{2}\hbar$. Note that in general an energy eigenket will not be an eigenket of L_z or S_z because these operators do not commute with H.

11.3 The free particle

For a free Klein–Gordon particle the energy and momentum eigenstates are given in the coordinate representation by the plane wave solutions to (11.4), namely,

$$\Psi(\mathbf{r}, t) = N(\mathbf{P})\exp[\mathrm{i}(\mathbf{P}.\mathbf{r} - Et)/\hbar]. \tag{11.58}$$

Here N is a normalization constant, **P** and E are the momentum and energy eigenvalues, and

$$E = \pm(\mathbf{P}^2c^2 + m_0^2c^4)^{1/2}. \tag{11.59}$$

The solutions (11.58) with negative energy are associated with the antiparticle states [4].

Similarly, for a free Dirac particle the eigenstates of energy and momentum are given by the plane wave solutions to (11.16)

$$\Psi(\mathbf{r}, t) = u(\mathbf{P})\exp[\mathrm{i}(\mathbf{P}.\mathbf{r} - Et)/\hbar]. \tag{11.60}$$

The column vector

$$u = \begin{bmatrix} u_1 \\ u_2 \\ u_3 \\ u_4 \end{bmatrix} \tag{11.61}$$

must satisfy

$$(c\boldsymbol{\alpha}.\mathbf{P} + \beta m_0c^2 - E)u = 0. \tag{11.62}$$

Equation (11.62) consists of the four linear equations

$$\begin{array}{llllllll}
(m_0c^2 - E)u_1 & & & + & cP_zu_3 & + & cP_-u_4 & = 0, \\
& & (m_0c^2 - E)u_2 & + & cP_+u_3 & - & cP_zu_4 & = 0, \\
cP_zu_1 & + & cP_-u_2 & & -(m_0c^2 + E)u_3 & & & = 0, \\
cP_+u_1 & - & cP_zu_2 & & & & -(m_0c^2 + E)u_4 & = 0,
\end{array} \tag{11.63}$$

where $P_\pm = P_x \pm iP_y$. Equations (11.63) have non-zero solutions for the u_i only if the determinant of the coefficients vanishes. This determinant is equal to $(E^2 - c^2\mathbf{P}^2 - m_0^2c^4)^2$, and hence the energy and momentum of a free Dirac particle also satisfy (11.59).

However, (11.63) cannot be solved for three of the u in terms of the fourth u. The reason is that H and $\mathbf{p}$ are not a c.s.c.o. (see below). From (11.63) one can only express u_3 and u_4 (say) in terms of u_1 and u_2. Corresponding to $u_2 = 0$ and $u_1 = 0$, respectively, we have the linearly independent solutions

$$u^{(1)}(\mathbf{P}) = N(\mathbf{P})\begin{bmatrix} E + m_0c^2 \\ 0 \\ cP_z \\ cP_+ \end{bmatrix}, \quad u^{(2)}(\mathbf{P}) = N(\mathbf{P})\begin{bmatrix} 0 \\ E + m_0c^2 \\ cP_- \\ -cP_z \end{bmatrix}, \tag{11.64}$$

where $N(\mathbf{P})$ is a normalization constant. These are solutions for $E > 0$: for $E < 0$ the solutions (11.64) vanish in the rest frame of the particle.

Similarly, corresponding to $u_4 = 0$ and $u_3 = 0$, respectively, we have the solutions for $E < 0$

$$u^{(3)}(\mathbf{P}) = N(\mathbf{P})\begin{bmatrix} -cP_z \\ -cP_+ \\ m_0c^2 - E \\ 0 \end{bmatrix}, \quad u^{(4)}(\mathbf{P}) = N(\mathbf{P})\begin{bmatrix} -cP_- \\ cP_z \\ 0 \\ m_0c^2 - E \end{bmatrix}. \tag{11.65}$$

As before, the negative energy states are associated with antiparticles [4].

The solutions (11.64) and (11.65) can also be obtained by starting with the Dirac equation in the rest frame

$$\beta m_0c^2\Psi = E\Psi, \tag{11.66}$$

where

$$\Psi = u\exp(-iEt/\hbar) \tag{11.67}$$

and $E = \pm m_0c^2$. For $E = m_0c^2$ the solutions to (11.66) have

$$u^{(1)}(0) = \begin{bmatrix} 1 \\ 0 \\ 0 \\ 0 \end{bmatrix}, \quad u^{(2)}(0) = \begin{bmatrix} 0 \\ 1 \\ 0 \\ 0 \end{bmatrix}. \tag{11.68}$$

Similarly, for $E = -m_0c^2$

$$u^{(3)}(0) = \begin{bmatrix} 0 \\ 0 \\ 1 \\ 0 \end{bmatrix}, \qquad u^{(4)}(0) = \begin{bmatrix} 0 \\ 0 \\ 0 \\ 1 \end{bmatrix}. \tag{11.69}$$

By applying the transformation (11.33) with

$$\mathbf{v} = -c^2\mathbf{P}/(c^2\mathbf{P}^2 + m_0^2c^4)^{1/2}$$

to (11.68) and (11.69), one obtains (11.64) and (11.65) [4].

It is clear from (11.49), (11.64), and (11.65) that the eigenstates

$$\Psi^{(\lambda)} = u^{(\lambda)}(\mathbf{P}) \exp[\mathrm{i}(\mathbf{P}\cdot\mathbf{r} - Et)/\hbar] \tag{11.70}$$

are not, in general, eigenstates of S_z. However, if the motion is along the z axis ($P_x = P_y = 0$)

$$S_z\Psi^{(\lambda)} = (-1)^{\lambda+1}\tfrac{1}{2}\hbar\Psi^{(\lambda)}. \tag{11.71}$$

Any linear combination of the positive energy solutions

$$\Psi = a\Psi^{(1)} + b\Psi^{(2)} \tag{11.72}$$

is an eigenstate of H and $\mathbf{p}$ with energy eigenvalue $E > 0$. Similarly for the negative energy solutions. Hence H and $\mathbf{p}$ are not a c.s.c.o. An additional operator that commutes with H is

$$I = c\mathbf{\Sigma}\cdot\mathbf{p}. \tag{11.73}$$

From (11.13)–(11.15)

$$I^2 = H^2 - m_0^2c^4, \tag{11.74}$$

so that the eigenvalues of I are

$$F = \pm cP, \tag{11.75}$$

where

$$P = (P_x^2 + P_y^2 + P_z^2)^{1/2}. \tag{11.76}$$

The eigenvalue equations

$$I\Psi^{(\lambda)} = F\Psi^{(\lambda)} \tag{11.77}$$

require

$$\begin{aligned} (P_z - P)u_1 + P_-u_2 = 0, \qquad P_+u_1 - (P_z + P)u_2 = 0, \\ (P_z - P)u_3 + P_-u_4 = 0, \qquad P_+u_3 - (P_z + P)u_4 = 0. \end{aligned} \tag{11.78}$$

Equations (11.63) and (11.78) together give u_2, u_3, and u_4 in terms of u_1. Substituting these in (11.61) we have

$$u(\mathbf{P}) = N(\mathbf{P}) \begin{pmatrix} (E + m_0c^2)(P + P_z) \\ (E + m_0c^2)P_+ \\ cP(P + P_z) \\ cPP_+ \end{pmatrix}. \tag{11.79}$$

With the normalization $u^\dagger u = 1$,

$$N(\mathbf{P}) = \tfrac{1}{2}[E(E + m_0c^2)P(P + P_z)]^{-1/2}. \tag{11.80}$$

One can also consider eigenstates of H, $\mathbf{J}^2$, and J_z. In the coordinate representation these eigenstates for a free particle are represented by spherical waves [8]. This completes our introduction to the Klein–Gordon and Dirac equations.

11.4 The Feynman–Gell-Mann reduction [11]

As we have seen, the Dirac equation consists of four coupled operator equations. The Feynman–Gell-Mann reduction applies to the Dirac equation for a charged particle in an external electromagnetic field; it enables one to work with a two-component (spinor) equation rather than with the four-component (bi-spinor) Dirac equation and it therefore simplifies the algebraic task of uncoupling the Dirac equation.

In (11.35) substitute

$$|\psi\rangle = c(-\mathrm{i}\gamma_\nu D_\nu + m_0c)\begin{pmatrix} |\phi\rangle \\ -|\phi\rangle \end{pmatrix} \tag{11.81}$$

and multiply out. Then

$$(\gamma_\mu\gamma_\nu D_\mu D_\nu + m_0^2c^2)\begin{pmatrix} |\phi\rangle \\ -|\phi\rangle \end{pmatrix} = 0. \tag{11.82}$$

Now

$$\gamma_\mu\gamma_\nu D_\mu D_\nu = \begin{pmatrix} (p_\mu - qA_\mu)^2 - q\hbar\boldsymbol{\sigma}.\mathbf{B} & \mathrm{i}c^{-1}q\hbar\boldsymbol{\sigma}.\mathbf{E} \\ \mathrm{i}c^{-1}q\hbar\boldsymbol{\sigma}.\mathbf{E} & (p_\mu - qA_\mu)^2 - q\hbar\boldsymbol{\sigma}.\mathbf{B} \end{pmatrix}, \tag{11.83}$$

where $\mathbf{E}$ and $\mathbf{B}$ are the electric and magnetic fields (see Appendix K). From (11.82) and (11.83)

$$\left[c^{-2}\left(\mathrm{i}\hbar\frac{\partial}{\partial t} - q\Phi\right)^2 - (\mathbf{p} - q\mathbf{A})^2 - m_0^2c^2 + q\hbar(\mathbf{B} + \mathrm{i}c^{-1}\mathbf{E}).\boldsymbol{\sigma}\right]|\phi\rangle = 0. \tag{11.84}$$

In order to solve the Dirac equation (11.35) it is sufficient to solve the two-component equation (11.84). The kets of the Dirac equation are given in terms of $|\phi\rangle$ by (11.81); that is, by

$$|\psi\rangle = \begin{pmatrix} [c\boldsymbol{\sigma}.(\mathbf{p} - q\mathbf{A}) + i\hbar(\partial/\partial t) - q\Phi + m_0c^2]|\phi\rangle \\ [c\boldsymbol{\sigma}.(\mathbf{p} - q\mathbf{A}) + i\hbar(\partial/\partial t) - q\Phi - m_0c^2]|\phi\rangle \end{pmatrix}. \tag{11.85}$$

For energy eigenstates one can replace $i\hbar\partial/\partial t$ with the energy eigenvalue E in (11.84) and (11.85).

The Dirac equation can be solved for a charged particle (i) in a uniform magnetic field, (ii) in a uniform electric field, (iii) in the field of a plane electromagnetic wave, (iv) in the field of a plane electromagnetic wave propagating parallel to a uniform magnetic field, (v) in parallel, uniform electric and magnetic fields, (vi) in perpendicular, uniform electric and magnetic fields, (vii) in a Coulomb field, and (viii) for certain cases in which the electromagnetic potentials have a particular dependence on the space coordinates [12]. In the next section we use the Feynman–Gell-Mann reduction to solve the Dirac equation for a charged particle in a uniform magnetic field.

11.5 Charged particle in a uniform magnetic field

Suppose $\Phi = 0$ and consider the magnetic field (2.181). Then for energy eigenstates (11.84) is

$$(\mathbf{D}^2 - \hbar qB\sigma_3)|\phi\rangle = c^{-2}(E^2 - m_0^2c^4)|\phi\rangle. \tag{11.86}$$

Because D_z commutes with $\mathbf{D}^2$ (see (2.182)) we can suppose that $|\phi\rangle$ is also an eigenstate of D_z with eigenvalue P_z. In terms of the dimensionless operators (2.187), (11.86) becomes

$$\tfrac{1}{2}(\tilde{D}_x^2 + \tilde{D}_y^2)|\phi\rangle = (2\hbar|q|B)^{-1}(E^2c^{-2} - m_0^2c^2 - P_z^2 + \hbar qB\sigma_3)|\phi\rangle. \tag{11.87}$$

Consider the linearly independent spinors

$$|\phi_{+1}\rangle = \begin{pmatrix} |\phi\rangle_{\mathrm{U}} \\ 0 \end{pmatrix} \quad \text{and} \quad |\phi_{-1}\rangle = \begin{pmatrix} 0 \\ |\phi\rangle_{\mathrm{L}} \end{pmatrix}, \tag{11.88}$$

that are eigenkets of σ_3 with eigenvalues $\sigma = +1$ and $\sigma = -1$ respectively. (Here U and L denote upper and lower elements.) For these (11.87) is

$$\tfrac{1}{2}(\tilde{D}_x^2 + \tilde{D}_y^2)|\phi_\sigma\rangle = (2\hbar|q|B)^{-1}(E^2c^{-2} - m_0^2c^2 - P_z^2 + \hbar qB\sigma)|\phi_\sigma\rangle. \tag{11.89}$$

Note that with $\sigma = 0$, eqn (11.89) is the Klein–Gordon equation (11.41) for a charged particle in a uniform magnetic field $(0, 0, B)$. Equation (11.89) is formally identical to the one-dimensional oscillator problem and therefore, as in the calculations given in Section 2.8, we find

$$E^2 = m_0^2c^4 + c^2P_z^2 + [n + \tfrac{1}{2} - \tfrac{1}{2}(\operatorname{sign} q)\sigma]2\hbar|q|Bc^2 \qquad (n = 0, 1, 2, \ldots). \tag{11.90}$$

For the Klein–Gordon problem the energy eigenvalues are given by (11.90) with $\sigma = 0$, and the eigenkets are

$$|\psi\rangle = |n\gamma\rangle \exp(-\mathrm{i}Et/\hbar). \tag{11.91}$$

Here we adopt the notation of Section 2.8: thus $|n\gamma\rangle$ is a normalized eigenket, at time $t = 0$, of the c.s.c.o. H^2 and $\Gamma\{D_z, \ldots\}$, and γ denotes the eigenvalues $\{P_z, \ldots\}$. Possible choices of Γ are discussed in Section 2.8.

For the Dirac problem the spinors (11.88) that are energy eigenstates are:

(i) If $q > 0$,

$$|\phi_{+1}\rangle = \begin{pmatrix} |n\gamma\rangle \\ 0 \end{pmatrix} \exp(-\mathrm{i}Et/\hbar) \tag{11.92}$$

and

$$\begin{aligned} |\phi_{-1}\rangle &= \begin{pmatrix} 0 \\ |n-1, \gamma\rangle \end{pmatrix} \exp(-\mathrm{i}Et/\hbar) \qquad (n \neq 0), \\ &= 0 \qquad (n = 0). \end{aligned} \tag{11.93}$$

(ii) If $q < 0$,

$$\begin{aligned} |\phi_{+1}\rangle &= \begin{pmatrix} |n-1, \gamma\rangle \\ 0 \end{pmatrix} \exp(-\mathrm{i}Et/\hbar) \qquad (n \neq 0), \\ &= 0 \qquad (n = 0), \end{aligned} \tag{11.94}$$

and

$$|\phi_{-1}\rangle = \begin{pmatrix} 0 \\ |n\gamma\rangle \end{pmatrix} \exp(-\mathrm{i}Et/\hbar). \tag{11.95}$$

The spinors (11.92)–(11.95) are solutions to (11.89) with energy

$$E = \pm(m_0^2c^4 + c^2P_z^2 + 2\hbar|q|Bc^2n)^{1/2} \tag{11.96}$$

(see (11.90)).

To calculate the Dirac eigenkets, first express (11.85) as

$$|\psi\rangle = \begin{bmatrix} \begin{pmatrix} E + m_0c^2 + cP_z & (\hbar|q|Bc^2)^{1/2}(\tilde{D}_x - \mathrm{i}\tilde{D}_y) \\ (\hbar|q|Bc^2)^{1/2}(\tilde{D}_x + \mathrm{i}\tilde{D}_y) & E + m_0c^2 - cP_z \end{pmatrix} |\phi\rangle \\ \begin{pmatrix} E - m_0c^2 + cP_z & (\hbar|q|Bc^2)^{1/2}(\tilde{D}_x - \mathrm{i}\tilde{D}_y) \\ (\hbar|q|Bc^2)^{1/2}(\tilde{D}_x + \mathrm{i}\tilde{D}_y) & E - m_0c^2 - cP_z \end{pmatrix} |\phi\rangle \end{bmatrix}. \tag{11.97}$$

(i) Suppose $q > 0$. With $|\phi\rangle = |\phi_{+1}\rangle$ and using the shift operations (2.192) and (2.193), where $\tilde{D}_\pm = 2^{-1/2}(\tilde{D}_x \mp \mathrm{i}\tilde{D}_y)$, (11.97) yields

$$|\psi_+\rangle = \begin{bmatrix} (E + m_0c^2 + cP_z)|n\gamma\rangle \\ (2\hbar qBc^2n)^{1/2}|n-1, \gamma\rangle \\ (E - m_0c^2 + cP_z)|n\gamma\rangle \\ (2\hbar qBc^2n)^{1/2}|n-1, \gamma\rangle \end{bmatrix} \exp(-\mathrm{i}Et/\hbar). \tag{11.98}$$

With $|\phi\rangle = |\phi_{-1}\rangle$, (11.97) gives

$$|\psi_-\rangle = \begin{bmatrix} (2\hbar qBc^2n)^{1/2}|n\gamma\rangle \\ (E + m_0c^2 - cP_z)|n-1,\gamma\rangle \\ (2\hbar qBc^2n)^{1/2}|n\gamma\rangle \\ (E - m_0c^2 - cP_z)|n-1,\gamma\rangle \end{bmatrix} \exp(-\mathrm{i}Et/\hbar) \quad (n \neq 0), \qquad (11.99)$$

$$= 0 \qquad (n = 0).$$

(ii) If $q < 0$, $\tilde{D}_\pm = 2^{-1/2}(\tilde{D}_x \pm \mathrm{i}\tilde{D}_y)$ and similar calculations lead to

$$|\psi_+\rangle = \begin{bmatrix} (E + m_0c^2 + cP_z)|n-1,\gamma\rangle \\ (2\hbar|q|Bc^2n)^{1/2}|n,\gamma\rangle \\ (E - m_0c^2 + cP_z)|n-1,\gamma\rangle \\ (2\hbar|q|Bc^2n)^{1/2}|n,\gamma\rangle \end{bmatrix} \exp(-\mathrm{i}Et/\hbar) \quad (n \neq 0), \qquad (11.100)$$

$$= 0 \qquad (n = 0).$$

and

$$|\psi_-\rangle = \begin{bmatrix} (2\hbar|q|Bc^2n)^{1/2}|n-1,\gamma\rangle \\ (E + m_0c^2 - cP_z)|n\gamma\rangle \\ (2\hbar|q|Bc^2n)^{1/2}|n-1,\gamma\rangle \\ (E - m_0c^2 - cP_z)|n\gamma\rangle \end{bmatrix} \exp(-\mathrm{i}Et/\hbar). \qquad (11.101)$$

The above kets can be normalized by including a factor $\frac{1}{2}(E^2 + cP_zE)^{-1/2}$ on the right-hand side of (11.98) and (11.99), and a factor $\frac{1}{2}(E^2 - cP_zE)^{-1/2}$ in (11.100) and (11.101).

Next we consider c.s.c.o.s for the above Klein–Gordon and Dirac problems. For the Klein–Gordon problem H^2, D_z, and X (eqn (2.203)) or d^2 (eqn (2.208)) are a c.s.c.o. Thus, for example, with the choice of vector potential (2.215) the coordinate-space wavefunctions $\psi_{n\gamma}(\mathbf{r}) = \langle \mathbf{r}|n\gamma\rangle$ are given by (2.238) or (2.250). Similarly, with the choice of vector potential (2.213), the operators H^2, p_y, and p_z are a c.s.c.o. and (2.227) are the coordinate-space wavefunctions.

For the Dirac problem H (eqn (11.20)), D_z, and X or d^2 are a set of commuting operators. However, they are not a c.s.c.o. because any linear combination

$$|\psi\rangle = a|\psi_+\rangle + b|\psi_-\rangle \qquad (11.102)$$

is an eigenket of these operators. Similarly, with the vector potential (2.213), H, p_y, and p_z are a commuting set of operators. They are not a c.s.c.o. and kets of the form (11.102) are eigenkets of these operators.

To obtain a c.s.c.o. an additional operator is required that is independent of and commutes with one of the above sets of commuting operators. One such operator is a generalization of (11.73), namely [13],

$$I = c\mathbf{\Sigma}.\mathbf{D} = c\begin{pmatrix} \boldsymbol{\sigma}.\mathbf{D} & 0 \\ 0 & \boldsymbol{\sigma}.\mathbf{D} \end{pmatrix}. \tag{11.103}$$

Clearly I commutes with the Hamiltonian

$$H = c\begin{pmatrix} m_0 c & \boldsymbol{\sigma}.\mathbf{D} \\ \boldsymbol{\sigma}.\mathbf{D} & -m_0 c \end{pmatrix}; \tag{11.104}$$

it also commutes with D_z and other constants of the motion such as X, d^2, and p_y (the latter for the choice of vector potential (2.213)). Now

$$H^2 = I^2 + m_0^2 c^4, \tag{11.105}$$

and hence if F is an eigenvalue of I

$$F = \pm(E^2 - m_0^2 c^4)^{1/2} \tag{11.106}$$

$$= \pm(c^2 P_z^2 + 2\hbar|q|Bc^2 n)^{1/2}. \tag{11.107}$$

Next we determine normalized eigenkets of a c.s.c.o. that includes H, D_z, and I. We consider $q < 0$. The eigenvalue equation

$$I|\psi\rangle = F|\psi\rangle, \tag{11.108}$$

where $|\psi\rangle$ is given by (11.100)–(11.102), requires (for $n \neq 0$)

$$(F - cP_z)a = (2\hbar|q|Bc^2 n)^{1/2} b \tag{11.109}$$

and the equivalent relation

$$(2\hbar|q|Bc^2 n)^{1/2} a = (F + cP_z)b. \tag{11.110}$$

Substituting (11.109) and (11.110) in $|\psi\rangle$ and normalizing we find

$$|\psi\rangle = C\begin{bmatrix} (E + m_0 c^2)(2\hbar|q|Bc^2 n)^{1/2}|n-1, \gamma\rangle \\ (E + m_0 c^2)(F - cP_z)|n\gamma\rangle \\ F(2\hbar|q|Bc^2 n)^{1/2}|n-1, \gamma\rangle \\ F(F - cP_z)|n\gamma\rangle \end{bmatrix} \exp(-\mathrm{i}Et/\hbar), \tag{11.111}$$

where

$$C = \tfrac{1}{2}[E(E + m_0 c^2)F(F - cP_z)]^{-1/2}. \tag{11.112}$$

There is one solution (11.111) for each of the values F corresponding to the $\pm$ sign in (11.106), except if $n = 0$ in which case only the lower sign in (11.106) should be used. (With $n = 0$, (11.111) is an eigenket of I with eigenvalue $F = -cP_z$ only.)

Finally, the coordinate-space wavefunctions corresponding to (11.111) can be written down by replacing $|n\gamma\rangle$ in (11.111) with $\psi_{n\gamma}(\mathbf{r}) = \langle \mathbf{r}|n\gamma\rangle$. Here $\psi_{n\gamma}(\mathbf{r})$ are the non-relativistic wavefunctions derived in Section 2.9. The explicit form of $\psi_{n\gamma}(\mathbf{r})$ depends on the choice of a c.s.c.o. and the vector potential $\mathbf{A}$:

(I) With the choice (2.213) for $\mathbf{A}$, and for the c.s.c.o. H, p_y, p_z, and I, $\psi_{n\gamma}$ is given by (2.227).

(II) For the c.s.c.o. H, D_z, I, and X, and the choice (2.215) for **A**, $\psi_{n\gamma}$ is given by (2.238).

(III) For the c.s.c.o. H, D_z, I, and d^2, and the choice (2.215) for **A**, $\psi_{n\gamma}$ is given by (2.250).

According to (11.106), the energy is degenerate in the sign of the eigenvalue F. It can be shown that for positive energy states the matter current and spin density are parallel (antiparallel) for positive (negative) values of F [13]. There are other constants of the motion that can be used in place of I in the c.s.c.o. given above [13].

References

1. See, for example, Rindler, W. (1982). *Introduction to special relativity*, Ch. 5. Oxford University Press.
2. See, for example, Durand, B. and Durand, L. (1983). Analytic solution of the relativistic Coulomb problem for a spinless Salpeter equation. *Physical Review*, D **28**, 396–406. Kissel-Phillip, M. and Schwartz, W.H.E. (1988). Relativity corrected Schrödinger equation. *Physical Review*, A **38**, 6027–33 and references therein.
3. This equation has a lengthy history: see, for example, Kragh, H. (1984). Equation with many fathers. The Klein–Gordon equation in 1926. *American Journal of Physics*, **52**, 1024–33.
4. See, for example, Baym, G. (1969). *Lectures on quantum mechanics*, Ch. 22. Benjamin, Reading, MA.
5. The substitution (11.8) is a particular case of the 'gauge principle'. See, for example, Aitchison, I.J.R. and Hey, A.J.G. (1982). *Gauge theories in particle physics*, Ch. 8. Adam Hilger, Bristol.
6. Dirac, P.A.M. (1928). The quantum theory of the electron. *Proceedings of the Royal Society of London*, **A117**, 610–24.
7. For other representations of the Dirac matrices see, for example, Corinaldesi, E. and Strocchi, F. (1963). *Relativistic wave mechanics*. North-Holland, Amsterdam.
8. See, for example, Messiah, A. (1962). *Quantum mechanics*, Vol. II. North-Holland, Amsterdam, and Green H.S. (1965). *Matrix mechanics*. Noordhoff, Groningen.
9. The momentum representation of the Dirac equation has been discussed in detail by Lévy, M. (1950). Wave equations in momentum space. *Proceedings of the Royal Society of London* **A204**, 145–69.
10. For a more detailed discussion of the spin angular momentum operator see Ryder, L.H. (1985). *Quantum field theory*, Ch. 2. Cambridge University Press and references therein.
11. Feynman, R.P. and Gell-Mann, M. (1958). Theory of the Fermi interaction. *Physical Review*, **109**, 193–8.
12. Numerous references to solutions of the Dirac equation are given by Nieto, M.M. and Taylor, P.L. (1985). A solution (Dirac electron in crossed, constant electric and magnetic fields) that has found a problem (relativistic quantized Hall effect). *American Journal of Physics*, **53**, 234–7.
13. Johnson, M.H. and Lippmann, B.A. (1950). Relativistic motion in a magnetic field. *Physical Review*, **77**, 702–5.

12

RELATIVISTIC SPHERICALLY SYMMETRIC PROBLEMS

In Section 12.1 we discuss some elementary properties of the Klein–Gordon and Dirac equations with a spherically symmetric electrostatic potential. In Section 12.2 a simple similarity transformation is applied to the Dirac equation. The main result of this section is (12.70), which applies to the Schrödinger, Klein–Gordon, and (transformed) Dirac equations with a Coulomb potential. Shift operators for energy in the eigenkets of (12.70) are derived (Section 12.3) and they are used to obtain energy eigenvalues (Section 12.4). Expectation values and coordinate-space wavefunctions for the Klein–Gordon and Dirac Coulomb problems are calculated in Sections 12.5 and 12.6. A relativistic (Dirac) oscillator is discussed in Sections 12.8–12.10.

12.1 Spherically symmetric potentials

Consider a particle of charge q and rest mass m_0 in a spherically symmetric electrostatic potential

$$A_\mu = \left(0, \frac{\mathrm{i}}{c}\Phi(r)\right). \tag{12.1}$$

Let $|\psi\rangle$ be an energy eigenket for this particle at time $t = 0$.*

For a spin 0 particle the Klein–Gordon equation (11.41) is

$$(c^2\mathbf{p}^2 + m_0^2c^4)|\psi\rangle = (E - q\Phi)^2|\psi\rangle. \tag{12.2}$$

It is convenient to express (12.2) in terms of the radial momentum operator p_r introduced in Chapter 5. Using the identity (5.23) in (12.2) we have

$$(c^2p_r^2 + c^2r^{-2}\mathbf{L}^2 + m_0^2c^4 + 2qE\Phi - q^2\Phi^2)|\psi\rangle = E^2|\psi\rangle. \tag{12.3}$$

Now $\mathbf{L}^2$ and L_z commute with the operator on the left-hand side of (12.3). Therefore, as in the non-relativistic limit, if the energy and angular momentum are a c.s.c.o. we can use the eigenkets $|Elm\rangle$ as basis kets (see Chapter 5).

Similarly, for a spin $\frac{1}{2}$ particle the Dirac equation (11.42) can be written

$$H|\psi\rangle = E|\psi\rangle, \tag{12.4}$$

where

$$H = c\boldsymbol{\alpha}.\mathbf{p} + \beta m_0c^2 + q\Phi. \tag{12.5}$$

We have seen in Section 11.2 that $\mathbf{J}^2$ and J_z commute with $c\boldsymbol{\alpha}.\mathbf{p} + \beta m_0c^2$. It

* The eigenket at time t is given by (1.111).

is obvious that $\mathbf{J}^2$ and J_z commute with $q\Phi(r)$. Hence H, $\mathbf{J}^2$, and J_z are commuting observables and we can consider energy eigenkets that are also eigenkets of $\mathbf{J}^2$ and J_z

$$\mathbf{J}^2|\psi\rangle = \hbar^2 j(j+1)|\psi\rangle \tag{12.6}$$

and

$$J_z|\psi\rangle = \hbar m|\psi\rangle. \tag{12.7}$$

The Hamiltonian (12.5) can be expressed in terms of the radial momentum operator p_r by using the operator identity (see Appendix L)

$$\boldsymbol{\alpha}.\mathbf{p} = \alpha_r p_r + ir^{-1}\alpha_r(\boldsymbol{\Sigma}.\mathbf{L} + \hbar), \tag{12.8}$$

where

$$\alpha_r = \boldsymbol{\alpha}.\hat{\mathbf{r}} = \begin{pmatrix} 0 & \boldsymbol{\sigma}.\hat{\mathbf{r}} \\ \boldsymbol{\sigma}.\hat{\mathbf{r}} & 0 \end{pmatrix}. \tag{12.9}$$

Hence*

$$H = c\alpha_r p_r + i\hbar c r^{-1}\alpha_r \beta K + \beta m_0 c^2 + q\Phi, \tag{12.10}$$

where

$$K = \beta(\hbar^{-1}\boldsymbol{\Sigma}.\mathbf{L} + 1) \tag{12.11}$$

is the Dirac operator. In block notation

$$K = \begin{pmatrix} \mathscr{K} & 0 \\ 0 & -\mathscr{K} \end{pmatrix}, \tag{12.12}$$

where

$$\mathscr{K} = \hbar^{-1}\boldsymbol{\sigma}.\mathbf{L} + 1 \tag{12.13}$$

is a 2×2 matrix. The dimensionless operator K is a measure of the alignment between the spin and orbital angular momentum of the particle: it plays an important role in the theory and we first discuss some of its properties.

It is shown in Appendix L that

$$[K, X] = 0, \tag{12.14}$$

where $X = \alpha_r, \beta, p_r, f(r), \mathbf{J}^2$, or J_z. Thus K commutes with the set of commuting operators H, $\mathbf{J}^2$, and J_z. Consider the eigenvalues in

$$K|\psi\rangle = k|\psi\rangle, \tag{12.15}$$

where k is the Dirac quantum number. Let

* Note that the Hermitian property of H requires p_r to be Hermitian.

$$|\phi\rangle = \begin{pmatrix} |\psi_1\rangle \\ |\psi_2\rangle \end{pmatrix} \quad \text{and} \quad |\chi\rangle = \begin{pmatrix} |\psi_3\rangle \\ |\psi_4\rangle \end{pmatrix}. \tag{12.16}$$

Then

$$|\psi\rangle = \begin{pmatrix} |\phi\rangle \\ |\chi\rangle \end{pmatrix}. \tag{12.17}$$

Substituting (12.12) and (12.17) in (12.15) we obtain

$$\mathscr{K}|\phi\rangle = k|\phi\rangle \tag{12.18}$$

and

$$\mathscr{K}|\chi\rangle = -k|\chi\rangle. \tag{12.19}$$

If we square (11.52) and use (11.51), (11.55), (12.11), and (11.22), we find

$$\begin{pmatrix} \mathscr{K} & 0 \\ 0 & \mathscr{K} \end{pmatrix} = \hbar^{-2}(\mathbf{J}^2 - \mathbf{L}^2 + \tfrac{1}{4}\hbar^2). \tag{12.20}$$

It follows from (12.6) and (12.17)–(12.20) that $|\phi\rangle$ and $|\chi\rangle$ must be eigenkets of $\mathbf{L}^2$. Consider first the eigenvalue equation for $|\phi\rangle$

$$\mathbf{L}^2|\phi\rangle = \hbar^2 l(l+1)|\phi\rangle. \tag{12.21}$$

The theory of combination of two angular momenta requires

$$j = l \pm \tfrac{1}{2}, \tag{12.22}$$

where $l \neq 0$ if $j = l - \frac{1}{2}$. Because $l = 0, 1, 2, \ldots$ (see Section 9.5),

$$j = \tfrac{1}{2}, \tfrac{3}{2}, \ldots. \tag{12.23}$$

From (12.6), (12.18), and (12.20)–(12.22) we find

$$\begin{aligned} k &= l + 1 \qquad \text{if } j = l + \tfrac{1}{2}, \\ &= -l \qquad\quad \text{if } j = l - \tfrac{1}{2}; \end{aligned}$$

that is,

$$k = (-1)^{j+l-1/2}(j + \tfrac{1}{2}) \qquad (= \pm 1, \pm 2, \ldots). \tag{12.24}$$

Since $|\chi\rangle$ is also an eigenket of $\mathbf{L}^2$ we write

$$\mathbf{L}^2|\chi\rangle = \hbar^2 l'(l'+1)|\chi\rangle. \tag{12.25}$$

For the quantum numbers l' similar calculations to the above yield

$$l' = l \pm 1 \qquad \text{for } j = l \pm \tfrac{1}{2}. \tag{12.26}$$

Thus $j = l \pm \frac{1}{2}$ corresponds to $j = l' \mp \frac{1}{2}$. We note that the Dirac ket (12.17) is not an eigenket of $\mathbf{L}^2$. The significance of the eigenvalue l is discussed in Section 12.7.

Consider the effect of the operator $\boldsymbol{\sigma}.\hat{\mathbf{r}}$ on a spinor such as $|\phi\rangle$ that satisfies the eigenvalue equations (12.18) and (12.21). Using the identity (L.6) we have

$$\mathscr{K}(\boldsymbol{\sigma}\cdot\hat{\mathbf{r}}|\phi\rangle) = -\boldsymbol{\sigma}.\hat{\mathbf{r}}(\mathscr{K}|\phi\rangle) = -k(\boldsymbol{\sigma}\cdot\hat{\mathbf{r}}|\phi\rangle). \tag{12.27}$$

Thus $\boldsymbol{\sigma}.\hat{\mathbf{r}}$ reverses the sign of k. Similarly, using the identities (4.213) and (L.6) we have

$$\begin{aligned} \mathbf{L}^2(\boldsymbol{\sigma}\cdot\hat{\mathbf{r}}|\phi\rangle) &= \hbar^2\mathscr{K}(\mathscr{K}-1)(\boldsymbol{\sigma}\cdot\hat{\mathbf{r}}|\phi\rangle) \\ &= \hbar^2\boldsymbol{\sigma}.\hat{\mathbf{r}}\mathscr{K}(\mathscr{K}+1)|\phi\rangle \\ &= \hbar^2 k(k+1)(\boldsymbol{\sigma}\cdot\hat{\mathbf{r}}|\phi\rangle) \\ &= \hbar^2(l+1)(l+2)(\boldsymbol{\sigma}\cdot\hat{\mathbf{r}}|\phi\rangle) \quad \text{if } j = l+\tfrac{1}{2}, \\ &= \hbar^2 l(l-1)(\boldsymbol{\sigma}\cdot\hat{\mathbf{r}}|\phi\rangle) \qquad \text{if } j = l-\tfrac{1}{2}. \end{aligned} \tag{12.28}$$

Hence $\boldsymbol{\sigma}.\hat{\mathbf{r}}$ raises l by one if $j = l + \frac{1}{2}$ and lowers l by one if $j = l - \frac{1}{2}$ (see also (12.213) and (12.214)).

The above results show that we can express (12.17) in the form

$$|\psi\rangle = \begin{pmatrix} \mathrm{i}|\psi_{Ejmkl}\rangle_{\mathrm{U}} \\ \boldsymbol{\sigma}.\hat{\mathbf{r}}|\psi_{Ejmkl}\rangle_{\mathrm{L}} \end{pmatrix}. \tag{12.29}$$

Here U and L denote the upper and lower elements of the bi-spinor $|\psi\rangle$, and the factor i has been included for later convenience. We will, for brevity, often omit the subscripts $Ejmkl$ in $|\psi\rangle_{\mathrm{U}}$ and $|\psi\rangle_{\mathrm{L}}$. For the ket (12.29) the normalization (11.38) and the identity (4.208) require

$${}_{\mathrm{U}}\langle\psi|\psi\rangle_{\mathrm{U}} + {}_{\mathrm{L}}\langle\psi|\psi\rangle_{\mathrm{L}} = 1. \tag{12.30}$$

The Dirac Hamiltonian can be written in 2×2 block form by substituting the block forms of α_r, β, and K (eqns (12.9), (11.22), and (12.12)) in (12.10)

$$H = \begin{pmatrix} m_0c^2 + q\Phi & \boldsymbol{\sigma}.\hat{\mathbf{r}}(cp_r + \mathrm{i}\hbar cr^{-1}\mathscr{K}) \\ \boldsymbol{\sigma}.\hat{\mathbf{r}}(cp_r + \mathrm{i}\hbar cr^{-1}\mathscr{K}) & -m_0c^2 + q\Phi \end{pmatrix}. \tag{12.31}$$

If we substitute (12.29) and (12.31) in (12.4), then multiply out and use the commutator (1.145) and the identities (L.6) and (4.208), we obtain the coupled equations

$$(\mathrm{i}cp_r - \hbar cr^{-1}\mathscr{K})|\psi\rangle_{\mathrm{U}} = (E + m_0c^2 - q\Phi)|\psi\rangle_{\mathrm{L}} \tag{12.32}$$

and

$$(\mathrm{i}cp_r + \hbar cr^{-1}\mathscr{K})|\psi\rangle_{\mathrm{L}} = -(E - m_0c^2 - q\Phi)|\psi\rangle_{\mathrm{U}}. \tag{12.33}$$

An important spherically symmetric potential for which the Klein–Gordon equation (12.3) and the Dirac equations (12.32) and (12.33) have been solved analytically is the Coulomb potential. In the following sections we discuss the application of operator methods to these two relativistic Coulomb problems.

12.2 Relativistic Coulomb problems

The Klein–Gordon equation (12.3) with the attractive Coulomb potential

$$q\Phi = -\frac{\hbar^2}{m_0 ar} \tag{12.34}$$

can be written

$$\left(c^2 p_r^2 - \frac{2\hbar c\gamma E}{r} + \frac{\hbar^2 c^2 [l(l+1) - \gamma^2]}{r^2} + m_0^2 c^4 - E^2\right)|\psi\rangle = 0. \tag{12.35}$$

Here

$$\gamma = \hbar/(m_0 ca) \tag{12.36}$$

and $|\psi\rangle$ is an eigenket of $\mathbf{L}^2$ and L_z with eigenvalues $\hbar^2 l(l+1)$ and $\hbar m$. For a hydrogen-like atom $a = 4\pi\varepsilon_0\hbar^2(m_0 q^2 Z)^{-1}$ is the Bohr radius divided by the atomic number Z, and $\gamma = Zq^2(4\pi\varepsilon_0\hbar c)^{-1} \approx Z/137$ is Z times the fine-structure constant.

For the Dirac equation we first perform a similarity transformation [1] that simplifies the application of operator methods to the Dirac–Coulomb problem. Consider the operator

$$T = a_1 + \mathrm{i}a_2\beta\alpha_r, \tag{12.37}$$

where a_1 and a_2 are real constants such that

$$a_1^2 - a_2^2 = 1. \tag{12.38}$$

(Note that T is Hermitian; however, it is not unitary because $TT^\dagger \neq 1$.) Then

$$T^{-1} = a_1 - \mathrm{i}a_2\beta\alpha_r. \tag{12.39}$$

In block notation

$$T = \begin{pmatrix} a_1 & \mathrm{i}a_2\boldsymbol{\sigma}.\hat{\mathbf{r}} \\ -\mathrm{i}a_2\boldsymbol{\sigma}.\hat{\mathbf{r}} & a_1 \end{pmatrix} \tag{12.40}$$

and

$$T^{-1} = \begin{pmatrix} a_1 & -\mathrm{i}a_2\boldsymbol{\sigma}.\hat{\mathbf{r}} \\ \mathrm{i}a_2\boldsymbol{\sigma}.\hat{\mathbf{r}} & a_1 \end{pmatrix}. \tag{12.41}$$

Under a similarity transformation the Dirac equation (12.4) becomes

$$H'|\psi'\rangle = E|\psi'\rangle, \tag{12.42}$$

where

$$H' = THT^{-1} \tag{12.43}$$

and

$$|\psi'\rangle = T|\psi\rangle. \tag{12.44}$$

Substituting (12.29) and (12.40) in (12.44) we find

$$|\psi'\rangle = \begin{pmatrix} \mathrm{i}|\psi'_{Ejmkl}\rangle_{\mathrm{U}} \\ \boldsymbol{\sigma}.\hat{\mathbf{r}}|\psi'_{Ejmkl}\rangle_{\mathrm{L}} \end{pmatrix}, \tag{12.45}$$

where

$$|\psi'_{Ejmkl}\rangle_{\mathrm{U}} = a_1|\psi_{Ejmkl}\rangle_{\mathrm{U}} + a_2|\psi_{Ejmkl}\rangle_{\mathrm{L}}, \tag{12.46}$$

$$|\psi'_{Ejmkl}\rangle_{\mathrm{L}} = a_2|\psi_{Ejmkl}\rangle_{\mathrm{U}} + a_1|\psi_{Ejmkl}\rangle_{\mathrm{L}}. \tag{12.47}$$

It is convenient to let

$$a_1 = \cosh\tfrac{1}{2}\theta, \qquad a_2 = \sinh\tfrac{1}{2}\theta. \tag{12.48}$$

Then

$$a_1^2 + a_2^2 = \cosh\theta \tag{12.49}$$

and

$$2a_1a_2 = \sinh\theta. \tag{12.50}$$

Substituting (12.31), (12.40), and (12.41) in (12.43), and multiplying out the matrices we obtain the transformed Hamiltonian

$$H' = \begin{pmatrix} m_0c^2\cosh\theta + \mathrm{i}\sinh\theta\, cp_r + q\Phi & \mathrm{i}\boldsymbol{\sigma}.\hat{\mathbf{r}}(-m_0c^2\sinh\theta - \mathrm{i}\cosh\theta\, cp_r + \hbar c r^{-1}\mathscr{K}) \\ \mathrm{i}\boldsymbol{\sigma}.\hat{\mathbf{r}}(-m_0c^2\sinh\theta - \mathrm{i}\cosh\theta\, cp_r + \hbar c r^{-1}\mathscr{K}) & -m_0c^2\cosh\theta - \mathrm{i}\sinh\theta\, cp_r + q\Phi \end{pmatrix}. \tag{12.51}$$

From (12.42), (12.45), and (12.51) we obtain the transformed equations

$$(m_0c^2\cosh\theta + \mathrm{i}\sinh\theta\, cp_r + q\Phi)|\psi'\rangle_{\mathrm{U}} - (m_0c^2\sinh\theta + \mathrm{i}\cosh\theta\, cp_r + \hbar c r^{-1}k)|\psi'\rangle_{\mathrm{L}} = E|\psi'\rangle_{\mathrm{U}} \tag{12.52}$$

and

$$(m_0c^2\sinh\theta + \mathrm{i}\cosh\theta\, cp_r - \hbar c r^{-1}k)|\psi'\rangle_{\mathrm{U}} - (m_0c^2\cosh\theta + \mathrm{i}\sinh\theta\, cp_r - q\Phi)|\psi'\rangle_{\mathrm{L}} = E|\psi'\rangle_{\mathrm{L}}. \tag{12.53}$$

In (12.52) and (12.53) we have used (12.18) and (12.19) to replace $\mathscr{K}$ with its eigenvalue and we have omitted the subscripts *Ejmkl*.

The above is for any spherically symmetric electrostatic potential. For the Coulomb potential (12.34) the transformed Dirac equations can be simplified by choosing [1]

$$\tanh\theta = \gamma k^{-1}. \tag{12.54}$$

In terms of the constants

$$\zeta = [(j+\tfrac{1}{2})^2 - \gamma^2]^{1/2} \tag{12.55}$$

and

$$\eta = (-1)^{j+l+1/2}, \tag{12.56}$$

we have

$$k = -\eta(\zeta^2 + \gamma^2)^{1/2}, \tag{12.57}$$

$$\cosh\theta = (1 + \gamma^2\zeta^{-2})^{1/2}, \tag{12.58}$$

and

$$\sinh\theta = -\eta\gamma\zeta^{-1}. \tag{12.59}$$

Equations (12.52)–(12.59) yield the transformed Dirac–Coulomb equations

$$(icp_r + \eta\hbar c\zeta r^{-1} - \eta\gamma E\zeta^{-1})|\psi'\rangle_U = [m_0c^2 + (1 + \gamma^2\zeta^{-2})^{1/2}E]|\psi'\rangle_L \tag{12.60}$$

and

$$(icp_r - \eta\hbar c\zeta r^{-1} + \eta\gamma E\zeta^{-1})|\psi'\rangle_L = [m_0c^2 - (1 + \gamma^2\zeta^{-2})^{1/2}E]|\psi'\rangle_U. \tag{12.61}$$

The components of the original Dirac ket are given in terms of the transformed components by the inverse of (12.46) and (12.47)

$$|\psi\rangle_U = a_1|\psi'\rangle_U - a_2|\psi'\rangle_L, \tag{12.62}$$

$$|\psi\rangle_L = -a_2|\psi'\rangle_U + a_1|\psi'\rangle_L, \tag{12.63}$$

where

$$a_1 = \frac{1}{\sqrt{2}}((1 + \gamma^2\zeta^{-2})^{1/2} + 1)^{1/2}, \qquad a_2 = -\frac{\eta}{\sqrt{2}}((1 + \gamma^2\zeta^{-2})^{1/2} - 1)^{1/2}. \tag{12.64}$$

An advantage of the transformed equations (12.60) and (12.61) is that they can be decoupled to yield equations that are similar to the Klein–Gordon equation (12.35) and the non-relativistic Schrödinger equation, namely,

$$\left(c^2p_r^2 - \frac{2\hbar c\gamma E}{r} + \frac{\hbar^2c^2\zeta(\zeta \pm \eta)}{r^2} + m_0^2c^4 - E^2\right)\begin{pmatrix}|\psi'\rangle_U \\ |\psi'\rangle_L\end{pmatrix} = 0. \tag{12.65}$$

Next we introduce some convenient notation. Define the dimensionless function of energy

$$\lambda = \left(\frac{\gamma^2E^2}{m_0^2c^4 - E^2}\right)^{1/2} \tag{12.66}$$

and the length

$$a_\lambda = \tfrac{1}{2}(\lambda^2 + \gamma^2)^{1/2}a. \tag{12.67}$$

For a Klein–Gordon particle define

$$\zeta = [(l + \tfrac{1}{2})^2 - \gamma^2]^{1/2} - \tfrac{1}{2}. \tag{12.68}$$

We also use the abbreviation $|\lambda\zeta\rangle$ to denote the ket $|\psi_{Elm}\rangle$ ($\equiv |Elm\rangle$) of a Klein–Gordon particle; for a Dirac particle we let $|\lambda\zeta\rangle$ denote the components

$$|\lambda\zeta\rangle_U \equiv |\psi'_{Ejmkl}\rangle_U, \qquad |\lambda\zeta\rangle_L \equiv |\psi'_{Ejmkl}\rangle_L \tag{12.69}$$

TABLE 12.1 The parameters λ, a_λ, ζ, and δ in (12.70)

	Klein–Gordon	Dirac	Schrödinger
λ	$\left(\dfrac{\gamma^2 E^2}{m_0^2c^4 - E^2}\right)^{1/2}$		$\left(\dfrac{\gamma^2 m_0 c^2}{2(m_0c^2 - E)}\right)^{1/2}$
a_λ	$\frac{1}{2}(\lambda^2+\gamma^2)^{1/2}a$		$\frac{1}{2}\lambda a$
ζ	$[(l+\frac{1}{2})^2 - \gamma^2]^{1/2} - \frac{1}{2}$	$[(j+\frac{1}{2})^2 - \gamma^2]^{1/2}$	l
δ	1	$\delta = \eta$ if $\lvert\lambda\zeta\rangle = \lvert\lambda\zeta\rangle_U$ $= -\eta$ if $\lvert\lambda\zeta\rangle = \lvert\lambda\zeta\rangle_L$, $\eta = (-1)^{j+l+1/2}$	1

of the transformed Dirac ket (12.45). In terms of these the Klein–Gordon equation (12.35) and the decoupled equations (12.65) are

$$\left(\frac{a_\lambda^2}{\hbar^2}p_r^2 + \frac{1}{4} - \frac{\lambda a_\lambda}{r} + \zeta(\zeta+\delta)\frac{a_\lambda^2}{r^2}\right)|\lambda\zeta\rangle = 0. \tag{12.70}$$

Here δ is defined in Table 12.1, where we have summarized the definitions of the other parameters in (12.70). The non-relativistic Schrödinger equation with a Coulomb potential is also a particular case of (12.70): the appropriate values of the parameters are given in Table 12.1, and $|\lambda\zeta\rangle$ denotes $|Elm\rangle$. (Here E includes the rest mass energy m_0c^2.)

For bound states ($|E| < m_0c^2$), λ and a_λ are real. We can, without loss of generality, define a_λ to be positive, as in Table 12.1; then $\lambda > 0$ for an attractive potential. The energy is given in terms of λ by

$$E = m_0c^2(1 - \tfrac{1}{2}\gamma^2\lambda^{-2}) \tag{12.71}$$

for the Schrödinger equation, and

$$E = m_0c^2(1 + \gamma^2\lambda^{-2})^{-1/2} \tag{12.72}$$

for the Klein–Gordon and Dirac equations. We note that in (12.70), a_λ and $\hbar a_\lambda^{-1}$ are scale factors for r and p_r.

The normalization condition for the Schrödinger and Klein–Gordon kets is

$$\langle\lambda\zeta|\lambda\zeta\rangle = 1. \tag{12.73}$$

For the Dirac equation the normalization (12.30) can be expressed in terms of the transformed kets by substituting (12.62) and (12.63), and using (12.58), (12.59), and the notation (12.69). Because ${}_U\langle\lambda\zeta|\lambda\zeta\rangle_L$ is real* we obtain

* See (12.60), (12.61), and (12.143).

$$(1+\gamma^2\zeta^{-2})^{1/2}({}_{\mathrm{U}}\langle\lambda\zeta|\lambda\zeta\rangle_{\mathrm{U}}+{}_{\mathrm{L}}\langle\lambda\zeta|\lambda\zeta\rangle_{\mathrm{L}})+2\eta\gamma\zeta^{-1}{}_{\mathrm{U}}\langle\lambda\zeta|\lambda\zeta\rangle_{\mathrm{L}}=1. \quad (12.74)$$

It is useful to evaluate the scalar products in (12.74). To do this we require two more relations between these products. Multiplying (12.60) and (12.61), respectively, on the left by ${}_{\mathrm{L}}\langle\psi'|$ and ${}_{\mathrm{U}}\langle\psi'|$ we find

$$[m_0c^2+(1+\gamma^2\zeta^{-2})^{1/2}E]_{\mathrm{L}}\langle\lambda\zeta|\lambda\zeta\rangle_{\mathrm{L}}=[(1+\gamma^2\zeta^{-2})^{1/2}E-m_0c^2]_{\mathrm{U}}\langle\lambda\zeta|\lambda\zeta\rangle_{\mathrm{U}}. \quad (12.75)$$

Similarly, multiplying (12.60) on the left by ${}_{\mathrm{U}}\langle\psi'|$ and using the expectation values (12.133) and (12.143) we obtain

$$[m_0c^2+(1+\gamma^2\zeta^{-2})^{1/2}E]_{\mathrm{U}}\langle\lambda\zeta|\lambda\zeta\rangle_{\mathrm{L}}=\eta\gamma E\zeta(\lambda^{-2}-\zeta^{-2})_{\mathrm{U}}\langle\lambda\zeta|\lambda\zeta\rangle_{\mathrm{U}}. \quad (12.76)$$

Equations (12.74)–(12.76) yield the useful results

$${}_{\mathrm{U}}\langle\lambda\zeta|\lambda\zeta\rangle_{\mathrm{U}}=\frac{1}{2(1+\gamma^2\lambda^{-2})^{1/2}}\left[\left(\frac{1+\gamma^2\zeta^{-2}}{1+\gamma^2\lambda^{-2}}\right)^{1/2}+1\right], \quad (12.77)$$

$${}_{\mathrm{L}}\langle\lambda\zeta|\lambda\zeta\rangle_{\mathrm{L}}=\frac{1}{2(1+\gamma^2\lambda^{-2})^{1/2}}\left[\left(\frac{1+\gamma^2\zeta^{-2}}{1+\gamma^2\lambda^{-2}}\right)^{1/2}-1\right], \quad (12.78)$$

and

$${}_{\mathrm{U}}\langle\lambda\zeta|\lambda\zeta\rangle_{\mathrm{L}}=-\frac{1}{2}\eta\gamma\frac{\gamma^2-\zeta^2}{\zeta(\lambda^2+\gamma^2)}. \quad (12.79)$$

From (12.78)

$$|\zeta\zeta\rangle_{\mathrm{L}}=0. \quad (12.80)$$

Note that for the components (12.62) and (12.63) of the original Dirac kets (12.77)–(12.79) yield the simple relations

$${}_{\mathrm{U}}\langle\psi|\psi\rangle_{\mathrm{U}}=\frac{1}{2}\left(1+\frac{E}{m_0c^2}\right), \quad (12.81)$$

$${}_{\mathrm{L}}\langle\psi|\psi\rangle_{\mathrm{L}}=\frac{1}{2}\left(1-\frac{E}{m_0c^2}\right), \quad (12.82)$$

and

$${}_{\mathrm{U}}\langle\psi|\psi\rangle_{\mathrm{L}}=-\frac{1}{2}\gamma^{-1}k\left(1-\frac{E}{m_0^2c^4}\right). \quad (12.83)$$

This completes our introductory remarks on relativistic Coulomb problems. In Sections 12.3–12.5 we present an operator analysis of (12.70). This analysis is based on shift operators for energy, and it is similar to that given in Section 5.6 for the non-relativistic Coulomb problem.

12.3 Shift operators for energy

We rewrite (12.70) as

$$O_\lambda|\lambda\zeta\rangle = -\zeta(\zeta+\delta)|\lambda\zeta\rangle, \tag{12.84}$$

where

$$O_\lambda = \hbar^{-2}r^2p_r^2 + \tfrac{1}{4}r^2a_\lambda^{-2} - \lambda r a_\lambda^{-1}. \tag{12.85}$$

We wish to derive shift operators $Q_\lambda^\pm$ that transform $|\lambda\zeta\rangle$ into $|\lambda \pm 1, \zeta\rangle$ [2]. Such operators must satisfy

$$Q_{\lambda\pm1}^\mp Q_\lambda^\pm = O_\lambda + F^\pm, \tag{12.86}$$

where $F^\pm$ are independent of r and p_r. As in Section 5.6 we perform the derivation of the factorizations (12.86) in three parts.

(i) Let

$$\tilde{O}_\lambda = u^2p_u^2 + \tfrac{1}{4}u^2 - \lambda u, \tag{12.87}$$

where u and p_u satisfy $[u, p_u] = \mathrm{i}$. Then

$$\tilde{Q}_{\lambda\pm1}^\mp \tilde{Q}_\lambda^\pm = \tilde{O}_\lambda + \lambda(\lambda \pm 1), \tag{12.88}$$

where

$$\tilde{Q}_\lambda^\pm = \pm\mathrm{i}up_u - \tfrac{1}{2}u + \lambda. \tag{12.89}$$

(ii) Define operators $\Delta_\lambda^\pm$ such that (cf. (5.114)–(5.116))

$$\Delta_\lambda^\pm \frac{r}{a_\lambda} = \frac{r}{a_{\lambda\pm1}}\Delta_\lambda^\pm, \tag{12.90}$$

$$\Delta_\lambda^\pm a_\lambda p_r = a_{\lambda\pm1}p_r\Delta_\lambda^\pm, \tag{12.91}$$

and

$$(\Delta_\lambda^\pm)^\dagger\Delta_\lambda^\pm = 1. \tag{12.92}$$

From (12.90)–(12.92)

$$(\Delta_\lambda^\pm)^\dagger = \Delta_{\lambda\pm1}^\mp. \tag{12.93}$$

In (12.87)–(12.89) let $u = ra_\lambda^{-1}$ and $p_u = \hbar^{-1}a_\lambda p_r$. Then operate on the left on both sides of (12.88) with $\Delta_{\lambda\pm1}^\mp\Delta_\lambda^\pm = 1$. This yields the desired factorizations

$$Q_{\lambda\pm1}^\mp Q_\lambda^\pm = O_\lambda + \lambda(\lambda \pm 1), \tag{12.94}$$

where

$$Q_\lambda^\pm = \Delta_\lambda^\pm T_\lambda^\pm \tag{12.95}$$

and

$$T_\lambda^\pm = \pm i\hbar^{-1} r p_r - \tfrac{1}{2} r a_\lambda^{-1} + \lambda. \tag{12.96}$$

(iii) The scaling operators $\Delta_\lambda^\pm$ can be derived from (12.90)–(12.92). The calculations are similar to those in Section 5.6 and the results are

$$\Delta_\lambda^\pm = \exp[(i\hbar^{-1} r p_r + \tfrac{1}{2}) \ln(a_\lambda / a_{\lambda \pm 1})]. \tag{12.97}$$

From (12.94) and (12.84)

$$Q_\lambda^\pm |\lambda\zeta\rangle = \alpha_{\lambda\zeta}^\pm |\lambda \pm 1, \zeta\rangle. \tag{12.98}$$

Consider first the special cases in which a shift operator annihilates a ket. Let both sides of (12.94) with $\lambda = \lambda'$ act on $|\lambda'\zeta\rangle$. Then if λ' is a solution to

$$\lambda'(\lambda' - 1) = \zeta(\zeta + \delta), \tag{12.99}$$

either

$$Q_{\lambda'}^- |\lambda'\zeta\rangle = 0, \tag{12.100}$$

or

$$Q_{\lambda'-1}^+ |\lambda' - 1, \zeta\rangle = 0, \tag{12.101}$$

or both of these must be true. (See also Section 12.7.)

Next we calculate the coefficients $\alpha_{\lambda\zeta}^\pm$. From (12.98)

$$\begin{aligned} \langle \lambda \pm 1, \zeta | \lambda \pm 1, \zeta\rangle |\alpha_{\lambda\zeta}^\pm|^2 &= \langle \lambda\zeta | (Q_\lambda^\pm)^\dagger Q_\lambda^\pm | \lambda\zeta\rangle \\ &= \langle \lambda\zeta | (T_\lambda^\pm)^\dagger T_\lambda^\pm | \lambda\zeta\rangle, \end{aligned} \tag{12.102}$$

where we have used (12.92). The adjoints of (12.96) are

$$(T_\lambda^\pm)^\dagger = T_\lambda^\mp \mp 1. \tag{12.103}$$

Substituting (12.96) and (12.103) in (12.102), and using (12.84) to eliminate quadratic terms in p_r, we find

$$\langle (T_\lambda^\pm)^\dagger T_\lambda^\pm \rangle = \langle -2i\hbar^{-1} r p_r \pm r a_\lambda^{-1} + \lambda^2 \mp \lambda - \zeta(\zeta + \delta)\rangle. \tag{12.104}$$

If the operator*

$$\Omega = \hbar^{-2} p_r^2 + \tfrac{1}{4} a_\lambda^{-2} - \lambda a_\lambda^{-1} r^{-1} + \zeta(\zeta + \delta) r^{-2} \tag{12.105}$$

in (12.70) is Hermitian, the expectation values of rp_r and r are given by (12.144) and (12.147). Using these values we obtain from (12.102) and (12.104)

$$\langle \lambda \pm 1, \zeta | \lambda \pm 1, \zeta\rangle |\alpha_{\lambda\zeta}^\pm|^2 = \lambda^{-1}(\lambda \pm 1)[\lambda(\lambda \pm 1) - \zeta(\zeta + \delta)]\langle \lambda\zeta | \lambda\zeta\rangle. \tag{12.106}$$

---

* Ω is a_λ^{-2} times the operator in (12.70). The form (12.105) is convenient for applying the Hellmann–Feynman theorem: see Section 12.5.

Hence

$$\alpha_{\lambda\zeta}^{\pm} = \exp(\pm i\varepsilon)\{\lambda^{-1}(\lambda \pm 1)[\lambda(\lambda \pm 1) - \zeta(\zeta + \delta)]\}^{1/2}\beta_{\lambda\zeta}^{\pm}, \quad (12.107)$$

where ε is a real constant (see Section 12.6) and

$$\beta_{\lambda\zeta}^{\pm} = (\langle\lambda\zeta|\lambda\zeta\rangle/\langle\lambda \pm 1, \zeta|\lambda \pm 1, \zeta\rangle)^{1/2}. \quad (12.108)$$

It follows from (12.106) that for normalizable kets ζ must be real. Hence (12.55) and (12.68) yield the restrictions $\gamma \le 1$ for the Dirac equation and $\gamma \le \frac{1}{2}$ for the Klein–Gordon equation; that is, $Z \le 137$ and $Z \le 137/2$. (If Z exceeds these values it is necessary to take into account modifications to the Coulomb potential by, for example, the finite nuclear size.) Note that (12.107) depend on the assumption that Ω is Hermitian (see Section 12.7).

For normalized kets of the Schrödinger and Klein–Gordon equations $\beta_{\lambda\zeta}^{\pm} = 1$. For the Dirac equation $\beta_{\lambda\zeta}^{\pm}$ are given by (12.108), (12.77), and (12.78), except that for $|\zeta + 1, \zeta\rangle_{\mathrm{L}}$ the lowering operation (12.98) is (see (12.80))

$$Q_{\zeta+1}^{-}|\zeta + 1, \zeta\rangle_{\mathrm{L}} = 0. \quad (12.109)$$

According to (12.107), both (12.100) and (12.101) are valid, and they can be written (see (12.102))

$$T_{\lambda'}^{-}|\lambda'\zeta\rangle = 0 \quad (12.110)$$

and

$$T_{\lambda'-1}^{+}|\lambda' - 1, \zeta\rangle = 0, \quad (12.111)$$

where λ' are the roots

$$\lambda_{\pm} = \tfrac{1}{2} \pm [\zeta(\zeta + \delta) + \tfrac{1}{4}]^{1/2} \quad (12.112)$$

of (12.99). Thus the possibility $\lambda' = 1$, that is contained in (12.98) and (12.107) for the lowering operation, applies only if $\zeta = 0$ or $-\delta$; that is, in the non-relativistic case. Equations (12.110) and (12.111) are discussed further in Section 12.7. We now consider some applications of (12.70) and of the shift operators derived above.

12.4 Energy eigenvalues

Consider bound states of an attractive Coulomb potential. Then $\lambda > 0$. Let $|\lambda\zeta\rangle$ be a normalized eigenket of the operator Ω. Suppose there exists such a ket with $\lambda \ge \lambda_+ + 1$, and consider the sequence

$$Q_{\lambda}^{-}|\lambda\zeta\rangle, Q_{\lambda-1}^{-}Q_{\lambda}^{-}|\lambda\zeta\rangle, \ldots. \quad (12.113)$$

The results obtained in the previous section show that there are three possibilities: either (12.113) terminates at $\lambda' = \lambda_+$, or at $\lambda' = \lambda_-$, or it does not terminate at all. We determine which of these possibilities ensures non-negativity of the norm in (12.113). We assume that Ω is Hermitian: then the kets in (12.113) will have non-negative norm provided (see (12.106) and (12.112))

$$F(\lambda) = \lambda^{-1}(\lambda - 1)(\lambda - \lambda_-)(\lambda - \lambda_+) \geq 0. \tag{12.114}$$

(i) For the Schrödinger equation $\lambda_+ = l + 1$ and $\lambda_- = -l$ from (12.112) and Table 12.1. If $l \geq 1$, $F < 0$ in $(1, l + 1)$. Hence (12.113) must terminate at λ_+, and the possible values of λ are

$$\lambda = N + \lambda_+ = N + l + 1 \qquad (N = 0, 1, 2, \ldots). \tag{12.115}$$

If $l = 0$, $F = (\lambda - 1)^2 \geq 0$ for all λ. But a normalizable s state with non-integral λ cannot exist because from such a state one could obtain a normalizable p state with non-integral λ by applying the operator C_z^+ (see (9.19)). Thus for s states (12.113) terminates at $\lambda_+ = 1$, and the values of λ are given by (12.115).

(ii) For the Klein–Gordon equation

$$\lambda_\pm = \tfrac{1}{2} \pm [(l + \tfrac{1}{2})^2 - \gamma^2]^{1/2}. \tag{12.116}$$

Because $\gamma \leq \frac{1}{2}$, then for $l \geq 1$, $F < 0$ in $(\lambda_-, 0)$ and $(1, \lambda_+)$. Consequently all sequences (12.113) that do not terminate at λ_+ will produce a ket for which $F < 0$. If $l = 0$, $F < 0$ in $(0, \frac{1}{2} - \frac{1}{2}(1 - 4\gamma^2)^{1/2})$ and $(\frac{1}{2} + \frac{1}{2}(1 - 4\gamma^2)^{1/2}, 1)$. These intervals are too small to conclude that (12.113) must terminate at λ_+. Normalizable s states of the Klein–Gordon equation with $\lambda \neq N + \lambda_+$ are called anomalous states. The existence of these states is discussed below (Section 12.7). It turns out that the operators p_r and Ω are not Hermitian with respect to anomalous states (see (12.139) and Section 12.7); for these operators to be Hermitian we must have $\lambda = N + \lambda_+$ for all l.

(iii) For the Dirac equation we start by considering

$$\begin{aligned} |\lambda\zeta\rangle &= |\lambda\zeta\rangle_L \qquad \text{if } j = l + \tfrac{1}{2}, \\ &= |\lambda\zeta\rangle_U \qquad \text{if } j = l - \tfrac{1}{2}. \end{aligned} \tag{12.117}$$

Then $\delta = 1$ (see Table 12.1), and

$$\lambda_\pm = \tfrac{1}{2} \pm \{\tfrac{1}{2} + [(j + \tfrac{1}{2})^2 - \gamma^2]^{1/2}\}. \tag{12.118}$$

Because $\gamma \leq 1$, then for $j \geq \frac{3}{2}$, $F < 0$ in $(1, \lambda_+)$ where $\lambda_+ > 2$. Hence (12.113) must terminate at λ_+. If $j = \frac{1}{2}$, $F < 0$ in $(-(1 - \gamma^2)^{1/2}, 0)$ and $(1, 1 + (1 - \gamma^2)^{1/2})$; consequently if $(1 - \gamma^2)^{1/2} > \frac{1}{2}$ (i.e. if $Z \leq 118$), eqn (12.113) must terminate at λ_+. However, if $Z > 118$ we cannot conclude from the above that (12.113) must terminate at λ_+. In fact, for $Z > 118$ and $j = \frac{1}{2}$ there exist normalizable states with $\lambda \neq N + \lambda_+$ (see Section 12.7).

If we restrict ourselves to $Z \leq 118$, the above results show that for the kets (12.117) with $\delta = 1$,

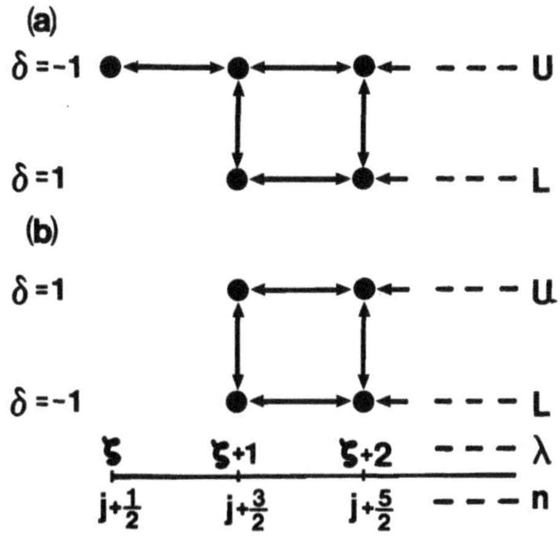

FIG. 12.1 Transformations of the Dirac kets $|\lambda\zeta\rangle_U$ and $|\lambda\zeta\rangle_L$ defined in (12.69): (a) for $j = l + \frac{1}{2}$, (b) for $j = l - \frac{1}{2}$. The transformations depicted by vertical arrows are between $|\lambda\zeta\rangle_L$ and $|\lambda\zeta\rangle_U$; they are effected by the operators in (12.60) and (12.61). The transformations depicted by horizontal arrows are between $|\lambda\zeta\rangle_U$ and $|\lambda \pm 1, \zeta\rangle_U$, etc.; they are effected by the shift operators $Q_\lambda^\pm$ in (12.98). The values of λ and n are from (12.121) and (12.126). In (b) the ket $|\zeta\zeta\rangle_L$ does not exist because of (12.80). (Reprinted from Ref. 2.)

$$\lambda = N + \lambda_+ = N + \zeta + 1 \qquad (N = 0, 1, 2, \ldots), \tag{12.119}$$

and that lowering with Q_λ^- terminates at $\lambda = \zeta + 1$. From these we can obtain kets for which $\delta = -1$ by using the transformations (12.60) and (12.61), and the lowering operator Q_λ^-. We see that the kets for which $\delta = -1$ have

$$\lambda = N + \zeta, \tag{12.120}$$

and that lowering with Q_λ^- terminates at $\lambda = \zeta$, except that $|\zeta\zeta\rangle_L$ does not exist (see (12.80)). (Note that with λ given by (12.119) or (12.120) the ratio of the norms in (12.75) is non-negative.) The transformations between Dirac kets are depicted in Fig. 12.1.

The above results for the Schrödinger, Klein–Gordon, and Dirac equations can be summarized: the normalizable eigenkets $|\lambda\zeta\rangle$ of the Hermitian operator Ω are those with

$$\lambda = N + \lambda_+ = N + \zeta + \tfrac{1}{2}(1 + \delta), \tag{12.121}$$

where $N = 0, 1, 2, \ldots$, except that for $N = 0$ and $j = l - \frac{1}{2}$ the lower component $|\zeta\zeta\rangle_L$ of the transformed Dirac ket does not exist.

The energy eigenvalues follow from (12.71), (12.72), and (12.121). For the Schrödinger equation

$$E = m_0 c^2 (1 - \tfrac{1}{2}\gamma^2 n^{-2}), \tag{12.122}$$

and for the Klein–Gordon equation

$$E = m_0 c^2 [1 + \gamma^2 \{n - l - \tfrac{1}{2} + [(l + \tfrac{1}{2})^2 - \gamma^2]^{1/2}\}^{-2}]^{-1/2}. \tag{12.123}$$

Here

$$n = N + l + 1 \qquad (N = 0, 1, 2, \ldots). \tag{12.124}$$

For the Dirac equation

$$E = m_0c^2[1 + \gamma^2\{n - j - \tfrac{1}{2} + [(j + \tfrac{1}{2})^2 - \gamma^2]^{1/2}\}^{-2}]^{-1/2}, \tag{12.125}$$

where

$$n = N + j + \tfrac{1}{2} \qquad (N = 0, 1, 2, \ldots) \tag{12.126}$$

and $n \neq j + \frac{1}{2}$ if $j = l - \frac{1}{2}$.

The eigenvalues (12.122) for the non-relativistic case are the Bohr levels (9.34) plus the rest mass energy. Also, the shift operators (12.95) with $\lambda = n$ and $a_\lambda = \frac{1}{2}na$ are those derived in Section 5.6 (see (5.120)).

A striking feature of the relativistic energy eigenvalues (12.123) and (12.125) is the breaking of the accidental degeneracy of the non-relativistic Coulomb problem (see Chapter 9). The formula (12.123) was obtained by Schrödinger, who first derived the Klein–Gordon equation [3]. Because the fine-structure splittings calculated from (12.123) for the hydrogen atom disagreed with the experimental values, Schrödinger rejected the Klein–Gordon equation for an electron [3]. Equation (12.125) was first derived in 1916 by Sommerfeld [4] who applied the old quantum theory to the action variables (9.156). According to this theory

$$J_r + J_\theta + J_\phi = n\hbar, \qquad J_r = n_r\hbar. \tag{12.127}$$

With the identification [5] $n_r = n - j - \frac{1}{2}$, eqns (9.155) and (12.127) yield (12.125). The reasons for this agreement between the old quantum theory and the Dirac theory have been discussed by Biedenharn [5]. The Dirac theory is an approximate treatment of the Coulomb problem for a spin $\frac{1}{2}$ particle. For the corrections that must be made to this theory, and a comparison with experimental results for a variety of physical systems, the reader should consult the book by Series [6].

The Klein–Gordon equation applies to a spinless particle such as a π meson. Thus to test the eigenvalue formula (12.123) use is made of pionic atoms like π^-Ti and π^-Fe. To minimize strong interaction effects it is necessary to use transitions from states with large n and l, whose overlap with the nucleus is small. These difficult experiments have been performed and, to order $(\gamma/n)^4$, the measured values confirm the formula (12.123) [7].

Some degeneracy remains in the Dirac eigenvalues (12.125). Each level with a given n and j, where $n \neq j + \frac{1}{2}$, is two-fold degenerate because E does not depend on $l = j \pm \frac{1}{2}$. Stated differently, E is independent of the sign of the Dirac quantum number k in (12.24). (If $n = j + \frac{1}{2}$ there is no degeneracy because the state with $l = j + \frac{1}{2}$ does not exist.) A schematic illustration of the spectrum for $j = \frac{1}{2}$ and $\frac{3}{2}$ is given in Fig. 12.2 where we have used the notation ${}^nl_j$ to label the levels.

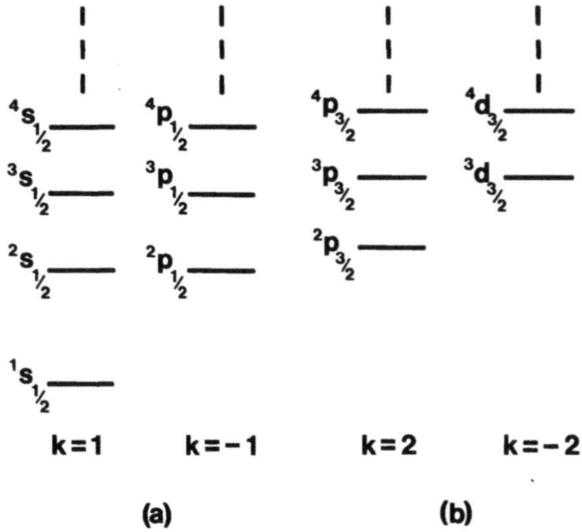

FIG. 12.2 Schematic illustration of the energy eigenvalues (12.125) of the Dirac–Coulomb problem: (a) for $j = \frac{1}{2}$, (b) for $j = \frac{3}{2}$. The levels are labelled by $^n l_j$ and values of the Dirac quantum number k (eqn (12.24)) are indicated.

An explanation of this degeneracy was given by Johnson and Lippmann [8], who considered the operator

$$I = \mathbf{\Sigma}.\hat{\mathbf{r}} + ia(\hbar c)^{-1}K\gamma_5(H - m_0c^2\beta), \qquad (12.128)$$

where $\gamma_5 \equiv \gamma_1\gamma_2\gamma_3\gamma_4 \equiv i\alpha_1\alpha_2\alpha_3$. They pointed out that I commutes with H and anticommutes with K. Thus I transforms a ket with energy E and Dirac quantum number $k = j + \frac{1}{2}$ into a ket with the same energy and $k = -(j + \frac{1}{2})$. If $n = j + \frac{1}{2}$ the ket with $k = j + \frac{1}{2}$ (i.e. $l = j - \frac{1}{2}$) is annihilated by I [9]. The Johnson–Lippmann operator can also be written

$$I = \mathbf{\Sigma}.\mathbf{A} - i\gamma a r^{-1}\gamma_5 K - a\hbar^{-2}\begin{pmatrix} 0 & 0 \\ 0 & \boldsymbol{\sigma} \end{pmatrix}.(\mathbf{L} \times \mathbf{p} - \mathbf{p} \times \mathbf{L}), \qquad (12.129)$$

where $\mathbf{A}$ is the Pauli–Lenz vector (9.10) and γ is given by (12.36). In the non-relativistic limit ($\gamma \ll 1$) the second term on the right-hand side of (12.129) is small. Similarly, the third term can be neglected because it acts on the lower components of a Dirac ket, and the latter are small in the non-relativistic limit. Thus in this limit the effect of I on a ket is the same as that of $\mathbf{\Sigma}.\mathbf{A}$.

The structure of the levels in Fig. 12.2 is suggestive of a supersymmetric system (Section 3.8), and it has been shown that the Dirac–Coulomb problem is, in fact, supersymmetric [10]. Corrections to the Dirac theory, such as those caused by interaction of the charged particle with vacuum fluctuations of the quantized radiation field [6], break the supersymmetry and consequently they remove the degeneracy in k.

12.5 Expectation values

In the following sections we continue to express our results in terms of the variables introduced in Section 12.2. Of these variables: (i) λ and λ_+ are given in terms of quantum numbers in Table 12.2; (ii) a_λ, ζ, and δ are defined in Table 12.1, and η is given by (12.56). Using these, any of the results obtained below can be expressed in terms of quantum numbers such as n, j, and l.

Equation (12.70) can be written

$$\Omega|\lambda\zeta\rangle = 0, \tag{12.130}$$

where Ω is given by (12.105). If Ω is Hermitian with respect to $|\lambda\zeta\rangle$, we can evaluate certain expectation values by applying the Hellmann–Feynman theorem and the hypervirial theorem to (12.130).

If we substitute (12.67) in (12.105) and treat the radius a as a continuous variable, the Hellmann–Feynman theorem yields

$$\langle \partial\Omega/\partial a\rangle = 0. \tag{12.131}$$

(Here, and in what follows, $\langle\ldots\rangle$ means $\langle\lambda\zeta|\ldots|\lambda\zeta\rangle$.) We then have

$$\left\langle \frac{1}{\lambda^2+\gamma^2}\frac{\mathrm{d}}{\mathrm{d}a}\frac{1}{a^2} - \frac{2\lambda}{r(\lambda^2+\gamma^2)^{1/2}}\frac{\mathrm{d}}{\mathrm{d}a}\frac{1}{a}\right\rangle = 0. \tag{12.132}$$

Hence

$$\langle r^{-1}\rangle = (2\lambda a_\lambda)^{-1}\langle\lambda\zeta|\lambda\zeta\rangle. \tag{12.133}$$

Similarly, if we treat λ as a continuous variable,

$$\langle\partial\Omega/\partial\lambda\rangle = 0. \tag{12.134}$$

Consider the ket $|\lambda'\zeta\rangle$ that satisfies (12.110), and the kets obtained from $|\lambda'\zeta\rangle$ by repeated application of raising operators Q_λ^+; that is, kets with $\lambda = N + \lambda_\pm$. Then (see (12.99))

$$\zeta(\zeta+\delta) = (\lambda - N)(\lambda - N - 1), \tag{12.135}$$

and (12.134) gives

$$(2\lambda_\pm - 1)a_\lambda^2\langle r^{-2}\rangle = (2a_\lambda^{-1})\frac{\partial a_\lambda}{\partial\lambda}\langle\lambda\zeta|\lambda\zeta\rangle - \left(\lambda\frac{\partial a_\lambda}{\partial\lambda} - a_\lambda\right)\langle r^{-1}\rangle. \tag{12.136}$$

Equations (12.133) and (12.136) yield

$$\langle r^{-2}\rangle = a_\lambda^{-2}[2\lambda(2\lambda_\pm - 1)]^{-1}\langle\lambda\zeta|\lambda\zeta\rangle. \tag{12.137}$$

From (12.133), (12.137), and (12.70)

$$\langle p_r^2\rangle = \tfrac{1}{4}\hbar^2 a_\lambda^{-2}\{1 - 2\lambda_\pm(\lambda_\pm - 1)[\lambda(2\lambda_\pm - 1)]^{-1}\}\langle\lambda\zeta|\lambda\zeta\rangle. \tag{12.138}$$

For the Klein–Gordon equation it is interesting to consider anomalous s states with $\lambda = N + \lambda_-$. For these (12.137) yields $\langle r^{-2}\rangle < 0$ because $\lambda_- < \frac{1}{2}$

TABLE 12.2 The parameters λ and λ_+ in terms of quantum numbers

	Klein–Gordon	Dirac	Schrödinger
λ	$n - l - \frac{1}{2} + [(l + \frac{1}{2})^2 - \gamma^2]^{1/2}$	$n - j - \frac{1}{2} + [(j + \frac{1}{2})^2 - \gamma^2]^{1/2}$	n
λ_+	$\frac{1}{2} + [(l + \frac{1}{2})^2 - \gamma^2]^{1/2}$	For $j = l \pm \frac{1}{2}$: $\lambda_+ = \frac{1}{2} \mp \frac{1}{2} + [(j + \frac{1}{2})^2 - \gamma^2]^{1/2}$ in $\lvert\lambda\zeta\rangle_U$ $= \frac{1}{2} \pm \frac{1}{2} + [(j + \frac{1}{2})^2 - \gamma^2]^{1/2}$ in $\lvert\lambda\zeta\rangle_L$	$l + 1$

(see (12.116)). Therefore, either Ω is not Hermitian or $|\lambda\zeta\rangle$ is not normalizable. Similarly, from (12.138)

$$\langle\lambda_-\zeta|p_r^2|\lambda_-\zeta\rangle = -\tfrac{1}{4}(1-4\gamma^2)^{-1/2}\hbar^2 a_\lambda^{-2}\langle\lambda_-\zeta|\lambda_-\zeta\rangle. \tag{12.139}$$

If $|\lambda_-\zeta\rangle$ is normalizable, the right-hand side of (12.139) is negative and hence p_r is not Hermitian with respect to $|\lambda_-\zeta\rangle$ (see also Section 12.7).

In Section 6.3 we applied the hypervirial theorem to the non-relativistic Coulomb problem. The application of this theorem to (12.130) can be carried out in a similar manner. We suppose that Ω is Hermitian with respect to $|\lambda\zeta\rangle$ and $W|\lambda\zeta\rangle$. Then (1.193) is valid with $|\psi\rangle = |\lambda\zeta\rangle$ and $A = \Omega$. Let $W = g(r)$ and $\mathrm{d}g/\mathrm{d}r = f(r)$. Then

$$\langle fp_r\rangle = \tfrac{1}{2}\mathrm{i}\hbar\langle \mathrm{d}f/\mathrm{d}r\rangle. \tag{12.140}$$

Similarly, with

$$W = f(r)p_r + \frac{1}{2}\mathrm{i}\hbar\frac{\mathrm{d}f}{\mathrm{d}r},$$

we find

$$\left\langle \frac{1}{4}\frac{\mathrm{d}^3 f}{\mathrm{d}r^3} + \left(-\frac{1}{4a_\lambda^2} + \frac{\lambda}{a_\lambda r} - \frac{\zeta(\zeta+\delta)}{r^2}\right)\frac{\mathrm{d}f}{\mathrm{d}r} + \left(\frac{\zeta(\zeta+\delta)}{r^3} - \frac{\lambda}{2a_\lambda r^2}\right)f\right\rangle = 0. \tag{12.141}$$

Several useful expectation values can be obtained from the last two equations. With $f = r^s$ (s a constant) (12.140) yields

$$\langle r^s p_r\rangle = \tfrac{1}{2}\mathrm{i}\hbar s\langle r^{s-1}\rangle. \tag{12.142}$$

Thus

$$\langle p_r\rangle = 0, \tag{12.143}$$

$$\langle rp_r\rangle = \tfrac{1}{2}\mathrm{i}\hbar\langle\lambda\zeta|\lambda\zeta\rangle, \tag{12.144}$$

$$\langle r^2 p_r\rangle = \mathrm{i}\hbar\langle r\rangle, \tag{12.145}$$

etc. Similarly, with $f = r^{s+1}$, (12.141) gives

$$(s+1)\langle r^s\rangle - 2(2s+1)\lambda a_\lambda\langle r^{s-1}\rangle - sa_\lambda^2[s^2 - 1 - 4\zeta(\zeta+\delta)]\langle r^{s-2}\rangle = 0. \tag{12.146}$$

With $s = 0$, (12.146) yields (12.133).* Substituting $s = 1, 2, \ldots$ in (12.146) we find

$$\langle r\rangle = a_\lambda\lambda^{-1}[3\lambda^2 - \zeta(\zeta+\delta)]\langle\lambda\zeta|\lambda\zeta\rangle, \tag{12.147}$$

---

* The proof of (12.133) by the Hellmann–Feynman theorem is more general than that based on the hypervirial theorem; the latter requires that Ω is Hermitian with respect to $rp_r|\lambda\zeta\rangle$, and this is not valid for all states $|\lambda\zeta\rangle$.

$$\langle r^2\rangle = 2a_\lambda^2[5\lambda^2 - 3\zeta(\zeta+\delta)+1]\langle\lambda\zeta|\lambda\zeta\rangle, \tag{12.148}$$

etc. Similarly, with $s = -1, -2, \ldots$ in (12.146) we obtain

$$\langle r^{-3}\rangle = \frac{\lambda}{2a_\lambda\zeta(\zeta+\delta)}\langle r^{-2}\rangle, \tag{12.149}$$

$$\langle r^{-4}\rangle = \frac{3\lambda^2 - \zeta(\zeta+\delta)}{\lambda a_\lambda[4\zeta(\zeta+\delta)-3]}\langle r^{-3}\rangle, \tag{12.150}$$

etc. Thus $\langle r^s\rangle$ for $s \le -3$ can be expressed in terms of the known expectation value $\langle r^{-2}\rangle$ (see (12.137)). The formulae derived above are restricted to their range of validity. For example, the parameters λ, a_λ, ζ, and δ must be such that the factors on the right-hand sides of (12.149) and (12.150) are not singular or negative.

For the Dirac equation the above expectation values are with respect to $|\lambda\zeta\rangle_\mathrm{U}$ and $|\lambda\zeta\rangle_\mathrm{L}$, the upper and lower elements of the transformed ket $|\psi'\rangle$ given by (12.46) and (12.47). With the aid of these we can calculate expectation values with respect to the Dirac ket $|\psi\rangle$. To illustrate this we calculate the expectation value of r. From (12.29), (4.208), (12.62), and (12.63)

$$\begin{aligned}\langle\psi|r|\psi\rangle &= (a_1^2+a_2^2)({}_\mathrm{U}\langle\lambda\zeta|r|\lambda\zeta\rangle_\mathrm{U} + {}_\mathrm{L}\langle\lambda\zeta|r|\lambda\zeta\rangle_\mathrm{L})\\ &\quad - 2a_1a_2({}_\mathrm{U}\langle\lambda\zeta|r|\lambda\zeta\rangle_\mathrm{L} + {}_\mathrm{L}\langle\lambda\zeta|r|\lambda\zeta\rangle_\mathrm{U}).\end{aligned} \tag{12.151}$$

The first two scalar products on the right-hand side of (12.151) are given by (12.147), (12.77), and (12.78). To calculate the other two scalar products in (12.151) we multiply (12.60) on the left by ${}_\mathrm{U}\langle\lambda\zeta|r$ and (12.61) on the left by ${}_\mathrm{L}\langle\lambda\zeta|r$, and subtract the resulting equations. A short calculation using (12.144) and (12.147) gives

$${}_\mathrm{U}\langle\lambda\zeta|r|\lambda\zeta\rangle_\mathrm{L} + {}_\mathrm{L}\langle\lambda\zeta|r|\lambda\zeta\rangle_\mathrm{U} = \tfrac{3}{2}a(1+\gamma^2\lambda^{-2})^{-1/2}\eta\gamma(\zeta - \lambda^2\zeta^{-1}). \tag{12.152}$$

Hence we obtain from (12.151)

$$\begin{aligned}\langle\psi|r|\psi\rangle &= \tfrac{1}{2}a(1+\gamma^2\lambda^{-2})^{-1/2}[3\lambda^2 - \zeta^2 + 2\gamma^2\\ &\quad - \eta\zeta(1+\gamma^2\zeta^{-2})^{1/2}(1+\gamma^2\lambda^{-2})^{1/2}].\end{aligned} \tag{12.153}$$

These algebraic methods can also be applied directly to the Dirac equation (12.4) [11], where (see (12.10) and (12.34))

$$H = c\alpha_r p_r + i\hbar c r^{-1}\alpha_r\beta k + \beta m_0c^2 - \gamma\hbar c r^{-1}. \tag{12.154}$$

If we treat the fine-structure constant and the Dirac quantum number as continuous variables in the Hellmann–Feynman theorem we have

$$\hbar c\langle\psi|r^{-1}|\psi\rangle = -\partial E/\partial\gamma \tag{12.155}$$

and

$$i\hbar c\langle\psi|r^{-1}\alpha_r\beta|\psi\rangle = \partial E/\partial k. \tag{12.156}$$

Substituting (12.121) in (12.72) and performing the above partial differentiations we obtain from (12.155) and (12.156) (note that N is a constant in these calculations)

$$\langle r^{-1}\rangle = (a\zeta\lambda^3)^{-1}(1+\gamma^2\lambda^{-2})^{-3/2}(\zeta\lambda+\gamma^2) \tag{12.157}$$

and

$$\langle r^{-1}\alpha_r\beta\rangle = \mathrm{i}\gamma a^{-1}\eta\lambda^{-3}(1+\gamma^2\zeta^{-2})^{1/2}(1+\gamma^2\lambda^{-2})^{-3/2}. \tag{12.158}$$

(In the rest of this section $\langle\ldots\rangle$ denotes the expectation value with respect to the Dirac ket, that is $\langle\psi|\ldots|\psi\rangle$.)

Next we apply the hypervirial theorem [11]. In (1.193) we substitute (12.154) and $W = r^s, p_r r^{s+1}, r^s\alpha_r, r^s\beta$, and $r^s\alpha_r\beta$. Hence we obtain the relations

$$s\langle r^{s-1}\alpha_r\rangle = 0, \tag{12.159}$$

$$\mathrm{i}\hbar ks\langle r^{s-1}\alpha_r\beta\rangle + m_0c^2(s+1)\langle r^s\beta\rangle - \gamma\hbar cs\langle r^{s-1}\rangle - E(s+1)\langle r^s\rangle = 0, \tag{12.160}$$

$$-2\mathrm{i}m_0c^2\langle r^s\alpha_r\beta\rangle + 2\hbar ck\langle r^{s-1}\beta\rangle + \hbar cs\langle r^{s-1}\rangle = 0, \tag{12.161}$$

$$-\tfrac{1}{2}\mathrm{i}\hbar cs\langle r^{s-1}\alpha_r\beta\rangle - E\langle r^s\beta\rangle - \gamma\hbar c\langle r^{s-1}\beta\rangle + m_0c^2\langle r^s\rangle = 0, \tag{12.162}$$

$$-\mathrm{i}E\langle r^s\alpha_r\beta\rangle - \mathrm{i}\gamma\hbar c\langle r^{s-1}\alpha_r\beta\rangle + \tfrac{1}{2}\hbar cs\langle r^{s-1}\beta\rangle + \hbar ck\langle r^{s-1}\rangle = 0. \tag{12.163}$$

The expectation values $\langle r^s\alpha_r\rangle$, $\langle r^s\beta\rangle$, $\langle r^s\alpha_r\beta\rangle$, and $\langle r^s\rangle$ for integral values of s can be calculated from (12.157)–(12.163). We give a few examples. With $s = 0$ in (12.160)

$$\langle\beta\rangle = E/m_0c^2. \tag{12.164}$$

With $s = 0$ in (12.163)

$$\begin{aligned}\mathrm{i}\langle\alpha_r\beta\rangle &= E^{-1}(\hbar ck\langle r^{-1}\rangle - \mathrm{i}\gamma\hbar c\langle r^{-1}\alpha_r\beta\rangle)\\ &= -\gamma\eta\zeta\lambda^{-2}(1+\gamma^2\zeta^{-2})^{1/2}(1+\gamma^2\lambda^{-2})^{-1}.\end{aligned} \tag{12.165}$$

Similarly, if we take $s = 1$ in (12.160) and (12.162), and eliminate $\langle r\beta\rangle$ from the resulting equations, we find

$$2E^{-1}(m_0^2c^4 - E^2)\langle r\rangle = 3\gamma\hbar c + \mathrm{i}\hbar cE^{-1}(m_0c^2 - kE)\langle\alpha_r\beta\rangle. \tag{12.166}$$

Equations (12.165) and (12.166) yield (12.153). Other examples and their physical applications are discussed in Ref. 11; extensions to non-Coulombic potentials are treated in Ref. 12.

From the expectation value of both sides of (12.4) with an attractive Coulomb potential we have

$$c\langle\boldsymbol{\alpha}.\mathbf{p}\rangle = \gamma\hbar c\langle r^{-1}\rangle, \tag{12.167}$$

where we have used (12.164). Equation (12.167) is a particular case of the virial theorem for the Dirac equation. This theorem can be proved by the simple scaling argument used for the Schrödinger equation in Section 1.10. Consider the Hamiltonian

$$\hat{H} = c\boldsymbol{\alpha}.\mathbf{p} + \beta m_0 c^2 + V(\xi \mathbf{r}), \tag{12.168}$$

where ξ is an arbitrary real parameter. Then

$$\left\langle \xi \frac{\partial \hat{H}}{\partial \xi}\bigg|_{\mathbf{r},\mathbf{p}} \right\rangle = \left\langle \xi \frac{\partial V(\xi \mathbf{r})}{\partial \xi} \right\rangle = \sum_{i=1}^{3} \left\langle r_i \frac{\partial V(\xi \mathbf{r})}{\partial r_i} \right\rangle, \tag{12.169}$$

where $\langle\ \rangle$ denotes an expectation value with respect to a normalizable eigenket of $\hat{H}$. In terms of the conjugate operators $\mathbf{r}' = \xi\mathbf{r}$ and $\mathbf{p}' = \xi^{-1}\mathbf{p}$,

$$\hat{H} = \xi c\boldsymbol{\alpha}.\mathbf{p}' + \beta m_0 c^2 + V(\mathbf{r}'). \tag{12.170}$$

Thus

$$\left\langle \xi \frac{\partial \hat{H}}{\partial \xi}\bigg|_{\mathbf{r}',\mathbf{p}'} \right\rangle = \xi c\langle \boldsymbol{\alpha}.\mathbf{p}' \rangle. \tag{12.171}$$

The left-hand sides of (12.169) and (12.171) are equal by the Hellmann–Feynman theorem (1.203). Thus setting $\xi = 1$ we obtain the virial theorem

$$c\langle \boldsymbol{\alpha}.\mathbf{p} \rangle = \sum_{i=1}^{3} \left\langle r_i \frac{\partial V(\mathbf{r})}{\partial r_i} \right\rangle. \tag{12.172}$$

For an attractive Coulomb potential (12.172) reduces to (12.167).

The above derivation can readily be generalized to include a vector potential and a Lorentz scalar potential and it can also be used to obtain the virial theorem for the Klein–Gordon equation [13].

12.6 Coordinate-space wavefunctions

In the coordinate representation the wavefunctions of the Schrödinger and Klein–Gordon equations are

$$\psi_{Elm}(\mathbf{r}) = \langle \mathbf{r} | Elm \rangle. \tag{12.173}$$

The transformed Dirac wavefunction is (see (12.45))

$$\psi'_{Ejm}(\mathbf{r}) = \begin{pmatrix} \mathrm{i}\psi'_{\mathrm{U}}(\mathbf{r}) \\ \boldsymbol{\sigma}.\hat{\mathbf{r}}\,\psi'_{\mathrm{L}}(\mathbf{r}) \end{pmatrix}, \tag{12.174}$$

where

$$\psi'_{\mathrm{U}}(\mathbf{r}) = \langle \mathbf{r} | \psi'_{Ejmkl} \rangle_{\mathrm{U}}, \qquad \psi'_{\mathrm{L}}(\mathbf{r}) = \langle \mathbf{r} | \psi'_{Ejmkl} \rangle_{\mathrm{L}}. \tag{12.175}$$

Consider the ground state; that is, the state with $\lambda = \lambda_+$ and energy

$$E = m_0 c^2 (1 + \gamma^2 \lambda_+^{-2})^{-1/2} \tag{12.176}$$

(see Section 12.4). For this state (12.110) expressed in the coordinate representation is

$$\left(r\frac{\partial}{\partial r} + \frac{r}{2a_{\lambda_+}} - \lambda_+ + 1 \right) \psi_{\lambda_+ \zeta}(\mathbf{r}) = 0. \tag{12.177}$$

Here $\psi_{\lambda_+\zeta}$ denotes the Schrödinger and Klein–Gordon wavefunctions (12.173) and the components (12.175) of the transformed Dirac wavefunctions with energy eigenvalue (12.176). In (12.177) we separate variables

$$\psi_{\lambda_+\zeta}(\mathbf{r}) = R_{\lambda_+\zeta}(r)V(\hat{\mathbf{r}}). \tag{12.178}$$

Then

$$R_{\lambda_+\zeta}(r) = C_{\lambda_+}\left(\frac{r}{a_{\lambda_+}}\right)^{\lambda_+-1} \exp\left(-\frac{r}{2a_{\lambda_+}}\right), \tag{12.179}$$

and normalization to unity requires

$$C_{\lambda_+} = a_{\lambda_+}^{-3/2}[\Gamma(2\lambda_+ + 1)]^{-1/2}. \tag{12.180}$$

From (12.179) we generate all normalized radial wavefunctions $R_{\lambda\zeta}$ by applying the energy shift operations (12.98). In these calculations we require the effect of the scaling operators on wavefunctions, namely*

$$\Delta_{\lambda}^{\pm}R_{\lambda\zeta}(r) = \left(\frac{a_\lambda}{a_{\lambda\pm1}}\right)^{3/2} R_{\lambda\zeta}\left(\frac{a_\lambda}{a_{\lambda\pm1}}r\right). \tag{12.181}$$

Using also (12.90), (12.91), (12.99), (12.107), and (12.112) we obtain the recurrence relation

$$R_{\lambda\zeta}(u) = e^{-i\varepsilon}\left(\frac{a_{\lambda-1}}{a_\lambda}\right)^{3/2}\left(\frac{N-1+\lambda_+}{N(N+\lambda_+)(N-1+2\lambda_+)}\right)^{1/2}$$
$$\times\left(u\frac{\mathrm{d}}{\mathrm{d}u} - \frac{1}{2}u + \lambda\right)R_{\lambda-1,\zeta}(u), \tag{12.182}$$

where $u = r/a_\lambda$ and $\lambda = N + \lambda_+$ $(N = 1, 2, \ldots)$. In writing down (12.182) we have, for convenience, assumed wavefunctions normalized to unity. Thus we have set $\beta_{\lambda\zeta}^{+} = 1$ in (12.107): for the radial Dirac functions the normalization is corrected to (12.77) and (12.78) by including an appropriate factor in the final result (the factor M in (12.192)).

Let

$$R_{\lambda\zeta}(u) = C_{\lambda_+}u^{\lambda_+-1}\mathrm{e}^{-(1/2)u}\phi_N(u). \tag{12.183}$$

Then (12.182) yields

$$\phi_N(u) = \gamma_N D_N \phi_{N-1}(u) \qquad (N = 1, 2, \ldots), \tag{12.184}$$

where

$$\gamma_N = \mathrm{e}^{-i\varepsilon}\left(\frac{a_{\lambda-1}}{a_\lambda}\right)^{3/2}\left(\frac{(N-1+\lambda_+)(N-1-2\lambda_+)}{N(N+\lambda_+)}\right)^{1/2} \tag{12.185}$$

* The proof of (12.181) is similar to that of (6.86) and (6.87).

and

$$D_N = 1 - \frac{1}{N - 1 + 2\lambda_+} u + \frac{1}{N - 1 + 2\lambda_+} u \frac{\mathrm{d}}{\mathrm{d}u}. \tag{12.186}$$

According to (12.179) and (12.183), $\phi_0(u) = 1$. Hence (12.184) gives

$$\phi_N(u) = \gamma_N \ldots \gamma_2 \gamma_1 P_N(u), \tag{12.187}$$

where

$$P_N(u) = D_N \ldots D_2 D_1 1. \tag{12.188}$$

The product of the γ in (12.187) is readily evaluated and we have

$$\gamma_N \ldots \gamma_2 \gamma_1 = \exp(-\mathrm{i}N\varepsilon)\left(\frac{a_{\lambda_+}}{a_{N+\lambda_+}}\right)^{3/2}\left(\frac{\lambda_+\Gamma(N + 2\lambda_+)}{N!(N + \lambda_+)\Gamma(2\lambda_+)}\right)^{1/2}. \tag{12.189}$$

Also, comparing (12.186) and (12.188) with (B.1) we see that

$$P_N(u) = {}_1F_1(-N; 2\lambda_+; u). \tag{12.190}$$

Substituting (12.187) and (12.180) in (12.183) we obtain

$$R_{\lambda\zeta}(r) = C_\lambda {}_1F_1\left(-N; 2\lambda_+; \frac{r}{a_\lambda}\right)\left(\frac{r}{a_\lambda}\right)^{\lambda_+ - 1} \exp\left(-\frac{r}{2a_\lambda}\right). \tag{12.191}$$

Here λ and N are defined in (12.121), and

$$C_\lambda = \exp(-\mathrm{i}N\varepsilon)a_\lambda^{-3/2}\left(\frac{\Gamma(N + 2\lambda_+)}{N!2\lambda[\Gamma(2\lambda_+)]^2}\right)^{1/2} M. \tag{12.192}$$

For Schrödinger and Klein–Gordon wavefunctions that are normalized to unity, $M = 1$. For the transformed Dirac wavefunctions the normalizations (12.77) and (12.78) require

$$M = \left(\frac{1}{2(1 + \gamma^2\lambda^{-2})^{1/2}}\right)^{1/2}\left[\left(\frac{1 + \gamma^2\zeta^{-2}}{1 + \gamma^2\lambda^{-2}}\right)^{1/2} \pm 1\right]^{1/2}, \tag{12.193}$$

where the upper (lower) sign applies to the upper (lower) wavefunction in (12.174).

The above solutions can readily be expressed in terms of quantum numbers such as n, j, and l (see Table 12.2). In the Schrödinger and Klein–Gordon wavefunctions $N = n - l - 1$. For the Schrödinger wavefunctions (12.191) yields the normalized radial wavefunctions $R_{nl}(r)$ given by (6.98).* Similarly, from (12.191) one can write down the radial Klein–Gordon wavefunctions $R_{nl}(r)$.

---

* The phase factor is that in (6.98) if we take $\varepsilon = 0$ in (12.192) and include a factor $(-1)^l$ in (12.179).

For the transformed Dirac wavefunctions with $j = l + \frac{1}{2}$, we have $N = n - j - \frac{1}{2}$ in the upper component and $N = n - j - \frac{3}{2}$ in the lower component; if $j = l - \frac{1}{2}$ these values of N are reversed (see Fig. 12.1 and eqn (12.121)). Thus for $j = l \pm \frac{1}{2}$ the transformed wavefunctions (12.174) are

$$\psi'_{njm}(\mathbf{r}) = \begin{pmatrix} \mathrm{i}R^{\pm}_{nj}(r)V(\hat{\mathbf{r}}) \\ \boldsymbol{\sigma}.\hat{\mathbf{r}}R^{\mp}_{nj}(r)V(\hat{\mathbf{r}}) \end{pmatrix}, \tag{12.194}$$

where

$$R^{\pm}_{nj}(r) = C^{\pm}_{nj}\,{}_1F_1\left(-N + \frac{1}{2} \mp \frac{1}{2}; 2\zeta + 1 \mp 1; \frac{r}{a_{nj}}\right)\left(\frac{r}{a_{nj}}\right)^{\zeta-(1/2)\mp(1/2)} \exp\left(-\frac{r}{2a_{nj}}\right), \tag{12.195}$$

$$C^{\pm}_{nj} = \exp[-\mathrm{i}(N - \tfrac{1}{2} \pm \tfrac{1}{2})\varepsilon]a_{nj}^{-3/2} \times \left(\frac{\Gamma(N + \frac{1}{2} \mp \frac{1}{2} + 2\zeta)}{(N - \frac{1}{2} \pm \frac{1}{2})!2(N + \zeta)[\Gamma(2\zeta + 1 \mp 1)]^2}\right)^{1/2} M^{\pm}_{nj}, \tag{12.196}$$

$$M^{\pm}_{nj} = \left(\frac{1}{2[1 + \gamma^2(N + \zeta)^{-2}]^{1/2}}\right)^{1/2}\left[\left(\frac{1 + \gamma^2\zeta^{-2}}{1 + \gamma^2(N + \zeta)^{-2}}\right)^{1/2} \mp \eta\right]^{1/2}, \tag{12.197}$$

$$a^{\pm}_{nj} = \tfrac{1}{2}[(N + \zeta)^2 + \gamma^2]^{1/2}a, \tag{12.198}$$

and $N = n - j - \frac{1}{2}$. The phase ε can be determined by expressing (12.60) and (12.61) in the coordinate representation and then substituting the limiting forms as $r \to 0$ of $R^{\pm}_{nj}$. This shows that $\varepsilon = \frac{1}{2}(1 - \eta)\pi$.

The solution $\psi_{njm}(\mathbf{r})$ to the original Dirac–Coulomb equation is given in terms of the above solutions by (12.29), (12.62), and (12.63) expressed in the coordinate representation; that is,

$$\psi_{njm}(r) = \begin{pmatrix} \mathrm{i}(a_1 R^{\pm}_{nj} - a_2 R^{\mp}_{nj})V(\hat{\mathbf{r}}) \\ \boldsymbol{\sigma}.\hat{\mathbf{r}}(-a_2 R^{\pm}_{nj} + a_1 R^{\mp}_{nj})V(\hat{\mathbf{r}}) \end{pmatrix}, \tag{12.199}$$

where a_1 and a_2 are given by (12.64). Each component of the Dirac wavefunction (12.199) consists of a combination of two confluent hypergeometric functions.

Finally, we determine the angular part of the wavefunctions $V(\hat{\mathbf{r}})$. For the Schrödinger and Klein–Gordon equations

$$\psi_{nlm}(\mathbf{r}) = R_{nl}(r)V(\hat{\mathbf{r}}), \tag{12.200}$$

and V is a normalized eigenfunction of $\mathbf{L}^2$ and L_z with eigenvalues $\hbar^2 l(l + 1)$ and $\hbar m$. Hence

$$V(\hat{\mathbf{r}}) = Y_{lm}(\hat{\mathbf{r}}), \tag{12.201}$$

where Y_{lm} is the spherical harmonic (4.129).

For the Dirac equation $V(\hat{\mathbf{r}})$ in (12.199) is the spinor

$$V(\hat{\mathbf{r}}) = \begin{pmatrix} V^{(1)}(\hat{\mathbf{r}}) \\ V^{(2)}(\hat{\mathbf{r}}) \end{pmatrix}. \tag{12.202}$$

Equations (12.199) and (12.29) show that $V(\hat{\mathbf{r}})$ is an eigenstate of $(\mathbf{L} + \frac{1}{2}\hbar\boldsymbol{\sigma})^2$, $L_z + \frac{1}{2}\hbar\sigma_z$, $\mathbf{L}^2$, and $\mathscr{K}$, with eigenvalues $\hbar^2 j(j+1)$, $\hbar m$, $\hbar^2 l(l+1)$, and k respectively. The eigenvalue equations for $\mathbf{L}^2$ and $L_z + \frac{1}{2}\hbar\sigma_z$ require

$$V(\hat{\mathbf{r}}) = \begin{pmatrix} a Y_{l,m-1/2}(\hat{\mathbf{r}}) \\ b Y_{l,m+1/2}(\hat{\mathbf{r}}) \end{pmatrix}, \tag{12.203}$$

where a and b are constants. The eigenvalue equation for $\mathscr{K}$ requires (see (12.13) and (4.190))

$$a\hbar(m + \tfrac{1}{2}) Y_{l,m-1/2} + bL_- Y_{l,m+1/2} = \hbar k a Y_{l,m-1/2} \tag{12.204}$$

and

$$aL_+ Y_{l,m-1/2} + b\hbar(-m + \tfrac{1}{2}) Y_{l,m+1/2} = \hbar k b Y_{l,m+1/2}. \tag{12.205}$$

Use of (4.30) in the last two equations yields

$$(k - m - \tfrac{1}{2})(a/b) = [l(l+1) - (m^2 - \tfrac{1}{4})]^{1/2} \tag{12.206}$$

and

$$(k + m - \tfrac{1}{2})(b/a) = [l(l+1) - (m^2 - \tfrac{1}{4})]^{1/2}. \tag{12.207}$$

In terms of the values of k given by (12.24), the right-hand sides of (12.206) and (12.207) are equal to $[(k - m - \frac{1}{2})(k + m - \frac{1}{2})]^{1/2}$. Hence

$$\frac{b}{a} = \pm \left(\frac{k - m - \frac{1}{2}}{k + m - \frac{1}{2}} \right)^{1/2} \tag{12.208}$$

for $j = l \pm \frac{1}{2}$. Normalization of V requires

$$|a|^2 + |b|^2 = 1. \tag{12.209}$$

The solutions to (12.208) and (12.209) are

$$a = \left(\frac{l \pm m + \frac{1}{2}}{2l + 1} \right)^{1/2} \exp(\mathrm{i}\alpha_\pm) \tag{12.210}$$

and

$$b = \pm \left(\frac{l \mp m + \frac{1}{2}}{2l + 1} \right)^{1/2} \exp(\mathrm{i}\alpha_\pm) \tag{12.211}$$

for $j = l \pm \frac{1}{2}$. The phases $\alpha_\pm$ are arbitrary real constants. Substituting (12.210) and (12.211) in (12.203), we obtain for $V(\hat{\mathbf{r}})$ the spinor spherical harmonics

$$V_{jml}^{\pm}(\hat{\mathbf{r}}) = \exp(\mathrm{i}\alpha_\pm)(2l + 1)^{-1/2} \begin{pmatrix} (l \pm m + \frac{1}{2})^{1/2} Y_{l,m-1/2}(\hat{\mathbf{r}}) \\ \pm(l \mp m + \frac{1}{2})^{1/2} Y_{l,m+1/2}(\hat{\mathbf{r}}) \end{pmatrix}, \tag{12.212}$$

for $j = l \pm \frac{1}{2}$.

With the aid of (4.83) and (4.84) it can be shown that

$$\boldsymbol{\sigma}.\hat{\mathbf{r}}V^{+}_{jml} = \exp[\mathrm{i}(\alpha_{+} - \alpha_{-})]\,V^{-}_{j,m,l+1} \qquad (12.213)$$

and

$$\boldsymbol{\sigma}.\hat{\mathbf{r}}V^{-}_{jml} = \exp[\mathrm{i}(\alpha_{-} - \alpha_{+})]\,V^{+}_{j,m,l-1}. \qquad (12.214)$$

These two equations verify that for the upper or lower component of the Dirac wavefunction, $\boldsymbol{\sigma}.\hat{\mathbf{r}}$ raises l by one if $j = l + \frac{1}{2}$ and lowers l by one if $j = l - \frac{1}{2}$; it also follows that $\boldsymbol{\sigma}.\hat{\mathbf{r}}$ reverses the sign of k (see Section 12.1).

12.7 Discussion

1. It is instructive to compare the operator analysis presented above with wave-mechanical properties of Coulomb problems. In the coordinate representation and assuming separation of variables, (12.70) becomes

$$\left(-a_\lambda^2\frac{\mathrm{d}^2}{\mathrm{d}r^2} - \frac{2a_\lambda^2}{r}\frac{\mathrm{d}}{\mathrm{d}r} + \frac{1}{4} - \frac{\lambda a_\lambda}{r} + \zeta(\zeta+\delta)\frac{a_\lambda^2}{r^2}\right)R_{\lambda\zeta}(r) = 0. \qquad (12.215)$$

For the Schrödinger equation the radial solutions have been discussed in Section 6.4. Here we consider the Klein–Gordon and Dirac equations. For these the (unnormalized) solutions of (12.215) that vanish at $r = \infty$ are [14]

$$\begin{aligned} R_{\lambda\zeta}\left(\frac{r}{a_\lambda}\right) = {} & \frac{\Gamma(1-2\lambda_-)}{\Gamma(\lambda_+ - \lambda)}\,{}_1F_1\left(\lambda_- - \lambda; 2\lambda_-; \frac{r}{a_\lambda}\right)\left(\frac{r}{a_\lambda}\right)^{\lambda_- - 1}\exp\left(-\frac{r}{2a_\lambda}\right) \\ & + \frac{\Gamma(1-2\lambda_+)}{\Gamma(\lambda_- - \lambda)}\,{}_1F_1\left(\lambda_+ - \lambda; 2\lambda_+; \frac{r}{a_\lambda}\right)\left(\frac{r}{a_\lambda}\right)^{\lambda_+ - 1}\exp\left(-\frac{r}{2a_\lambda}\right), \end{aligned} \qquad (12.216)$$

where $\lambda_\pm$ are given by (12.112).

One can readily show that*

$$Q_\lambda^- R_{\lambda\zeta}\left(\frac{r}{a_\lambda}\right) = -(\lambda - \lambda_-)(\lambda - \lambda_+)R_{\lambda-1,\zeta}\left(\frac{r}{a_{\lambda-1}}\right) \qquad (12.217)$$

and

$$Q_{\lambda-1}^+ R_{\lambda-1,\zeta}\left(\frac{r}{a_{\lambda-1}}\right) = (\lambda - \lambda_-)(\lambda - \lambda_+)R_{\lambda\zeta}\left(\frac{r}{a_\lambda}\right), \qquad (12.218)$$

where

$$Q_\lambda^\pm = \left[\exp\left(r\frac{\mathrm{d}}{\mathrm{d}r} + \frac{3}{2}\right)\ln\left(\frac{a_\lambda}{a_{\lambda\pm1}}\right)\right]\left(\pm r\frac{\mathrm{d}}{\mathrm{d}r} - \frac{r}{2a_\lambda} + \lambda \pm 1\right). \qquad (12.219)$$

* Equations (12.217) and (12.218) can be proved with the aid of eqns (10) on p. 258 of Ref. 14.

When $\lambda = \lambda_\pm$, eqns (12.217) and (12.218) are the differential forms of (12.100) and (12.101).

The quadratic integrability of the solutions (12.216) depends on their behaviour as $r \to 0$. The confluent hypergeometric functions in (12.216) are well behaved for all values of λ as $r \to 0$ [14]. Because $\lambda_- < \lambda_+$, the first term in (12.216) dominates as $r \to 0$, and

$$R_{\lambda\zeta} \sim r^{\lambda_- - 1}. \tag{12.220}$$

Hence if $\lambda_- \leq -\frac{1}{2}$, $R_{\lambda\zeta}$ is quadratically integrable only if the first term in (12.216) is zero. This restricts the values of $\lambda_+ - \lambda$ to the poles of $\Gamma(\lambda_+ - \lambda)$, namely,

$$\lambda = N + \lambda_+ \qquad (N = 0, 1, 2, \ldots). \tag{12.221}$$

With these values of λ the solutions (12.216) are, when normalized, identical to the Klein–Gordon wavefunctions and the transformed Dirac wavefunctions given by (12.191).

There are two cases in which $\lambda_- > -\frac{1}{2}$: for the Klein–Gordon equation with $l = 0$, and for the Dirac equation with $j = \frac{1}{2}$ and $Z > 118$ (see Section 12.4). The corresponding solutions (12.216) are quadratically integrable for all values of λ: the solutions for $\lambda \neq N + \lambda_+$ are the radial coordinate-space solutions associated with the anomalous states mentioned in Section 12.4 [15].

Consider next the Hermitian property of the differential form of Ω, namely the operator in (12.215). This operator is Hermitian if the expectation value of the 'potential' in (12.215) is finite [16]; that is,

$$|\langle -\lambda a_\lambda r^{-1} + \zeta(\zeta + \delta)a_\lambda^2 r^{-2}\rangle| < \infty, \tag{12.222}$$

which requires

$$r^{1/2} R_{\lambda\zeta} \to 0 \tag{12.223}$$

as $r \to 0$. The solutions with λ given by (12.221) satisfy (12.223), but the anomalous solutions referred to above do not. Another feature of these anomalous solutions is that they do not have a satisfactory non-relativistic limit. Consider, for example, a spinless particle. For the anomalous 1s state $\lambda = \lambda_-$ and in the limit $\gamma \to 0$, $\lambda_- \approx \gamma^2$. Therefore $E \approx \gamma m_0 c^2$ (see (12.72)). This behaviour occurs because in the non-relativistic limit the anomalous states of the Klein–Gordon equation are not solutions of the Schrödinger–Coulomb equation [17].

2. In the coordinate representation the Dirac wavefunction for a spherically symmetric potential is

$$\psi_{Ejm}(\mathbf{r}) = \begin{pmatrix} ir^{-1}F(r)V_{jml}^{\pm}(\hat{\mathbf{r}}) \\ r^{-1}G(r)\boldsymbol{\sigma}\cdot\hat{\mathbf{r}}V_{jml}^{\pm}(\hat{\mathbf{r}}) \end{pmatrix}, \tag{12.224}$$

where $V_{jml}^{\pm}$ are the spinors (12.212), and F and G are the solutions to the coupled equations (see (12.32) and (12.33))

$$\hbar c\left(\frac{\mathrm{d}}{\mathrm{d}r}-\frac{k}{r}\right)F=(E+m_0c^2-q\Phi)G \tag{12.225}$$

and

$$\hbar c\left(\frac{\mathrm{d}}{\mathrm{d}r}+\frac{k}{r}\right)G=-(E-m_0c^2-q\Phi)F. \tag{12.226}$$

For non-Coulombic potentials these equations must be solved numerically [18].

The Dirac wavefunction (12.224) is, of course, not an eigenstate of $\mathbf{L}^2$. To understand the significance of the quantum number l, consider the parity operator [19]

$$P=\beta\mathscr{P}. \tag{12.227}$$

Here β is the matrix (11.22) and $\mathscr{P}$ anticommutes with $\mathbf{r}$ and $\mathbf{p}$. Clearly P commutes with the Hamiltonian (12.5) and with $\mathbf{J}^2$ and J_z. The effect of P on a wavefunction is

$$P\psi_{Ejm}(\mathbf{r})=\beta\psi_{Ejm}(-\mathbf{r})=(-1)^l\psi_{Ejm}(\mathbf{r}), \tag{12.228}$$

where in the last step we have used (12.224) and (4.131). Hence ψ_{Ejm} is an eigenstate of parity with eigenvalue $(-1)^l$.

12.8 A relativistic (Dirac) oscillator

Relativistic quantum-mechanical oscillators have important applications, particularly in particle physics [20]. An interesting system of this kind is associated with a Dirac Hamiltonian that is linear in both $\mathbf{p}$ and $\mathbf{r}$, namely

$$H=c\boldsymbol{\alpha}.(\mathbf{p}-\mathrm{i}m_0\omega\beta\mathbf{r})+\beta m_0c^2. \tag{12.229}$$

Here ω is a positive constant. For reasons that will become apparent below the system described by the corresponding Dirac equation

$$[c\boldsymbol{\alpha}.(\mathbf{p}-\mathrm{i}m_0\omega\beta\mathbf{r})+\beta m_0c^2]|\Psi\rangle=\mathrm{i}\hbar\frac{\partial}{\partial t}|\Psi\rangle \tag{12.230}$$

is referred to as a Dirac oscillator [21]. The Dirac oscillator has simple algebraic properties and it possesses exact solutions. Before discussing these we make some introductory remarks.

Using (11.22), (11.28), and (12.17) we see that for energy eigenkets (12.230) consists of the coupled equations

$$c\boldsymbol{\sigma}.(\mathbf{p}-\mathrm{i}m_0\omega\mathbf{r})|\phi\rangle=(E+m_0c^2)|\chi\rangle \tag{12.231}$$

and

$$c\boldsymbol{\sigma}.(\mathbf{p} + \mathrm{i}m_0\omega\mathbf{r})|\chi\rangle = (E - m_0c^2)|\phi\rangle. \tag{12.232}$$

These equations can be decoupled: multiplying (12.231) on the left by $c\boldsymbol{\sigma}.(\mathbf{p} + \mathrm{i}m_0\omega\mathbf{r})$ and using (12.232), (4.206), and the identity

$$(\boldsymbol{\sigma}.\mathbf{p})(\boldsymbol{\sigma}.\mathbf{r}) - (\boldsymbol{\sigma}.\mathbf{r})(\boldsymbol{\sigma}.\mathbf{p}) = -2\mathrm{i}\boldsymbol{\sigma}.\mathbf{L} - 3\mathrm{i}\hbar$$

(see (4.210) and (1.138)), we obtain

$$\left(\frac{1}{2m_0}\mathbf{p}^2 + \frac{1}{2}m_0\omega^2\mathbf{r}^2 - 2\frac{\omega}{\hbar}\mathbf{L}.\mathbf{S} - \frac{3}{2}\hbar\omega\right)|\phi\rangle = (2m_0c^2)^{-1}(E^2 - m_0^2c^4)|\phi\rangle, \tag{12.233}$$

where $\mathbf{S} = \frac{1}{2}\hbar\boldsymbol{\sigma}$. Similarly

$$\left(\frac{1}{2m_0}\mathbf{p}^2 + \frac{1}{2}m_0\omega^2\mathbf{r}^2 + 2\frac{\omega}{\hbar}\mathbf{L}.\mathbf{S} + \frac{3}{2}\hbar\omega\right)|\chi\rangle = (2m_0c^2)^{-1}(E^2 - m_0^2c^4)|\chi\rangle. \tag{12.234}$$

Consider the non-relativistic limit. For positive energy solutions the norm of $|\chi\rangle$ is small in this limit (see (12.254)) and the Dirac equation reduces to (12.233). Let $E = m_0c^2 + \varepsilon$. Then $E^2 - m_0^2c^4 \approx 2m_0c^2\varepsilon$ and hence ε is the eigenvalue of the operator on the left-hand side of (12.233). This operator corresponds to the Hamiltonian of a three-dimensional isotropic harmonic oscillator with constant spin–orbit coupling.

It is also interesting to write (12.230) in a covariant form analogous to (11.35) [22, 23]. In terms of the frame-dependent quantity $u_\mu = \mathrm{i}m_0\omega(\mathbf{0}, 1)$ and the matrices $\sigma_{\mu\nu} = \frac{1}{2}\mathrm{i}[\gamma_\mu, \gamma_\nu]$, the interaction term in (12.230) is

$$-\mathrm{i}m_0c\omega\boldsymbol{\alpha}.\beta\mathbf{r} = \mathrm{i}m_0c\omega\beta\boldsymbol{\alpha}.\mathbf{r} = \beta\tfrac{1}{2}c\sigma_{\mu\nu}F_{\mu\nu}, \tag{12.235}$$

where

$$F_{\mu\nu} = x_\nu u_\mu - x_\mu u_\nu. \tag{12.236}$$

Substituting (12.235) in (12.230) and proceeding as in the derivation of (11.17) we obtain the desired form [22, 23]

$$(\mathrm{i}\gamma_\mu p_\mu + m_0c + \tfrac{1}{2}\sigma_{\mu\nu}F_{\mu\nu})|\Psi\rangle = 0. \tag{12.237}$$

Here the interaction term resembles that for a Dirac particle with an anomalous magnetic moment in an electromagnetic field with field tensor [24]

$$F_{\mu\nu} = \frac{\partial A_\nu}{\partial x_\mu} - \frac{\partial A_\mu}{\partial x_\nu}. \tag{12.238}$$

Comparing (12.238) and (12.236) we see that the effective four-vector potential is [23]

$$A_\mu = \tfrac{1}{2}(x_\mu(u_\lambda x_\lambda) - \tfrac{1}{2}x_\lambda^2 u_\mu). \tag{12.239}$$

Similarly, the (frame-dependent) 'magnetic' and 'electric' fields associated with

(12.238) are [23]

$$\mathbf{B} = 0, \qquad \mathbf{E} = m_0 c\omega \mathbf{r}. \tag{12.240}$$

Thus the interaction in (12.229) occurs as if it were due to an infinite uniform charge density, and it is this aspect that is of interest in models of quark confinement [23].

Next we determine a c.s.c.o. for the Dirac oscillator. It is straightforward to show that $\mathbf{J} = \mathbf{L} + \mathbf{S}$ commutes with the interaction term in (12.229); because $\mathbf{J}$ also commutes with the Hamiltonian for a free Dirac particle (Section 11.2) it follows that $\mathbf{J}$ commutes with (12.229). Much of the theory discussed in Section 12.1 applies here as well: the Dirac operator K commutes with (12.229) and the quantum number k is given by (12.24). The parity operator P (eqn (12.227)) commutes with (12.229) and the parity eigenvalues are equal to $(-1)^l$. The operators H, $\mathbf{J}^2$, J_z, $\mathbf{S}^2$, and P are a c.s.c.o. (see Section 12.10).

The eigenvalue equations for the elements $|\phi\rangle$ and $|\chi\rangle$ in (12.17) are (see (12.233), (12.234), (12.6), (12.7), (12.18), (12.19), and (12.21))

$$\mathscr{H}|njm\zeta\rangle = \hbar\omega(n + 1 + \delta)|njm\zeta\rangle, \tag{12.241}$$

$$\bar{\mathbf{J}}^2|njm\zeta\rangle = \hbar^2 j(j+1)|njm\zeta\rangle, \tag{12.242}$$

$$\bar{J}_z|njm\zeta\rangle = \hbar m|njm\zeta\rangle, \tag{12.243}$$

$$\mathbf{L}^2|njm\zeta\rangle = \hbar^2\zeta(\zeta+1)|njm\zeta\rangle. \tag{12.244}$$

Here*

$$\mathscr{H} = \frac{1}{2m_0}\mathbf{p}^2 + \frac{1}{2}m_0\omega^2\mathbf{r}^2, \tag{12.245}$$

$$\bar{\mathbf{J}} = \mathbf{L} + \tfrac{1}{2}\hbar\boldsymbol{\sigma}, \tag{12.246}$$

n is the dimensionless parameter

$$n = (2m_0c^2\hbar\omega)^{-1}(E^2 - m_0^2c^4) + k - 1, \tag{12.247}$$

and $|njm\zeta\rangle$ denotes either $|\phi\rangle$ or $|\chi\rangle$. If $|njm\zeta\rangle = |\phi\rangle$,

$$\delta = \tfrac{1}{2}, \qquad \zeta = l; \tag{12.248}$$

if $|njm\zeta\rangle = |\chi\rangle$,

$$\delta = -\tfrac{1}{2}, \qquad \zeta = l - \eta, \tag{12.249}$$

where η is given by (12.56). For later use we note that the positive and negative energy eigenvalues are given in terms of n and j by (see (12.247) and (12.24))

$$E_{nj} = \pm m_0c^2\{1 + 2\hbar\omega(m_0c^2)^{-1}[n + 1 + \eta(j + \tfrac{1}{2})]\}^{1/2}. \tag{12.250}$$

---

* Note that $\mathscr{H}$ and $\bar{\mathbf{J}}$ are in 2×2 block form.

The operators $\mathscr{H}$, $\bar{\mathbf{J}}^2$, $\bar{J}_z$, and $\mathbf{L}^2$ are a set of commuting operators for the elements $|\phi\rangle$ and $|\chi\rangle$ of the Dirac ket. In the next section we determine shift operators for the quantum numbers n, j, m, and ζ. We will need the normalization conditions for $|\phi\rangle$ and $|\chi\rangle$. The normalization (11.38) of the Dirac ket requires

$$\langle\phi|\phi\rangle + \langle\chi|\chi\rangle = 1. \tag{12.251}$$

Multiplying (12.231) and (12.232) on the left by $\langle\chi|$ and $\langle\phi|$, respectively, and using the Hermitian properties of $\mathbf{r}$, $\mathbf{p}$, and $\boldsymbol{\sigma}$, we have

$$(E + m_0c^2)\langle\chi|\chi\rangle = (E - m_0c^2)\langle\phi|\phi\rangle. \tag{12.252}$$

Hence

$$\langle\phi|\phi\rangle = \tfrac{1}{2}(1 + m_0c^2/E) \tag{12.253}$$

$$\langle\chi|\chi\rangle = \tfrac{1}{2}(1 - m_0c^2/E). \tag{12.254}$$

We see that in the non-relativistic limit the norm of $|\chi\rangle$ is small for positive energy solutions, while the norm of $|\phi\rangle$ is small for negative energy solutions.

12.9 Shift operators for the Dirac oscillator [25]

The operator $\mathscr{H}$ in (12.245) is the same as the Hamiltonian for a non-relativistic oscillator, and this suggests that to construct shift operators for the eigenkets $|njm\zeta\rangle$ we can use results for the non-relativistic oscillator in an angular momentum basis (Section 8.1). Thus we consider the dimensionless operator (see (8.12))

$$\mathbf{D}(\omega) = \hbar^{-1}(\mathbf{a}^\dagger \times \mathbf{L} - i\mathbf{a}^\dagger K), \tag{12.255}$$

where

$$\mathbf{a}^\dagger = \frac{1}{(2m_0\hbar|\omega|)^{1/2}}(-i\mathbf{p} + m_0\omega\mathbf{r}) \tag{12.256}$$

is the boson creation operator and K is to be determined.*

Note that (12.255) corresponds to the choice $\mathbf{v} = -i\hbar^{-1}\mathbf{a}^\dagger$ in (4.44) for U^+, and thus we can use results from Sections 4.4 and 8.1 to write down certain commutators involving $\mathbf{D}(\omega)$. If

$$[\mathscr{H}, K] = 0 \tag{12.257}$$

then (see (8.21))

$$[\mathscr{H}, \mathbf{D}(\omega)] = \hbar\omega\mathbf{D}(\omega). \tag{12.258}$$

Also, if

* In writing down (12.255) we have, for later convenience, omitted a constant factor in (8.12). Also, the modulus sign in (12.256) is advantageous: see (12.264).

$$[\mathbf{L}^2, K] = 0 \tag{12.259}$$

and (see (4.47) with $\Omega = 0$)

$$K^2 - \hbar K - \mathbf{L}^2 = 0, \tag{12.260}$$

then (see (4.51))

$$[\mathbf{L}^2, \mathbf{D}(\omega)] = 2\hbar \mathbf{D}(\omega) K. \tag{12.261}$$

The solution (see (4.213))

$$K = -\boldsymbol{\sigma}.\mathbf{L} \tag{12.262}$$

to (12.260) satisfies (12.257) and (12.259).*

In what follows we suppose that K is given by (12.262); thus

$$\mathbf{D}(\omega) = \hbar^{-1}[\mathbf{a}^\dagger \times \mathbf{L} + \mathrm{i}\mathbf{a}^\dagger(\boldsymbol{\sigma}.\mathbf{L})]. \tag{12.263}$$

Then

$$\mathbf{D}(-\omega) = -\hbar^{-1}[\mathbf{a} \times \mathbf{L} + \mathrm{i}\mathbf{a}(\boldsymbol{\sigma}.\mathbf{L})] \tag{12.264}$$

where

$$\mathbf{a} = \frac{1}{(2m_0\hbar|\omega|)^{1/2}}(\mathrm{i}\mathbf{p} + m_0\omega\mathbf{r}) \tag{12.265}$$

is the boson annihilation operator.

Because $\bar{\mathbf{J}}$ commutes with (12.262), we see from (4.45) and (12.246) that $\mathbf{D}(\omega)$ is a vector operator with respect to $\bar{\mathbf{J}}$

$$[\bar{J}_i, D_j(\omega)] = \mathrm{i}\hbar\varepsilon_{ijk}D_k(\omega). \tag{12.266}$$

It is also straightforward to show that

$$[\mathscr{K}, \mathbf{D}(\omega)] = -\mathbf{D}(\omega). \tag{12.267}$$

From the commutators (12.258), (12.261), (12.266), (12.267) and from (12.18), (12.19), (12.241)–(12.244) we obtain the 12 shift operations

$$D_\mu(\pm\omega)|njm\zeta\rangle_{\mathrm{U}} = \left(\frac{1 + m_0c^2/E_{nj}}{1 + m_0c^2/E_{n\pm1,j+\eta}}\right)^{1/2} \alpha_\mu^\pm(njm)|n \pm 1, j + \eta, m + \mu, \zeta + \eta\rangle_{\mathrm{U}} \tag{12.268}$$

and

$$D_\mu(\pm\omega)|njm\zeta\rangle_{\mathrm{L}} = \left(\frac{1 - m_0c^2/E_{nj}}{1 - m_0c^2/E_{n\pm1,j-\eta}}\right)^{1/2} \beta_\mu^\pm(njm)|n \pm 1, j - \eta, m + \mu, \zeta - \eta\rangle_{\mathrm{L}}. \tag{12.269}$$

---

* Other solutions to (12.260) are (4.49); however, with these (12.267) is not satisfied.

Here $\mu = 0$ or ± 1 and $D_{\pm 1} = D_x \pm \mathrm{i}D_y$ and $D_0 = D_z$. The coefficients $\alpha_\mu^\pm$ and $\beta_\mu^\pm$ are calculated below. The energy-dependent factors in (12.268) and (12.269) have been included to take account of the normalization conditions (12.253) and (12.254), and the subscripts U and L indicate whether $|njm\zeta\rangle$ is the upper or lower element ($|\phi\rangle$ or $|\chi\rangle$) in (12.17).

Using (12.268), (12.269), and orthogonality of the kets we obtain an additional 12 shift operations

$$D_\mu^\dagger(\pm\omega)|njm\zeta\rangle_{\mathrm{U}} = \left(\frac{1 + m_0c^2/E_{nj}}{1 + m_0c^2/E_{n\mp1,j-\eta}}\right)^{1/2} [\alpha^\pm_{-\mu}(n \mp 1, j - \eta, m + \mu)]^* \times |n \mp 1, j - \eta, m + \mu, \zeta - \eta\rangle_{\mathrm{U}} \tag{12.270}$$

and

$$D_\mu^\dagger(\pm\omega)|njm\zeta\rangle_{\mathrm{L}} = \left(\frac{1 - m_0c^2/E_{nj}}{1 - m_0c^2/E_{n\mp1,j+\eta}}\right)^{1/2} [\beta^\pm_{-\mu}(n \mp 1, j + \eta, m + \mu)]^* \times |n \mp 1, j + \eta, m + \mu, \zeta + \eta\rangle_{\mathrm{L}}, \tag{12.271}$$

where $D_{\pm1}^\dagger = D_x^\dagger \pm \mathrm{i}D_y^\dagger$ and $D_0^\dagger = D_z^\dagger$. Explicit expressions for $\mathbf{D}^\dagger$ are given below.

We now consider some algebraic properties of the vector operators $\mathbf{D}(\pm\omega)$. Like their non-relativistic counterparts (Section 8.1) $\mathbf{D}(\pm\omega)$ can be factorized into a product of a vector operator and a scalar operator, and this factorization can be performed in two different ways. The first such factorization is

$$\mathbf{D}(\pm\omega) = (2m_0\hbar^3\omega)^{-1/2}\mathbf{U}R(\pm\omega), \tag{12.272}$$

where

$$\mathbf{U} = \hat{\mathbf{r}} \times \mathbf{L} + \mathrm{i}\hat{\mathbf{r}}(\boldsymbol{\sigma}.\mathbf{L}) \tag{12.273}$$

and

$$R(\omega) = -\mathrm{i}p_r - r^{-1}\boldsymbol{\sigma}.\mathbf{L} + m_0\omega r. \tag{12.274}$$

The second factorization is

$$\mathbf{D}(\pm\omega) = \mp\mathrm{i}(m_0\omega/2\hbar^3)^{1/2}\mathbf{V}P(\pm\omega), \tag{12.275}$$

where

$$\mathbf{V} = \hat{\mathbf{p}} \times \mathbf{L} + \mathrm{i}\hat{\mathbf{p}}(\boldsymbol{\sigma}.\mathbf{L}), \tag{12.276}$$

$$P(\omega) = \mathrm{i}r_p - p^{-1}\boldsymbol{\sigma}.\mathbf{L} + (m_0\omega)^{-1}p, \tag{12.277}$$

and r_p is the radial operator (7.2). Both p_r and r_p are Hermitian with respect to $|njm\zeta\rangle$ (see Section 12.10).

The factorizations of the adjoints are

$$\mathbf{D}^\dagger(\pm\omega) = (2m_0\hbar^3\omega)^{-1/2}R^\dagger(\pm\omega)\mathbf{U}^\dagger \tag{12.278}$$

$$= \pm\mathrm{i}(m_0\omega/2\hbar^3)^{1/2}P^\dagger(\pm\omega)\mathbf{V}^\dagger. \tag{12.279}$$

Equations (12.272), (12.275), (12.278), and (12.279) show that the operators $\mathbf{D}$ and $\mathbf{D}^\dagger$ are bifactorizable.

The vector operators $\mathbf{U}$ and $\mathbf{V}$ also have useful factorizations:

$$\mathbf{U} = \boldsymbol{\sigma}.\hat{\mathbf{r}}(\boldsymbol{\sigma} \times \mathbf{L}), \tag{12.280}$$

$$\mathbf{V} = \boldsymbol{\sigma}.\hat{\mathbf{p}}(\boldsymbol{\sigma} \times \mathbf{L}). \tag{12.281}$$

Thus

$$\mathbf{U}^\dagger = (\boldsymbol{\sigma} \times \mathbf{L})\boldsymbol{\sigma}.\hat{\mathbf{r}}, \tag{12.282}$$

$$\mathbf{V}^\dagger = (\boldsymbol{\sigma} \times \mathbf{L})\boldsymbol{\sigma}.\hat{\mathbf{p}}. \tag{12.283}$$

The operators R, P, $\mathbf{U}$, and $\mathbf{V}$ introduced above factorize functions of the commuting operators $\mathscr{H}$, $\boldsymbol{\sigma}.\mathbf{L}$, $\mathbf{L}^2$, and $\bar{J}_z$. For example,

$$R^\dagger(\pm\omega)R(\pm\omega) = 2m_0(\mathscr{H} \pm \tfrac{1}{2}\hbar\omega \mp \omega\boldsymbol{\sigma}.\mathbf{L}), \tag{12.284}$$

$$\begin{aligned} U_0^\dagger U_0 &= (\boldsymbol{\sigma} \times \mathbf{L})_0^2 \\ &= \mathbf{L}^2 - \bar{J}_z^2 + \tfrac{1}{4}\hbar^2, \end{aligned} \tag{12.285}$$

$$\begin{aligned} (U_{\pm 1})^\dagger U_{\pm 1} &= [(\boldsymbol{\sigma} \times \mathbf{L})_{\pm 1}]^\dagger(\boldsymbol{\sigma} \times \mathbf{L})_{\pm 1} \\ &= (\boldsymbol{\sigma}.\mathbf{L} \mp \bar{J}_z)^2 - \tfrac{1}{4}\hbar^2, \end{aligned} \tag{12.286}$$

where in the last two calculations we have used (12.280) and $(\boldsymbol{\sigma}.\hat{\mathbf{r}})^2 = 1$. Similar results apply for P and $\mathbf{V}$.

With the aid of these algebraic properties, it is simple to calculate the coefficients in (12.268)–(12.271): using (12.272), (12.280), and (12.282) we have

$$[D_\mu(\pm\omega)]^\dagger D_\mu(\pm\omega) = (2m_0\hbar^3\omega)^{-1}R^\dagger(\pm\omega)[(\boldsymbol{\sigma} \times \mathbf{L})_\mu]^\dagger(\boldsymbol{\sigma} \times \mathbf{L})_\mu R(\pm\omega).$$

Using (12.284)–(12.286) and noting that the scalar operators R commute with the right-hand sides of (12.285) and (12.286), we obtain

$$[D_0(\pm\omega)]^\dagger D_0(\pm\omega) = \hbar^{-3}(\omega^{-1}\mathscr{H} \pm \tfrac{1}{2}\hbar \mp \boldsymbol{\sigma}.\mathbf{L})(\mathbf{L}^2 - \bar{J}_z^2 + \tfrac{1}{4}\hbar^2) \tag{12.287}$$

and

$$[D_\mu(\pm\omega)]^\dagger D_\mu(\pm\omega) = \hbar^{-3}(\omega^{-1}\mathscr{H} \pm \tfrac{1}{2}\hbar \mp \boldsymbol{\sigma}.\mathbf{L})[(\boldsymbol{\sigma}.\mathbf{L} - \mu\bar{J}_z)^2 - \tfrac{1}{4}\hbar^2], \tag{12.288}$$

where $\mu = \pm 1$. The expectation values of the right-hand sides of (12.287) and (12.288) can be written down using (12.18), (12.19), and (12.241)–(12.244). Thus we find

$$\alpha_\mu^\pm(njm) = \theta_\mu^\pm(n \pm \eta j \pm \tfrac{1}{2}\eta \pm \tfrac{3}{2} + \tfrac{3}{2})\gamma_\mu(jm\eta) \tag{12.289}$$

$$\beta_\mu^\pm(njm) = \phi_\mu^\pm(n \mp \eta j \mp \tfrac{1}{2}\eta \pm \tfrac{3}{2} + \tfrac{1}{2})\gamma_\mu(j, m, -\eta), \tag{12.290}$$

where $\theta_\mu^\pm$ and $\phi_\mu^\pm$ are phase factors (see Section 12.10) and

$$\gamma_0(jm\eta) = [(j + \tfrac{1}{2}\eta + \tfrac{1}{2})^2 - m^2]^{1/2} \tag{12.291}$$

$$\gamma_{\pm 1}(jm\eta) = [(\eta j + \tfrac{1}{2}\eta + 1 \pm m)^2 - \tfrac{1}{4}]^{1/2}. \tag{12.292}$$

The vector operators $\mathbf{D}(\omega)$ change both the energy and the angular momentum. It is useful to construct also shift operators that change only the energy; these are

$$[D_\mu(\mp\omega)]^\dagger D_\mu(\pm\omega) \qquad (\mu = 0, \pm 1).$$

Using the factorizations (12.272), (12.278), (12.285), and (12.286) we find that

$$[D_0(\mp\omega)]^\dagger D_0(\pm\omega) = Q^\pm(\mathbf{L}^2 - J_z^2 + \tfrac{1}{4}\hbar^2) \tag{12.293}$$

$$[D_\mu(\mp\omega)]^\dagger D_\mu(\pm\omega) = Q^\pm[(\boldsymbol{\sigma}.\mathbf{L} - J_z)^2 - \tfrac{1}{4}\hbar^2] \qquad (\mu = \pm 1), \tag{12.294}$$

where

$$Q^\pm = \pm i\mathbf{r}.\mathbf{p} - m_0\omega\mathbf{r}^2 + \omega^{-1}\mathscr{H} \pm \tfrac{3}{2}\hbar \tag{12.295}$$

$$= \pm i\mathbf{p}.\mathbf{r} + (m_0\omega)^{-1}\mathbf{p}^2 - \omega^{-1}\mathscr{H} \mp \tfrac{3}{2}\hbar. \tag{12.296}$$

Using (12.293) and the shift operators (12.268)–(12.271) with $\mu = 0$, we obtain

$$Q^\pm|njm\zeta\rangle_{\mathrm{U}} = \eta\left(\frac{1 + m_0c^2/E_{nj}}{1 + m_0c^2/E_{n\pm 2,j}}\right)^{1/2}[(n - \eta j - \tfrac{1}{2}\eta + 1 \pm 1)$$

$$\times (n + \eta j + \tfrac{1}{2}\eta + 2 \pm 1)]^{1/2}|n \pm 2, j, m, \zeta\rangle_{\mathrm{U}} \tag{12.297}$$

and

$$Q^\pm|njm\zeta\rangle_{\mathrm{L}} = \eta\left(\frac{1 - m_0c^2/E_{nj}}{1 - m_0c^2/E_{n\pm 2,j}}\right)^{1/2}[(n - \eta j - \tfrac{1}{2}\eta + 1 \pm 1)$$

$$\times (n + \eta j + \tfrac{1}{2}\eta \pm 1)]^{1/2}|n \pm 2, j, m, \zeta\rangle_{\mathrm{L}}, \tag{12.298}$$

where we have set the phase factors equal to η (see Section 12.10).

The shift operations (12.297) and (12.298) can also be derived by the method used in Section 12.3 for relativistic Coulomb problems: one rewrites the eigenvalue equation (12.241) in the form (12.84) (with $\lambda = n$) and then factorizes the operator O_λ [26].

Finally we consider the 12 phase factors $\theta_\mu^\pm(njm)$ and $\phi_\mu^\pm(njm)$ that appear in the coefficients (12.289) and (12.290). These factors cannot all be chosen independently because the shift operators are related by identities: thus (12.293), (12.294), and the choice of η for the phase factors in (12.297) and (12.298) require

$$[\theta_\mu^\mp(n \pm 2, j, m)]^*\theta_\mu^\pm(njm) = \eta. \tag{12.299}$$

Also, $|njm\zeta\rangle_{\mathrm{U}}$ can be transformed into $|n \pm 1, j + \eta, m + \mu, \zeta + \eta\rangle_{\mathrm{L}}$ in two different ways, namely by applying $D_\mu(\pm\omega)$ and then (12.231), or by applying (12.231) and then $D_\mu^\dagger(\mp\omega)$. Comparing the results we find

$$[\phi_{-\mu}^{\mp}(n \pm 1, j + \eta, m + \mu)]^* = -\theta_\mu^{\pm}(njm). \qquad (12.300)$$

Equations (12.299) and (12.300) show that only three of the phase factors can be chosen independently. For example, if we choose

$$\theta_0^+ = -\eta\mathrm{i} \quad \text{and} \quad \theta_{\pm 1}^+ = \pm\mathrm{i}, \qquad (12.301)$$

then $\theta_0^- = -\mathrm{i}$, $\theta_{\pm 1}^- = \pm\eta\mathrm{i}$, $\phi_0^+ = -\mathrm{i}$, $\phi_{\pm 1}^+ = \mp\eta\mathrm{i}$, $\phi_0^- = -\eta\mathrm{i}$, and $\phi_{\pm 1}^- = \mp\mathrm{i}$.

12.10 Solutions for the Dirac oscillator

In this section we consider some applications of the results derived above. First we use the shift operations (12.297) and (12.298) to determine the possible value of the quantum number n. Non-negativity of the norm in (12.297) and (12.298) requires that the lowering operations must terminate at a minimum value $n' = j + \frac{1}{2}\eta$, except that for $j = l + \frac{1}{2}$, $n' = j + \frac{3}{2}$ in $|njm\zeta\rangle_{\mathrm{L}}$. The raising operations in (12.297) and (12.298) do not terminate, and if $|n'jm\zeta\rangle$ is normalizable so are kets with

$$n = 2N + j + \tfrac{1}{2}\eta \qquad (N = 0, 1, 2, \ldots). \qquad (12.302)$$

For $j = l + \frac{1}{2}$, $|njm\zeta\rangle_{\mathrm{L}}$ does not exist if $N = 0$ (i.e. $n = j - \frac{1}{2}$). The energy eigenvalues of the Dirac oscillator are given by (12.250) and (12.302). The degeneracies of these eigenvalues have been discussed in terms of supersymmetric quantum mechanics [27] and in terms of a symmetry Lie algebra [21].

From a normalized ket $|njm\zeta\rangle_{\mathrm{U}}$ (or $|njm\zeta\rangle_{\mathrm{L}}$) the shift operators $\mathbf{D}$, $\mathbf{D}^\dagger$, and the operators in the coupled equations (12.231) and (12.232) will generate all the kets of the Dirac oscillator. Some of these transformations are depicted in Fig. 12.3, which is similar to the corresponding diagram for the Dirac–Coulomb problem (Fig. 12.1).

Expectation values of integer powers of r can be obtained by operator methods; these calculations are similar to those for the non-relativistic oscillator (Section 6.2) and they are left as an exercise for the reader. The matrix elements of $\mathbf{r}$ can be written down using the shift operations (12.268)–(12.271), and the results are given in Table 12.3. From these one can evaluate the matrix elements of $x^{n_1}y^{n_2}z^{n_3}$ where n_i are non-negative integers.

Finally we consider wavefunctions of the Dirac oscillator. These wavefunctions can be obtained by using the shift operators for energy and angular momentum that were derived in Section 12.9. The calculations are similar to those given for the non-relativistic oscillator in Sections 6.4, 7.6, and 8.5, and we will simply write down the results.

With the choice (12.301) of phases the Dirac ket (12.17) appears in the coordinate representation as

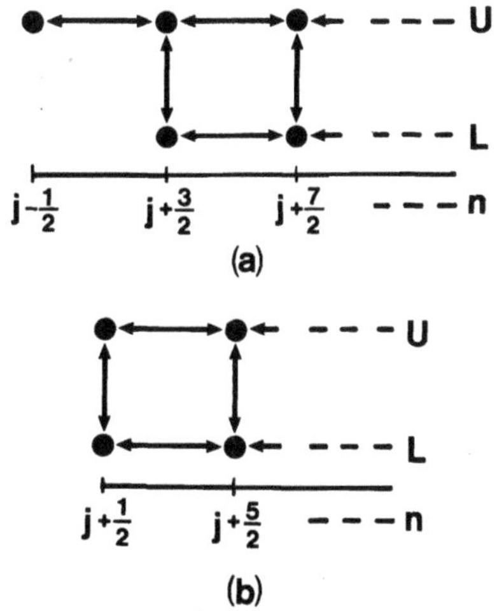

FIG. 12.3 Some transformations of the kets $|njm\zeta\rangle_U$ and $|njm\zeta\rangle_L$ for the Dirac oscillator: (a) for $j = l + \frac{1}{2}$, (b) for $j = l - \frac{1}{2}$. The transformations indicated by the vertical arrows are between $|njm\zeta\rangle_U$ and $|njm\zeta\rangle_L$; they are effected by the operators in (12.231) and (12.232). The horizontal arrows depict the transformations (12.297) and (12.298). (Reprinted from Ref. 25.)

$$\Psi^{\pm}(\mathbf{r}, t) = \begin{pmatrix} \mathrm{i}F_{Nl}(r)\, V^{\pm}_{jml}(\hat{\mathbf{r}}) \\ G_{Nl}(r)\, V^{\mp}_{jml'}(\hat{\mathbf{r}}) \end{pmatrix} \exp(-\mathrm{i}Et/\hbar) \tag{12.303}$$

for $j = l \pm \frac{1}{2}$. Here

$$F_{Nl}(r) = [\tfrac{1}{2}(1 + m_0c^2/E)]^{1/2} C_{Nl} R_{Nl}, \tag{12.304}$$

$$G_{Nl}(r) = (\mathrm{sign}\ E)[\tfrac{1}{2}(1 - m_0c^2/E)]^{1/2} C_{N'l'} R_{N'l'}, \tag{12.305}$$

where

$$R_{Nl} = {}_1F_1\left(-N; l + \frac{3}{2}; \frac{r^2}{a^2}\right)\left(\frac{r^2}{a^2}\right)^{l/2} \exp\left(-\frac{r^2}{2a^2}\right), \tag{12.306}$$

$$C_{Nl} = \eta^N \left(\frac{2^{l-N+2}(2N + 2l + 1)!!}{a^3\pi^{1/2} N![(2l + 1)!!]^2}\right)^{1/2}, \tag{12.307}$$

$a = (\hbar/m_0\omega)^{1/2}$, $l' = l - \eta$, $N' = N - \frac{1}{2}(1 - \eta)$, $N = 0, 1, \ldots$ is the radial quantum number, E is given by (12.250) and (12.302), and l is a non-negative integer. The factors sign E in (12.305) and η^N in (12.307) are required by the coupled equations (12.231) and (12.232). The square-root factors in (12.304) and (12.305) are required by the normalizations (12.253) and (12.254). The angular functions in (12.303) are the spinor spherical harmonics (12.212) with phases $\alpha_{\pm} = 0$. One can readily show that p_r is Hermitian with respect to the wavefunctions (12.303) (see Appendix F).

The wavefunction in the momentum representation is obtained by making the substitutions $r \to (m_0\omega)^{-1}p$ and $\hat{\mathbf{r}} \to \hat{\mathbf{p}}$ in (12.303)–(12.306), and replacing

TABLE 12.3 The non-zero matrix elements of **r** for the Dirac oscillator. In these formulae $|njm\zeta\rangle_U$ and $|njm\zeta\rangle_L$ are the elements $|\phi\rangle$ and $|\chi\rangle$, respectively, in (12.17), and $\mu = 0$ or ± 1. The energy eigenvalues E_{nj} are given by (12.250) and (12.302), and the coefficients $\alpha_\mu^\pm$ and $\beta_\mu^\pm$ by (12.289) and (12.290)

$$_U\langle n \pm 1, j + \eta, m + \mu, \zeta + \eta | r_\mu | njm\zeta\rangle_U = \pm d_+(1 + m_0c^2/E_{n\pm 1, j+\eta})^{1/2}\alpha_\mu^\pm(njm)$$

$$_U\langle n \pm 1, j - \eta, m + \mu, \zeta - \eta | r_\mu | njm\zeta\rangle_U = \pm d_+(1 + m_0c^2/E_{n\pm 1, j-\eta})^{1/2}[\alpha_{-\mu}^\mp(n \pm 1, j - \eta, m + \mu)]^*$$

$$_L\langle n \pm 1, j - \eta, m + \mu, \zeta + \eta | r_\mu | njm\zeta\rangle_L = \mp d_-(1 - m_0c^2/E_{n\pm 1, j-\eta})^{1/2}\beta_\mu^\pm(njm)$$

$$_L\langle n \pm 1, j + \eta, m + \mu, \zeta + \eta | r_\mu | njm\zeta\rangle_L = \mp d_-(1 - m_0c^2/E_{n\pm 1, j+\eta})^{1/2}[\beta_{-\mu}^\mp(n \pm 1, j + \eta, m + \mu)]^*$$

where

$$d_\pm = \mathrm{i}\{2(2m_0\omega/\hbar)^{1/2}[\eta(2j + 1) \mp 1]\}^{-1}(1 \pm m_0c^2/E_{nj})^{1/2}$$

$a^{-3/2}$ in (12.307) with* $(\hbar/a)^{-3/2}(-\eta i)^l$. The radial operator r_p is Hermitian with respect to these wavefunctions (see Appendix F).

It is instructive to consider the effect of the factorized operators (12.272) on the coordinate-space wavefunctions (12.303). The effect of the factor $R(\omega)$ on the upper and lower elements of (12.303) is the same as that of the first-order differential operator

$$R_j(\omega) = -\hbar\frac{\partial}{\partial r} \pm \hbar\eta\left(j+\frac{1}{2}\right)r^{-1} + m_0\omega r \qquad (12.308)$$

obtained by replacing $\boldsymbol{\sigma}\cdot\mathbf{L}$ with its eigenvalues (see (12.18), (12.19), and (12.24)) and p_r by (5.35). In (12.308) the upper (lower) sign applies if R_j acts on the upper (lower) elements in (12.303). The R_j change N and l in the radial functions F_{Nl} and G_{Nl}. The factor $\mathbf{U}$ in (12.272) operates on the spinor spherical harmonics, and the resulting shift operations are

$$U_\mu V^+_{jml} = \theta^+_\mu\gamma_\mu(jm\eta)V^+_{j-1,m+\mu,l-1}, \qquad U_\mu V^-_{jml} = \theta^-_\mu\gamma_\mu(jm\eta)V^-_{j+1,m+\mu,l+1}. \qquad (12.309)$$

Here $V^\pm_{jml}$ are given by (12.212) with $\alpha_\pm = 0$, $\gamma^\pm_\mu$ by (12.291) and (12.292), and the phase factors $\theta^\pm_\mu$ correspond to the choice (12.301). Similar remarks apply to the effect of the factorized operators (12.275) on the momentum-space wavefunctions.

References

1. Transformations of the Dirac–Coulomb equation have been discussed by several authors: see Biedenharn, L.C. (1962). Remarks on the relativistic Kepler problem. *Physical Review*, **126**, 845–51; Su, J.Y. (1985). Simplified solutions of the Dirac–Coulomb equation. *Physical Review*, A **32**, 3251–7; Wong, M.K.F. (1986). Infeld–Hull factorization and the simplified solution of the Dirac–Coulomb equation. *Physical Review*, A **34**, 1559–60.
2. de Lange, O.L. (1989). An operator analysis for the Schrödinger, Klein–Gordon, and Dirac equations with a Coulomb potential. *Journal of Mathematical Physics*, **30**, 858–66.
3. See, for example, Ref. 3, Chapter 11.
4. Sommerfeld, A. (1916). Zur Quantentheorie der Spektrallinien. *Annalen der Physik*, **51**, 1–94.
5. Biedenharn, L.C. (1983). The 'Sommerfeld Puzzle' revisited and resolved. *Foundations of Physics*, **13**, 13–34.
6. Series, G.W. (1957). *The spectrum of atomic hydrogen*. Oxford University Press. More recent results are reviewed by Kugel, H.W. and Murnick, D.E. (1977). The Lamb shift in hydrogenic ions. *Reports on Progress in Physics*, **40**, 297–343.
7. Delker, L. *et al.* (1979). Experimental verification of the relativistic fine-structure term of the Klein–Gordon equation in pionic titanium atoms. *Physical Review Letters*, **42**, 89–92. Wang, K.C. *et al.* (1980). Experimental determination of the

* The phase factor $(-\eta i)^l$ is required by the factors $\mp i$ in (12.275).

relativistic fine-structure splitting in pionic Ti and Fe atoms. *Physical Review*, A **22**, 1072–77.

8. Johnson, M.H. and Lippmann, B.A. (1950). Relativistic Kepler problem. *Physical Review*, **78**, 329. See also Biedenharn, L.C. Ref. 1.
9. See Baym, G. (1969). *Lectures on quantum mechanics*, Ch. 23. Benjamin, Reading, MA.
10. Sukumar, C.V. (1985). Supersymmetry and the Dirac equation for a central Coulomb field. *Journal of Physics A: Math. Gen.*, **18**, L697–701.
11. Epstein, J.H. and Epstein, S.T. (1962). Some applications of hypervirial theorems to the calculation of average values. *American Journal of Physics*, **30**, 266–8. For physical applications, see also Armstrong, L. (1971). *Theory of the hyperfine structure of free atoms*. Wiley, New York, and Sobel'man, I.I. (1972). *Introduction to the theory of atomic spectra*, Ch. 7. Pergamon, Oxford.
12. Goldman, S.P. and Drake, G.W.F. (1982). Relativistic sum rules and integral properties of the Dirac equation. *Physical Review*, A **25**, 2877–81.
13. Dongpei, Zhu (1986). Proof of the quantum virial theorem. *American Journal of Physics*, **54**, 267–70.
14. Erdélyi, A., Magnus, W., Oberhettinger, F., and Tricomi, F.G. (1953). *Higher transcendental functions*, Vol. 1, Ch. 3. McGraw-Hill, New York.
15. Armstrong, B.H. (1963). Anomalous solutions to the Dirac and Schrödinger equations for the Coulomb potential. *Physical Review*, **130**, 2506–18.
16. Prugovečki, E. (1981). *Quantum mechanics in Hilbert space*, 2nd edn, pp. 48–9. Academic Press, New York.
17. Armstrong, B.H. and Power, E.A. (1963). Acceptable solutions and boundary conditions for the Schrödinger and Dirac equations. *American Journal of Physics*, **31**, 262–8.
18. See, for example, Malli, G.L. (ed.) (1983). *Relativistic effects in atoms, molecules, and solids*. Plenum, New York.
19. See, for example, Messiah, A. (1962). *Quantum mechanics*, Vol. II, Ch. 20. North-Holland, Amsterdam.
20. See Kim, Y.S. and Noz, M.E. (1978). Relativistic harmonic oscillators and hadronic structures in the quantum-mechanics curriculum. *American Journal of Physics*, **46**, 484–8 and references therein.
21. The Dirac oscillator has a fairly lengthy history. See Quesne, C. and Moshinsky, M. (1990). Symmetry Lie algebra of the Dirac oscillator. *Journal of Physics A: Math. Gen.*, **23**, 2263–72.
22. Moreno, M. and Zentella, Z. (1989). Covariance, CPT and the Foldy–Wouthuysen transformation for the Dirac oscillator. *Journal of Physics A: Math. Gen.*, **22**, L821–5.
23. Benítez, J. et. al. (1990). Solution and hidden supersymmetry of a Dirac oscillator. *Physical Review Letters*, **64**, 1643–5.
24. See, for example, Bethe, H.A. and Jackiw, R. (1968). *Intermediate quantum mechanics*, 2nd edn. Benjamin, New York.
25. de Lange, O.L. (1991). Shift operators for a Dirac oscillator. *Journal of Mathematical Physics*, **32**, 1296–300.
26. de Lange, O.L. (1991). Algebraic properties of the Dirac oscillator. *Journal of Physics A: Math. Gen.*, **24**, 667–77.
27. Ui, H. and Takeda, G. (1984). Does accidental degeneracy imply a symmetry group?. *Progress of Theoretical Physics*, **72**, 266–84. Balantekin, A.B. (1985). Accidental degeneracies and supersymmetric quantum mechanics. *Annals of Physics*, **164**, 277–87.

APPENDICES

Appendix A

We prove the identity (2.53). Consider the function

$$F(z) = \exp(zA)\exp(zB)\exp[-z(A+B)], \tag{A.1}$$

where the operators A and B are independent of z. Then

$$\frac{\mathrm{d}F}{\mathrm{d}z} = \exp(zA)[A, \exp(zB)]\exp[-z(A+B)]. \tag{A.2}$$

If $[B, [A, B]] = 0$, we can use induction to show that

$$[A, B^n] = n[A, B]B^{n-1} \qquad (n = 1, 2, \ldots).$$

Therefore

$$[A, \exp(zB)] = z[A, B]\exp(zB). \tag{A.3}$$

If $[A, [A, B]] = 0$, it follows from (A.2) and (A.3) that

$$\frac{\mathrm{d}F}{\mathrm{d}z} = z[A, B]F. \tag{A.4}$$

The solution to (A.4) is

$$F = \exp(\tfrac{1}{2}z^2[A, B]). \tag{A.5}$$

Equations (A.1) and (A.5) yield (2.53).

Appendix B

We prove the following recurrence relations for hypergeometric functions:

$$\left(1 - \frac{x}{N+B} + \frac{x}{N+B}\frac{\mathrm{d}}{\mathrm{d}x}\right) {}_1F_1(-N; B; x)$$

$$= {}_1F_1(-(N+1); B; x), \tag{B.1}$$

$$\left(1 - \frac{N+B-1}{(B-1)(B-2)}x + \frac{x}{B-1}\frac{\mathrm{d}}{\mathrm{d}x}\right) {}_1F_1(-N; B; x)$$

$$= {}_1F_1(-(N+1); B-2; x), \tag{B.2}$$

$$\left(1 - \frac{N+B+1}{N+C}x + \frac{N+B+1}{B(N+C)}(1-x)x\frac{\mathrm{d}}{\mathrm{d}x}\right) {}_2F_1(-N, B; C; x)$$

$$= {}_2F_1(-(N+1), B+1; C; x). \tag{B.3}$$

Here N is a non-negative integer. The function

$$ {}_1F_1(-N; B; x) = \sum_{j=0}^{N} a_j(N, B)x^j, \tag{B.4} $$

with

$$ a_j(N, B) = \frac{(-1)^j}{j!} \frac{\Gamma(N+1)}{\Gamma(N+1-j)} \frac{\Gamma(B)}{\Gamma(B+j)}, \tag{B.5} $$

is a confluent hypergeometric function, and

$$ {}_2F_1(-N, B; C; x) = \sum_{j=0}^{N} a_j(N, B, C)x^j, \tag{B.6} $$

with

$$ a_j(N, B, C) = a_j(N, C)\frac{\Gamma(B+j)}{\Gamma(B)}, \tag{B.7} $$

is a hypergeometric function.

The left-hand side of (B.1)–(B.3) has the form

$$ \sum_{j=0}^{N+1} b_j x^j, \tag{B.8} $$

where $b_0 = 1$.

B.1 *Proof of* (B.1)

Substituting (B.4) and (B.5) on the left-hand side of (B.1) we obtain

$$ b_j = \left(1 + \frac{j}{N+B}\right)a_j(N, B) - \frac{1}{N+B}a_{j-1}(N, B) $$

if $1 \le j \le N$, and

$$ b_{N+1} = -\frac{1}{N+B}a_N(N, B). $$

Substituting (B.5) in these equations we find

$$ b_j = a_j(N+1, B) \qquad (1 \le j \le N) $$

$$ b_{N+1} = a_{N+1}(N+1, B). $$

Hence (B.8) is equal to the right-hand side of (B.1).

B.2 *Proof of* (B.2)

Here

$$ b_j = \left(1 + \frac{j}{B-1}\right)a_j(N, B) - \frac{N+B-1}{(B-1)(B-2)}a_{j-1}(N, B) $$

if $1 \le j \le N$, and

$$b_{N+1} = -\frac{N+B-1}{(B-1)(B-2)}a_N(N, B).$$

Substituting (B.5) in these equations yields

$$b_j = a_j(N+1, B-2) \qquad (1 \le j \le N)$$

$$b_{N+1} = a_{N+1}(N+1, B-2).$$

Hence (B.8) is equal to the right-hand side of (B.2).

B.3 Proof of (B.3)

Here

$$b_j = \left(1 + j\frac{N+B+1}{B(N+C)}\right)a_j(N, B, C)$$
$$-\frac{(N+B+1)(B+j-1)}{(N+C)B}a_{j-1}(N, B, C),$$

if $1 \le j \le N$, and

$$b_{N+1} = -\frac{(N+B+1)(N+B)}{(N+C)B}a_N(N, B, C).$$

Substituting (B.7) in these equations gives

$$b_j = a_j(N+1, B+1, C) \qquad (1 \le j \le N)$$

$$b_{N+1} = a_{N+1}(N+1, B+1, C).$$

Hence (B.8) is equal to the right-hand side of (B.3).

When expressed in the coordinate or momentum representation, many of the shift operations derived in the text yield recurrence relations that are special cases of (B.1)–(B.3). Thus these equations are often used to determine the polynomial that appears as a factor in certain wavefunctions (see Chapters 3–7, 10, and 12).

Appendix C

We prove certain results that are stated in Chapter 4.

1. $$[L_x, L_y] = [yp_z - zp_y, zp_x - xp_z] = y[p_z, z]p_x + p_y[z, p_z]x$$
$$= -i\hbar yp_x + i\hbar xp_y = i\hbar L_z. \tag{C.1}$$

   Similarly, $[L_y, L_z] = i\hbar L_x$ and $[L_z, L_x] = i\hbar L_y$. This proves (4.4).

2. Let $\mathbf{v}$ be a vector operator. Then

$$[\mathbf{L}^2, v_j] = L_i^2 v_j - v_j L_i^2 = L_i(v_j L_i + i\hbar\varepsilon_{ijk}v_k) - v_j L_i^2$$
$$= v_j L_i^2 + i\hbar\varepsilon_{ijk}v_k L_i + i\hbar\varepsilon_{ijk}(v_k L_i + i\hbar\varepsilon_{ikl}v_l) - v_j L_i^2,$$

   where we have used (4.4) and (4.32). Thus

$$[\mathbf{L}^2, v_j] = 2i\hbar\varepsilon_{ijk}v_k L_i - \hbar^2\varepsilon_{ijk}\varepsilon_{ikl}v_l = 2i\hbar(\mathbf{v}\times\mathbf{L})_j + 2\hbar^2 v_j, \tag{C.2}$$

where we have used (1.154) and (1.156). This proves (4.33).

Also

$$\begin{aligned}[L_i, (\mathbf{v}\times\mathbf{L})_j] &= \varepsilon_{jkl}(L_i v_k L_l - v_k L_l L_i)\\ &= i\hbar\varepsilon_{jkl}(\varepsilon_{ikm}v_m L_l - \varepsilon_{lin}v_k L_n)\\ &= i\hbar(-\delta_{jk}\delta_{in} + \delta_{jn}\delta_{ik})v_k L_n,\end{aligned}$$

where we have used (4.4), (4.32), and (1.157). Thus

$$[L_i, (\mathbf{v}\times\mathbf{L})_j] = i\hbar\varepsilon_{ijl}\varepsilon_{knl}v_k L_n = i\hbar\varepsilon_{ijk}(\mathbf{v}\times\mathbf{L})_k, \tag{C.3}$$

which is (4.39).

3. $$[(\mathbf{v}\times\mathbf{L})\times\mathbf{L}]_i = \varepsilon_{ijk}(\mathbf{v}\times\mathbf{L})_j L_k = \varepsilon_{ijk}\varepsilon_{jmn}v_m L_n L_k = -v_i\mathbf{L}^2 + v_m L_i L_m, \tag{C.4}$$

where we have used (1.157). Now

$$v_m L_i L_m = v_m L_m L_i + i\hbar\varepsilon_{imn}v_m L_n = (\mathbf{v}.\mathbf{L})L_i + i\hbar(\mathbf{v}\times\mathbf{L})_i. \tag{C.5}$$

Thus if $\mathbf{v}.\mathbf{L} = 0$, then

$$[(\mathbf{v}\times\mathbf{L})\times\mathbf{L}]_i = -v_i\mathbf{L}^2 + i\hbar(\mathbf{v}\times\mathbf{L})_i. \tag{C.6}$$

This proves (4.42).

4. $$[L_i, f(r)] = i\hbar\varepsilon_{ijk}r_j[p_k, f(r)] = \hbar^2\varepsilon_{ijk}r_j\frac{\partial f(r)}{\partial r_k} = \hbar^2 r^{-1}\frac{\mathrm{d}f}{\mathrm{d}r}\varepsilon_{ijk}r_j r_k = 0 \tag{C.7}$$

because

$$\varepsilon_{ijk}r_j r_k = \tfrac{1}{2}(\varepsilon_{ijk} + \varepsilon_{ikj})r_k r_j = 0. \tag{C.8}$$

Similarly,

$$[L_i, f(\mathbf{r}.\mathbf{p})] = 0, \qquad [L_i, f(p)] = 0. \tag{C.9}$$

Thus any differentiable function of r, $\mathbf{r}.\mathbf{p}$, or p is a scalar.

5. $$[L_i, r_j] = \varepsilon_{ikl}(r_k p_l r_j - r_j r_k p_l) = \varepsilon_{ikl}r_k[p_l, r_j] = i\hbar\varepsilon_{ijk}r_k. \tag{C.10}$$

Similarly,

$$[L_i, p_j] = i\hbar\varepsilon_{ijk}p_k. \tag{C.11}$$

Thus $\mathbf{r}$ and $\mathbf{p}$ are vector operators.

6. $$\mathbf{r}.\mathbf{L} = r_i L_i = \varepsilon_{ijk}r_i r_j p_k = 0 \tag{C.12}$$

(see (C.8)). This proves (4.65). Similarly,

$$\mathbf{p}.\mathbf{L} = 0. \tag{C.13}$$

Appendix D

We prove the four identities (4.59) and (4.60). These can be written

$$U_\lambda^{\pm}(L_z + \lambda\hbar \mp \lambda K^{\pm}) = U_0^{\pm}L_\lambda, \tag{D.1}$$

where $\lambda = \pm 1$ and $L_\lambda = L_x + i\lambda L_y$. In the left-hand side of (D.1) substitute (4.50) and (4.57), and multiply out. Then use (4.54) and the commutation relations (4.4), (4.31), and (4.32). This gives

$$U_\lambda^\pm(L_z + \lambda\hbar \mp \lambda K^\pm) = [\pm i(v_x L_y L_x - v_y L_x^2) + v_z K^\pm L_x \mp \lambda(v_x L_y^2 - v_y L_x L_y) + i\lambda v_z K^\pm L_y] \mp \lambda \mathbf{v}.\mathbf{L}(L_x + i\lambda L_y). \tag{D.2}$$

The term in square brackets is equal to $U_0^\pm L_\lambda$. The remaining term is zero (see (4.41)).

Appendix E

We use the hypervirial theorem (Section 1.10) to calculate the three expectation values in (4.62) and (4.69). From (4.32), (4.33), and (4.43),

$$[\mathbf{L}^2, \hat{z}^2] = 4i\hbar\hat{z}(\hat{\mathbf{r}} \times \mathbf{L})_z + 6\hbar^2\hat{z}^2 - 2\hbar^2 \tag{E.1}$$

and

$$[\mathbf{L}^2, \hat{z}(\hat{\mathbf{r}} \times \mathbf{L})_z] = 2i\hbar(\hat{\mathbf{r}} \times \mathbf{L})_z^2 + 2\hbar^2\hat{z}(\hat{\mathbf{r}} \times \mathbf{L})_z - 2i\hbar\hat{z}^2\mathbf{L}^2. \tag{E.2}$$

The expectation values of the left-hand sides of (E.1) and (E.2) are zero and hence we obtain two equations relating three expectation values. A third such equation is provided by the operator identity

$$(\hat{\mathbf{r}} \times \mathbf{L})_z^2 = (1 - \hat{z}^2)(\mathbf{L}^2 - L_z^2) - \hat{z}^2 L_z^2. \tag{E.3}$$

Solving these three equations we find

$$\langle \hat{z}^2 \rangle = \frac{2[l(l+1) - m^2] - 1}{4l(l+1) - 3}, \tag{E.4}$$

$$\langle \hat{z}(\hat{\mathbf{r}} \times \mathbf{L})_z \rangle = i\hbar \frac{l(l+1) - 3m^2}{4l(l+1) - 3}, \tag{E.5}$$

$$\langle (\hat{\mathbf{r}} \times \mathbf{L})_z^2 \rangle = \hbar^2 \frac{2l(l+1)[l(l+1) - m^2 - 1] + 3m^2}{4l(l+1) - 3}. \tag{E.6}$$

Because $\hat{\mathbf{p}}$ is also a vector operator satisfying (4.41), the same expectation values are obtained when $\hat{\mathbf{r}}$ is replaced by $\hat{\mathbf{p}}$ in (E.4)–(E.6). In the above the averages are with respect to $|lm\rangle$. We note that these results remain valid if the averages are with respect to the kets $|Elm\rangle$ of eqns (5.11)–(5.13). (This property is used in the derivation of (8.27).)

Appendix F

We determine the conditions under which the radial operators p_r (eqn (5.20)) and r_p (eqn (7.2)) are Hermitian* with respect to $|El\rangle$.

I. p_r is Hermitian with respect to $|El\rangle$ if

$$\langle E'l'|p_r^\dagger - p_r|El\rangle = 0. \tag{F.1}$$

* Note that p_r and r_p are not Hermitian with respect to their eigenstates; they do not have real eigenvalues and they are not observables [1].

From (5.21), $rp_r = \mathbf{r}.\mathbf{p} - i\hbar$. Thus $p_r^\dagger r = \mathbf{p}.\mathbf{r} + i\hbar = \mathbf{r}.\mathbf{p} - 2i\hbar = p_r r$ and hence

$$\langle E'l'|(p_r^\dagger - p_r)r|E''l''\rangle = 0. \tag{F.2}$$

To obtain (F.1) from (F.2) requires an additional condition, namely

$$\langle E''l''|r^{-1}|El\rangle < \infty \tag{F.3}$$

for all $|E''l''\rangle$ and $|El\rangle$. In the coordinate representation (F.3) is

$$\int_0^\infty R^*_{E''l''}(r)r^{-1}R_{El}(r)r^2\,\mathrm{d}r < \infty, \tag{F.4}$$

and for square-integrable wavefunctions this requires

$$rR_{El}(r) \to 0 \quad \text{as} \quad r \to 0. \tag{F.5}$$

It is instructive to give an alternative derivation of (F.5). For the differential operator (5.35) an integration by part yields

$$\int_0^\infty R^*_{E'l'}(p_r R_{El})r^2\,\mathrm{d}r = \int_0^\infty \{[p_r + i\hbar\delta(r)]R_{E'l'}\}^* R_{El} r^2\,\mathrm{d}r.$$

Thus the adjoint of (5.35) is [1]

$$p_r^\dagger = p_r + i\hbar\delta(r), \tag{F.6}$$

and (F.5) is the condition for p_r to be Hermitian with respect to $R_{El}(r)$.

We note that (F.5) is also the condition for the equivalence of the Schrödinger equation in Cartesian coordinates and the Schrödinger equation in spherical polar coordinates [2]. For the three-dimensional isotropic harmonic oscillator and bound states of the Coulomb potential (F.5) is satisfied (see Section 6.4). In Section 5.4 we have assumed that p_r is Hermitian with respect to $\langle El|$ and $R_l^\pm|El\rangle$. This requires instead of (F.5) the more restrictive condition $r^{1/2}R_{El}(r) \to 0$ as $r \to 0$, which is associated with finiteness of the average effective potential in (5.25) [1].

II. A similar analysis applies to the operator r_p. From (7.2), $pr_p = \mathbf{p}.\mathbf{r} + i\hbar$ and

$$\langle E'l'|r_p^\dagger - r_p|El\rangle = 0 \tag{F.7}$$

if

$$\langle E''l''|p^{-1}|El\rangle < \infty \tag{F.8}$$

for all $|E''l''\rangle$ and $|El\rangle$. In the momentum representation (F.8) is

$$\int_0^\infty \bar{R}^*_{E''l''}(p)p^{-1}\bar{R}_{El}(p)p^2\,\mathrm{d}p < \infty, \tag{F.9}$$

and for square-integrable wavefunctions this requires

$$p\bar{R}_{El}(p) \to 0 \quad \text{as} \quad p \to 0. \tag{F.10}$$

For the isotropic harmonic oscillator and bound states of the Coulomb potential (F.10) is satisfied (see Section 7.6).

References

1. Liboff, R.L., Nebenzahl, I., and Fleischmann, H.H. (1973). On the radial momentum operator. *American Journal of Physics*, **41**, 976–80. The second term in

eqn (6) of this reference has an incorrect sign. The use of Dirac δ functions in expressions for generalized momentum operators was first discussed by Robinson, P.D., and Hirschfelder, J.O. (1963). Generalized momentum operators in quantum mechanics. *Journal of Mathematical Physics*, **4**, 338–47.

2. Dirac, P.A.M. (1958). *Quantum mechanics*, 4th edn, Section 38. Oxford University Press.

Appendix G

1. Proof of the operator identity (5.23):

$$\mathbf{L}^2 = \varepsilon_{ijk} r_j p_k \varepsilon_{ilm} r_l p_m = r_j p_k r_j p_k - r_j p_k r_k p_j. \tag{G.1}$$

(See (1.157).) Using (1.65)–(1.67) and (1.138) in (G.1) we have

$$\mathbf{L}^2 = r_j^2 p_k^2 - r_j p_j r_k p_k + i\hbar r_j p_k = \mathbf{r}^2\mathbf{p}^2 - [(\mathbf{r}.\mathbf{p})^2 - i\hbar\mathbf{r}.\mathbf{p}]. \tag{G.2}$$

But

$$\mathbf{r}^2 p_r^2 = \mathbf{r}^2\left(\frac{1}{r}\mathbf{r}.\mathbf{p} - \frac{i\hbar}{r}\right)\left(\frac{1}{r}\mathbf{r}.\mathbf{p} - \frac{i\hbar}{r}\right) = \mathbf{r}.\mathbf{p}(\mathbf{r}.\mathbf{p} - i\hbar), \tag{G.3}$$

where we have used (1.141). Equations (G.2) and (G.3) yield (5.23).

2. Proof of the operator identity (7.4). In the same manner as the calculations leading to (G.2) one can show that

$$\mathbf{L}^2 = \mathbf{p}^2\mathbf{r}^2 - [(\mathbf{p}.\mathbf{r})^2 + i\hbar\mathbf{p}.\mathbf{r}]. \tag{G.4}$$

Also,

$$\mathbf{p}^2 r_p^2 = \mathbf{p}^2\left(\frac{1}{p}\mathbf{p}.\mathbf{r} + \frac{i\hbar}{p}\right)\left(\frac{1}{p}\mathbf{p}.\mathbf{r} + \frac{i\hbar}{p}\right) = \mathbf{p}.\mathbf{r}(\mathbf{p}.\mathbf{r} + i\hbar), \tag{G.5}$$

where we have used (1.142). Equations (G.4) and (G.5) yield (7.4.).

Appendix H

We prove (7.69). If $f(p, r_p)$ is a power series in p and r_p then (1.100) and (7.95) yield

$$\langle Elm|f(p, r_p)|Elm\rangle = \int \bar{\psi}^*_{Elm}(\mathbf{p}) f\left(p, i\hbar\frac{d}{dp} + \frac{i\hbar}{p}\right)\bar{\psi}_{Elm}(\mathbf{p})\, d^3p, \tag{H.1}$$

where $d^3p = p^2\, dp\, d\Omega$. Substituting (7.96) in (H.1) and using the normalization of $Y_{lm}(\hat{\mathbf{p}})$ to perform the integration over solid angle, we have

$$\langle Elm|f(p, r_p)|Elm\rangle = \int_0^\infty p^2 \bar{R}^*_{El}(p) f\left(p, i\hbar\frac{d}{dp} + \frac{i\hbar}{p}\right)\bar{R}_{El}(p)\, dp. \tag{H.2}$$

Similarly, using the expansion

$$\langle El| = \int_0^\infty \langle El|p\rangle\langle p|\, dp,$$

(7.98), and

$$\langle p|r_p|El\rangle = \mathrm{i}\hbar\left(\frac{\mathrm{d}}{\mathrm{d}p}+\frac{1}{p}\right)\bar{R}_{El}(p),$$

we see that $\langle El|f(p, r_p)|El\rangle$ is also equal to the right-hand side of (H.2). Hence we obtain (7.69).

Appendix I

We prove the operator identities

$$\mathbf{U}^{\pm}R^{\pm}(\omega) = \mathbf{F}^{\pm}(\omega) \pm \mathrm{i}\mathbf{G}^{\pm}(\omega), \tag{I.1}$$

where $R^{\pm}$, $\mathbf{U}^{\pm}$, $\mathbf{F}^{\pm}$, and $\mathbf{G}^{\pm}$ are given by (8.2), (8.8), (8.10), and (8.11) respectively. First express $R^{\pm}$ and $\mathbf{U}^{\pm}$ in terms of $\mathbf{r}$, $\mathbf{p}$, and S:

$$R^{\pm}(\omega) = \pm \mathrm{i}r^{-1}\mathbf{r}.\mathbf{p} - \hbar(S \mp \tfrac{1}{2})r^{-1} - M\omega r, \tag{I.2}$$

$$\mathbf{U}^{\pm} = \pm \mathrm{i}r^{-1}\mathbf{r}(\mathbf{r}.\mathbf{p}) \mp \mathrm{i}r\mathbf{p} + \hbar r^{-1}\mathbf{r}(S \pm \tfrac{1}{2}). \tag{I.3}$$

In (I.3) we have used the identity

$$\mathbf{r} \times \mathbf{L} = \mathbf{r}(\mathbf{r}.\mathbf{p}) - \mathbf{r}^2\mathbf{p}. \tag{I.4}$$

Now multiply out the products $\mathbf{U}^{\pm}R^{\pm}(\omega)$ and use the commutators (1.141), (1.143), (1.149), (1.150), the identity (G.2), and

$$(S \pm \tfrac{1}{2})(S \mp \tfrac{1}{2}) = \hbar^{-2}\mathbf{L}^2. \tag{I.5}$$

After some manipulation we find

$$\begin{aligned}\mathbf{U}^{\pm}R^{\pm}(\omega) = &-[\mathbf{r}\mathbf{p}^2 - \mathbf{p}(\mathbf{r}.\mathbf{p})] - M\hbar\omega\mathbf{r}(S \pm \tfrac{1}{2})\\ &\pm \mathrm{i}\{-M\omega[\mathbf{r}(\mathbf{r}.\mathbf{p}) - \mathbf{r}^2\mathbf{p}] + \hbar\mathbf{p}(S \mp \tfrac{1}{2})\}.\end{aligned} \tag{I.6}$$

In (I.6) substitute (I.4) and

$$\mathbf{p} \times \mathbf{L} = \mathbf{r}\mathbf{p}^2 - \mathbf{p}(\mathbf{r}.\mathbf{p}) + \mathrm{i}\hbar\mathbf{p}. \tag{I.7}$$

Hence we obtain (I.1).

Appendix J

1. Proof of the operator identities

$$a\hbar^{-1}\mathbf{U}^{\pm}R^{\pm} = \hbar\mathbf{A}(S \pm \tfrac{1}{2}) \pm \mathrm{i}\mathbf{F}, \tag{J.1}$$

where $\mathbf{U}^{\pm}$, $R^{\pm}$, $\mathbf{A}$, and $\mathbf{F}$ are given by (8.8), (9.4), (9.10), and (9.11).

In (I.3), $\mathbf{U}^{\pm}$ are expressed in terms of $\mathbf{r}$, $\mathbf{p}$, and S. Similarly,

$$R^{\pm} = \pm \mathrm{i}r^{-1}\mathbf{r}.\mathbf{p}(S \pm \tfrac{1}{2}) - \hbar^{-1}\mathbf{L}^2r^{-1} + \hbar a^{-1}. \tag{J.2}$$

Multiplying out the products $\mathbf{U}^{\pm}R^{\pm}$ and using the commutators (1.143) and (1.150) we find

$$\mathbf{U}^{\pm}R^{\pm} = \{-r^{-2}\mathbf{r}[(\mathbf{r}.\mathbf{p})^2 - i\hbar\mathbf{r}.\mathbf{p} + \mathbf{L}^2] + \mathbf{p}(\mathbf{r}.\mathbf{p}) + \hbar^2 a^{-1}r^{-1}\mathbf{r}\}(S \pm \tfrac{1}{2})$$
$$\pm\, i\hbar a^{-1}[r^{-1}\mathbf{r}(\mathbf{r}.\mathbf{p}) - r\mathbf{p} + a\hbar^{-2}\mathbf{p}\mathbf{L}^2]. \quad \text{(J.3)}$$

Use of the identities (G.2), (I.7), and (9.12) shows that the operator in braces in (J.3) is equal to $\hbar^2 a^{-1}\mathbf{A}$. Use of the identity (I.4) shows that the second term in (J.3) is equal to $\pm i\hbar a^{-1}\mathbf{F}$.

2. Proof of the factorizations (9.25): (i) Multiply out the products $(P^{\pm})^{\dagger}P^{\pm}$, where $P^{\pm}$ and $(P^{\pm})^{\dagger}$ are given by (9.22) and (9.24). (ii) Use the identity (7.4) to express r_p^2 in terms of $\mathbf{r}^2$ and $p^{-2}\mathbf{L}^2$. (iii) Note that $\mathbf{r}^2$ commutes with $\mathbf{p}^2 - 2MH$. (iv) Use the identities $\hbar^{-2}\mathbf{L}^2 = (S \pm \tfrac{1}{2})^2 \mp (S \pm \tfrac{1}{2})$, and the commutators $[r_p, p] = i\hbar$ and $[r_p, p^{-1}] = -i\hbar p^{-2}$. This yields

$$(P^{\pm})^{\dagger}P^{\pm} = \hbar^{-2}\mathbf{r}^2(\mathbf{p}^2 - 2MH)^2 + 8MH(S \pm \tfrac{1}{2})^2$$
$$\pm\, 2\{2MH + p^2 + i\hbar^{-1}[pr_p, 2MH]\}(S \pm \tfrac{1}{2}). \quad \text{(J.4)}$$

Now

$$[pr_p, 2MH] = [\mathbf{p}.\mathbf{r}, \mathbf{p}^2 - 2\hbar^2 a^{-1}r^{-1}]$$
$$= \mathbf{p}.[\mathbf{r}, \mathbf{p}^2] - 2\hbar^2 a^{-1}[\mathbf{p}, r^{-1}].\mathbf{r}$$
$$= i\hbar(2MH + \mathbf{p}^2).$$

(In the last step we have used (1.147) and (1.150).) Thus the brace in (J.4) is zero. The first term in (J.4) is equal to $4\hbar^2 a^{-2}$ (see (7.19)). Hence we obtain (9.25).

3. Proof of the operator identities

$$\mathbf{C}^{\pm} = \pm i\tfrac{1}{2}a\hbar^{-1}\mathbf{V}^{\pm}P^{\pm}, \quad \text{(J.5)}$$

where $\mathbf{C}^{\pm}$, $\mathbf{V}^{\pm}$, and $P^{\pm}$ are given by (9.9), (8.31), and (9.22).

The proof of (J.5) is rather cumbersome. First express $P^{\pm}$ in terms of $\mathbf{r}$, $\mathbf{p}$, and S:

$$P^{\pm} = \mp i\hbar^{-1}p^{-1}\mathbf{r}.\mathbf{p}(\mathbf{p}^2 - 2MH) + p^{-1}(\mathbf{p}^2 + 2MH)(S \pm \tfrac{1}{2}) \mp 2p^{-1}(\mathbf{p}^2 - 2MH). \quad \text{(J.6)}$$

Note that $\mathbf{L}^2$, and hence S, commutes with all the operators in (J.6). Consider first the product of $\mathbf{V}^{\pm}$ with the second term on the right-hand side of (J.6). Writing $\mathbf{p}^2 + 2MH = 2\mathbf{p}^2 - 2\hbar^2 a^{-1}r^{-1}$, multiplying out, and using (9.9)–(9.11) yields

$$i\hbar^{-1}a\mathbf{V}^{\pm}p^{-1}(\mathbf{p}^2 + 2MH)(S \pm \tfrac{1}{2}) = \pm 2\hbar[\mathbf{C}^{\pm} - \hbar(r^{-1}\mathbf{r} \pm i\mathbf{V}^{\pm}p^{-1}r^{-1})$$
$$\times (S \pm \tfrac{1}{2})] - 2i\hbar r^{-1}\mathbf{r} \times \mathbf{L}. \quad \text{(J.7)}$$

Solving (J.7) for $\mathbf{C}^{\pm}$ and using (J.6) we have

$$\mathbf{C}^{\pm} = \pm i\tfrac{1}{2}a\hbar^{-2}\mathbf{V}^{\pm}P^{\pm} + \mathbf{Z}^{\pm}, \quad \text{(J.8)}$$

where

$$\mathbf{Z}^{\pm} = \hbar(r^{-1}\mathbf{r} \pm i\mathbf{V}^{\pm}p^{-1}r^{-1})(S \pm \tfrac{1}{2}) \pm ir^{-1}\mathbf{r} \times \mathbf{L}$$
$$-\, \tfrac{1}{2}a\hbar^{-2}\mathbf{V}^{\pm}p^{-1}(\mathbf{r}.\mathbf{p} - 2i\hbar)(\mathbf{p}^2 - 2MH). \quad \text{(J.9)}$$

In (J.9) we substitute

$$\mathbf{V}^{\pm} = \mp ip^{-1}\mathbf{p}(\mathbf{r}.\mathbf{p}) \pm ip\mathbf{r} + \hbar p^{-1}\mathbf{p}(S \mp \tfrac{5}{2}),$$

$$\mathbf{p}^2 - 2MH = 2\hbar^2 a^{-1} r^{-1},$$

the identities (I.4) and (G.2), and we use the commutators (1.143), (1.149), (1.150), and (1.153). This shows that $\mathbf{Z}^{\pm} = 0$.

4. Proof of the commutator (9.76). Equation (9.76) can be written as

$$\mathbf{A} \times \mathbf{A} = -2i\hbar^{-3} Ma^2 H\mathbf{L}. \tag{J.10}$$

To prove (J.10) first express the Pauli–Lenz vector in terms of $\mathbf{r}$ and $\mathbf{p}$: from (9.10), (9.12), (I.7), and (1.138)

$$\mathbf{A} = a\hbar^{-2}(-\mathbf{r}p^2 + \mathbf{p}\mathbf{p}.\mathbf{r} + 3i\hbar\mathbf{p}) + r^{-1}\mathbf{r}. \tag{J.11}$$

To evaluate the product $\mathbf{A} \times \mathbf{A}$ substitute (J.11) and multiply out. Then use $\mathbf{r} \times \mathbf{r} = \mathbf{p} \times \mathbf{p} = 0$ and the fact that $\mathbf{L} = \mathbf{r} \times \mathbf{p} = -\mathbf{p} \times \mathbf{r}$ commutes with the scalars $\mathbf{p}.\mathbf{r}$, $\mathbf{p}^2$, and functions of r (see Chapter 4). Hence we find

$$\mathbf{A} \times \mathbf{A} = a^2\hbar^{-4}([\mathbf{p}.\mathbf{r}, \mathbf{p}^2] - 3i\hbar\mathbf{p}^2)\mathbf{L} + a\hbar^{-2}([r^{-1}, \mathbf{p}.\mathbf{r}] + 3i\hbar r^{-1})\mathbf{L}. \tag{J.12}$$

Use of the commutators (1.142), (1.138), and (1.143), and (9.1) shows that (J.12) is (J.10).

Appendix K

1. Proof of the identity (11.83). Let $\{A, B\} \equiv AB + BA$. Then

$$\begin{aligned}\gamma_\mu\gamma_\nu D_\mu D_\nu &= \tfrac{1}{2}\gamma_\mu\gamma_\nu(\{D_\mu, D_\nu\} + [D_\mu, D_\nu]) \\ &= \tfrac{1}{2}\{\gamma_\mu, \gamma_\nu\}D_\mu D_\nu + \tfrac{1}{2}\gamma_\mu\gamma_\nu[D_\mu, D_\nu] \\ &= D_\mu^2 + \tfrac{1}{2}\gamma_\mu\gamma_\nu[D_\mu, D_\nu],\end{aligned} \tag{K.1}$$

where we have used (11.19). Also

$$\begin{aligned}[D_\mu, D_\nu] &= [p_\mu - qA_\mu, p_\nu - qA_\nu] = -q[p_\mu, A_\nu] + q[p_\nu, A_\mu] \\ &= iq\hbar\left(\frac{\partial A_\nu}{\partial x_\mu} - \frac{\partial A_\mu}{\partial x_\nu}\right).\end{aligned} \tag{K.2}$$

In the last step we have used (11.36) and (1.136). From (K.2)

$$\gamma_\mu\gamma_\nu[D_\mu, D_\nu] = \tfrac{1}{2}iq\hbar[\gamma_i, \gamma_j]\left(\frac{\partial A_j}{\partial x_i} - \frac{\partial A_i}{\partial x_j}\right) + iq\hbar[\gamma_i, \gamma_4]\left(\frac{\partial A_4}{\partial x_i} - \frac{\partial A_i}{\partial x_4}\right). \tag{K.3}$$

Now

$$\begin{aligned}[\gamma_i, \gamma_j] &= -\beta\alpha_i\beta\alpha_j + \beta\alpha_j\beta\alpha_i = \alpha_i\alpha_j - \alpha_j\alpha_i \\ &= \begin{pmatrix}\sigma_i\sigma_j - \sigma_j\sigma_i & 0 \\ 0 & \sigma_i\sigma_j - \sigma_j\sigma_i\end{pmatrix} \\ &= 2i\varepsilon_{ijk}\Sigma_k,\end{aligned} \tag{K.4}$$

where Σ_k are defined in (11.49). In the above calculations we have used (11.18), (11.13), (11.14), (11.28), and (4.204). Similarly

$$[\gamma_i, \gamma_4] = 2\mathrm{i}\alpha_i. \tag{K.5}$$

Also

$$\frac{\partial A_4}{\partial x_i} - \frac{\partial A_i}{\partial x_4} = -\frac{\mathrm{i}}{c}E_i. \tag{K.6}$$

From (K.1), (K.3)–(K.6), (2.179), and (1.156) we obtain

$$\gamma_\mu\gamma_\nu D_\mu D_\nu = D_\mu^2 - q\hbar\mathbf{\Sigma}.\mathbf{B} + \mathrm{i}c^{-1}q\hbar\boldsymbol{\alpha}.\mathbf{E}, \tag{K.7}$$

which is (11.83)

Appendix L

1. Proof of the identity (12.8):

$$(\boldsymbol{\alpha}.\mathbf{r})(\boldsymbol{\alpha}.\mathbf{p}) = \begin{pmatrix} 0 & \boldsymbol{\sigma}.\mathbf{r} \\ \boldsymbol{\sigma}.\mathbf{r} & 0 \end{pmatrix}\begin{pmatrix} 0 & \boldsymbol{\sigma}.\mathbf{p} \\ \boldsymbol{\sigma}.\mathbf{p} & 0 \end{pmatrix} = \mathbf{r}.\mathbf{p} + \mathrm{i}\mathbf{\Sigma}.\mathbf{L}, \tag{L.1}$$

where we have used (11.28), (4.211), and (11.49). Multiplying (L.1) on the left by $r^{-1}\boldsymbol{\alpha}.\hat{\mathbf{r}}$ we have

$$(\boldsymbol{\alpha}.\hat{\mathbf{r}})^2\boldsymbol{\alpha}.\mathbf{p} = \alpha_r p_r + \mathrm{i}r^{-1}\alpha_r(\mathbf{\Sigma}.\mathbf{L} + \hbar), \tag{L.2}$$

where p_r and α_r are given by (5.20) and (12.9). But

$$(\boldsymbol{\alpha}.\hat{\mathbf{r}})^2 = \begin{pmatrix} 0 & (\boldsymbol{\sigma}.\hat{\mathbf{r}})^2 \\ (\boldsymbol{\sigma}.\hat{\mathbf{r}})^2 & 0 \end{pmatrix} = 1, \tag{L.3}$$

because of (4.208). From (L.2) and (L.3) we obtain (12.8).

2. Proof of the commutation relations (12.14):

(i)
$$[K, \alpha_r] = \begin{pmatrix} 0 & \mathscr{K}\boldsymbol{\sigma}.\hat{\mathbf{r}} + \boldsymbol{\sigma}.\hat{\mathbf{r}}\mathscr{K} \\ -\mathscr{K}\boldsymbol{\sigma}.\hat{\mathbf{r}} - \boldsymbol{\sigma}.\hat{\mathbf{r}}\mathscr{K} & 0 \end{pmatrix}. \tag{L.4}$$

But

$$\begin{aligned}\mathscr{K}\boldsymbol{\sigma}.\hat{\mathbf{r}} &= (\hbar^{-1}\boldsymbol{\sigma}.\mathbf{L} + 1)\boldsymbol{\sigma}.\hat{\mathbf{r}} \\ &= \hbar^{-1}\sigma_i\sigma_j L_i r^{-1} r_j + \boldsymbol{\sigma}.\hat{\mathbf{r}} \\ &= \hbar^{-1}(-\sigma_j\sigma_i + 2\delta_{ij})L_i r^{-1} r_j + \boldsymbol{\sigma}.\hat{\mathbf{r}} \\ &= -\hbar^{-1}\sigma_j\sigma_i L_i r^{-1} r_j + \boldsymbol{\sigma}.\hat{\mathbf{r}},\end{aligned} \tag{L.5}$$

because $\mathbf{L}.\hat{\mathbf{r}} = 0$. Then from (4.206), (1.156), (C.7) and (C.10)

$$\mathscr{K}\boldsymbol{\sigma}.\hat{\mathbf{r}} = -\hbar^{-1}(\boldsymbol{\sigma}.\hat{\mathbf{r}})(\boldsymbol{\sigma}.\mathbf{L}) - \boldsymbol{\sigma}.\hat{\mathbf{r}} = -\boldsymbol{\sigma}.\hat{\mathbf{r}}\mathscr{K}. \tag{L.6}$$

From (L.4) and (L.6)

$$[K, \alpha_r] = 0. \tag{L.7}$$

(ii) From (12.12) and (11.22) it is obvious that

$$[K, \beta] = 0. \tag{L.8}$$

(iii) Because $\mathbf{L}$, $\boldsymbol{\Sigma}$, and β commute with p_r and $f(r)$,

$$[K, p_r] = 0, \qquad [K, f(r)] = 0. \tag{L.9}$$

(iv) Each term in $\mathbf{J}^2 = \mathbf{L}^2 + 2\mathbf{L}.\mathbf{S} + \mathbf{S}^2$ commutes with β and $\boldsymbol{\Sigma}.\mathbf{L} = 2\hbar^{-1}\mathbf{S}.\mathbf{L}$. Thus

$$[K, \mathbf{J}^2] = 0. \tag{L.10}$$

(v) $[K, J_z] = [\beta(\hbar^{-1}\boldsymbol{\Sigma}.\mathbf{L} + 1), J_z] = 2\beta\hbar^{-2}[\mathbf{S}.\mathbf{L}, L_z + S_z] = 0.$ (L.11)

AUTHOR INDEX

SUBJECT INDEX

The manufacturer's authorised representative in the EU for product safety is Oxford University Press España S.A. of el Parque Empresarial San Fernando de Henares, Avenida de Castilla, 2 – 28830 Madrid (www.oup.es/en or product.safety@oup.com). OUP España S.A. also acts as importer into Spain of products made by the manufacturer.

www.ingramcontent.com/pod-product-compliance
Ingram Content Group UK Ltd.
Pitfield, Milton Keynes, MK11 3LW, UK
UKHW020924170926
475595UK00031B/103
* 9 7 8 0 1 9 8 5 3 9 6 1 2 *